MATHEMATICS AND STATISTICS FOR
THE BIO-SCIENCES

MATHEMATICS AND ITS APPLICATIONS
Series Editor: G. M. BELL
Emeritus Professor of Mathematics, King's College London, University of London

STATISTICS, OPERATIONAL RESEARCH
AND COMPUTATIONAL MATHEMATICS Section
Editor: B. W. CONOLLY,
Emeritus Professor of Mathematics (Operational Research), Queen Mary College, University of London

Mathematics and its applications are now awe-inspiring in their scope, variety and depth. Not only is there rapid growth in pure mathematics and its applications to the traditional fields of the physical sciences, engineering and statistics, but new fields of application are emerging in biology, ecology and social organization. The user of mathematics must assimilate subtle new techniques and also learn to handle the great power of the computer efficiently and economically.

The need for clear, concise and authoritative texts is thus greater than ever and our series endeavours to supply this need. It aims to be comprehensive and yet flexible. Works surveying recent research will introduce new areas and up-to-date mathematical methods. Undergraduate texts on established topics will stimulate student interest by including applications relevant at the present day. The series will also include selected volumes of lecture notes which will enable certain important topics to be presented earlier than would otherwise be possible.

In all these ways it is hoped to render a valuable service to those who learn, teach, develop and use mathematics.

Mathematics and its Applications
Series Editor: G. M. BELL
Professor of Mathematics, King's College London, University of London

series continued at back

MATHEMATICS AND STATISTICS FOR THE BIO-SCIENCES

G. EASON, M.Sc., Ph.D., F.I.M.A., C.W. COLES, B.A., D.Phil.
and G. GETTINBY, B.Sc., Ph.D.
Department of Mathematics, University of Strathclyde

ELLIS HORWOOD
NEW YORK   LONDON   TORONTO   SYDNEY   TOKYO   SINGAPORE

First published in 1980
Reprinted in 1982, 1986 (with corrections, and an additional section in Chapter 22
on the Poisson distribution), 1989 and 1992.
ELLIS HORWOOD LIMITED
Market Cross House, Cooper Street,
Chichester, West Sussex, PO19 1EB, England

A division of
Simon & Schuster International Group
A Paramount Communications Company

© Ellis Horwood Limited, 1992

Printed and bound in Great Britain
by Hartnolls, Bodmin

British Library Cataloguing in Publication Data

A catalogue record for this book is available from the British Library

ISBN 0–13–560541–5   Pbk

Library of Congress Cataloging-in-Publication Data

Available from the publishers

Table of Contents

Authors' Preface

In recent years there has been a rapid growth in the use of mathematical and statistical techniques to describe a wide variety of processes encountered in the biosciences. Increasingly, the use of mathematical models has led to a greater understanding of the biological processes. This trend is certain to continue in the forseeable future and it will become essential for bioscience students to be familiar with these techniques. Moreover, many researchers in the biosciences will find their training in mathematics and statistics inadequate and require a more complete treatment of the techniques available. This book provides the essential requirements of such people and covers all the topics needed in a first course in mathematics or a first course in statistics for bioscience students. It also introduces more advanced topics that final year undergraduates, postgraduates and research workers may require. No comparable text is thought to be as self-contained or cover so much material needed by bioscience students.

The material is based on lectures given to bioscience students at the University of Strathclyde for a number of years. The book is aimed at students entering universities or colleges with a pass in the H-grade mathematics course in the Scottish Certificate of Education but, in view of the revision material included, it should be suitable for students with O-grade or O-level qualifications. The presentation of proofs is at a level appropriate to bioscience students and excessive rigour has been avoided. A 'traditional' approach is adopted, but in such a way that students with a background of 'modern' mathematics should experience no difficulty. References for further reading are included when it is thought useful. Numerous worked examples illustrating the application of the mathematics and statistics to biology, but not requiring detailed knowledge of that subject, are included. In addition a large range of problems, complete with answers, are available for the reader to work through.

Topics covered are the binomial theorem and series, curve fitting, matrices and solution of equations, differentiation and integration techniques, logarithmic and exponential functions, differential equations, numerical integration, probability, parametric and non-parametric statistics. The first chapter is devoted to revision of school work and there are a number of appendices covering material directly related to the text.

Acknowledgements

It is a pleasure to acknowledge the permission granted by the Universities of Edinburgh and Strathclyde to use questions from their examination papers. Dr. Munn of the University of Edinburgh made available a number of examples, as did a number of colleagues. Statistical tables have been reproduced with the kind permission of Dr. R. J. Henery. Our thanks are due particularly to Miss Tove Korstvedt for her unfailing patience in typing the manuscript.

CHAPTER 1

Basic mathematics

1.1 INTRODUCTION

This chapter contains a review of the basic concepts of arithmetic and algebra that are considered to be essential to the later chapters. A summary of the various sets of numbers is followed by some examples involving elementary algebra. The chapter concludes with a discussion of logarithms and a summary of essential results in trigonometry. The reader is assumed to be familiar with much of this material but because of the sequential nature of mathematics it is essential that these topics are fully understood in order to give a firm foundation for later work.

1.2 SETS OF NUMBERS

In order to perform mathematical operations it is essential to have a complete understanding of the sets of numbers which are fundamental to these operations. The number systems required here will now be summarised. **Real numbers**, such as 1, 2.13, -4.2, 7/9, π, $\sqrt{2}$ etc are the numbers which we use in elementary mathematics. They may be thought of as distances, measured from an origin O at zero, which lie along a straight line (the **real line**). It is conventional to place positive numbers to the right of O and negative numbers to the left (see Fig. 1.1).

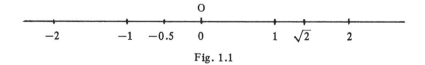

Fig. 1.1

The **integers** are the numbers 0, ± 1, ± 2, etc. They are particular points on the real line on either side of O. Successive integers are one unit apart.

The **whole numbers** are the integers to the right of and including O, namely 0, 1, 2, 3 etc.

The **natural numbers** 1, 2, 3 etc. are the integers to the right of O and are the numbers with which we count. Included in the natural numbers are the **prime numbers** that are divisible by 1 and themselves only.

If p is an integer and q is a natural number and if p and q have no common factor then p/q is a **rational number**. Integers are rational numbers with $q = 1$. For example, $-\frac{3}{4}$ is a rational number since -3 is an integer and 4 is a natural number, and there is no common factor.

The term **fraction** is sometimes used to describe numbers of the form p/q where p is an integer and q is a natural number. The numbers p and q may now have common factors. For example, $-\frac{6}{8}$ is a fraction since -6 is an integer and 8 is a natural number. There is a common factor 2 so that $-\frac{6}{8} = -\frac{3}{4}$ which is a rational number.

Not all real numbers may be expressed as rational numbers. Numbers such as $\pi = 3.14159 \ldots$ and $\sqrt{2} = 1.41421 \ldots$, which are called **irrational numbers**, can only be written down approximately as numbers, and it is usual in calculations to leave them as π, $\sqrt{2}$ etc. A numerical value may then be substituted to the required accuracy.

In addition to the set of real numbers and its various subsets listed here, other sets of numbers exist. The most common of these is the set of **complex numbers** such as $2 + 3i$ where $i^2 = -1$. This is a natural extension to the set of real numbers since by introducing this set it is possible to write down solutions of equations such as $x^2 = -1$. A complete discussion of the properties of complex numbers is beyond the scope of this book; however, a brief discussion of some of their properties is included in Chapter 3. Many books devote sections to them, see for example Smyrl (1978).

1.3 EQUALITY AND INEQUALITY

As already indicated, real numbers may be thought of as being distributed along a line. In order to perform operations with them we need certain symbols that enable us to compare two numbers.

If a and b are real numbers then:

$a = b$ means a is **equal** to b

$a > b$ means a is **greater than** b (a lies to the right of b on the real line).

$a < b$ means a is **less than** b (a lies to the left of b on the real line).

$a \geqslant b$ means a is **greater than or equal** to b.

$a \leqslant b$ means a is **less than or equal** to b.

$a \neq b$ means a is **not equal** to b (either $a < b$ or $a > b$).

$a \cong b$ or $a \simeq b$ mean a is **approximately equal** to b. (Thus $\sqrt{2} \simeq 1.4142$).

If the same operation is performed on each side of an inequality the direction of the inequality remains unchanged except when multiplying by a negative number when the inequality changes direction. This is illustrated by

$$5 > 1$$
$$10 > 2 \text{ (multiply by 2)}$$
$$-10 < -2 \text{ (multiply by } -1)$$
$$0 < 8 \text{ (add 10)}$$
$$0 < 1 \text{ (divide by 8)}.$$

When both sides of an inequality are positive it is sometimes useful to square them. This does not change the direction of the inequality. For example, $5 > 2$ on squaring becomes $25 > 4$.

1.4 SIGNIFICANT FIGURES AND DECIMAL PLACES

It is not always possible (or necessary) to state the value of a number exactly. In order to give a measure of the accuracy of a number we make use of two descriptions.

(i) The number of **significant figures** (SF) is determined by counting all digits to the right of and including the first non-zero digit.

(ii) The number of **decimal places** (D) is the number of digits occurring after the decimal point.

For example: 2.3   has 2 SF, 1 D.

2.4028 has 5 SF, 4 D.

0.003   has 1 SF, 3 D.

We are often required to give the result of a calculation to a specified accuracy. This process is called **rounding-off**. To round-off a number to n SF (or m D) we start with the first non-zero digit (or start at the decimal point) and count n (or m) digits to the right. If the remaining digits amount to more than half a unit in the $n(m)^{\text{th}}$ place we add one to the $n(m)^{\text{th}}$ digit; if they amount to less than half a unit we do nothing. If the remaining part is exactly half a unit then we round the n (or m)th digit to the nearest even digit.

For example: 1.470588 is 1.4706 to 5 SF (4 D),

1.471   to 4 SF (3 D),

1.47   to 3 SF (2 D).

0.01625 is 0.0162 to 3 SF (4 D).

7.6135   is 7.614 to 4 SF (3 D).

The true value of a number which is known to be rounded lies within half a unit of the final digit so that $2.475 \leqslant 2.48 \leqslant 2.485$ and $2.485 < 2.49 < 2.495$.

1.5 THE MODULUS OF A REAL NUMBER

The **modulus** or **absolute value** of a real number gives its magnitude, regardless of sign. In terms of numbers on the real line the modulus is the distance of the number from O and is denoted by placing a vertical line on either side of the number.

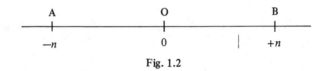

Fig. 1.2

Referring to Fig. 1.2, if $n > 0$ then OA represents $-n$ and OB represents $+n$.

The modulus of $+n = |+n| = n =$ length of OB > 0.

The modulus of $-n = |-n| = n =$ length of OA $=$ length of OB > 0.

The numbers are equidistant from O and have the same magnitude or modulus.

For instance $|7| = |-7| = 7$.

Alternatively the modulus of n may be thought of as the positive square root of n^2 so that $|n| = \sqrt{n^2}$.

Example 1: Determine all values of x such that $|x| \leqslant 2$.

If x is positive and $|x| \leqslant 2$, then $0 < x \leqslant 2$ (referring to Fig. 1.2, with $n = 2$, the points on the real line lie between B and the origin).

If x is negative and $|x| \leqslant 2$, then $-2 \leqslant x < 0$ (referring to Fig. 1.2, with $n = 2$, the points on the real line lie between A and the origin).

Combining these results gives $-2 \leqslant x \leqslant 2$ (since 0 must be included).

Example 2: Determine all values of x such that $|x + 1| > 1$.

If $(x + 1)$ is positive and $|x + 1| > 1$ then $x + 1 > 1$, and $x > 0$.

If $(x + 1)$ is negative and $|x + 1| > 1$ then $x + 1 < -1$, and $x < -2$.

Combining these results gives possible values of x as $x < -2$ or $x > 0$.

A section of the real line is called an **interval**. The interval $a \leqslant x \leqslant b$ is said to be **closed** since the end points are included and is denoted by $[a, b]$. When the end points are not included so that $a < x < b$ the interval is said to be **open** and is denoted by (a, b). Example 1 with $a = -2, b = 2$ is an example of a closed interval. The intervals in example 2 are open.

1.6 FACTORIAL

The product of all the integers from 1 to n where n is a whole number is called n **factorial** denoted by n!

Thus
$$3! = 3.2.1 = 6 ,$$
$$4! = 4.3.2.1 = 24 ,$$
$$n! = n(n-1)(n-2)\ldots 2.1 = n\{(n-1)!\}.$$

In some situations it is convenient to introduce 0!. This is defined by $0! = 1$.
Factorials for large numbers are not usually evaluated.

1.7 RECIPROCAL

The **reciprocal** of a real number a is $b = 1/a$ provided $a \neq 0$. An alternative notation for the reciprocal is a^{-1} and it should be remembered that $1/a$ and a^{-1} are the same. The reciprocal should not be confused with the functional inverse which will be discussed in Section 3.3.

Infinity, denoted by ∞, is larger than any given number. Similarly $-\infty$ is less than any given number. Technically infinity is not a number as it does not obey the laws of algebra.

The reciprocal of a very large positive number is small and positive, and the reciprocal of a very large negative number is small and negative. Similarly the reciprocal of a small positive number is a large positive number and the reciprocal of a small negative number is a large negative number. It is therefore sometimes convenient to think of the reciprocal of $\pm\infty$ as zero and the reciprocal of 0 as $\pm\infty$ (more information being required to determine the sign).

A quantity a is **proportional** to another quantity b ($a \propto b$) if $a = kb$ where k is a constant. a is **inversely proportional** to b ($a \propto 1/b$) if $a = k/b$. This particular concept arises in many biological and other applications.

The simplest model describing the behaviour of a population of animals, bacteria, etc., assumes that the rate of growth R of the population is proportional to the number n present in the population so that

$$R \propto n$$
and $\quad R = kn ,$

where k is a constant.

A more complicated model assumes that R is proportional to the product of n with $(a - n)$ where a is a constant so that

$$R \propto n(a-n)$$
and $\quad R = kn(a-n) ,$

where k is a constant.

Protein is sometimes assumed to disintegrate in such a way that the mass m is inversely proportional to $(a + t)$ where a is a constant and t is a measure of time. Then

$$m \propto \frac{1}{a+t}$$

and

$$m = \frac{k}{a+t} \ ,$$

where k is a constant.

1.8 RULES OF ARITHMETIC

It is necessary to adopt certain conventions or rules in performing arithmetical and algebraic operations in order to avoid ambiguity. Thus it is necessary to introduce brackets in order to distinguish between

$$8 + 6 \times 3 = \ 8 + 18 = 26 \tag{1.1}$$

and

$$(8 + 6) \times 3 = 14 \times 3 = 42 \ . \tag{1.2}$$

The convention is adopted that $a \times b = ab$.

The priority of operations is:

(i) Brackets (innermost first) so that in (1.2) 8 is added to 6 before multiplying by 3.

(ii) Multiplication and division; these have equal priority, so that in (1.1) 6 is multiplied by 3 before being added to 8.

(iii) Addition and subtraction; these have equal priority, so that $8 + 6 - 3 = 14 - 3 = 11$, is identical with $8 + 6 - 3 = 8 + 3 = 11$.

It is convenient to introduce two further symbols at this stage. If a result α implies a result β (but not the reverse) then $\alpha \Rightarrow \beta$, whereas if a result α implies a result β and the reverse is also true then $\alpha \Leftrightarrow \beta$. The basic rules are summarised by

$$a + b + c = a + (b + c) = (a + b) + c,$$

$$a + (-a) = 0,$$

$$a + b = a + c \Leftrightarrow b = c,$$

$$-(-a) = a,$$

$$-(a + b) = -a - b, \ -(a - b) = -a + b,$$

$$a + x = b \Leftrightarrow x = b - a,$$

$$a(b + c) = ab + ac,$$

$$-a(b + c) = -ab - ac,$$

$$\frac{1}{1/a} = a, \ a \neq 0,$$

$$a/b = 0 \Rightarrow a = 0 \text{ if } b \neq 0,$$

$$\frac{a}{b} \times \frac{c}{d} = \frac{ac}{bd} = \frac{a}{b} \bigg/ \frac{d}{c} \ ,$$

$$\frac{ac}{ab} = \frac{c}{b}, \ a \neq 0,$$

$$\frac{a+b}{c} = \frac{a}{c} + \frac{b}{c} \ ,$$

$$\frac{a}{b} + \frac{c}{d} = \frac{ad+bc}{bd} \ .$$

1.9 INDICES

If a quantity is written in the form A^b then b is the **index**. It is usually a rational number, but not necessarily so. Examples involving the use of indices are

$$3^2 \ = 3 \times 3 \qquad = 9,$$
$$2^3 \ = 2 \times 2 \times 2 = 8,$$
$$3^{0.5} = 3^{\frac{1}{2}} = \sqrt{3}.$$

In the last of these $3^{\frac{1}{2}}$ is the **square root** of 3. In general the qth **root** of A is denoted by $A^{1/q}$ or $\sqrt[q]{A}$. These are alternative notations for the same quantity. In performing arithmetical operations indices have priority over multiplication and division.

Rules for operating with indices are:

(i) $A^b \times A^c = A^{b+c}$,

(ii) $(A^b)^c = A^{b \times c} = A^{bc}$,

(iii) $A^0 = 1, A \neq 0$,

(iv) $A^{-b} = 1/A^b$ (follows from (i) and (iii)).

It should be noted that in (i) the indices are added and in (ii) they are multiplied. Particular care must be taken to distinguish between these two rules.

On squaring both sides of $x = a$ there results

$$x^2 = (a)^2 = a^2 \ .$$

Similarly if $x = -a$

$$x^2 = (-a)^2 = a^2 \ .$$

If the equation $x^2 = a^2$ is to be solved it is essential to write down both solutions so that $x = \pm a$. For example if $x^2 = 3$ then

$$x = +\sqrt{3} \cong 1.732 \quad \text{or} \quad x = -\sqrt{3} \cong -1.732 \ .$$

However, the convention is adopted that if $x = \sqrt{3}$ the positive square root only is taken.

By using the properties of indices it is possible to write extremely large (or small) numbers in a compact form. For example

$$1\ 200\ 000\ =\ 1.2 \times 10^6,$$
$$0.000\ 001\ 2\ =\ 1.2 \times 10^{-6}.$$

Example 1: Evaluate (a) $9^{\frac{1}{2}}$, (b) $4^3 \times 4^{1.5}$, (c) $(3^2)^3$, (d) $(3^3)^2$, (e) $1/3^{-2}$.

(a) $9^{\frac{1}{2}} = \sqrt{9} = 3$ (positive square root only).

(b) $4^3 \times 4^{1.5} = 4^{3+1.5} = 4^{4.5} = 4 \times 4 \times 4 \times 4 \times 2 = 512 = 2^9$.

(c) $(3^2)^3 = 9^3 = 729 = 3^6$.

(d) $(3^3)^2 = (27)^2 = 729 = 3^6$.

(e) $1/3^{-2} = 3^2 = 9$ (using rule (iv)).

{Rule (ii) indicates that both (c) and (d) have the value $3^{2 \times 3} = 3^6$ so that it was not necessary to evaluate the results separately}.

Example 2: A biological population, initially of size n, doubles its size every day. Assuming that the process is continuous, demonstrate that at any time the size of the population is $n2^t$ where t is the age of the population measured in days.

At $t = 0$, the initial population size is n,

at $t = 1$, the end of day 1, the size is $2n$,

at $t = 2$, the end of day 2, the size is $4n = 2^2 n$,

at $t = 3$, the end of day 3, the size is $8n = 2^3 n$,

at $t = 4$, the end of day 4, the size is $16n = 2^4 n$.

This process can be continued for t a whole number. All the results are of the form that give a population size $n2^t$.

{This result has not been proved rigorously; it can be shown to be correct}.

1.10 SOME WORKED EXAMPLES

This section consists of a selection of examples that illustrate the use of the rules of Sections 1.8 and 1.9. It is assumed that these are essentially revision examples, but they involve basic concepts that are necessary for later sections.

Example 1: Simplify (a) $\dfrac{p}{q} + 1$, (b) $\dfrac{p+q}{\dfrac{p}{q}+1}$, (c) $\dfrac{a^2-b^2}{a+b}$.

(a) $\dfrac{p}{q} + 1 = \dfrac{p}{q} + \dfrac{q}{q} = \dfrac{p+q}{q} = (p+q)\,q^{-1}$.

(b) $\dfrac{p+q}{\dfrac{p}{q}+1} = \dfrac{p+q}{(p+q)q^{-1}} = q\,\dfrac{(p+q)}{p+q} = q$.

(c) $\dfrac{a^2 - b^2}{a+b} = \dfrac{(a-b)\,(a+b)}{(a+b)} = (a-b)$.

{Note that $(a^2 - b^2)$ factorises unlike $(a^2 + b^2)$ which does not have real factors}.

Example 2: Solve for x: $ax - b(x+1) = 2$.

$$ax - bx - b = 2$$
$$x(a-b) = 2 + b$$
$$x = \frac{2+b}{a-b}, \quad a \neq b \ .$$

Note that in solving an equation of this type for x the most efficient procedure is to collect all the terms in x on one side of the equation and the remaining terms on the other side.

Example 3: Expand $\left(x + \dfrac{1}{x}\right)^2$.

$$\left(x + \frac{1}{x}\right)^2 = \left(x + \frac{1}{x}\right)\left(x + \frac{1}{x}\right) = x^2 + 2x\left(\frac{1}{x}\right) + \left(\frac{1}{x}\right)^2$$

$$= x^2 + 2 + \frac{1}{x^2} \ .$$

Example 4: Determine the values of the constants a and b such that $a(1 + bx^2)^{\frac{1}{2}} \equiv (3 + x^2)^{\frac{1}{2}}$. {The symbol $\equiv$ means identical with (in all respects)}.

Here the coefficient of x^2 on the right-hand side (RHS) is unity so that to obtain an identity we must make the coefficient on the left-hand side (LHS) unity also.

$$1 + bx^2 = b\left(\frac{1}{b} + x^2\right)$$

$$(1 + bx^2)^{\frac{1}{2}} = \{b\left(\frac{1}{b} + x^2\right)\}^{\frac{1}{2}} = b^{\frac{1}{2}}\left(\frac{1}{b} + x^2\right)^{\frac{1}{2}}$$

$$a(1 + bx^2)^{\frac{1}{2}} = ab^{\frac{1}{2}} \left(\frac{1}{b} + x^2\right)^{\frac{1}{2}} \equiv (3 + x^2)^{\frac{1}{2}}$$

if $\dfrac{1}{b} = 3$, that is, $b = 1/3$ *and* $ab^{\frac{1}{2}} = 1$, that is, $a = \sqrt{3}$.

1.11 LOGARITHMS

The basic concept used in working with logarithms to base 10 is that of expressing a real positive number A in the form $A = 10^m$. By writing another number B as 10^n we have

$$AB = (10^m)(10^n) = 10^{m+n}$$

and multiplication of A and B is transformed by this process into the addition of their indices m and n.

The **logarithm to base 10** of $A(= 10^m)$ is defined to be the index m so that $\log_{10} A = m$. Tables of logarithms to base 10 (**common logarithms**) have been constructed giving the values of the logarithms of numbers between 1 and 10. Thus if

$$A = 10^m, \ B = 10^n, \ C = AB = 10^{m+n},$$

$$\log_{10} A = m, \ \log_{10} B = n, \ \log_{10} C = m + n \ .$$

The values of m and n may be read from the tables, the number $m + n$ calculated and the value of C determined by using tables of **anti-logarithms** (these are tables from which the value of a number may be determined when its logarithm is given). By using logarithms difficult arithmetic is simplified and prior to the introduction of calculating machines complicated arithmetical calculations were performed using logarithms to base 10.

This basic concept may be generalised to logarithms to any base, say a. The base must be a positive number. If A, B are positive numbers with

$$A = a^M \ \text{and} \ B = a^N$$

then $M = \log_a A$, $N = \log_a B$ and a is the base of these logarithms.

The laws of indices give the following results for logarithms:

$$a^0 = 1 \ \text{so that} \ \log_a 1 = 0 \ ,$$

$$a^1 = a \ \text{so that} \ \log_a a = 1 \ ,$$

$$\log_a(AB) = \log_a A + \log_a B = M + N \ ,$$

$$\log_a(A/B) = \log_a A - \log_a B = M - N \ ,$$

$$\log_a(A^c) = c \log_a A = cM, \ c \ \text{any real number.}$$

If $A > B$ then for $a > 1$, $M > N$ and $\log_a A > \log_a B$.

The two most commonly occurring bases are 10 and the irrational number $e = 2.71828 \ldots$. The number e gives rise to natural or Napierian logarithms with $\log_e x$ denoted by $\ln x$. These occur naturally in calculus and have many practical applications (see Chapter 12).

It is possible to establish relationships between different bases of logarithms. If we write a number A in the form

$$A = a^M = b^N$$

then
$$\log_a A = M = N \log_a(b)$$
$$\log_b A = N = M \log_b(a) .$$

Combining these,

$$\log_a A = \log_b A \, \log_a(b),$$
$$\log_b A = \log_a A \, \log_b(a) .$$

These are particularly useful in connecting logarithms to the bases 10 and e.

Example: Given that $\log_{10}(3.978) = 0.5997$ evaluate

(a) $\log_{10}(397.8)$, (b) $\log_{10}(0.3978)$.

(a) $397.8 = 3.978 \times 100 = 3.978 \times 10^2$.

Taking logarithms to base 10 gives
$$\log_{10}(397.8) = \log_{10}(3.978) + 2 = 2.5997.$$

(b) $0.3978 = 3.978 \times 10^{-1}$,
$$\log_{10}(0.3978) = \log_{10}(3.978) - 1 = -1 + 0.5997 = -0.4003.$$

{ It is conventional when working with logarithms to base 10 to write

$-1 + 0.5997$ in the form $\bar{1}.5997$}.

1.12 GRAPHS

One way of showing the dependence of one variable y on another variable x is by means of a graph. This is a particularly useful method when no precise mathematical relationship is known, for example in displaying the results of an experiment.

An **origin** O is chosen and **orthogonal** (perpendicular) axes OX and OY (**Cartesian axes**) are drawn; conventionally with the axis of the **independent variable** x horizontal and the axis of the **dependent variable** y vertical. Positive values of the coordinate x are taken to the right of O and positive values of the coordinate y are taken above O. The **origin** is the point with coordinates $(0,0)$ with the coordinates of a point denoted by (x, y). Plotting the position of points arising from a set of experimental observations or from a numerical evaluation of an equation gives a pictorial representation of the dependence of y on x.

Example 1: Plot a graph corresponding to the equation $y = \log_{10}x$.

In order to plot y against x we need a table of values. A set of log tables contains the following values.

x	1.0	1.5	2.0	2.5	3.0	4.0	5.0	7.5	10.0
y	0	0.176	0.301	0.398	0.477	0.602	0.699	0.875	1.0

Additional values may be obtained by using the properties of logarithms. For example

$$\log_{10}25 = \log_{10}(2.5 \times 10) = 1.398.$$
$$\log_{10}(\tfrac{1}{2}) = -\log_{10}2 = -0.301.$$

The variation of y with x is shown in Fig. 1.3.

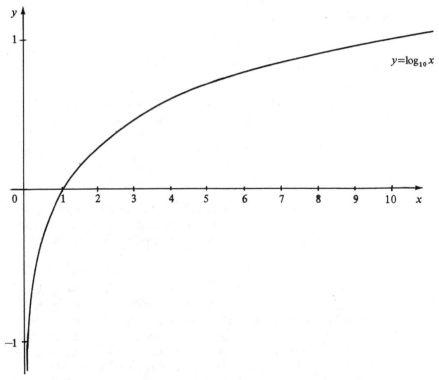

Fig. 1 .3

Several properties of this curve, in particular its shape, are common to all graphs of the form $y = \log_a x$ with $a > 1$. These properties are that no part of the curve lies to the left of the y-axis; $\log_a 1 = 0$; $\log_a a = 1$; y becomes increasingly large as x increases; y approaches $-\infty$ as x approaches zero.

Example 2: In an experiment to determine the change in body temperature resulting from the administration of a new drug the following table was obtained.

Time (hours) x	0	1	2	3	4	5
Temperature °C y	36.81	37.23	38.28	37.87	37.72	37.50

In this example $36.5 < y < 39.0$ only, so that it is pointless plotting y with a scale starting at the origin. Only that section of the vertical axis lying between 36.5 and 39.0 need be shown, see Fig. 1.4.

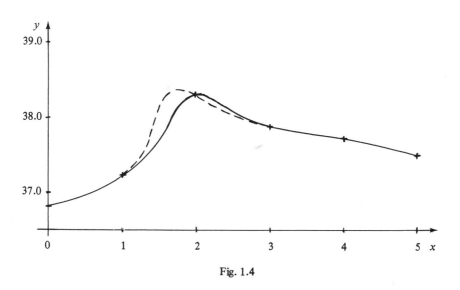

Fig. 1.4

Although the table gives values for isolated points only, it is normal practice to assume a continuous dependence of y on x and to draw a smooth curve connecting the points represented by the solid line of Fig. 1.4. In this case the number of readings is insufficient to be certain that this curve is accurate between $x = 1$ and $x = 3$, and alternative curves are equally convincing as, for example, the broken line. To determine the true curve, a repeated experiment, with smaller intervals of time between the readings, is desirable.

1.13 THE TRIGONOMETRIC RATIOS

For most practical purposes it is convenient to measure angles in degrees with one right angle divided into 90°. In calculus and many other mathematical theories it is necessary to measure angles in radians.

Suppose that the centre of a wheel which can roll in a horizontal straight line moves a distance R, the radius of the wheel, from O to O′ (see Fig. 1.5). The point of the wheel originally in contact with the ground moves to P′ and a point Q is now in contact with the ground with the arc QP′ = R. The angle $\alpha = Q\hat{O}'P'$ is defined to have a magnitude of **one radian**. Consequently

2π radians is equivalent to 360°,

1 radian is equivalent to $180°/\pi \cong 57.3°$,

1 degree is equivalent to $\pi/180$ radians.

As the wheel continues to rotate, the angle through which OP turns will also continue to increase. Consequently, in order to describe the motion of the wheel, it is necessary to measure angles which may be larger than 360° (2π radians), and it is convenient to define the trigonometric ratios to allow for any size of angle.

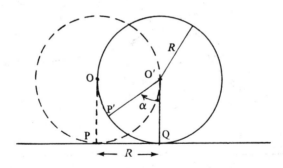

Fig. 1.5

Choose an origin O with Cartesian axes OX, OY so that the coordinates of a point P are (x, y). Referring to Fig. 1.6 the distance of P from O is found, using Pythagoras' theorem, to be $r = (x^2 + y^2)^{\frac{1}{2}} > 0$ (x and y may be positive or negative). The (positive) angle $\theta = A\hat{O}P$ is measured from OX in an anti-clockwise direction. It is sometimes convenient to use the quantities r and θ as alternative coordinates, called **polar coordinates**, to define the position of the point P. They are given in terms of the coordinates x and y by the definitions of $\sin \theta$ and $\cos \theta$ which now follow.

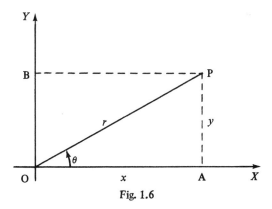

Fig. 1.6

The trigonometric ratios are defined by:

$$\sin\theta = \frac{AP}{OP} = \frac{y}{r}, \qquad \cos\theta = \frac{OA}{OP} = \frac{x}{r},$$

$$\tan\theta = \frac{AP}{OA} = \frac{y}{x}, \qquad \cot\theta = \frac{1}{\tan\theta} = \frac{x}{y},$$

$$\sec\theta = \frac{1}{\cos\theta} = \frac{r}{x}, \qquad \mathrm{cosec}\,\theta = \frac{1}{\sin\theta} = \frac{r}{y}.$$

These results hold for all values of x, y and $\theta > 0$. Since r is greater than or equal to both $|x|$ and $|y|$ then $|\sin\theta| \leqslant 1$, $|\cos\theta| \leqslant 1$, $|\sec\theta| \geqslant 1$, $|\mathrm{cosec}\,\theta| \geqslant 1$, whereas $\tan\theta$ and $\cot\theta$ can take any real value. Other relationships are

$$\tan\theta = \frac{\sin\theta}{\cos\theta}, \qquad \cot\theta = \frac{\cos\theta}{\sin\theta}.$$

A negative angle is measured clockwise from OX and may be thought of as arising from a reflection of a point P in the x-axis in which x and r are unchanged and y and θ change signs. Then

$$\sin(-\theta) = -\sin\theta, \; \cos(-\theta) = \cos\theta, \tan(-\theta) = -\tan\theta.$$

By considering negative and positive values of x and y it is found that all trigonometric ratios are positive in the first quandrant; sine and cosecant only in the second quadrant, etc., so that:

sine positive	all positive
tangent positive	cosine positive

Certain angles occur frequently in mathematical exercises, and the values of their trigonometric ratios are particularly important. Values for the angles

$0(y = 0, x = r)$ and $\pi/2$ $(x = 0, y = r)$ are obtained easily. Values for $\pi/4$ and for $\pi/6$ and $\pi/3$ may be obtained by considering appropriate right-angled triangles and making use of Pythagoras' theorem.

degrees	0	30	45	60	90
radians	0	$\pi/6$	$\pi/4$	$\pi/3$	$\pi/2$
cos	1	$\sqrt{3}/2$	$1/\sqrt{2}$	$1/2$	0
sin	0	$1/2$	$1/\sqrt{2}$	$\sqrt{3}/2$	1
tan	0	$1/\sqrt{3}$	1	$\sqrt{3}$	∞

The values of x and y corresponding to the angles θ and $\theta + 2\pi$ are identical (for the same value of r) so that all trigonometric ratios repeat their values (they are said to be periodic). Once values are available for $0 \leqslant \theta \leqslant 2\pi$ they are available for all values of θ, and graphs may be plotted. The graphs of $\sin \theta$ and $\cos \theta$ are shown in Fig. 1.7. These two graphs have exactly the same smooth shape but one is displaced relative to the other. Fig. 1.8 shows the graph of $\tan \theta$. In this case there are infinite discontinuities at $\theta = -\pi/2, \pi/2, 3\pi/2$ etc., (when $\cos \theta$ is zero). In all cases the periodic repetition is clear.

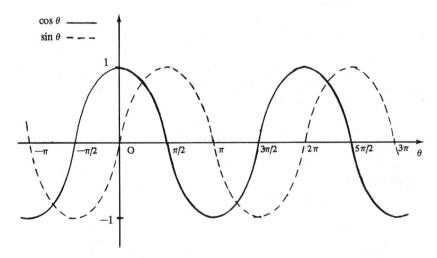

Fig. 1.7

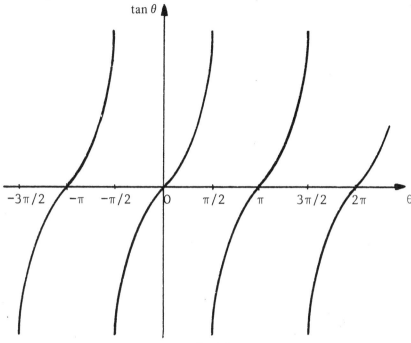

Fig. 1.8

1.14 TRIGONOMETRIC FORMULAE

A large number of trigonometric formulae exist relating the various trigonometric ratios defined in the previous section. A list of those most commonly occurring is given in Appendix A. It is beyond the scope of this book to present proofs here; proofs may be found in any text on trigonometry. It is desirable however to become familiar with the use of these results, and a number of examples will now be discussed. Note that in proving these results there is a steady progression from one side of the identity to the other.

Example 1: Prove that $\sin^2\theta + \cos^2\theta = 1$.

From the definitions of $\sin\theta$ and $\cos\theta$

$$\sin^2\theta + \cos^2\theta = \frac{y^2}{r^2} + \frac{x^2}{r^2} = 1$$

since $x^2 + y^2 = r^2$.

Example 2: Prove the formula for $\tan(A + B)$ using the results for $\sin(A + B)$ and $\cos(A + B)$.

$$\tan(A + B) = \frac{\sin(A + B)}{\cos(A + B)} = \frac{\sin A \cos B + \cos A \sin B}{\cos A \cos B - \sin A \sin B} \ .$$

Divide the numerator and denominator by $\cos A \cos B$ to obtain

$$\tan (A + B) = \frac{\dfrac{\sin A \cos B}{\cos A \cos B} + \dfrac{\cos A \sin B}{\cos A \cos B}}{\dfrac{\cos A \cos B}{\cos A \cos B} - \dfrac{\sin A \sin B}{\cos A \cos B}}$$

$$= \frac{\tan A + \tan B}{1 - \tan A \tan B}.$$

Example 3: Prove that $\dfrac{1 + \cos x}{\sin x} + \dfrac{\sin x}{1 + \cos x} = 2 \operatorname{cosec} x.$

Start with the left-hand side (the more complicated side) and express it over a common denominator

$$\text{LHS} = \frac{(1 + \cos x)^2 + \sin^2 x}{\sin x \, (1 + \cos x)}$$

$$= \frac{1 + 2 \cos x + \cos^2 x + \sin^2 x}{\sin x \, (1 + \cos x)}$$

$$= \frac{2(1 + \cos x)}{\sin x \, (1 + \cos x)} \quad (\text{using } \cos^2 x + \sin^2 x = 1)$$

$$= 2/\sin x = 2 \operatorname{cosec} x.$$

Example 4: Prove that $\dfrac{\sin 3A + \sin A}{\cos 3A + \cos A} = \tan 2A.$

This example involves the use of results (A 24) and (A 25)[†]. Again start with the LHS to obtain

$$\text{LHS} = \frac{2 \sin 2A \cos A}{2 \cos 2A \cos A} = \tan 2A.$$

1.15 THE INVERSE TRIGONOMETRIC FUNCTIONS

In Section 1.13 when the trigonometric ratios were first introduced the quantity

$$\operatorname{cosec} x = 1/\sin x = (\sin x)^{-1} \tag{1.3}$$

was introduced. This is the **multiplicative inverse** of $\sin x$ since $\operatorname{cosec} x \sin x = 1$. Similarly $\sec x$ and $\cot x$ are the multiplicative inverses of $\cos x$ and $\tan x$ respectively. It is necessary to introduce brackets when writing the multiplicative inverse in (1.3) in order to avoid confusion with the inverse function since the

† Appendix A.

expressions $\sin^{-1}x$, $\cos^{-1}x$ and $\tan^{-1}x$ are used to denote the **inverse trigono-metric functions.** This is an unfortunate mathematical notation and tends to create additional difficulties initially. However, its use is now universal so that it would be fruitless to introduce an alternative notation.

The notation $y = \sin^{-1}x$ is used to denote that angle between $-\pi/2$ and $+\pi/2$ whose sine is x. The equation $x = \sin y$ is implied, but the graph of $y = \sin^{-1}x$ consists of that part of the graph of $x = \sin y$ for which $-\pi/2 \leqslant y \leqslant \pi/2$ only, see Fig. 1.9. The restriction imposed on y ensures that there is only one value of y, the **principal value,** for any given value of x. This restriction is to avoid ambiguity since there are many values of y for which the equation $x = \sin y$ holds.

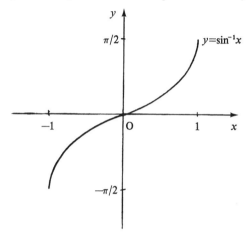

Fig. 1.9

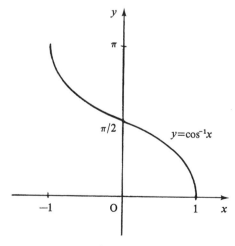

Fig. 1.10

The notation $y = \cos^{-1}x$ denotes that angle between 0 and π whose cosine is x. The equation $x = \cos y$ is implied but $y = \cos^{-1}x$ consists of that part of the graph of $x = \cos y$ for which $0 \leqslant y \leqslant \pi$, see Fig. 1.10.

The notation $y = \tan^{-1}x$ denotes that angle between $-\pi/2$ and $\pi/2$ whose tangent is x. Again $x = \tan y$ is implied but the graph of $y = \tan^{-1}x$ consists of that section of the graph of $x = \tan y$ for which $-\pi/2 \leqslant y \leqslant \pi/2$, see Fig. 1.11.

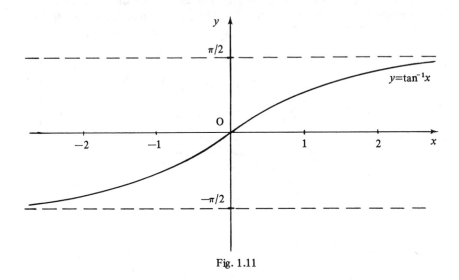

Fig. 1.11

Example: Find the principal value of θ and all values of θ such that (i) $\sin \theta = \frac{1}{2}$, (ii) $\cos \theta = -1/\sqrt{2}$, (iii) $\tan \theta = 1$.

(i) When $\sin \theta = \frac{1}{2}$ the principal value is given by

$$\theta = \sin^{-1}(\tfrac{1}{2}) \text{ with } -\pi/2 \leqslant \theta \leqslant \pi/2.$$

The appropriate value of θ is $\pi/6$.

Since $\sin(\pi - \theta) = \sin \theta$, $\sin\left(\dfrac{5\pi}{6}\right) = \frac{1}{2}$ also.

The values $\pi/6$ and $5\pi/6$ are the two values of θ lying between $-\pi$ and π for which $\sin \theta = \frac{1}{2}$. The general solutions for θ are

$$\theta = \pi/6 + 2n\pi, \text{ where } n \text{ is an integer,}$$

$$\theta = 5\pi/6 + 2m\pi, \text{ where } m \text{ is an integer.}$$

(ii) When $\cos \theta = -1/\sqrt{2}$ the principal value is given by $\theta = \cos^{-1}(-1/\sqrt{2})$ with $0 \leqslant \theta \leqslant \pi$. The appropriate value is $\theta = 3\pi/4$.

Since $\cos(-\theta) = \cos\theta$, $\cos\left(-\dfrac{3\pi}{4}\right) = -1/\sqrt{2}$ also. Thus $-3\pi/4$ and $3\pi/4$ are the two values of θ lying between $-\pi$ and π for which $\cos\theta = -1/\sqrt{2}$. General solutions are

$\theta = 3\pi/4 + 2n\pi$, where n is an integer,

$\theta = -3\pi/4 + 2m\pi$, where m is an integer.

(iii) When $\tan\theta = 1$ the principal value of θ is $\pi/4$. In this case, since $\tan(\pi + \theta) = \tan\theta$, it is possible to express the general solution in the form

$\theta = \pi/4 + n\pi$, where n is an integer.

1.16 SUMMARY AND FURTHER READING

This chapter has been concerned, on the whole, with reviewing results which the reader had encountered previously. Topics of fundamental importance to later work are the factorial notation, proportionality, indices, logarithms, graphical representation and some of the results from trigonometry.

For further information on these topics readers are referred to Smyrl (1978), Sparks and Rees (1937).

PROBLEMS

1. Determine values of x which satisfy the following:

    (i) $x + 1 > 0$,

    (ii) $x - 2 > 0$,

    (iii) $1 - 2x \leqslant 2$,

    (iv) $(x + 1)(x - 2) > 0$,

    (v) $\dfrac{x + 1}{x - 2} < 0$.

2. Express the following numbers in decimal form to three decimal places:

    (i) 7/8,

    (ii) 17/8,

    (iii) 11/9,

    (iv) 35/3,

    (v) 1/30,

    (vi) 7/80.

3. Express the numbers in question 2 to three significant figures.

4. Evaluate the following when $x = -2, 0, 1/2, 2$:
    (i) $|x + 1|$,
    (ii) $|x - 1|$,
    (iii) $|2 - x|$,
    (iv) $||x + 1| - |x - 1|| + |2 - x|$,
    (v) $||x + 1| + |x - 1|| - |2 - x|$.

5. Determine all values of x such that:
    (i) $|x| > 3$,
    (ii) $|x + 1| \leqslant 2$,
    (iii) $|x - 1| < 2$,
    (iv) $|2x + 1| = 2$,
    (v) $|3x - 1| \geqslant 2$.

6. Evaluate the following:
    (i) $4!/2!$,
    (ii) $6! \, 4!/(5! \, 3!)$,
    (iii) $(5! + 4!)/3!$,
    (iv) $5! + 4!/3!$,
    (v) $5!/3! + 4!$,
    (vi) $(n + 1)!/(n - 1)!$.

7. The rate of growth R of a plant in a certain time interval is proportional to $(4 + 3t + t^2)^{\frac{1}{2}}$, where t is the time. Express this statement as an equation. If $R = 6$ when $t = 0$ show that $R = 3(4 + 3t + t^2)^{\frac{1}{2}}$.

8. The volume V of a gas at constant temperature is inversely proportional to the pressure P. Express this statement as an equation.

9. The pressure P of a gas is proportional to the temperature T and also inversely proportional to the volume V. Express this statement as an equation.

10. The volume V and surface area S of a sphere are proportional to the cube and square of the radius r respectively. Express these statements as equations and deduce that $V^2 \propto S^3$.

11. The volume V of a sphere is proportional to the cube of the radius r. Deduce that if the radius of a sphere is doubled in magnitude then its volume is increased by a factor 8.

12. Evaluate the following if $p = 2, q = 3, r = -4$:

   (i) $pq + r$,

   (ii) pq/r,

   (iii) $(p + q)/r$,

   (iv) $p + q/r$,

   (v) $\dfrac{p - q}{\dfrac{1}{p} + \dfrac{1}{q}}$,

   (vi) $\dfrac{\dfrac{1}{p} - \dfrac{1}{r}}{p + r}$.

13. Express the following numbers in the form $2^p \times 3^q$:

   (i) $\dfrac{6^2 \times 8}{12^3}$,

   (ii) $\dfrac{6^3 \times 8^2}{12^4}$,

   (iii) $\dfrac{6^4 \times 12}{4^5}$,

   (iv) $6^{10}\, 4^9\, 12^{-8}$.

14. Express the cube root of 200 divided by the square root of 40 in the form $2^p \times 5^q$.

15. Solve the following equations for x:

   (i) $(x + a)b = c$,

   (ii) $x + ab = c$,

   (iii) $x/a + b = c$,

   (iv) $a/x + b = c$,

   (v) $a/x + b = c/x$,

   (vi) $(a + x)b + c = x(a + 1)$,

(vii) $ax + a(b + x) = c$,

(viii) $ax - c(b + x) = a$,

(ix) $(a + b)/x + c = (a - b)/x$,

(x) $(a - b)x - c = (b - c)x - d$.

16. Find the values of the constants a and b in the following identities:

(i) $\dfrac{1}{1 + 4x^2} \equiv \dfrac{a}{b + x^2}$,

(ii) $(1 - 9x^2)^{\frac{1}{2}} \equiv a(b - x^2)^{\frac{1}{2}}$,

(iii) $(1 + 3x)^2 \equiv a(b + x)^2$,

(iv) $(1 + 3x)^{\frac{1}{2}} \equiv a(b + x)^{\frac{1}{2}}$,

(v) $\dfrac{1}{(1 - 9x^2)^{\frac{1}{2}}} \equiv \dfrac{a}{(b - x^2)^{\frac{1}{2}}}$.

17. Given that $\log_{10} 4.78 = 0.6794$ evaluate:

(i) $\log_{10} 47.8$,

(ii) $\log_{10} 0.478$,

(iii) $\log_{10} 4780$,

(iv) $\log_{10} 0.0478$.

18. Given that $\log_{10} 6.85 = 0.8357$ write down the numbers whose log to base 10 is:

(i) 1.8357,

(ii) 3.8357,

(iii) $\overline{1}.8357$,

(iv) $\overline{3}.8357$,

(v) -0.1643.

19. If $\log_{10} 2 = x$ and $\log_{10} 3 = y$, find, in terms of x and y, the values of:

(i) $\log_{10} 6$,

(ii) $\log_{10}(4/3)$,

(iii) $\log_{10} 72$,

(iv) $\log_{10}(1/36)$.

20. Given that $\log_{10} y = \alpha \log_{10} x + \beta$ where α, β are constants and that $y = 10$ when $x = 0.1$ and $y = 100$ when $x = 1$ determine α and β.

21. By taking logs find the smallest positive integer n for which $4^n < 5^{n-1}$.

22. A biological population, initially of size 1000, doubles its size every day. The size of the population after n generation times is $N = 10^3 \, 2^n$. Find, using logarithms, the number of generations that must elapse before the size exceeds (i) 10^5 and (ii) 10^{10}.

23. The volume V of timber in a given tree increases by 5% each year so that $V = a(1.05)^t$ where t is the time in years and a is the volume at time $t = 0$.

    (i) Calculate the volume of the tree after 5 years.

    (ii) Find the time taken for the tree to double its volume.
    Give both answers to three significant figures.

24. In the study of hearing the loudness L is expressed in terms of the intensity I by the equation

    $$L = 10 \log_{10}(I/I_0)$$

    where I_0 is approximately 10^{-12} Watt/m^2. Express I in terms of L and determine I when $L = 60$.

25. Plot the graph of the exponential function $y = e^x$ using the values in the table

x	−2.0	−1.5	−1.0	−0.5	0	0.5	1.0	1.5	2.0
y	0.1353	0.2231	0.3679	0.6065	1.0	1.649	2.718	4.482	7.389

26. Protein is synthesised in such a way that the mass m of protein at time t is given by

    $$m = 500 + 50t + 4t^2 .$$

    Plot the graph of m against t for $0 \leqslant t \leqslant 6$.

27. The height h of a plant at time t is given in the table

t	0	1.0	2.0	3.0	4.0	5.0	6.0	7.0
h	2.00	2.83	3.74	4.69	5.66	6.63	7.62	8.60

    Plot a graph of h against t.

28. Use the results for $\sin(A - B)$ and $\cos(A - B)$ to prove that

    $$\tan(A - B) = \frac{\tan A - \tan B}{1 + \tan A \tan B} .$$

    By taking $A = \pi/3$ and $B = \pi/4$ deduce that $\tan(\pi/12) = 2 - \sqrt{3}$.

29. Prove that:

(i) $(1 - \cos A)(1 + \cos A) = \sin^2 A,$

(ii) $(\cos A + \sin A)^2 = 1 + \sin 2A,$

(iii) $\dfrac{1 - \cos A}{\sin A} + \dfrac{\sin A}{1 - \cos A} = 2 \operatorname{cosec} A,$

(iv) $\dfrac{\sin A + \sin B}{\cos A + \cos B} = \tan\tfrac{1}{2}(A + B),$

(v) $\dfrac{\sin A - \sin B}{\cos A - \cos B} = -\cot\tfrac{1}{2}(A + B).$

30. Find, in radian measure, the principal value of θ when

(i) $\cos \theta = 1/2,$

(ii) $\cos \theta = 1/\sqrt{2},$

(iii) $\tan \theta = -1,$

(iv) $\tan \theta = \sqrt{3},$

(v) $\cos \theta = -1/2,$

(vi) $\sin \theta = -1/2,$

(vii) $\cot \theta = -1,$

(viii) $\sec \theta = 2.$

31. Find, in radians, all values of θ when

(i) $\tan \theta = -1,$

(ii) $\cos \theta = 1/2,$

(iii) $\sin \theta = -1/2.$

32. Use the relationships

$$\tan(A + B) = \frac{\tan A + \tan B}{1 - \tan A \tan B},$$

and

$$\tan(A - B) = \frac{\tan A - \tan B}{1 + \tan A \tan B},$$

to show that

$$\tan^{-1} 1 + \tan^{-1}\frac{1}{2} = \tan^{-1} 3,$$

$$\tan^{-1} 1 - \tan^{-1}\frac{1}{2} = \tan^{-1}\frac{1}{3}.$$

33. Plot graphs of

   (i) $y = |\sin x|$,

   (ii) $y = |\tan x|$,

   (iii) $y = |\cos 2x|$.

34. The bio-rhythmical replication of two types of cells, y_1 and y_2, in response to antibodies are described by the equations

$$y_1 = 2a + a \cos (\pi t/3) \ ,$$

$$y_2 = 2a - a \cos \{\pi(t - 1)/5\} \ .$$

   Plot the corresponding graphs for $t \geqslant 0$ and deduce from your graphs that y_1 and y_2 both take maximum values when $t = 6$.

35. (i) In three successive years, the size of a population increases by 20%, increases by 20%, and decreases by 25%. Find the overall percentage change.

   (ii) The size of a population, initially N_0, increases by 7.15% each year. Write down an expression for N_n, the size of the population after n years.
   Use logarithms to find the least value of n, a positive integer, for which N_n exceeds $6N_0$.   (Edinburgh, 1978)

36. Find (showing all your working) the value of

$$\frac{4! + 5!}{6! + 7!} \ .$$

   Express this value in standard form, that is, in the form $A \times 10^n$, where $1 \leqslant A < 10$ and n is an integer.   (Edinburgh, 1978)

37. (i) In a certain bacterial culture, each cell divides 30 minutes after its creation; there are 6×10^3 cells initially. If the initial cells are newly created:

   (a) How many cells are there after two hours?

   (b) After how many more hours does the number of cells exceed 10^6?

   (ii) In two successive years, a population increases in size by 50% and 20% respectively, and in the following two years it decreases in size by 10% and 50%. Show that there is an overall decrease in the population size, which is equivalent to a decrease of approximately 5% for each of the four years.   (Edinburgh, 1977)

38.   (i) How do you know, without calculation, that $\log_5 7 > 1$?

      (ii) Given that

$$\log_b N = \log_a N / \log_a b \ ,$$

use common logarithms to show that $\log_5 7 = 1.21$ (to 2 decimal places).

(Edinburgh, 1977)

39.   (i) Initially, a bacterial culture consists of x cells. Each of these cells, and each of its descendants, divides 20 minutes after its creation. How many cells are there after 2 hours?

      (ii) Express

$$\frac{8.4 \times 10^2}{(8 \times 10^6) \times (1.2 \times 10^{-3})}$$

in the form $A \times 10^n$, where n is an integer and $1 \leqslant A < 10$.

      (iii) A population is initially of size N. In the first year, the population increases in size by 10%, and in the following year it decreases in size by 10%. What is the overall percentage change in population size?

(Edinburgh, 1977)

40.   (i) Explain why $4 < \log_2 20 < 5$ without first finding the value of $\log_2 20$. By using the definition of a logarithm, prove that

$$\log_2 A = \log_{10} A / \log_{10} 2 \ .$$

      (ii) Evaluate $\log_2 20$ correct to 2 decimal places.   (Edinburgh, 1976)

41.   A population increases at the rate of 2% per annum for 40 years. Show that the overall increase is about 121%.

A population increases by $x\%$ in one year, and then by $3x\%$ in the following year. Show that the size of the population at the end of the second year is less than if there had been an increase of $2x\%$ in each of these two years.   (Edinburgh, 1975)

CHAPTER 2

Sequences and series

2.1 INTRODUCTION

A number of simple series that occur in applications of mathematics to biology are discussed in this chapter. The particular topics discussed are sequences, the arithmetic series, the geometric series and the binomial theorem. The general ideas involved in the consideration of the binomial theorem are particularly relevant to the later chapters on probability and statistics.

2.2 SEQUENCES AND SERIES

A **sequence** $\{u_n\}$ is an ordered set of symbols of the form

$$u_1, u_2, \ldots, u_n, \ldots \, , \qquad\qquad (2.1)$$

where n is an integer, and is such that each value of u is associated in turn with each natural number.

Thus u_1 is associated with the number 1, u_2 with the number 2, and in general the term u_n is associated with the number n. Sequences take many forms; some simple examples are:

$\{2n\}$ for which $\quad u_1 = 2, u_2 = 4, u_3 = 6, \ldots, u_n = 2n, \ldots$

$\{x^n\}$ which denotes $x, x^2, x^3, \ldots, x^n, \ldots$

$\{n^2\}$ which denotes $1, 4, 9, \ldots, n^2, \ldots$

$\{\dfrac{1}{n}\}$ which denotes $1, \dfrac{1}{2}, \dfrac{1}{3}, \ldots, \dfrac{1}{n}, \ldots$

In all of these examples the one-to-one correspondence with the natural numbers is clear.

Given a sequence such as (2.1) it is possible to form a second sequence denoted by $\{S_n\}$ with terms S_n that are the sums of the first n terms of the sequence $\{u_n\}$. That is

$$\left.\begin{aligned}
S_1 &= u_1 , \\
S_2 &= u_1 + u_2 = S_1 + u_2 , \\
S_3 &= u_1 + u_2 + u_3 = S_2 + u_3 , \\
S_n &= u_1 + u_2 + \ldots + u_{n-1} + u_n = S_{n-1} + u_n .
\end{aligned}\right\} \quad (2.2)$$

Thus the first four terms in the sequence $\{S_n\}$ derived from the sequence $\{2n\}$ are

$$\begin{aligned}
S_1 &= u_1 = 2 , \\
S_2 &= u_1 + u_2 = 2 + 4 = 6 , \\
S_3 &= u_1 + u_2 + u_3 = 2 + 4 + 6 = 12 , \\
S_4 &= u_1 + u_2 + u_3 + u_4 = 2 + 4 + 6 + 8 = 20 .
\end{aligned}$$

For any given sequence $\{u_n\}$ it is possible to write down an **infinite series** S where

$$S = u_1 + u_2 + u_3 + \ldots + u_n + \ldots \quad (2.3)$$

is formed by adding together the infinity of terms from the sequence $\{u_n\}$. When S has a definite value it is also referred to as the **sum of the infinite series**. Clearly S is closely related to the sequence $\{S_n\}$. The terms S_n are called the **partial sums** of the series S since S_n denotes the sum of the first n terms of S. The terms S_n are sometimes referred to as **finite series** since their structure is similar to that of S and they contain a finite number of terms.

The symbol Σ is used in mathematics as an abbreviated notation for 'the sum of'. Thus $\sum_{k=1}^{3} \frac{1}{k+1}$ means the sum of all terms of the form $1/(k+1)$ taking k as successive integers from 1 to 3 inclusive so that

$$\sum_{k=1}^{3} \frac{1}{k+1} = \frac{1}{2} + \frac{1}{3} + \frac{1}{4} .$$

Using this notation it is possible to write equations such as those in (2.2) in the compact forms

$$S_3 = u_1 + u_2 + u_3 = \sum_{k=1}^{3} u_k ,$$

$$S_n = u_1 + u_2 + \ldots + u_n = \sum_{k=1}^{n} u_k ,$$

and the series S given by (2.3) may be written

$$S = \sum_{k=1}^{\infty} u_k .$$

Note that $S_n = \sum\limits_{k=1}^{n} u_k = \sum\limits_{t=1}^{n} u_t$,

and that S_n contains n terms (u_1 to u_n inclusive).

Example: A parasite's egg completes 1/10 of its development for every day that it is kept at 20 °C and 1/20 of its development for every day at 15°C. Determine the sequence of development fractions (the fraction of complete development) for each day if the egg is kept at 20°C and 15°C on alternate days, commencing with a day at 20°C. If the egg hatches when the development fraction sum reaches one, after how many days will this occur?

The development fraction on day 1 is $\dfrac{1}{10}$,

the development fraction on day 2 is $\dfrac{1}{20}$,

the development fraction on day 3 is $\dfrac{1}{10}$,

so that the sequence is $\left\{ \dfrac{1}{10}, \dfrac{1}{20}, \dfrac{1}{10}, \dfrac{1}{20}, \dfrac{1}{10}, \cdots \right\}$.

The development fraction sum on day 1 is $\dfrac{1}{10}$,

the development fraction sum on day 2 is $\dfrac{1}{10} + \dfrac{1}{20} = \dfrac{3}{20}$,

the development fraction sum on day 3 is $\dfrac{1}{10} + \dfrac{1}{20} + \dfrac{1}{10} = \dfrac{5}{20}$,

so that the sequence of development fraction sums is

$$\left[\frac{2}{20}, \frac{3}{20}, \frac{5}{20}, \frac{6}{20}, \frac{8}{20}, \frac{9}{20}, \frac{11}{20}, \frac{12}{20}, \frac{14}{20}, \frac{15}{20}, \frac{17}{20}, \frac{18}{20}, \frac{20}{20}, \frac{21}{20} \cdots \right] ,$$

and the egg hatches on day 13.

2.3 THE ARITHMETIC SERIES

The sequence $\{u_n\}$ where

$$u_1 = a,$$
$$u_2 = a + d,$$
$$u_3 = a + 2d,$$
$$u_n = a + (n-1)d, \tag{2.4}$$

which is constructed by taking the first term u_1 equal to a and forming successive terms by adding the same quantity d (the **common difference**) to the preceding term is called an **arithmetic progression** (AP). A second sequence $\{S_n\}$ may be constructed as in (2.2) from $\{u_n\}$ by summing the terms of the sequence $\{u_n\}$. Thus

$$S_2 = u_1 + u_2 = a + (a+d) = 2a+d$$

and in general

$$S_n = a + (a+d) + (a+2d) + \ldots + \{a + (n-1)d\} \ . \qquad (2.5)$$

The quantity S_n, called the **arithmetic series** of order n, is the sum of the first n terms of the arithmetic progression whose first term is a and whose common difference is d. Equation (2.5) is not a convenient expression from which to make calculations, and a more suitable form will now be obtained.

The expression for S_n, (2.5), may be written in natural order (2.5) or in reverse order so that

$$S_n = a + (a+d) + \ldots + \{a + (n-2)d\} + \{a + (n-1)d\}$$

and $\qquad S_n = \{a + (n-1)d\} + \{a + (n-2)d\} + \ldots + (a+d) + a \ .$

Adding these expressions term by term results in

$$2S_n = \{2a + (n-1)d\} + \{2a + (n-1)d\} + \ldots + \{2a + (n-1)d\}$$

so that there are now n terms each of magnitude $\{2a + (n-1)d\}$. Thus

$$2S_n = \{2a + (n-1)d\} \ ,$$
$$S_n = \tfrac{1}{2}n\{2a + (n-1)d\} \ . \qquad (2.6)$$

Note that in deriving this result no restriction has been placed on the values or signs of a and d. It is not realistic to consider the infinite series that corresponds to (2.6) since its sum cannot be finite as n approaches infinity.

Example 1: Find the sum of (a) the first five natural numbers, (b) the first n natural numbers.

(a) The first five natural numbers $1, 2, 3, 4, 5$ constitute the sequence $\{u_n\}$ in the previous notation with $a = 1$, $d = 1$ and $u_n = 1 + (n-1).1 = n$. Setting $n = 5$ in (2.6) gives

$$S_5 = \tfrac{1}{2}\times 5\{2\times 1 + 4\times 1\} = 15 = 1 + 2 + 3 + 4 + 5 \ .$$

(b) The general result is obtained by taking $a = 1$ and $d = 1$ in (2.6) so that

$$S_n = \tfrac{1}{2}n\{2 + n - 1\} = \tfrac{1}{2}n(n+1) \ .$$

Example 2: If the sum of the first ten terms of an arithmetic series and the tenth term of the series both have the value $-5/2$, determine the value of the first term and the common difference.

In this example the values of n and S_n are given but a and d are unknown.

The information that we are given provides two equations from which to determine a and d. With $n = 10$, (2.6) gives

$$S_{10} = 5\{2a+9d\} = -5/2$$

so that $\qquad\qquad 2a+9d = -1/2$.   (2.7)

Equation (2.4) gives the 10th term as

$$u_{10} = a + 9d = -5/2$$

so that $\qquad a = -5/2 - 9d$.   (2.8)

This may be substituted into (2.7) to give

$$-5 - 18d + 9d = -1/2$$
$$- 9d = 5 - 1/2 = 9/2$$
$$d = -1/2 \ .$$

From (2.8) $\quad a = -5/2 + 9/2 = 2$.

Hence the first term is 2 and the common difference $-\frac{1}{2}$.

Example 3: A population is divided into $n(>1)$ age groups each containing a members such that during the following year each group has a different capacity for reproduction. The first group is unable to reproduce, the second group increases the population by d members, the third group by $2d$ members and so on. Find the total number in the population at the end of the year if there are no deaths.

At the end of the year the population resulting from group 1 is a,

from group 2 is $a + d$,

from group 3 is $a + 2d$,

. . . . .

from group n is $a + (n-1)d$.

The total of the population is

$$a + (a+d) + (a+2d) + \ldots + \{a + (n-1)d\}$$

which is an arithmetic series (see (2.5)) with sum, from (2.6)

$$S_n = \tfrac{1}{2}n\{2a + (n-1)d\} \ .$$

2.4 THE GEOMETRIC SERIES

The sequence $\{u_n\}$ defined by

$$u_1 = a,$$
$$u_2 = ar,$$
$$u_3 = ar^2,$$
$$u_n = ar^{n-1},$$ (2.9)

which is constructed by taking the first term u_1 equal to a and by forming successive terms equal to the previous term multiplied by the **common ratio** r is called a **geometric progression** (GP). The sequence $\{S_n\}$ which is constructed by taking the sum of the first n terms in $\{u_n\}$ is

$$S_1 = u_1 = a \ ,$$
$$S_2 = u_1 + u_2 = a(1+r) \ ,$$
$$S_3 = u_1 + u_2 + u_3 = a(1+r+r^2) \ ,$$
$$S_n = a(1 + r + r^2 + \ldots + r^{n-1}) = \sum_{k=1}^{n} ar^{k-1} \ . \qquad (2.10)$$

The quantity S_n gives the sum of the first n terms of the **geometric series** whose first term is a and which has common ratio r.

A more convenient form than (2.10) for S_n is obtained by writing

$$S_n = a + ar + ar^2 + \ldots + ar^{n-1}$$
$$rS_n = \quad\;\; ar + ar^2 + \ldots + ar^{n-1} + ar^n \ .$$

Subtracting these equations results in

$$(1-r)S_n = a - ar^n$$

since terms cancel in pairs. Then

$$S_n = \frac{a(1-r^n)}{1-r} = \frac{a(r^n-1)}{r-1} \ . \qquad (2.11)$$

These expressions are valid for all finite values of a and r (positive or negative) except $r = 1$. The case $r = 1$ is trivial since (2.10) gives $S_n = na$ in that case.

As n become very large the value u_n given by (2.9) becomes increasingly large in magnitude if $|r| > 1$ and decreases in magnitude if $|r| < 1$. It can be shown rigorously that if $|r| < 1$ then the infinite series

$$S = a + ar + ar^2 + \ldots + ar^n + \ldots = \sum_{k=1}^{\infty} ar^{k-1} \ ,$$

has the sum

$$S = \frac{a}{1-r} \ . \qquad (2.12)$$

This value for S is obtained from (2.11) by letting n become increasingly large as the magnitude of r^n decreases to zero for $|r| < 1$.

Example 1: Find the sum of the series defined by $S_n = 1 + \frac{1}{2} + (\frac{1}{2})^2 + \ldots + (\frac{1}{2})^{n-1}$ when (a) $n = 4$, (b) $n = 8$. Calculate the sum to infinity.

Here $a = 1, r = \frac{1}{2}$ so that from (2.11)

(a) $S_4 = \dfrac{1 - (\frac{1}{2})^4}{1 - \frac{1}{2}} = 2(1 - \dfrac{1}{16}) = \dfrac{15}{8} = 1.875 = 1 + \frac{1}{2} + \frac{1}{4} + \frac{1}{8}$.

(b) $S_8 = \dfrac{1 - (\frac{1}{2})^8}{1 - \frac{1}{2}} = 2(1 - \dfrac{1}{256}) = \dfrac{255}{128} \cong 1.9922$.

Using (2.12) $S = \dfrac{1}{1 - \frac{1}{2}} = 2.0$.

This example illustrates that the values of S_n approach S rapidly even for r as large as $\frac{1}{2}$.

Example 2: Each hour a cell gives birth by dividing into two new cells. Write down the sequence corresponding to the number of cells present at the beginning of each hour and determine the total number of cells that have existed to the start of the 10th hour.

At the beginning of hour 1 there is 1 cell,

at the beginning of hour 2 there are 2 cells,

at the beginning of hour 3 there are $2 \times 2 = 2^2 = 4$ cells,

at the beginning of hour 4 there are $4 \times 2 = 2^3 = 8$ cells,

. . . .

at the beginning of hour n there are 2^{n-1} cells.

The total number of cells that have existed up to the beginning of the 10th hour is

$$S_{10} = 1 + 2 + 2^2 + \ldots + 2^9 .$$

This is a geometric series with $a = 1, r = 2, n = 10$ so that from (2.11)

$$S_{10} = \dfrac{1(2^{10} - 1)}{2 - 1}$$

$$= 2^{10} - 1 = 1024 - 1 = 1023 .$$

Example 3: Under prescribed conditions the aperture of a skin pore shrinks by a factor of $1/10$ for each unit of time. After how many complete time intervals will the aperture first be less than half its initial size?

Let the aperture have initial size A.

At the end of time unit 1 the aperture size is $A - \dfrac{1}{10}A = \dfrac{9}{10}A$,

at the end of time unit 2 the aperture size is $\left(\dfrac{9}{10}\right)^2 A$,

. . .

at the end of time unit n the aperture size is $\left(\dfrac{9}{10}\right)^n A$.

It is required that

$$\left(\frac{9}{10}\right)^n A < \frac{1}{2}A$$

so that, $2 < (10/9)^n$, since $A > 0$.

Taking logarithms to base 10 gives

$$\log_{10} 2 < n \log_{10}(10/9) = n\{\log_{10}10 - \log_{10}9\}$$

$$0.3010 < n\{1 - 0.9542\} = 0.0458n .$$

Hence $n > \dfrac{0.3010}{0.0458} \cong 6.57$

and the aperture is first less than half its initial size at the end of time unit 7.

2.5 THE BINOMIAL THEOREM FOR INTEGRAL INDEX

This section is concerned with the expansion of $(1+x)^n$, where n is a positive integer, in a series of terms each of which is a power of x multiplied by a quantity depending on n. For small values of n it is possible to obtain the required result by multiplying out. Examples are

$$
\begin{aligned}
n &= 0, \ (1+x)^0 = 1, \\
n &= 1, \ (1+x)^1 = 1 + x, \\
n &= 2, \ (1+x)^2 = (1+x)(1+x)^1 = 1 + 2x + x^2 , \\
n &= 3, \ (1+x)^3 = (1+x)(1+x)^2 = 1 + 3x + 3x^2 + x^3 , \quad (2.13) \\
n &= 4, \ (1+x)^4 = (1+x)(1+x)^3 = 1 + 4x + 6x^2 + 4x^3 + x^4 , \\
n &= 5, \ (1+x)^5 = (1+x)(1+x)^4 = 1 + 5x + 10x^2 + 10x^3 \\
&\qquad\qquad\qquad\qquad\qquad\qquad\qquad\qquad\quad + 5x^4 + x^5 .
\end{aligned}
$$

Clearly $(1+x)^6$ is obtained by multiplying the expansion of $(1+x)^5$ by $(1+x)$, and this process could be continued indefinitely.

The expansions on the right-hand side of (2.13) have coefficients which correspond to the symmetrical array

$$
\begin{array}{ccccccccccc}
 & & & & & 1 & & & & & \\
 & & & & 1 & & 1 & & & & \\
 & & & 1 & & 2 & & 1 & & & \\
 & & 1 & & 3 & & 3 & & 1 & & \\
 & 1 & & 4 & & 6 & & 4 & & 1 & \\
1 & & 5 & & 10 & & 10 & & 5 & & 1
\end{array}
\qquad (2.14)
$$

known as **Pascal's triangle**. Each number in Pascal's triangle, with the exception of the two 1's on each row, is the sum of the two numbers immediately above it. Thus the next row is

$$1 \quad (1+5) \quad (5+10) \quad (10+10) \quad (10+5) \quad (5+1) \quad 1$$

that is $1 \quad 6 \quad 15 \quad 20 \quad 15 \quad 6 \quad 1$.

Multiplying out, as in (2.13), or using Pascal's triangle (2.14) is not a satisfactory method of obtaining the expansion of $(1+x)^n$ for large values of n. These processes lead us to the **binomial theorem** when n is a positive integer in the form

$$(1+x)^n = 1 + \frac{n}{1}x + \frac{n(n-1)}{1.2}x^2 + \frac{n(n-1)(n-2)}{1.2.3}x^3 + \ldots + x_n$$

$$= \binom{n}{0}x^0 + \binom{n}{1}x^1 + \binom{n}{2}x^2 + \binom{n}{3}x^3 + \ldots + \binom{n}{n}x^n$$

$$= \sum_{r=0}^{n} \binom{n}{r}x^r . \tag{2.15}$$

The symbol $\binom{n}{r}$ denotes the **binomial coefficient** for n and r given by

$$\binom{n}{r} = \frac{n!}{r!(n-r)!} , \quad r = 0, 1, 2, \ldots, n ,$$

or $$\binom{n}{r} = \frac{n(n-1)(n-2)\ldots(n-r+1)}{1.2.3\ldots r}, \quad r = 1, 2, 3, \ldots, n . \tag{2.16}$$

The first form for $\binom{n}{r}$ is valid for $r = 0$ since $0! = 1$ by definition. A useful property of the binomial coefficient is

$$\binom{n}{n-r} = \frac{n!}{(n-r)!r!} = \binom{n}{r} . \tag{2.17}$$

This property of the coefficients corresponds to the symmetry in the expansions (2.13) and Pascal's triangle (2.14). The binomial coefficient is sometimes denoted by ^nC_r (see Chapter 20).

A more general form for the binomial theorem when n is a positive integer is

$$(y+z)^n = y^n + \frac{n}{1}y^{n-1}z + \frac{n(n-1)}{1.2}y^{n-2}z^2 + \ldots + \frac{n}{1}yz^{n-1} + z^n$$

$$= \binom{n}{0}y^n z^0 + \binom{n}{1}y^{n-1}z^1 + \binom{n}{2}y^{n-2}z^2 + \ldots + \binom{n}{n-1}y^1 z^{n-1}$$

$$+ \binom{n}{n}y^0 z^n$$

$$= \sum_{r=0}^{n} \binom{n}{r}y^{n-r} z^r . \tag{2.18}$$

The expansions (2.15) and (2.18) appear reasonable in view of (2.13) and (2.14) and may be proved rigorously. A typical term (the $(r+1)^{\text{th}}$ term) in (2.18) is given by

$$\binom{n}{r} y^{n-r} z^r. \tag{2.19}$$

Note that in this expression the sum of the indices of y and z is $(n-r) + r = n$.

Example 1: Evaluate $\binom{18}{12}$.

Equation (2.17) with $n = 18$ and $r = 12$ gives

$$\binom{18}{12} = \binom{18}{6} = \frac{18.17.16.15.14.13}{1.2.3.4.5.6} = 18\,564 \ .$$

It is usual when expanding by the binomial theorem for large values of n to leave the coefficients in the form $\binom{n}{r}$, or in factorial form, rather than evaluating them explicitly, since the numbers may become very large.

Example 2: Use the binomial theorem to expand (a) $(1+x)^5$, (b) $(1-y)^5$.
    (a) Equation (2.15) gives

$$(1+x)^5 = \binom{5}{0} + \binom{5}{1}x + \binom{5}{2}x^2 + \binom{5}{3}x^3 + \binom{5}{4}x^4 + \binom{5}{5}x^5$$

where $$\binom{5}{0} = \binom{5}{5} = \frac{5!}{0!5!} = 1$$

$$\binom{5}{1} = \binom{5}{4} = \frac{5!}{1!4!} = 5$$

$$\binom{5}{2} = \binom{5}{3} = \frac{5!}{2!3!} = 10 \ .$$

This now gives

$$(1+x)^5 = 1 + 5x + 10x^2 + 10x^3 + 5x^4 + x^5 \tag{2.20}$$

as in (2.13).

    (b) The expansion for $(1-y)^5$ may be obtained immediately by replacing x in (2.20) by $-y$ since

$$(1+x)^5 = (1-y)^5 = 1 + 5(-y) + 10(-y)^2 + 10(-y)^3 + 5(-y)^4$$
$$+ (-y)^5$$
$$= 1 - 5y + 10y^2 - 10y^3 + 5y^4 - y^5 \ .$$

Example 3: Determine the coefficient of x in the expansion of $\left(x - \dfrac{2}{x}\right)^{11}$.

It is not necessary to write out the full expansion in order to obtain the information requested. The expression $\left(x - \dfrac{2}{x}\right)^{11}$ is of the form $(y+z)^{11}$ provided that $y = x$ and $z = -2/x$ so that a typical term is given by (2.19) with $n = 11$. It has the form

$$\binom{11}{r} y^{11-r} z^r = \binom{11}{r} (x)^{11-r} \left(-\frac{2}{x}\right)^r$$

$$= \binom{11}{r} (-2)^r x^{11-r-r} . \tag{2.21}$$

The whole number r has not been specified in writing down (2.21). It is determined by requiring (2.21) to correspond to the term in x; that is, the index of x in (2.21) is to be made equal to unity so that

$$11 - 2r = 1$$

$$r = 5 .$$

The expression (2.21) with $r = 5$ gives

$$\binom{11}{5} (-2)^5 x$$

so that the coefficient of x is $-\binom{11}{5} 2^5$.

Example 4: The volume of a spherical cell of radius r is given by $V = \dfrac{4}{3} \pi r^3$.

Show that if r is increased by 1% the resulting increase in V is approximately 3%.

After an increase of 1% in r the new radius is $r + \dfrac{r}{100} = r\left(1 + \dfrac{1}{100}\right)$ and the new volume V' is given by

$$V' = \frac{4}{3}\pi r^3\left(1 + \frac{1}{100}\right)^3 = V\left(1 + \frac{1}{100}\right)^3 .$$

Clearly it is possible to evaluate $\left(\dfrac{101}{100}\right)^3$ by a calculator or logarithms but an approximate answer may be obtained by using the binomial theorem for $(1+x)^3$ and taking $x = \dfrac{1}{100} = 10^{-2}$. It is found that

$$\left(1 + \frac{1}{100}\right)^3 = 1 + 3.10^{-2} + 3.10^{-4} + 10^{-6}$$

and since 10^{-4} and 10^{-6} are much smaller than 10^{-2} an approximate value is given by

$$\left(1 + \frac{1}{100}\right)^3 \cong 1 + \frac{3}{100}$$

so that $\quad V' \cong V\left(1 + \frac{3}{100}\right)$

and the volume has increased by approximately 3%.

2.6 THE BINOMIAL THEOREM FOR GENERAL INDEX

The binomial theorem for positive integer index which was discussed in the previous section is valid for all values of x, although example 4 of Section 2.5 illustrates that for small values of x an approximate value may be obtained without taking all the terms. This process of approximation can be very useful in certain physical applications and it is helpful to have a suitable expansion process for $(1+x)^\alpha$ when α is a real number. Usually α is a rational number.

The appropriate series expansion which is stated here without proof is similar to the expansion for positive integer values of the index and is

$$(1+x)^\alpha = 1 + \frac{\alpha}{1!}x + \frac{\alpha(\alpha-1)}{2!}x^2 + \frac{\alpha(\alpha-1)(\alpha-2)}{3!}x^3$$
$$+ \frac{\alpha(\alpha-1)(\alpha-2)(\alpha-3)}{4!}x^4 + \ldots \qquad (2.22)$$

provided that $|x| < 1$. This result can be proved rigorously, but the proof is more difficult than that for integer values and will not be given here. Note that the expansion contains an infinite number of terms when α is not a whole number, and the restriction $|x| < 1$ is necessary to ensure that the terms decrease in value sufficiently quickly as the number of terms increases. Since it is not possible to write down all the terms in the expansion it is usual to present the first few terms, the number of terms being chosen to give the required accuracy.

In order to expand $(y+z)^\alpha$ when α is a real number it is necessary to write

$$(y+z) = y\left(1 + \frac{z}{y}\right)$$

so that $\quad (y+z)^\alpha = y^\alpha\left(1 + \frac{z}{y}\right)^\alpha = y^\alpha(1+x)^\alpha$

where $x = z/y$. The expansion (2.22) may now be used.

Example 1: Use the binomial theorem to expand $(1+x)^{\frac{1}{2}}$ up to the term in x^3 and hence determine $(1.05)^{\frac{1}{2}}$ to 4 places of decimals.

Here $\alpha = \frac{1}{2}$ so that (2.22) gives

$$(1+x)^{\frac{1}{2}} = 1 + \frac{\frac{1}{2}}{1!}x + \frac{\frac{1}{2}(-\frac{1}{2})}{2!}x^2 + \frac{(\frac{1}{2})(-\frac{1}{2})(-\frac{3}{2})}{3!}x^3 + \ldots$$

$$= 1 + \frac{1}{2}x - \frac{1}{8}x^2 + \frac{1}{16}x^3 + \ldots \ldots$$

An approximate value for $(1.05)^{\frac{1}{2}}$ is obtained by setting $x = 0.05$ in the expansion for $(1+x)^{\frac{1}{2}}$. It is found that

$$(1.05)^{\frac{1}{2}} \cong 1 + \frac{1}{2}(0.05) - \frac{1}{8}(0.05)^2 + \frac{1}{16}(0.05)^3$$

$$= 1 + 0.025 - 0.0003125 + 0.0000078125$$

$$= 1.0247 \quad \text{to four places of decimals.}$$

Example 2: Use the binomial theorem to expand $1/(1-y)$ up to the term in y^4.

Here $\quad \dfrac{1}{1-y} = (1-y)^{-1}$

which is of the form $(1+x)^\alpha$ provided that $x = -y$ and $\alpha = -1$. Equation (2.22) gives

$$(1-y)^{-1} = 1 + \frac{(-1)}{1!}(-y) + \frac{(-1)(-2)}{2!}(-y)^2 + \frac{(-1)(-2)(-3)}{3!}(-y)^3$$

$$+ \frac{(-1)(-2)(-3)(-4)}{4!}(-y)^4 + \ldots \ldots$$

$$= 1 + y + y^2 + y^3 + y^4 + \ldots \ldots$$

This expression is related to the infinite geometric series (2.12), and indicates that the binomial expansion is correct for that series.

Example 3: Expand $\dfrac{1}{(4-x)^{\frac{1}{2}}}$ up to the term in x^3.

The quantity $(4-x)$ is not of the form $(1+y)$ but may be expressed in that form by writing

$$4 - x = 4\left(1 - \frac{x}{4}\right)$$

so that $\quad (4-x)^{\frac{1}{2}} = 2\left(1 - \frac{x}{4}\right)^{\frac{1}{2}} = 2(1+y)^{\frac{1}{2}}$

with $y = -x/4$. Then

$$\frac{1}{(4-x)^{\frac{1}{2}}} = \frac{1}{2\left(1 - \frac{x}{4}\right)^{\frac{1}{2}}} = \frac{1}{2}(1+y)^{-\frac{1}{2}}.$$

Equation (2.22) with $\alpha = -\frac{1}{2}$ gives

$$(1+y)^{-\frac{1}{2}} = 1 + \frac{\left(-\frac{1}{2}\right)}{1!}y + \frac{\left(-\frac{1}{2}\right)\left(-\frac{3}{2}\right)}{2!}y^2 + \frac{\left(-\frac{1}{2}\right)\left(-\frac{3}{2}\right)\left(-\frac{5}{2}\right)}{3!}y^3 + \ldots$$

$$= 1 - \frac{1}{2}y + \frac{3}{8}y^2 - \frac{5}{16}y^3 + \ldots$$

so that replacing y by $-x/4$ results in

$$\frac{1}{2}\left(1 - \frac{x}{4}\right)^{-\frac{1}{2}} = \frac{1}{2}\left(1 - \frac{x}{8} + \frac{3x^2}{128} - \frac{5x^3}{1024} + \ldots\right).$$

Example 4: The increase in weight of a young animal over a 30 day period is modelled by the equation $W = W_0 (1+x)^{T/30}$ where W is the current weight, W_0 is the initial weight, x is the fractional increase in weight for 30 days, and T the number of days. If $W_0 = 50$ kg and $x = 1/10$ determine the weight after (a) 30 days and (b) 10 days. If this rate of growth continues, estimate the weight of the animal after (c) 45 days and (d) 60 days.

(a) When $T = 30$

$$W = 50\left(1 + \frac{1}{10}\right)^1 = 50\left(\frac{11}{10}\right) = 55 \text{ kg} .$$

(b) When $T = 10$

$$W = 50\left(1 + \frac{1}{10}\right)^{1/3}$$

and using the binomial theorem

$$W \cong 50\left\{1 + \frac{\left(\frac{1}{3}\right)}{1!}\left(\frac{1}{10}\right) + \frac{\left(\frac{1}{3}\right)\left(-\frac{2}{3}\right)}{2!}\left(\frac{1}{10}\right)^2\right\}$$

$$= 50\left\{1 + \frac{1}{30} - \frac{1}{900}\right\}$$

$$\cong 50 \{1 + 0.033 - 0.001\}$$

$$= 51.6 \text{ kg.}$$

(c) When $T = 45$
$$W = 50\left(1 + \frac{1}{10}\right)^{3/2}$$

and using the binomial theorem

$$W \cong 50\left\{1 + \frac{\left(\frac{3}{2}\right)}{1!}\left(\frac{1}{10}\right) + \frac{\left(\frac{3}{2}\right)\left(\frac{1}{2}\right)}{2!}\left(\frac{1}{10}\right)^2 + \frac{\left(\frac{3}{2}\right)\left(\frac{1}{2}\right)\left(-\frac{1}{2}\right)}{3!}\left(\frac{1}{10}\right)^3\right\}$$

$$= 50\left\{1 + \frac{3}{20} + \frac{3}{800} - \frac{1}{16000}\right\}$$

$$\cong 50\{1 + 0.15 + 0.004\}$$

$$= 57.7 \text{ kg.}$$

(d) When $T = 60$
$$W = 50\left(1 + \frac{1}{10}\right)^2$$

$$= 50\left(\frac{11}{10}\right)^2$$

$$= 60.5 \text{ kg.}$$

Note that the answers to (a) and (d) are exact, those to (b) and (c) are approximate owing to the use of the binomial theorem and approximations used in the arithmetic.

PROBLEMS

1. Write down the first four terms of the sequence $\{u_n\}$ where

   (i) $u_n = 3n - 1$,

   (ii) $u_n = n^3$,

   (iii) $u_n = x^{2n}$,

   (iv) $u_n = 1/n^2$,

   (v) $u_n = \dfrac{n - 1}{n + 1}$.

2. The ages of animals to be entered in an experiment are chosen using the sequence $\{5 \times 2^{n-1}\}$ where $n = 1, 2, 3, 4$. Write down the terms of the sequence.

3. An animal population, initially of size N, doubles its size each year. Write down the terms of the sequence giving the population size at the end of each of the first four years.

4. The density d of fungal fruiting bodies which grow from the roots of a tree increase with the distance r from the centre of the tree according to the relationship

$$d = kr^{3/4},$$

where k is a constant. Write down the sequence of values of d corresponding to $r = 1, 2, 3, 4$. For what value of r is $d > 3k$?

5. An animal population consists of N breeding adults. Each year the population increases by a number equal to the number of breeding adults. The young animals start to breed when two years old. Write down the terms of the sequence giving the population size at the end of each of the first four years. (The numbers arising in this sequence are the Fibonacci numbers, see Chapter 16).

6. Find the sum of the arithmetic series for which

(i) $a = 3, d = 3/2, n = 11$,

(ii) $a = 15, d = -2, n = 12$,

(iii) $a = -12, d = 2, n = 13$.

7. Find the sum of (a) the first six even numbers, (b) the first n even numbers. For what value of n does the sum in (b) equal 90? For what value of n, does the sum first exceed 1000?

8. The third term in an arithmetic series is 9 and the eleventh term is 25. Find the sum of the first eleven terms of the series.

9. The first term of an arithmetic series is 2, the nth term is -16, and the sum of the first n terms is -49. Find the value of n.

10. Find the sum of the geometric series for which

(i) $a = 2, r = 3, n = 4$,

(ii) $a = 1, r = 1/3, n = 4$,

(iii) $a = 1, r = -1/3, n = 4$.

11. Find the sum to infinity of the geometric series for which

(i) $a = 1, r = 1/3$,

(ii) $a = 1, r = -1/3$.

12. The first term of a geometric series is 4 and the fourth term is 32. Determine the sum of the first ten terms of the series.

13. The first term of a geometric series is 4 and the fifth term is 64. Find the two possible values for the common ratio and determine the sum of the first five terms in each case.

14. The sum to infinity of a geometric series is 6 and the first term is 2. Determine the common ratio.

15. Each hour a cell divides into two new cells. If there are ten cells present initially determine the number present at the beginning of the sixth hour and the total number of cells that have existed to the start of the sixth hour.

16. During a certain growth phase a plant increases in size by 10% each week, Determine the percentage increase in the size over a 4-week period (to the nearest integer).

17. The biomass of a tree increases by 3% each year. Calculate the percentage increase in the biomass of the tree at the end of year 7. After how many years will the tree be 50% larger?

18. During a certain period the area of a healing wound is observed to shrink by a factor of 1/8 each day. Determine the fraction of the wound that has healed at the end of day 3. After how many complete days will the wound first be less than half of its initial size?

19. Yeast is grown in a sugar solution in such a way that the weight of yeast increases by 5% every hour. If the initial weight is 1 gm the weight after t hours is $w = (1.05)^t$. Calculate the weight after

   (i) 20 minutes,

   (ii) 30 minutes,

   (iii) 90 minutes,

   (iv) 2 hours.

20. The ages of animals entered in a certain experiment are chosen to form a geometric sequence. There are five groups of animals of which the youngest group is aged 16 months and the oldest 81 months. Determine the ages of the other groups.

21. The mean number of eggs n laid per female fly per day is related approximately to the number of flies per square centimetre d by

   $$n = 55(0.9)^d .$$

Calculate the mean number of eggs laid per female fly per day when $d = 3, 5, 10, 40$. Give your answer to the nearest whole number.

22. Evaluate the binomial coefficients

(i) $\binom{8}{6}$,   (ii) $\binom{9}{4}$,

(iii) $\binom{9}{5}$,   (iv) $\binom{50}{48}$,

(v) $\binom{6}{0}$,   (vi) $\binom{n}{n-1}$.

23. Use the binomial theorem to expand

(i) $(1 + x)^6$,

(ii) $(1 + 2x)^6$,

(iii) $(1 - x)^6$,

(iv) $(2 - x)^6$.

Use the first three terms of (iv) to estimate $(1.95)^6$.

24. Use the binomial theorem to expand

(i) $(x + y)^4$,

(ii) $\left(x + \dfrac{1}{x}\right)^4$,

(iii) $\left(x - \dfrac{1}{x^2}\right)^4$,

(iv) $\left(2x^2 - \dfrac{3}{x}\right)^4$,

(v) $(1 + x + y)^4$,

(vi) $\left(1 + x - \dfrac{1}{x}\right)^4$.

25. Determine the coefficient of

(i) x^4 in $(1 + x)^7$,

(ii) x^5 in $(1 - x)^7$,

(iii) x^9 in $(2 - x)^{12}$,

(iv) x^3 in $(1 + 2x)^{20}$,

(v) x^8 in $\left(x^2 + \dfrac{1}{x}\right)^{10}$,

(vi) x^0 in $\left(\dfrac{1}{x} - 2x\right)^8$,

(vii) x^{-3} in $\left(\dfrac{3}{x^3} - \dfrac{x^2}{2}\right)^6$.

26. A sum of money is invested at a compound interest rate of 10% per annum. Use the binomial theorem to calculate the percentage increase in the value of the investment after ten years. Check your answer by using logarithms.

27. The length of each side of a cube is increased by 2%. Show that the increase in volume is approximately 6%.

28. Use the binomial theorem to find the first four terms, and give the range of x for which the full expansion is valid, for:

    (i) $(1 + x)^{1/3}$,

    (ii) $(1 - x)^{1/3}$,

    (iii) $(1 + 2x)^{1/2}$,

    (iv) $(1 + 2x)^{-1/2}$,

    (v) $(4 - x)^{1/2}$,

    (vi) $(4 - x)^{-1/2}$,

    (vii) $(1 + x)^{-1}$,

    (viii) $(1 - x)^{-2}$.

Use the expansion obtained in (v) to evaluate $(3.8)^{1/2}$ to four places of decimals.

29. The weight in kilograms of a young animal over a 20-day period is given by

$$W = 10\left(1 + \dfrac{1}{8}\right)^{T/20}$$

where T is the time in days. Use the binomial theorem to determine the weight of the animal to four significant figures

    (i) at time $T = 0$,

    (ii) after 20 days,

    (iii) after 15 days.

Use the same formula to estimate the weight after

(iv) 40 days,

(v) 30 days.

30. The weight of a person suffering from a disease decreased from 100 kg to 90 kg over a 20-day period. The weight W after T days was given by

$$W = 100\left(1 - \frac{1}{10}\right)^{T/20}.$$

Use the binomial theorem to estimate to three significant figures the weight after

(i) 2 days,

(ii) 10 days,

(iii) 15 days.

31. It is observed that the cannibal population of an island has a natural net increase of 5% per annum but that 25 individuals disappear mysteriously just before the festivities to mark the end of each year. By summing an appropriate geometric series find, in terms of r and n_0, the population n_r after r years if the initial population is n_0.   (Edinburgh, 1978)

32. In a certain colony the number of male and female rabbits are equal. At the end of the sixth month after birth, and at the end of every third month thereafter, each female rabbit produces two female and two male rabbits, which themselves produce offspring similarly from six months onwards. If the colonly consists initially of one male and one female newly born rabbit, show that after nine months there are five females and after twelve months eleven females.   (Edinburgh, 1978)

33. (i) Write down the first three terms in the binomial expansion of $(1+x)^7$.

    (ii) The size of a population is initially N_0 and increases at the rate of 1% each day. Show that the size has increased by just over 7% after 7 days.

    Write down an expression for the size of the population after t days, and show that the population has doubled in about 70 days.

    (iii) A second population behaves differently. It increases in size by 50% in the first year and then decreases by 50% in the second year. Find the overall percentage increase or decrease in size. (Edinburgh, 1976)

CHAPTER 3

Polynomials

3.1 INTRODUCTION

This chapter is concerned with the behaviour of one of the simplest mathematical entities, a polynomial of a single variable. Polynomials are particularly important in the experimental sciences since they often give a simple theoretical description of experimental results. A definition is given in Section 3.2, and the properties of some polynomials are then considered before presenting results for the general case. The chapter concludes with a discussion of interpolation.

3.2 DEFINITION

A **polynomial** p_n, of degree n in the single variable x, is an expression of the form

$$p_n = a_0 + a_1x + a_2x^2 + \ldots + a_{n-1}x^{n-1} + a_nx^n$$

$$= \sum_{r=1}^{n} a_rx^r , \qquad a_n \neq 0 ,$$

where the coefficients $a_0, a_1, a_2, \ldots, a_n$ are real constants. The **degree**, n, of the polynomial is determined by the highest power of x present so that the coefficient a_n is not zero. Any of the coefficients $a_0, a_1, \ldots, a_{n-1}$ may be zero. The finite series (but not the infinite series) encountered in Section 2.5 are examples of polynomials.

Some examples are

$$\text{degree } 0, p_0 = a_0, \qquad\qquad a_0 \neq 0 ;$$
$$\text{degree } 1, p_1 = a_0 + a_1x, \qquad a_1 \neq 0 ;$$
$$\text{degree } 2, p_2 = a_0 + a_1x + a_2x^2, a_2 \neq 0 .$$

In particular the expressions $1 + 2x$, x^2, $3 + 4x^2$, $x - 3x^3$, $1 + x^6$ are all polynomials in x. The polynomial of degree zero, $p_0 = $ constant, is of little interest. The polynomial of degree one, the equation of a straight line, is particularly important since use is often made of its properties in fitting experimental data, see Chapter 19. Examples involving polynomials of degree two are: (i) the rate of growth of a population which is often assumed to be a polynomial in N of the form $cN - kN^2$ where c, k are positive constants and N is the size of the population; (ii) the distance an object falls from rest under the effect of gravity which is given by $\frac{1}{2}gt^2$ where g is the acceleration due to gravity and t is the time.

The properties of some simple polynomials will now be examined.

3.3 POLYNOMIAL OF DEGREE ONE

The most general polynomial of degree one is

$$p_1 = a_0 + a_1 x , \qquad a_1 \neq 0 ,$$

where a_0 and a_1 are constants. This is most commonly encountered in the form of the linear equation

$$y = mx + c \qquad\qquad (3.1)$$

where m and c are constants, and it is this particular form that will be examined here.

The graph corresponding to equation (3.1) is a straight line that cuts the y-axis ($x = 0$) in the point A where $y = c$ and the x-axis ($y = 0$) in the point B where $x = -c/m$, see Fig. 3.1. Thus the intercept on the y-axis is c and the slope of the line is given by $\tan \theta = m = $ constant (from the triangle OAB).

It is often the case that the equation of the polynomial is not immediately given in the form (3.1) but the values of c and m are to be determined from given information. A straight line in the plane OXY is uniquely determined by either (i) the gradient of the line and one point that lies on the line or (ii) two different points that lie on it.

The equation of the straight line of given slope m that passes through the point C(x_1, y_1) (see Fig. 3.1) is of the form (3.1). The value of c is determined by the requirement that C lies on the line so that

$$y_1 = mx_1 + c,$$
$$c = y_1 - mx_1 .$$

Equation (3.1) then gives

$$y = mx + y_1 - mx_1$$

and finally

$$y - y_1 = m(x - x_1) . \qquad\qquad (3.2)$$

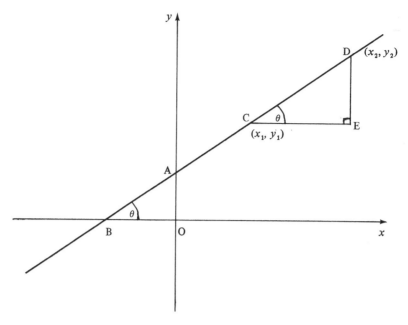

Fig. 3.1

Example 1: Determine the equation of the line with slope 3 that passes through the point $(1, -1)$. Find where this line cuts the x-axis.

Here $m = 3, x_1 = 1, y_1 = -1$ so that (3.2) gives

$$y + 1 = 3(x - 1) = 3x - 3 ,$$

$$y = 3x - 4 .$$

The line cuts the x-axis when $y = 0$ and

$$x = 4/3 .$$

In order to determine the equation of the straight line that passes through the points $C(x_1, y_1)$ and $D(x_2, y_2)$, see Fig. 3.1, construct the lines CE and DE parallel to the x and y axes respectively. Then

$$CE = x_2 - x_1 ,$$

$$ED = y_2 - y_1 ,$$

and from the triangle CDE

$$m = \tan \theta = \frac{ED}{CE} = \frac{y_2 - y_1}{x_2 - x_1} .$$

Equation (3.2) now gives the equation of the line in the form

$$y - y_1 = \frac{(y_2 - y_1)(x - x_1)}{x_2 - x_1} \tag{3.3}$$

so that

$$y = y_1 + \frac{(y_2 - y_1)(x - x_1)}{x_2 - x_1} . \tag{3.4}$$

An alternative form of (3.3) that indicates the symmetry of the relationship is

$$\frac{y - y_1}{y_2 - y_1} = \frac{x - x_1}{x_2 - x_1} . \tag{3.5}$$

In this form of the equation of a straight line it is immaterial which of the given points is chosen to be (x_1, y_1) and which (x_2, y_2).

Example 2: Obtain the equation of the straight line passing through the points $(1, 2)$ and $(3, 1)$ in the form $y = mx + c$.

In (3.4) write $x_1 = 1, y_1 = 2, x_2 = 3, y_2 = 1$ so that

$$y = 2 + \frac{(1 - 2)(x - 1)}{3 - 1}$$

$$= 2 - \tfrac{1}{2}(x - 1) = 2 - \tfrac{1}{2}x + \tfrac{1}{2} ,$$

$$= -\tfrac{1}{2}x + \tfrac{5}{2} .$$

In this example $m = -\tfrac{1}{2}, c = \tfrac{5}{2}$.

3.4 POLYNOMIAL OF DEGREE TWO

The most general polynomial of degree two is

$$p_2 = a_0 + a_1 x + a_2 x^2 , \qquad a_2 \neq 0 ,$$

where a_0, a_1, a_2 are constants. To avoid the use of subscripts the polynomial considered here is

$$y = ax^2 + bx + c , \qquad a \neq 0 , \tag{3.6}$$

where a, b, c are constants.

The curve corresponding to (3.6) cuts the y-axis ($x = 0$) once in the point $(0, c)$. It cuts the x-axis when $y = 0$ and, from (3.6),

$$ax^2 + bx + c = 0 . \tag{3.7}$$

To determine those points (if any) where the curve cuts the x-axis it is necessary to solve the **quadratic equation** (3.7). When the roots of (3.7) are real it is possible to express the equation as the product of two real linear factors. In the particular case when the roots are rational numbers it is usually possible to factorise (3.7) by inspection since if

$$ax_0^2 + bx_0 + c = 0$$

then $x = x_0$ is a root of the equation and $x - x_0$ is a factor. Some examples are

$$x^2 - 1 = (x - 1)(x + 1),$$
$$x^2 - 2x + 1 = (x - 1)(x - 1) = (x - 1)^2,$$
$$x^2 + x - 2 = (x - 1)(x + 2),$$
$$2x^2 - x - 1 = (x - 1)(2x + 1),$$
$$6x^2 + x - 12 = (3x - 4)(2x + 3).$$

When the roots of (3.7) are not rational numbers it is not usually possible to recognise factors, and it is then necessary to make use of the general formula for the roots of (3.7) derived in Section 13.2.1 and given by

$$x = \frac{-b \pm (b^2 - 4ac)^{\frac{1}{2}}}{2a}. \tag{3.8}$$

When $b^2 - 4ac > 0$ the roots given by (3.8) are real (rational or irrational numbers) and distinct and (3.7) has two real linear factors. When $b^2 - 4ac = 0$ the two roots are equal and have the value $-b/(2a)$. Equation (3.7) then has two repeated linear factors and becomes a perfect square. When $b^2 - 4ac < 0$ there are no real solutions of (3.7) since the square root of a negative number does not have a real value. It has been found convenient for both theoretical and practical reasons to invent a number system that permits the two roots of (3.7) to be defined. These invented numbers are called **complex numbers**. By introducing the symbol i which is such that $i^2 = -1$ it is possible to write down the two roots of any quadratic equation. For example the roots of $x^2 + 1 = 0$ are $x = \pm i$, and $x^2 + 1$ factorises into $(x + i)(x - i)$; the roots of $x^2 + 4 = 0$ are $x = \pm 2i$, and $x^2 + 4 = (x + 2i)(x - 2i)$.

Example 1: Solve the equation $f(x) = x^2 - 2x - 2 = 0$ and hence express $f(x)$ as the product of two linear factors.

Here, on comparison with (3.7), $a = 1, b = -2, c = -2$ so that from (3.8)

$$x = \frac{2 \pm (4 + 8)^{\frac{1}{2}}}{2} = 1 \pm \sqrt{3}.$$

The two real roots are $x = 1 + \sqrt{3}$ and $x = 1 - \sqrt{3}$ so that the linear factors are

$$(x - 1 - \sqrt{3}) \text{ and } (x - 1 + \sqrt{3}).$$

The quadratic expression may then be written

$$f(x) = x^2 - 2x - 2 = (x - 1 - \sqrt{3})(x - 1 + \sqrt{3}).$$

Example 2: Solve the equation $f(x) = x^2 - 2x + 5 = 0$ and hence express $f(x)$ as the product of two linear factors.

Here $a = 1, b = -2, c = 5$ so that (3.8) gives

$$x = \frac{2 \pm (4 - 20)^{\frac{1}{2}}}{2} = 1 \pm (-4)^{\frac{1}{2}} = 1 \pm 2i \ .$$

The two roots are $x = 1 + 2i$ and $x = 1 - 2i$ and the associated factors are

$$(x - 1 - 2i) \text{ and } (x - 1 + 2i) \ .$$

The quadratic expression becomes

$$f(x) = (x - 1 - 2i)(x - 1 + 2i) \ .$$

3.4.1 Properties of complex numbers

A complex number z may be written in the standard form $z = a + ib$ where a and b are real numbers and $i^2 = -1$. The number a is the **real part** and the number b the **imaginary part.** In order to manipulate complex numbers the following rules must be obeyed:

Equality, $a + ib = c + id$ if and only if $a = c$ and $b = d$.

Addition, $(a + ib) + (c + id) = (a + c) + i(b + d)$.

Multiplication, $c(a + ib) = ca + i(cb)$.

$$(a + ib)(c + id) = (ac - bd) + i(ad + bc).$$

In these results a, b, c and d are real numbers.

It is seen that in manipulating complex numbers the basic rules of algebra set out in Chapter 1 remain unchanged; this is illustrated by the definition of multiplication. In the new number system the number zero is obtained when both a and b are zero.

A number closely related to $z = a + ib$ is the **complex conjugate** $\bar{z} = a - ib$.

Then $z + \bar{z} = 2a$ is real,

$z - \bar{z} = 2ib$ is imaginary,

$z\bar{z} = (a + ib)(a - ib) = a^2 + b^2$ is real and positive.

In order to divide two complex numbers use is made of the complex conjugate since

$$\frac{a + ib}{c + id} = \frac{(a + ib)(c - id)}{(c + id)(c - id)} = \frac{(ac + bd)}{c^2 + d^2} + \frac{i(bc - ad)}{c^2 + d^2} \ ,$$

which is now in the standard form.

Example 3: Express the number $\dfrac{1 - i}{1 + 2i} - \dfrac{1 + i}{2 - i}$ in the form $a + ib$.

Multiplying the numerator and denominator of each term by the complex conjugate of the denominator gives

$$\frac{(1 - i)(1 - 2i)}{(1 + 2i)(1 - 2i)} - \frac{(1 + i)(2 + i)}{(2 - i)(2 + i)} = \frac{(-1 - 3i)}{5} - \frac{(1 + 3i)}{5}$$

$$= -\frac{2}{5} - \frac{6}{5}i \ .$$

A complex number $z = x + iy$ may be represented geometrically in the xy-plane by the point P(x, y), see Fig. 3.2. The x-coordinate represents the real part of the number and the y-coordinate the imaginary part. Such a representation is called an **Argand diagram**.

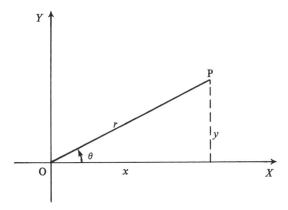

Fig. 3.2

Complex numbers may also be expressed in terms of trigonometric functions by

$$z = x + iy = r(\cos \theta + i \sin \theta) \ .$$

This is referred to as the **polar form** of the complex number. Equating real and imaginary parts gives

$$x = r \cos \theta \text{ and } y = r \sin \theta \ .$$

Squaring and adding these equations results in

$$x^2 + y^2 = r^2(\cos^2\theta + \sin^2\theta) = r^2 \ .$$

The quantity

$$r = (x^2 + y^2)^{\frac{1}{2}} \geqslant 0$$

is called the **modulus** of the complex number, written $r = \text{mod } z = |z|$, and gives the magnitude of the number. The angle θ given by

$$\cos \theta = x/r \text{ and } \sin \theta = y/r$$

is called the **argument** of the complex number, written $\theta = \arg z$. The **principal value** of θ is such that $-\pi < \theta \leqslant \pi$. Referring to Fig. 3.2 it is seen that r and θ give a geometrical representation of z in terms of polar coordinates (see Section 1.13), hence the term polar form. Note that $|z| = r$ is the distance of P from the origin.

Example 4: Express in polar form the complex numbers

(a) $1 + i\sqrt{3}$, (b) $-1 + i\sqrt{3}$, (c) $1 - i\sqrt{3}$.

(a) $1 + i\sqrt{3} = r(\cos\theta + i\sin\theta)$

gives $1 = r\cos\theta$ and $\sqrt{3} = r\sin\theta$ so that

$$r^2 = 1 + 3 = 4 ,$$
$$r = 2 .$$

The angle θ is determined by

$$\cos\theta = 1/2 \text{ and } \sin\theta = \sqrt{3}/2$$

so that the principal value is $\theta = \pi/3$.

$$1 + i\sqrt{3} = 2\{\cos(\pi/3) + i\sin(\pi/3)\} .$$

(b) $-1 + i\sqrt{3} = r(\cos\theta + i\sin\theta)$,

$$-1 = r\cos\theta \text{ and } \sqrt{3} = r\sin\theta$$

and again $r = 2$. In this case θ is determined by

$$\cos\theta = -1/2 \text{ and } \sin\theta = \sqrt{3}/2$$

so that the principal value is $2\pi/3$.

$$-1 + i\sqrt{3} = 2\{\cos(2\pi/3) + i\sin(2\pi/3)\} .$$

(c) In this case $r = 2$ also and θ is such that

$$\cos\theta = 1/2 \text{ and } \sin\theta = -\sqrt{3}/2$$

so that the principal value is $-\pi/3$.

$$1 - i\sqrt{3} = 2\{\cos(-\pi/3) + i\sin(-\pi/3)\} .$$

Many of the properties of complex numbers discussed here will be used in Chapters 15 and 16 where certain differential equations and recurrence equations are discussed.

3.4.2 Curve sketching

When the two roots of (3.7) given by (3.8) are real they give the points at which the graph of (3.6) cuts the x-axis. When $b^2 - 4ac > 0$ the roots are real and distinct and the curve cuts the x-axis at two distinct points. When $b^2 - 4ac = 0$ the two roots are equal and the curve corresponding to (3.6) touches the x-axis at the point $(-b/(2a), 0)$. Finally, when $b^2 - 4ac < 0$ there are no real solutions of (3.7) and the curve given by (3.6) does not cut the x-axis.

Further information may be obtained concerning the shape of the curve given by (3.6). When $a > 0$ the value of y must be positive for large values of x (both positive and negative) since ax^2 then makes the greatest contribution to y. In this case the curve has a **minimum value** and the curve is dish-shaped (it 'holds water'). When $a < 0$ the value of y obtained for large values of $|x|$ must be negative. The variable y has a **maximum value** and the curve is shaped like an inverted dish (it 'sheds water'). It can be shown that the curve is symmetrical about the line $x = -b/(2a)$ for all values of a and that the minimum (or maximum) value of y occurs for this value of x.

Example 5: Plot the curves given by (a) $y = x^2 - 2x + 3$, (b) $y = x^2 - 2x + 1$, (c) $y = x^2 - 2x - 1$.

In all cases $a = 1 > 0$, $b = -2$, $b/(2a) = -1$. All curves are symmetrical about the line $x = 1$ and y has a minimum value for that value of x.

(a) $y = x^2 - 2x + 3 = (x - 1)^2 + 2$.

Here $c = 3$, $b^2 - 4ac = -8 < 0$ so that the curve does not cut the x-axis. It is the top curve in Fig. 3.3.

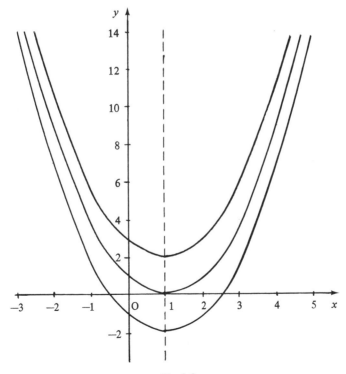

Fig. 3.3

(b) $y = x^2 - 2x + 1 = (x - 1)^2$.

Here $c = 1$, $b^2 - 4ac = 0$ so that the curve touches the x-axis when $x = 1$. It is the middle curve in Fig. 3.3.

(c) $y = x^2 - 2x - 1 = (x - 1)^2 - 2$.

Now $c = -1$, $b^2 - 4ac = 8 > 0$ so that the curve cuts the x-axis in the two points $x = 1 \pm \sqrt{2}$ (from (3.8)). It is the bottom curve in Fig. 3.3.

Example 6: Plot the curves given by (a) $y = -x^2 + 2x + 1$, (b) $y = -x^2 + 2x - 1$, (c) $y = -x^2 + 2x - 3$.

In all cases $a = -1 < 0$, $b = 2$, $b/(2a) = -1$. The three curves are symmetrical about the line $x = 1$ and y has a maximum value for that value of x.

(a) $y = -x^2 + 2x + 1 = -(x - 1)^2 + 2$.

Here $c = 1$, $b^2 - 4ac = 8 > 0$ so that the curve cuts the x-axis in the two points $x = 1 \pm \sqrt{2}$ (from (3.8)). It is the top curve in Fig. 3.4.

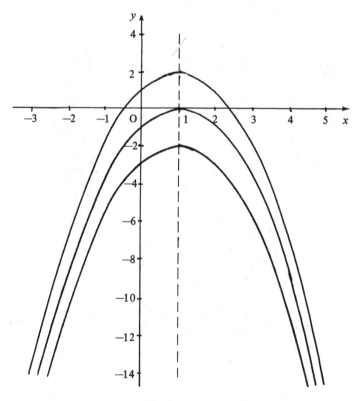

Fig. 3.4

(b) $y = -x^2 + 2x - 1 = -(x - 1)^2$.

Here $c = -1$, $b^2 - 4ac = 0$ so that the curve touches the x-axis when $x = 1$. It is the middle curve in Fig. 3.4.

(c) $y = -x^2 + 2x - 3 = -(x - 1)^2 - 2$.

Here $c = -3$, $b^2 - 4ac = -8 < 0$ so that the curve does not cut the x-axis. It is the bottom curve in Fig. 3.4.

If the two roots of (3.7) are denoted by α and β then

$$ax^2 + bx + c \equiv a(x - \alpha)(x - \beta) \equiv a\{x^2 - (\alpha + \beta)x + \alpha\beta\} .$$

These expressions are identical so that equating coefficients it is found that

$$\alpha + \beta = -b/a , \qquad \alpha\beta = c/a .$$

In certain circumstances these results are helpful in making accurate calculations.

Example 7: Solve the equation $x^2 - 24x + 1 = 0$.

The formula (3.8) gives

$$x = 12 \pm \sqrt{143} \cong 12 \pm 11.958$$

so that, using this method, the roots are $\alpha \cong 23.958$ and $\beta \cong 0.042$. The first root is obtained by adding together numbers of comparable magnitudes, and it is very accurate. The second root is calculated by subtracting two numbers of almost the same magnitude so that the difference is very small and not very accurate owing to the loss of significant figures. A more accurate value for β may be found by using the relationship that $\alpha\beta = 1$ for this equation. Thus

$$\beta = 1/\alpha = 1/23.958 = 0.04174 .$$

The roots are $x = 23.958$ and $x = 0.04174$.

3.5 POLYNOMIAL OF DEGREE THREE

The most general polynomial of degree three may be written in the form

$$y = ax^3 + bx^2 + cx + d , \qquad a \neq 0 , \tag{3.9}$$

where a, b, c, d are constants. The curve corresponding to (3.9) cuts the y-axis when $y = d$ and cuts the x-axis when

$$ax^3 + bx^2 + cx + d = 0 . \tag{3.10}$$

Equation (3.10) is a **cubic equation** for x which may be shown to have either

(i) three distinct real roots, for example

$$x^3 - 2x^2 - x + 2 = (x + 1)(x - 1)(x - 2) \text{ with roots } -1, 1, 2;$$

or   (ii) two distinct real roots, one of which is repeated, for example

$$x^3 - x^2 - x + 1 = (x + 1)(x - 1)^2 \text{ with roots } -1, 1, 1;$$

or (iii) one real triple root, for example
$$x^3 - 3x^2 + 3x - 1 = (x-1)^3 \text{ with roots } 1, 1, 1;$$

or (iv) one real root and two complex roots, for example
$$x^3 - x^2 + x - 1 = (x-1)(x^2+1) \text{ with real root } 1, \text{ complex roots } -i, i.$$

There is no simple method of solving a cubic equation which does not factorise by inspection comparable to the formula (3.8) for a quadratic equation. Since the equation has at least one real root it is sometimes possible to recognise a root and then factorise the equation. This method gives exact values for the roots but is not always applicable. An alternative approach is to plot the curve (3.9) accurately and find the points where the curve cuts the x-axis. Approximate values for the roots of (3.10) may always be found by using this method but the values are not normally very accurate. A third method, sometimes used in conjunction with the graphical method, is to evaluate the root numerically by applying Newton's method which is discussed in Chapter 8.

Example 1: Verify that $x = 1$ is a root of the equation $2x^3 + x^2 - 5x + 2 = 0$ and find the other roots.

Substitute $x = 1$ into the LHS of the equation to obtain
$$2 + 1 - 5 + 2 = 0 .$$

Thus $x = 1$ is a root and the cubic equation may be written
$$2x^3 + x^2 - 5x + 2 = (x - 1)(2x^2 + 3x - 2)$$
$$= (x - 1)(2x - 1)(x + 2) = 0 .$$

Solutions are $x = 1, x = \tfrac{1}{2}, x = -2$.

Example 2: Plot the graph of $y = x^3 + x^2 - 12x + 8$ and hence obtain approximate values for the roots of the equation $x^3 + x^2 - 12x + 8 = 0$.

In order to plot the graph it is necessary to construct a table of values.

x	-5	-4	-3	-2	-1	0	1	2	3	4
y	-32	8	26	28	20	8	-2	-4	8	40

Fig. 3.5 gives a plot of these values from which it is seen that the curve cuts the x-axis at points close to $x = -4.2$, $x = 0.7$, $x = 2.7$. These values of x represent approximate solutions of the cubic equation.

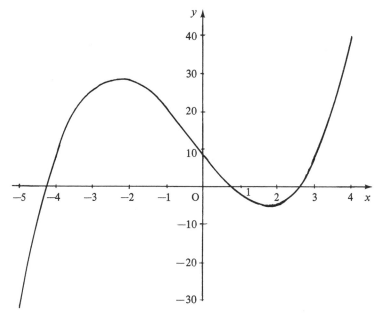

Fig. 3.5

3.6 SOLUTION OF POLYNOMIAL EQUATIONS

The methods available for determining real roots of polynomial equations of degree higher than three are essentially the same as those for equations of degree three. Given an equation of the form

$$p_n = a_0 + a_1 x + a_2 x^2 + \ldots + a_n x^n = 0, a_n \neq 0, n > 3 , \qquad (3.11)$$

values of x for which $p_n = 0$ may be obtained by factorisation in certain cases, by plotting an accurate graph of $y = p_n$ and finding those values of x for which $y = 0$ or by numerical methods. If p_n is zero when $x = x_0$ then $x = x_0$ is a root of (3.11) and $(x - x_0)$ is a factor. Using this information it is sometimes possible to find roots which are rational numbers. If the n roots of (3.11) are denoted by $\alpha_1, \alpha_2, \alpha_3, \ldots, \alpha_n$ it can be shown that the sum of the roots

$$\alpha_1 + \alpha_2 + \alpha_3 + \ldots + \alpha_n = -a_{n-1}/a_n , \qquad (3.12)$$

and the product of the roots

$$\alpha_1 \ \alpha_2 \ \alpha_3 \ \ldots \ \alpha_n = (-1)^n \ a_0/a_n . \qquad (3.13)$$

These two results are generalisations of the results for a quadratic obtained in Section (3.4) and may be proved by similar methods. If all roots are distinct it is possible to show that when n is even the equation has an even number of real roots. When n is odd the equation has an odd number of real roots so that there is at least one real root. Thus an equation of degree four has 0, 2 or 4 real roots; an equation of degree five has 1, 3, or 5 real roots.

Since the roots of $p_n = 0$ occur when the curve corresponding to $y = p_n$ cuts the x-axis, the value of y changes sign near a root. This is a useful aid in locating the approximate position of roots. Examples of this are found in example 2 of Section 3.5. From the table of values y changes sign (the curve cuts the x-axis, see Fig. 3.5) between $x = -5$ and $x = -4$, between $x = 0$ and $x = 1$ and between $x = 2$ and $x = 3$. There is (at least) one root in each of these intervals.

By substituting $z = x^r$ it is possible to reduce the degree of a polynomial when the polynomial contains powers of x^r only. This process is best illustrated by means of examples.

Example: Solve the equations (a) $x^4 - 3x^2 + 2 = 0$, (b) $x^4 - x^2 - 2 = 0$, (c) $x^4 + 3x^2 + 2 = 0$.

All equations are of degree four and will have 0, 2 or 4 real roots. The polynomials contain powers of x^2 alone so that it is possible to substitute $z = x^2$.

(a) Setting $z = x^2$ gives

$$z^2 - 3z + 2 = 0 ,$$

$$(z - 1)(z - 2) = 0$$

so that roots are $z = 1$ and $z = 2$. Then

$$x^2 = z = 1 \text{ gives } x = \pm 1$$

and $x^2 = z = 2$ gives $x = \pm\sqrt{2}$. There are four real roots for x.

(b) Substituting $z = x^2$ gives

$$z^2 - z - 2 = 0 ,$$

$$(z - 2)(z + 1) = 0$$

so that $z = 2$ and $z = -1$ are roots. Then

$$x^2 = z = 2 \text{ gives } x = \pm\sqrt{2} .$$

and $x^2 = z = -1$ gives $x = \pm i$. There are two real roots and two imaginary roots.

(c) Let $z = x^2$ to obtain

$$z^2 + 3z + 2 = 0 ,$$

$$(z + 1)(z + 2) = 0$$

so that $x^2 = z = -1$ and $x^2 = z = -2$. Neither of these values gives real roots and there are no real roots of this equation. The imaginary roots are $x = \pm i$ and $x = \pm i\sqrt{2}$.

3.7 MANIPULATION OF POLYNOMIALS

Occasionally it is necessary to carry out the algebraic operations of addition, subtraction, multiplication and division of polynomials. The operation of addition (subtraction) is straightforward in that the coefficients of corresponding powers of x are added (subtracted). Multiplication and division may be carried out, using the method of **detached coefficients**, in a manner very similar to the corresponding arithmetical operations. The process is best illustrated by means of examples.

Example 1: Multiply $x^3 - 4x^2 + 1$ by $x^2 - x - 2$.

Set out the coefficients of the powers of x in columns (as for arithmetical multiplication).

	x^5	x^4	x^3	x^2	x^1	x^0
line 1			1	-4	0	1
line 2				1	-1	-2
line 3			-2	8	0	-2
line 4		-1	4	0	-1	
line 5	1	-4	0	1		
line 6	1	-5	2	9	-1	-2

Lines 1 and 2 respectively contain the coefficients of the two given polynomials; lines 3, 4 and 5 contain the coefficients resulting from the multiplication of line 1 by -2, -1 and 1 respectively (the values in line 2); and line 6 contains the final coefficients obtained by summing the columns. The resulting polynomial is $x^5 - 5x^4 + 2x^3 + 9x^2 - x - 2$.

Example 2: Divide $2x^4 + 3x^3 - 2x + 1$ by $x^2 + 2x - 1$.

Set out the coefficients of powers of x in columns as for arithmetical long division to obtain

	x^2	x^1	x^0	) x^4	x^3	x^2	x^1	x^0
line 1						2	-1	4
line 2	1	2	-1	) 2	3	0	-2	1
line 3				2	4	-2		
line 4					-1	2	-2	
line 5					-1	-2	1	
line 6						4	-3	1
line 7						4	8	-4
line 8							-11	5

Line 2 contains the coefficients of the divisor first and the polynomial being divided second; line 3 contains the products of the coefficients of the divisor with 2, the number that ensures that the highest power of x (x^4 here) may be made zero; line 4 is the difference between lines 2 and 3 with the co-efficient of the next power of x brought down. The process is continued until we are left with a polynomial of lower degree than the divisor. The result of the division is given in line 1 with the remainder in line 8. In the present example

$$\frac{2x^4 + 3x^3 - 2x + 1}{x^2 + 2x - 1} = 2x^2 - x + 4 + \frac{-11x + 5}{x^2 + 2x - 1}.$$

Note, if the divisor is a factor of the polynomial being divided then the remainder is zero.

3.8 INTERPOLATION

It is often convenient to present values of a complicated mathematical expression or the results obtained from an experiment in the form of a table. Typically a variable y depends on a prescribed variable x, and values of y corresponding to (usually) equally spaced values of x are given for $a \leqslant x \leqslant b$. It is sometimes necessary to determine the value of y corresponding to a value of x lying between a and b but not given in the table. This process is called **interpolation**. There are several methods of interpolating between two given values that make use of the properties of polynomials, but the simplest, and the only one considered here, is **linear interpolation**.

Given neighbouring values of x in the table x_1 and x_2 ($x_2 > x_1$) and corresponding values of y, y_1 and y_2, then the value of y corresponding to a value of x lying between x_1 and x_2 is given approximately by

$$y = y_1 + \frac{(y_2 - y_1)(x - x_1)}{(x_2 - x_1)}. \tag{3.14}$$

(The point (x, y) is assumed to lie on a straight line connecting the points (x_1, y_1) and (x_2, y_2) and (3.14) is equation (3.4) of Section 3.3).

It is sometimes desirable to determine the value of y corresponding to a value of x not lying between a and b. This process is called **extrapolation**. The value of y may be calculated by using a linear extension in much the same way as for interpolation, but highly inaccurate values of y are sometimes obtained. Interpolation usually leads to accurate values of y provided that the values of x are reasonably closely spaced.

Example 1: The following table is an extract from a table of values of $y = e^x$, the exponential function. Determine approximate values of y corresponding to (a) $x = 0.54$ and (b) $x = 0.673$.

x	$y = e^x$
0.5	1.649
0.6	1.822
0.7	2.014

(a) Here $x_1 = 0.5, y_1 = 1.649; x_2 = 0.6, y_2 = 1.822$.

Setting $x = 0.54$ equation (3.14) gives

$$y = 1.649 + \frac{(0.173)(0.04)}{0.1}$$

$$= 1.649 + 0.069$$

$$= 1.718.$$

The correct value of y is 1.716.

{Note that x is 4/10 of $x_2 - x_1$ so that the corresponding y is 4/10 of $y_2 - y_1$}.

(b) Here $x_1 = 0.6, y_1 = 1.822; x_2 = 0.7, y_2 = 2.014$.

Setting $x = 0.673$ equation (3.14) gives

$$y = 1.822 + \frac{(0.192)(0.073)}{0.1}$$

$$= 1.822 + 0.140$$

$$= 1.1962.$$

Example 2: In an experiment to determine the change in body temperature resulting from the administration of a new drug the following table was obtained

Time (hours) x	0	1	2	3	4	5
Temperature $^\circ$C y	36.81	37.23	38.28	37.87	37.72	37.50

Determine the approximate body temperature after $2\frac{3}{4}$ hours have elapsed. {See example 2 of Section 1.12}.

Here $x_1 = 2, y_1 = 38.28; x_2 = 3, y_2 = 37.87; x = 2.75$.

Equation (3.14) results in

$$y = 38.28 + \frac{(-0.41)\,(0.75)}{1}$$

$$= 38.28 - 0.31$$

$$= 37.97 .$$

{ The point $(2.75, 37.97)$ lies close to the curves of Fig. 1.4 in Chapter 1}.

3.9 SUMMARY

The topics discussed in this chapter concerning the graphical representation of polynomials and the solution of polynomial equations are fundamental to many situations in which experimental results are approximated by such functions. The final section on interpolation is particularly important whenever it is necessary to deduce values for a function that do not themselves appear in a table.

Additional information on polynomial equations is to be found in Turnbull (1947).

PROBLEMS

1. Obtain an equation for the straight lines which pass through:

    (i) $(0, 0)$ with slope 2,

    (ii) $(1, -2)$ with slope -3,

    (iii) $(-1, 2)$ with slope -3,

    (iv) $(0, 0)$ and $(1, 1)$,

    (v) $(1, -2)$ and $(-1, 2)$,

    (vi) $(1, 0)$ and $(0, -2)$,

    (vii) $(1, 2)$ and $(0, 5)$,

    (viii) $(1, 2)$ and $(-2, 2)$,

    (ix) $(2, 1)$ and $(2, -1)$.

2. Find an equation for the line L which passes through $(1, -3)$ and is parallel to the line $y + 2x = 4$. Show that L cuts the y-axis in the point $(0, -1)$.

3. Show that the line through the points $(0, 3)$ and $(4, -1)$ meets the line through $(-3, 0)$ and $(-1, 1)$ in the point $(1, 2)$.

4.   (i) Show that the Centigrade temperature C and the Fahrenheit tempera-
     ture F are related by the equation $9C = 5F - 160$. Use the informa-
     tion that when $C = 0$, $F = 32$ and when $C = 100$, $F = 212$.

     (ii) What is the value of C when $F = 0$?

     (iii) Determine the temperature at which the Centigrade and Fahrenheit
     readings are the same.

5.   During a 30-day period a plant grows from a height of 10 cm to a height
     of 34 cm. Assuming that the growth is linear and that $t = 0$ when $l = 10$
     where t is the time in days and l the height in cm show that $5l = 4t + 50$.
     What will be the height of the plant after 20 days?

6.   The mean number f of fledglings in a bird sanctuary is related approxi-
     mately to the number n of breeding adults by the equation

     $$f = 5 - 0.05n, \quad 1 \leqslant n < 100 .$$

     Plot the graph of f against n. In two successive years the number of breeding
     adults rises from 20 to 40. Calculate the percentage decrease in the mean
     number of fledglings.

7.   Factorise, and hence solve, the equations:

     (i) $x^2 - 3x + 2 = 0$,

     (ii) $x^2 + 3x + 2 = 0$,

     (iii) $2x^2 - 5x + 2 = 0$,

     (iv) $6x^2 + 13x - 5 = 0$,

     (v) $6x^2 - x - 12 = 0$.

8.   Use the formula for the solution of a quadratic equation to solve:

     (i) $x^2 + 2x - 1 = 0$,

     (ii) $x^2 + 2x - 2 = 0$,

     (iii) $2x^2 + 3x - 1 = 0$,

     (iv) $3x^2 - 2x - 2 = 0$.

9.   Express as the product of two linear factors:

     (i) $x^2 + 2x - 1$,

     (ii) $x^2 + 2x - 2$,

     (iii) $2x^2 + 3x - 1$.

10. An object projected vertically upwards at time $t = 0$ with a velocity of 14 metres/second reaches a height, y, in metres given by

$$y = 14t - 4.9t^2 \ .$$

(i) Find the time that elapses before the object returns to its point of projection.

(ii) Find the values of t for which $y = 5$.

(iii) Find t when $y = 10$; what is the significance of this value of y?

11. Solve the equations:

(i) $x^2 + 9 = 0$,

(ii) $x^2 - 2x + 2 = 0$,

(iii) $x^2 + 2x + 5 = 0$,

(iv) $2x^2 - 2x + 1 = 0$,

(v) $2x^2 + 2x + 3 = 0$.

12. Express as the product of two linear factors:

(i) $x^2 - 2x + 2$,

(ii) $x^2 + 2x + 5$,

(iii) $2x^2 + 2x + 3$.

13. Express in polar form with $r > 0$ and $-\pi < \theta \leqslant \pi$:

(i) $1 + i$,

(ii) $-1 + i$,

(iii) $-1 - i\sqrt{3}$,

(iv) $\dfrac{1 + i}{1 - i}$.

14. Express in the form $a + ib$ the complex numbers:

(i) $(1 - i)(2 + i)$,

(ii) $\dfrac{1 - 2i}{1 + i}$,

(iii) $\dfrac{2 - i}{1 - i} + 3 - 2i$,

(iv) $\dfrac{1 + 2i}{1 - i} - \dfrac{3 + i}{2i}$,

(v) $\dfrac{1 - i}{1 - 3i} + \dfrac{2 - i}{3 + i}$.

15. Plot the curves given by:

   (i) $y = x^2 + 2x - 3$,

   (ii) $y = -x^2 - 2x + 3$,

   (iii) $y = 2x^2 - 4x + 3$,

   (iv) $y = -2x^2 - 4x - 5$.

   In all cases determine the point at which the curve has a finite maximum or minimum value.

16. Plot $y = x^2 + 4x$ and $y = x + 4$ on the same figure. Determine their points of intersection (i) graphically and (ii) algebraically.

17. Solve, by factorisation, the equations

   (i) $x^3 + 4x^2 + x - 6 = 0$,

   (ii) $2x^3 - 3x^2 - 8x - 3 = 0$,

   (iii) $x^3 - 3x - 2 = 0$,

   (iv) $8x^3 - 4x^2 - 2x + 1 = 0$,

   (v) $x^3 + 3x^2 + 3x + 1 = 0$.

18. Obtain one root by inspection and hence solve

   (i) $x^3 + x^2 - 3x + 1 = 0$,

   (ii) $x^3 - 2x^2 - 2x + 1 = 0$,

   (iii) $x^3 - x^2 + 4x - 4 = 0$,

   (iv) $x^3 - 1 = 0$,

   (v) $x^3 + 1 = 0$.

19. Plot the graphs of $y = x^3$ and $y = 5x + 2$ on the same figure. Hence obtain approximate values for the roots of the equation

   $$x^3 - 5x - 2 = 0 .$$

20. Plot $y = x^3 - 3x^2$ and $y = 2 - 3x$ on the same figure. Determine their point of intersection (i) graphically and (ii) algebraically.

21. Solve, by factorisation, the equations

   (i) $x^4 - 5x^3 + 5x^2 + 5x - 6 = 0$,

   (ii) $x^4 - 1 = 0$,

   (iii) $x^4 - 2x^2 + 1 = 0$,

(iv) $x^4 - 2x^3 + 2x - 1 = 0,$

(v) $x^5 - 3x^4 - 5x^3 + 15x^2 + 4x - 12 = 0.$

22. Find the real roots of the following equations:

(i) $x^4 - 5x^2 + 6 = 0,$

(ii) $x^6 - 9x^3 + 8 = 0,$

(iii) $x^8 - 17x^4 + 16 = 0,$

(iv) $x^4 - x^2 - 12 = 0,$

(v) $x^4 + 7x^2 + 12 = 0.$

23. Use the method of detached coefficients to evaluate

(i) $(x^2 - 2x + 1)(x^3 + 3x - 2),$

(ii) $(2x^3 - 3x^2 + 2x + 1)(-2x^2 + x - 3),$

(iii) $\dfrac{3x^4 + 2x^3 - 1}{x^2 - 2x + 3},$

(iv) $\dfrac{2x^3 - 3x^2 + 2x + 1}{2x^2 - x - 3}.$

24. Values of sin x and cos x, for x measured in radians, are given in the following table. Determine approximate values of sin x and cos x when x equals (i) 1.02, (ii) 1.075, (iii) 1.083.

x	sin x	cos x
1.00	0.8415	0.5403
1.05	0.8674	0.4976
1.10	0.8912	0.4536

25. The surface area A, in square metres, and weight W, in kilograms, of children is given by

W	5	6	7	8
A	0.307	0.345	0.381	0.415

Estimate the surface area of a child of weight 6.3 kilograms.

26. The pulse rates p of people of height h (metres) are given by

h	0.6	0.8	1.0	1.2	1.4	1.6	1.8
p	124	107	95	86	80	74	69

Plot the graph of p against h and estimate (i) graphically and (ii) by interpolation the approximate pulse rate of a person of height 0.9 metres.

27. The following table gives values of y for a certain three values of x:

x	4	16	25
y	1	2	2.5 .

(i) Show (by calculation) that y is not a linear function of x.

(ii) When $u = \log_{10}x$ and $v = \log_{10}y$, tabulate — correct to two decimal places only — the values of u, v corresponding to the given values of x, y.

(iii) Deduce that v could be a linear function of u.

(iv) If (iii) is valid, find — correct to one decimal place — the value of y when $x = 10$.   (Edinburgh, 1978)

28. In a biochemical reaction, certain variables v and w are related by an equation of the form

$$v = \frac{w}{aw + b},$$

where a and b are constants. If $x = 1/w$ and $y = 1/v$, show that y is a linear function of x.

Three pairs of values w, v are given below. By finding the corresponding values of x, y, verify (by calculation, not by plotting) that the three points (x, y) lie (almost exactly) on a line whose slope is about 1.67.

w	2.32	4.52	38.5
v	0.95	1.42	2.64 .

(Edinburgh, 1974)

CHAPTER 4

Functions

4.1 INTRODUCTION

The material covered in this chapter is concerned with some aspects of functions. Function and inverse function are defined, and there follows a brief discussion of continuity and limits. The ideas mentioned here are basic to any situation in which one variable depends on another and are essential prerequisites for many of the later chapters.

4.2 FUNCTIONS

Given two variable numbers x and y such that with each value of x there is associated one value, or several values, of y then y is said to be a **function** of x. The set of values of x is called the **domain** of the function and the corresponding set of values of y is called the **range** of the function. A function is said to be **single-valued** if there is only one value of y for each value of x. It is conventional to denote single-valued functions by writing

$$y = f(x), \quad a \leqslant x \leqslant b .$$

This is read 'y is equal to a function f of x for values of x in the domain $a \leqslant x \leqslant b$'. The domain sometimes includes a and b as above, sometimes not. Frequently no domain is specified and in that case the domain is chosen to be those values of x for which the function takes real values. The range is determined once the domain is given. Some simple examples are

FUNCTION	DOMAIN	RANGE
$y = x^3$,	$-1 < x \leqslant 3$,	$-1 < y \leqslant 9$,
$y = x^3$,	all x ,	all y ,
$y = x^2$,	all x ,	$0 \leqslant y$,
$y = \log_{10} x$,	$0 < x$,	all y ,
$y = (1+x)^{\frac{1}{2}}$,	$-1 \leqslant x$,	$0 \leqslant y$,
$y = (1-x)^{\frac{1}{2}}$,	$x \leqslant 1$,	$0 \leqslant y$.

The variable x is referred to as the independent variable and y is known as the dependent variable. In the majority of theoretical discussions the relationship between x and y is an equation (like $y = x^3$) but in many practical situations the relationship is given by a table of values, resulting from an experiment (see Chapter 19). In most of the cases encountered here the relationship is single-valued and y is said to be an **explicit function** of x. Other situations exist in which the equation connecting x and y is not solved for y; y is then said to be an **implicit function** of x. Some implicit relationships may be made explicit by solving for y; in other cases it is not possible to solve the equation. Examples of implicit equations are

$$y^2 = 1 + x^2 ,$$
$$y^3 + xy - x^4 = 4 ,$$
$$\sin y + \tan x = \log_{10} y + x .$$

The first of these implicit equations gives two values of y for each value of x since

$$y = \pm(1+x^2)^{\frac{1}{2}} .$$

Hence we think of it as two explicit, single-valued equations

$$y_1 = +(1+x^2)^{\frac{1}{2}} \quad \text{and} \quad y_2 = -(1+x^2)^{\frac{1}{2}} .$$

The domain of both functions is all values of x, but the range of y_1 is $1 \leqslant y_1$ and that of y_2 is $y_2 \leqslant -1$. The other two examples cannot be made explicit.

Example: If $f(x) = x^2 + 4x - 1$ find (a) $f(2)$, (b) $f(-2)$, (c) $f(-t)$, (d) $f(\sin t)$.

(a) $f(2)$ is found by replacing x by 2 to obtain $f(2) = 4 + 8 - 1 = 11$.

(b) $f(-2)$ is found by replacing x by -2 to obtain $f(-2) = 4 - 8 - 1 = -5$.

(c) $f(-t)$ is found by replacing x by $-t$ to obtain $f(-t) = t^2 - 4t - 1$.

(d) $f(\sin t)$ is found by replacing x by $\sin t$ to obtain

$$f(\sin t) = \sin^2 t + 4 \sin t - 1 .$$

Part (d) of this example illustrates a situation in which the required value of the function is itself given by another function. In general if $y = f(x)$ and $x = g(t)$ (that is, x is some function g of t) then it is possible to write

$$y = f(x) = f\{g(t)\} = F(t)$$

and y is an implied function F of the variable t. It is not always practicable to determine $F(t)$ explicitly.

4.3 INVERSE FUNCTIONS

It is sometimes desirable when y is given as a function of x to be able to express x as a function of y so that the roles of x and y are interchanged. The normal procedure is to solve the equation $y = f(x)$ for x whenever possible to obtain $x = f^{-1}(y)$, where the index -1 denotes the functional inverse, not the multiplicative inverse discussed in Chapter 1. Usually the symbols x and y are then interchanged so that $y = f^{-1}(x)$. If possible a single-valued form is determined. Examples of this type of relationship are the trigonometric functions and their inverses discussed in Chapter 1. It can be shown that the functions $y = f(x)$ and $y = f^{-1}(x)$ are symmetrical with respect to the line $y = x$ and also that $f\{f^{-1}(x)\} = x = f^{-1}\{f(x)\}$.

Example: Determine the inverse functions of (a) $2x+1$, (b) x^3, (c) x^2+1.

(a) Here $y = f(x) = 2x+1$.

Solve for x to obtain $x = \frac{1}{2}(y-1) = f^{-1}(y)$.

Interchange x and y so that $y = \frac{1}{2}(x-1) = f^{-1}(x)$. Graphs of $f(x)$ and $f^{-1}(x)$ are shown in Fig. 4.1. The symmetry with respect to $y = x$ is clear.

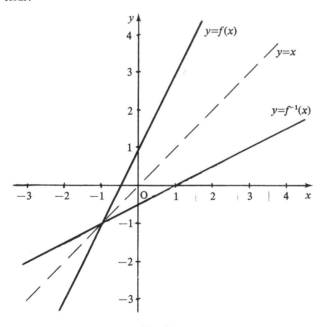

Fig. 4.1

(b) Here $y = f(x) = x^3$

so that $x = y^{1/3} = f^{-1}(y)$.

Interchanging x and y results in $y = x^{1/3} = f^{-1}(x)$. Graphs of $f(x)$ and $f^{-1}(x)$ and the line $y = x$ are shown in Fig. 4.2.

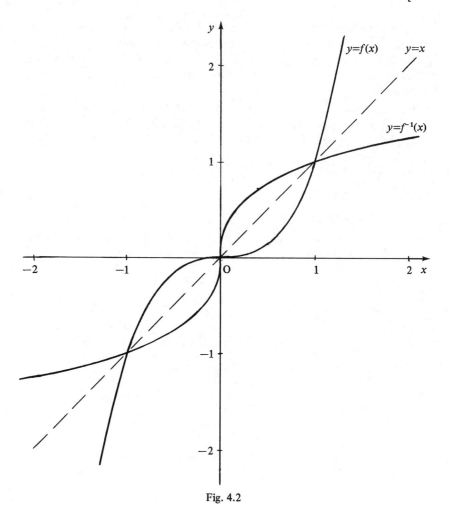

Fig. 4.2

(c) Here $y = f(x) = x^2 + 1$ with domain all values of x and range $1 \leqslant y$.
Solving for x gives

$$x^2 = y - 1 \ ,$$

so that suitable single-valued expressions are

$$x = (y-1)^{\frac{1}{2}} = f_1^{-1}(y) \text{ and } x = -(y-1)^{\frac{1}{2}} = f_2^{-1}(y) \ .$$

Interchanging x and y gives

$$y = (x-1)^{\frac{1}{2}} = f_1^{-1}(x) \text{ and } y = -(x-1)^{\frac{1}{2}} = f_2^{-1}(x) \ .$$

The function of x, $f_1^{-1}(x)$, has domain $1 \leqslant x$ and range $0 \leqslant y$ whereas $f_2^{-1}(x)$ has domain $1 \leqslant x$ and range $y \leqslant 0$. Graphs of $f(x), f_1^{-1}(x), f_2^{-1}(x)$ and $y = x$ are shown in Fig. 4.3. Note that $f_2^{-1}(x)$ corresponds to that part of $f(x)$ for which $x \leqslant 0$.

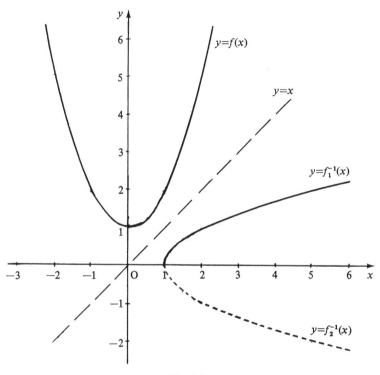

Fig. 4.3

4.4 CONTINUITY

The function of x

$$y = f(x), \qquad a \leqslant x \leqslant b ,$$

is **continuous** at $x = c$ where $a < c < b$ provided that the function has a definite value V at $x = c$ and the value of $f(x)$ approaches V as x approaches c from either side. A function is **continuous in an interval** $a < x < b$ if it is continuous at all points of that interval. The graph of a continuous function contains no jumps or breaks in the appropriate interval. A function is said to be **discontinuous** at points where there is a break in the graph. A function is said to be **undefined** at $x = c$ whenever it has no meaningful value at the point. Usually at such a point $f(c) = 0/0$ or ∞/∞ or $0.\infty$ (note that $0.0 = 0$ and $\infty.\infty = \infty$). In the next section an attempt will be made to assign a definite value to a function at points where it is not defined. Points of discontinuity, if present, usually occur at (a) the end points of the domain, or (b) where there is a change in the definition of the function, or (c) when an attempt is made to divide by zero, or (d) where the function is not defined.

It will be assumed here that a function exists for those values of x for which the function takes real values unless it is clearly indicated otherwise. This approach gives rise to a very simple treatment of continuity which is adequate for most practical purposes although reference to a text on mathematical analysis such as Protter and Morrey (1964) will indicate that this approach is inadequate in certain situations. In particular one function that will not be discussed here, but which is sometimes used to describe the behaviour of a nerve impulse, is assigned the value zero at all time except at the instant of the impulse. Practical examples of discontinuous functions are (a) the size of a population which increases or decreases by unit amounts and (b) the intensity of light in a room when a light switch is turned on (or off).

Example: Determine the points of discontinutiy, if any, of $f(x)$ where (a) $f(x) = x$, (b) $f(x) = |x|$, (c) $f(x) = (\sin x)/x$, (d) $f(x) = \tan x$,

(e) $f(x) = \begin{cases} 1, x > 0, \\ 0, x = 0, \\ -1, x < 0, \end{cases}$ (f) $f(x) = (x^2-1)^{\frac{1}{2}}$.

(a) $y = f(x) = x$ is continuous for all x (see Fig. 4.4).

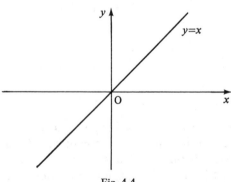

Fig. 4.4

(b) $y = f(x) = |x|$ is also continuous for all x although there is a change of character at $x = 0$ (see Fig. 4.5).

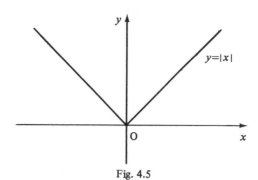

Fig. 4.5

(c) $y = f(x) = \dfrac{\sin x}{x}$ is continuous for all values of x except $x = 0$ when

$y = 0/0$. At $x = 0$ the function is (at present) undefined although as will be seen in Section 4.6 it is in fact continuous at that point.

(d) $y = f(x) = \tan x = \dfrac{\sin x}{\cos x}$. This function is discontinuous whenever

$\cos x = 0$, that is when $x = \pm\frac{1}{2}\pi, \pm\frac{3}{2}\pi, \pm\frac{5}{2}\pi$ etc. Otherwise it is continuous. (For a graph, see Fig. 1.8.)

(e) This function is continuous for all values of x except $x = 0$ when there is a jump in the value of y (see Fig. 4.6).

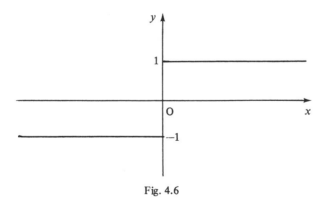

Fig. 4.6

(f) $y = f(x) = (x^2-1)^{\frac{1}{2}}$ has domain $x \leqslant -1$ and $1 \leqslant x$ and range $0 \leqslant y$. It is continuous for all values of x in its domain and is discontinuous between $x = -1$ and $x = +1$ where it is undefined (see Fig. 4.7).

Note that it is often helpful to sketch a graph of the function.

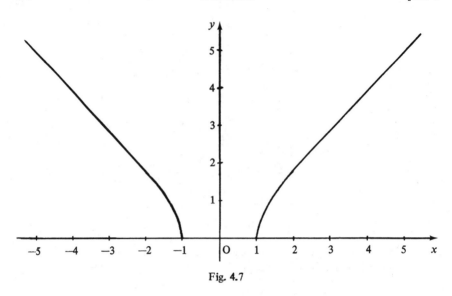

Fig. 4.7

4.5 LIMITS

Part (c) of the example in the previous section, $f(x) = \sin x/x$, is an example in which $y = f(x)$ is undefined at $x = 0$ owing to obtaining the value $y = 0/0$. Many examples exist in which a situation of this type arises, and an attempt will be made in this section to assign values to such expressions. A particular example will first be considered, and some general results will then be stated.

Consider the function $y = f(x) = \dfrac{x^2-1}{x-1}$ which is continuous (and equal to $x+1$) everywhere except at the point $x = 1$ where $y (= 0/0)$ is undefined. A table of values of the function as x approaches 1 results in

x	−1.0	0	0.9	0.99	0.999
y	0	1.0	1.9	1.99	1.999

x	3.0	2.0	1.1	1.01	1.001
y	4.0	3.0	2.1	2.01	2.001

and it is possible to calculate a finite value of y for any given value of x as close to $x = 1.0$ as is desired. A graph of the function is shown in Fig. 4.8 from which

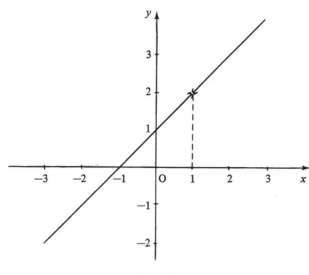

Fig. 4.8

it is clear that as x approaches 1.0 (from either side) the value of y approaches 2.0. This is supported by the values in the table, and consequently it appears reasonable to assign the value 2.0 to $f(1.0)$. We say that the value of $f(x)$ approaches 2 as x approaches 1, written $f(x) \rightarrow 2$ as $x \rightarrow 1$, or alternatively, the **limit** of $f(x)$ as x **tends to** 1 is 2, written

$$\lim_{x \to 1} \frac{x^2-1}{x-1} = 2 \ .$$

This value of $f(1)$ is the same as the value of $(x+1)$ when $x = 1$, and the limit may be evaluated by writing

$$\lim_{x \to 1} \frac{x^2-1}{x-1} = \lim_{x \to 1} (x+1) = 2 \ .$$

In general $f(x)$ tends to a **limiting value** l as $x \rightarrow c$, written $\lim_{x \to c} f(x) = l$, if the value of $f(x)$ approaches l as x approaches c, with x taken as close to c as we wish, either above or below c.

In calculating the value of a limit to the function $f(x)$ at $x = c$ always check that $f(c)$ is undefined; if it exists then $\lim_{x \to c} f(x) = f(c)$.

The values of c and l may be finite (including zero) or infinite. The limit of an algebraic expression may usually be calculated by rearranging the expression.

It can be shown that if $\lim_{x \to c} f(x) = F$ and $\lim_{x \to c} g(x) = G$ then

(i) $\lim_{x \to c} \{f(x) \pm g(x)\} = F \pm G$,

(ii) $\lim\limits_{x \to c} \{f(x)\, g(x)\} = FG,$

(iii) $\lim\limits_{x \to c} \{K\, f(x)\} = KF,$ where K is a constant,

(iv) $\lim\limits_{x \to c} \left\{ \dfrac{f(x)}{g(x)} \right\} = \dfrac{F}{G},$ provided $G \neq 0.$

Example: Determine the values of:

(a) $\lim\limits_{x \to 3} \dfrac{x^2-9}{x^2-x-6}$, (b) $\lim\limits_{x \to 0} \dfrac{x^2+2}{2x^2+1}$, (c) $\lim\limits_{x \to 0} \dfrac{1-(1-x)^{\frac{1}{2}}}{x}$, (d) $\lim\limits_{x \to \infty} \dfrac{x^2+2}{2x^2+1}$.

(a) $\lim\limits_{x \to 3} \dfrac{x^2-9}{x^2-x-6} = \lim\limits_{x \to 3} \dfrac{(x-3)\,(x+3)}{(x-3)\,(x+2)}$

$\qquad\qquad = \lim\limits_{x \to 3} \dfrac{x+3}{x+2} = \dfrac{6}{5}.$

(b) In this example the value of $f(x) = (x^2+2)/(2x^2+1)$ is defined at $x = 0$ so that $f(0) = \dfrac{2}{1} = 2.$

(c) In this example it is not possible to factorise the numerator (which $\to 0$ as $x \to 0$) and it is necessary to make some other adjustment of the function. The 'awkward' component is $(1-x)^{\frac{1}{2}}$, and in order to rationalise the numerator it is necessary to multiply numerator and denominator by $\{1 + (1-x)^{\frac{1}{2}}\}$. Then

$\lim\limits_{x \to 0} \dfrac{\{1 - (1-x)^{\frac{1}{2}}\}}{x} = \lim\limits_{x \to 0} \dfrac{\{1 - (1-x)^{\frac{1}{2}}\}\{1 + (1-x)^{\frac{1}{2}}\}}{x\{1 + (1-x)^{\frac{1}{2}}\}}$

$\qquad\qquad = \lim\limits_{x \to 0} \dfrac{\{1 - (1-x)\}}{x\{1 + (1-x)^{\frac{1}{2}}\}}$

$\qquad\qquad = \lim\limits_{x \to 0} \dfrac{x}{x\{1 + (1-x)^{\frac{1}{2}}\}}$

$\qquad\qquad = \lim\limits_{x \to 0} \dfrac{1}{1 + (1-x)^{\frac{1}{2}}} = \tfrac{1}{2}.$

Note that $\{1 + (1-x)^{\frac{1}{2}}\}$ is non-zero as $x \to 0$ so that it is not necessary to change this term where it appears in the denominator.

(d) In this example $f(x) \to \infty/\infty$ as $x \to \infty$. The most satisfactory procedure with examples of this type is to take $t = 1/x$ so that as $x \to \infty$ we have $t \to 0$. It is found that

$$\lim_{x \to \infty} \frac{x^2+2}{2x^2+1} = \lim_{t \to 0} \frac{\dfrac{1}{t^2} + 2}{\dfrac{2}{t^2} + 1}$$

$$= \lim_{t \to 0} \frac{1 + 2t^2}{2 + t^2} = \tfrac{1}{2} \, .$$

Note that in order to remove the terms containing $1/t^2$ it is necessary to multiply the numerator and the denominator by t^2.

4.6 A TRIGONOMETRIC LIMIT

In order to carry out certain calculations at a later stage it is necessary to assign a value to the quantity $\sin x/x$ as x approaches zero. A limiting process is required since both numerator and denominator vanish when x is zero.

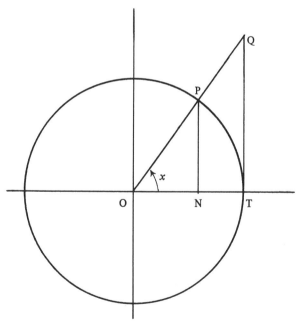

Fig. 4.9

Referring to Fig. 4.9, we consider a circle of radius R and centre O. Take a base radius OT and construct another radial line OP to obtain an acute angle $T\hat{O}P = x$, positive in the anti-clockwise direction. The line PN is perpendicular to OT, and QT is the tangent at T. When radian measure is adopted for x it is found that

$$PN \quad = R \sin x,$$
$$\text{arc PT} = Rx,$$
$$QT \quad = R \tan x,$$

so that

$$\text{area of triangle OPT} = \tfrac{1}{2}OT.PN = \tfrac{1}{2}R^2 \sin x,$$
$$\text{area of sector} \quad OPT = \tfrac{1}{2}R^2 x, \; .$$
$$\text{area of triangle OQT} = \tfrac{1}{2}OT.QT = \tfrac{1}{2}R^2 \tan x.$$

The geometry of the figure indicates that, for $x > 0$,

area of triangle OPT < area of sector OPT < area of triangle OQT, so that

$$\tfrac{1}{2}R^2 \sin x < \tfrac{1}{2}R^2 x < \tfrac{1}{2}R^2 \tan x$$

$$\sin x < \quad x \quad < \tan x = \sin x / \cos x$$

and since $\quad \sin x > 0$ for $0 < x < \tfrac{1}{2}\pi$

$$1 < \frac{x}{\sin x} < \frac{1}{\cos x} \; . \tag{4.1}$$

Inverting the terms in the inequality (4.1) results in

$$1 > \frac{\sin x}{x} > \cos x \; . \tag{4.2}$$

It can be shown that the results (4.1) and (4.2) also hold when $-\tfrac{1}{2}\pi < x < 0$. As x approaches zero (through positive or negative values) so $\cos x$ approaches 1 and the quantities $x/\sin x$ in (4.1) and $\sin x/x$ in (4.2) are trapped between two quantities each of which becomes 1 in the limit. As a result it is found that

$$\lim_{x \to 0} \frac{x}{\sin x} = 1 \tag{4.3}$$

and $\quad \lim_{x \to 0} \frac{\sin x}{x} = 1 \; . \tag{4.4}$

The assumption that radian measure is used for x is an essential ingredient in obtaining the result (4.4). If x is measured in degrees then the area of the sector OPT is not $\tfrac{1}{2}R^2 x$ and the final limit is not unity but $\pi/180$. The simple forms obtained for (4.3) and (4.4) by using radian measure are attractive, and it is for this reason that radian measure is adopted in the later chapters dealing with calculus.

Example: Evaluate (a) $\lim_{x \to 0} \dfrac{\tan x}{x}$, (b) $\lim_{x \to 0} \dfrac{1 - \cos x}{x}$, (c) $\lim_{x \to 0} \dfrac{\sin 3x}{x}$.

(a) $\lim\limits_{x\to 0} \dfrac{\tan x}{x} = \lim\limits_{x\to 0} \dfrac{\sin x}{x \cos x}$

$$= \lim\limits_{x\to 0} \left(\dfrac{\sin x}{x}\right)\left(\dfrac{1}{\cos x}\right) = 1 \ .$$

(b) $\lim\limits_{x\to 0} \dfrac{1-\cos x}{x} = \lim\limits_{x\to 0} \dfrac{2 \sin^2(\frac{1}{2}x)}{x}$

$$= \lim\limits_{x\to 0} \left\{\dfrac{2 \sin(\frac{1}{2}x)}{x}\right\} \{\sin(\tfrac{1}{2}x)\} \ .$$

Substitute $t = \frac{1}{2}x$, and since $t \to 0$ as $x \to 0$ it is found that

$$\lim\limits_{x\to 0} \dfrac{1-\cos x}{x} = \lim\limits_{t\to 0} \left(\dfrac{\sin t}{t}\right) \{\sin t\} = 1\times 0 = 0 \ .$$

(c) In this example substitute $t = 3x$ so that, since $t \to 0$ as $x \to 0$,

$$\lim\limits_{x\to 0} \dfrac{\sin 3x}{x} = \lim\limits_{t\to 0} \dfrac{3 \sin t}{t} = 3 \ .$$

PROBLEMS

1. State the domain of x (where none is given) and determine the range of y for:

    (i) $y = 1 - x$, $1 \leqslant x$,

    (ii) $y = 1 + x$, $-2 < x \leqslant 1$,

    (iii) $y = x^2 - 1$,

    (iv) $y = |x|$,

    (v) $y = |1 + x|$, $x < 0$,

    (vi) $y = \sin x$,

    (vii) $y = \tan x$, $0 \leqslant x < \pi/2$,

    (viii) $y = (2 - x)^{\frac{1}{2}}$,

    (ix) $y = x^2/(x^2 + 1)$,

    (x) $y = x/(x + 1)$.

2. The rate of growth v of a biological population is given in terms of its size n by

$$v = kn(a - n)$$

where a, k are positive constants.

    (i) If the domain of n is $0 < n < a$ find the range of v.

    (ii) If the domain of n is $a < n$ find the range of v.

3. If $f(x) = x^2 - 2x + 3$ find (i) $f(1)$, (ii) $f(-2)$, (iii) $f(2t)$, (iv) $f(-t^2)$, (v) $f(\cos x)$.

4. The size of a population is given approximately by $N(t) = 1000 + 50t^2$ where t is the time measured in hours. Determine $N(0), N(2), N(4)$.

5. The mass $m(t)$ of protein disintegrates according to the formula

$$m(t) = \frac{\alpha}{\beta + t}$$

    where t is the time. Calculate $m(0)$, $m(\beta)$ and evaluate $\lim_{t \to \infty} m(t)$.

6. Given the implicit relation $y^2 = x$, find two explicit relationships for y. State the domain of x and range of y in each case.

7. Given $y^2 = 4 + x^2$ find two explicit forms for y. State the domain of x and the range of y in each case.

8. Given the implicit relation $x^2 + y^2 = 4$, find two explicit forms for y. State the domain of x and the range of y in each case.

9. The velocity v of blood flowing in a cylindrical artery of radius R is given approximately by

$$v = f(r) = \alpha(R^2 - r^2) ,$$

    where r is the distance from the centre of the artery and α is a constant. State the domain of r and the range of v and express r in terms of v. Evaluate $f(R), f(\tfrac{1}{2}R), f(0)$.

10. Determine the inverse functions $f^{-1}(x)$ for each of the following functions $f(x)$; in all cases sketch $y = f(x)$ and $y = f^{-1}(x)$ showing their symmetry with respect to the line $y = x$.

    (i) $f(x) = 1 - 2x$,

    (ii) $f(x) = 1/x, \quad x \neq 0$,

    (iii) $f(x) = (1 + x)/(1 - x), \quad x \neq 1$,

    (iv) $f(x) = 1 - 8x^3$,

    (v) $f(x) = 1 - x^2$,

    (vi) $f(x) = (1 + x)^{\frac{1}{2}}$.

11. By taking logarithms to base 10 determine the inverse function of $f(x) = \alpha 10^x$, where α is a constant.

12. Determine the points of discontinuity, if any, of the following functions
     $f(x)$; in all cases sketch $y = f(x)$:

     (i) $f(x) = x^3$,

     (ii) $f(x) = |x^3|$,

     (iii) $f(x) = (x + 1)/(x - 1)$,

     (iv) $f(x) = (x^2 + 4)/(x^2 - 4)$,

     (v) $f(x) = (x^2 - 4)/(x^2 + 4)$,

     (vi) $f(x) = (\cos x)/x$,

     (vii) $f(x) = \cot x$,

     (viii) $f(x) = (x^2 - 4)^{\frac{1}{2}}$.

13. Determine the points of discontinuity, if any, of the following functions
     $f(x)$; in all cases sketch $y = f(x)$:

     (i) $f(x) = \begin{cases} x + 3, & x \leqslant 1, \\ 7 - 3x, & x > 1, \end{cases}$

     (ii) $f(x) = \begin{cases} 2x + 1, & x \leqslant 1, \\ 5 - x, & x > 1, \end{cases}$

     (iii) $f(x) = \begin{cases} 3x - 1, & x < -1, \\ 5x + 1, & x \geqslant -1, \end{cases}$

     (iv) $f(x) = \begin{cases} 4x/\pi^2, & x \leqslant \pi/2, \\ (\sin x)/x, & x > \pi/2, \end{cases}$

     (v) $f(x) = \begin{cases} -x, & x < -1, \\ 1/(x^2 - 2), & -1 \leqslant x \leqslant 1, \\ -x, & 1 < x. \end{cases}$

14. The size N of an animal population is given at time t by the function

     $N(t) = \begin{cases} 10, & 0 \leqslant t \leqslant 4, \\ 32, & 4 < t \leqslant 6, \\ 28, & 6 < t \leqslant 10, \\ 39, & 10 < t \leqslant 12, \\ 15, & 12 < t \leqslant 14, \\ 0, & 14 < t, \end{cases}$

     where t is measured in weeks. Plot a graph of N against t. Describe what
     happens at times $t = 4, 12, 14$.

15. Evaluate the following limits:

(i) $\lim\limits_{x \to 1} \dfrac{x^2 - 1}{x^2 + x - 2}$,

(ii) $\lim\limits_{x \to 1} \dfrac{x^2 - 1}{x^2 + x - 3}$,

(iii) $\lim\limits_{x \to 2} \dfrac{x^2 - 5x + 6}{x^2 - 4}$,

(iv) $\lim\limits_{x \to 1} \dfrac{x^3 - 1}{x - 1}$,

(v) $\lim\limits_{x \to 2} \dfrac{x^2 - 5x + 4}{x^2 + 4}$,

(vi) $\lim\limits_{x \to 0} \dfrac{(1 + x)^2 - (1 - x)^2}{x}$,

(vii) $\lim\limits_{x \to 3} \dfrac{27 - x^3}{x^2 - 3x}$.

16. Evaluate the following limits:

(i) $\lim\limits_{x \to -1} \dfrac{1 + 3x + 2x^2}{1 - x^2}$,

(ii) $\lim\limits_{x \to 1} \dfrac{2 - 3x + x^2}{1 - x^2}$,

(iii) $\lim\limits_{x \to 0} \dfrac{2 + 3x - x^2}{1 - x^2}$,

(iv) $\lim\limits_{x \to 0} \dfrac{x^3 + x}{3x^2 + 1}$,

(v) $\lim\limits_{h \to 0} \dfrac{(x + h)^2 - x^2}{h}$,

(vi) $\lim\limits_{h \to 0} \dfrac{1}{h} \{2(x + h)^2 - (x + h) - 2x^2 + x\}$,

(vii) $\lim\limits_{x \to 0} \dfrac{(1 + x)^3 + (1 - x)^3 - 2}{x^2}$.

17. Evaluate the following limits:

(i) $\lim\limits_{x \to \infty} \dfrac{x^2 + 1}{x^2 - 1}$,

(ii) $\lim\limits_{x\to\infty} \dfrac{x^2 - 1}{x^2 + 1}$,

(iii) $\lim\limits_{x\to\infty} \dfrac{2x + 1}{x^2 + x - 3}$,

(iv) $\lim\limits_{x\to\infty} \dfrac{3x^2 + 2x + 1}{x - 3}$,

(v) $\lim\limits_{x\to\infty} \dfrac{3x^2 + 2x + 1}{x^2 + x - 3}$,

(vi) $\lim\limits_{x\to\infty} \dfrac{x - x^3}{3x^3 + 1}$,

(vii) $\lim\limits_{x\to\infty} \dfrac{5x^2 + 1}{3x^2 + 3}$.

18. Evaluate the following limits:

(i) $\lim\limits_{x\to 0} \dfrac{1 - (1 + x)^{\frac{1}{2}}}{x}$,

(ii) $\lim\limits_{x\to 0} \dfrac{1 - (1 + x^2)^{\frac{1}{2}}}{x}$,

(iii) $\lim\limits_{x\to 3} \dfrac{(4x - 3)^{\frac{1}{2}} - x}{x - 3}$,

(iv) $\lim\limits_{x\to 1} \dfrac{(x^2 + 3)^{\frac{1}{2}} - 2}{x - 1}$,

(v) $\lim\limits_{x\to 1} \dfrac{1 - x}{(3x^2 + 1)^{\frac{1}{2}} + 2}$,

(vi) $\lim\limits_{x\to 0} \dfrac{(1 - x^2)^{\frac{1}{2}} - (1 + x^2)^{\frac{1}{2}}}{x^2}$,

(vii) $\lim\limits_{h\to 0} \dfrac{1}{h}\{(x + h)^{\frac{1}{2}} - x^{\frac{1}{2}}\}$,

(viii) $\lim\limits_{h\to 0} \dfrac{1}{h}\{(x + h)^{-\frac{1}{2}} - x^{-\frac{1}{2}}\}$.

19. Evaluate the following limits:

(i) $\lim\limits_{x\to -\infty} \dfrac{x^2 + 1}{x^2 - 1}$,

(ii) $\lim_{x \to -\infty} \dfrac{x + 1}{x - 1}$,

(iii) $\lim_{x \to \infty} \dfrac{x + 1}{(x^2 + 1)^{\frac{1}{2}}}$,

(iv) $\lim_{x \to \infty} \{(x^2 + 1)^{\frac{1}{2}} - x\}$,

(v) $\lim_{x \to \infty} x\{(x^2 + 1)^{\frac{1}{2}} - x\}$.

20. Evaluate the following limits:

(i) $\lim_{x \to 0} \dfrac{\sin 4x}{x}$,

(ii) $\lim_{x \to 0} \dfrac{\sin 3x}{\sin x}$,

(iii) $\lim_{x \to 0} \dfrac{x \cos x}{\sin 2x}$,

(iv) $\lim_{x \to 0} x \cot x$,

(v) $\lim_{x \to 0} \dfrac{1 - \cos x}{x^2}$,

(vi) $\lim_{x \to 0} \dfrac{1 - \cos^2 x}{x^2}$,

(vii) $\lim_{h \to 0} \dfrac{1}{h} \{\sin(x + h) - \sin x\}$,

(viii) $\lim_{h \to 0} \dfrac{1}{h} \{\cos(x + h) - \cos x\}$.

21. The size N of a biological population is related to the time $t (\geqslant 0)$ by the equation

$$10^t = \frac{aN}{b - N} ,$$

where a and b are positive constants. Determine N in terms of t and show that $\lim_{t \to \infty} N = b$.

CHAPTER 5

Differentiation

5.1 INTRODUCTION

The way in which the height of a plant changes with time during its life is illustrated in Fig. 5.1. Changes in height are markedly different during the different stages of growth. During its early life OA the height changes slowly as the root system develops, in middle life AB there is vigorous growth and the height changes rapidly, and then in later life BC the height again changes slowly as the plant concentrates on reproduction.

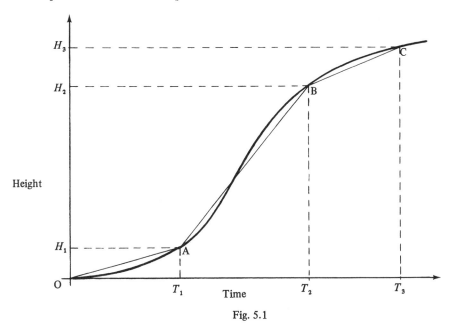

Fig. 5.1

In order to compare the various stages in the plant's life it is useful to measure the **average rate of growth.** From O to A the average rate of growth R_1 is given by

$$R_1 = \frac{\text{increase in height}}{\text{time taken}} = \frac{H_1}{T_1} = \text{slope of the straight line OA},$$

so that if the height is measured in centimetres and the time in weeks R_1 is measured in centimetres per week. From A to B the average rate of growth R_2 is given by

$$R_2 = \frac{\text{increase in height}}{\text{time taken}} = \frac{H_2 - H_1}{T_2 - T_1} = \text{slope of the straight line AB},$$

and from B to C the average growth rate R_3 is given by

$$R_3 = \frac{H_3 - H_2}{T_3 - T_2} = \text{slope of the straight line BC}.$$

The values of R_1, R_2 and R_3 will differ significantly, with R_2 being the greatest. The growth of a plant is a continuous process, and it would be possible to calculate its average rate of growth over shorter or longer periods than those indicated.

The concept of calculating the rate of change of a quantity is the basic objective of the process of differentiation discussed in this chapter. Many illustrations exist both within and outside the biological sciences. The rate of growth of an individual, the growth of a population, temperature changes during cooling, radioactive decay, changes in chemical reactions, the velocity of a car, are all illustrations.

In the remainder of this chapter the process of differentiation is defined and results are obtained for some commonly occurring functions. The basic rules of differentiation are then developed in preparation for application to particular problems.

5.2 THE DERIVATIVE

In order to discuss the process of differentiation we consider the function $y = f(x)$ which is assumed to be continuous. Referring to Fig. 5.2, which may be taken to illustrate the behaviour of a general function, the average rate of change of y between the two points $P(x_0, y_0)$ and $Q(x_0 + h, y_0 + k)$ is R where

$$R = \frac{\text{increase in the value of } y \text{ between P and Q}}{\text{increase in the value of } x \text{ between P and Q}}$$

$$= \frac{(y_0 + k) - y_0}{(x_0 + h) - x_0} = \frac{k}{h},$$

and since $y = f(x)$ for all points on the curve we may also write

$$R = \frac{f(x_0 + h) - f(x_0)}{h}. \tag{5.1}$$

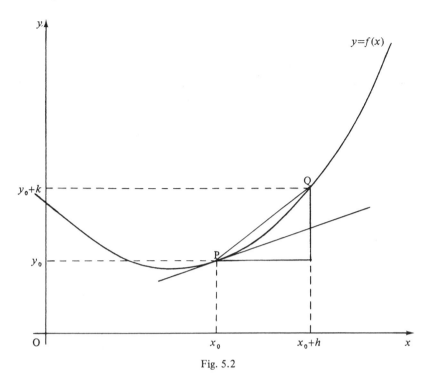

Fig. 5.2

This average rate of change clearly depends on the particular point P, in effect the value of x_0, and the particular end point Q, in effect the value of h, as well as the given function. The quantity R measures the slope of the chord PQ.

In many physical situations it is useful to make use of the idea of an **instantaneous rate of change**. For example, referring to Fig. 5.1, the average rate of growth of the plant between A and B may give a very crude estimate of the growth rate at a particular time such as T_2. To obtain an improved estimate a much shorter time interval is necessary. For the general function $f(x)$ the instantaneous rate of change is obtained by allowing h (and k) to approach zero whilst x_0 (and y_0) is kept constant. In effect in Fig. 5.2 Q approaches P and the chord PQ approaches the tangent to the curve at P. This amounts to a limiting process with the instantaneous rate of change r given by

$$r = \lim_{h \to 0} \left\{ \frac{f(x_0 + h) - f(x_0)}{h} \right\},$$

the value of R as $h \to 0$. Since this rate of change may be calculated for any point x_0 for which $f(x)$ is defined it is conventional to drop the subscript 0 and to write

$$r = \lim_{h \to 0} \left\{ \frac{f(x + h) - f(x)}{h} \right\} = f'(x) \qquad (5.2)$$

whenever this limit exists. Now x represents any value in the domain of $f(x)$ and $r = f'(x)$ is a function of the variable x. The quantity r gives the **gradient** of the function at P(x, y) (the **slope of the tangent**) and also measures the (instantaneous) rate of change of y with respect to x at P, usually referred to as the **rate of change**. The notation $f'(x)$, used to denote the rate of change of $f(x)$ or the **derivative** of $f(x)$ at the point x, is suitable when it is clear that the rate of change of $f(x)$ with x is under consideration. Situations sometimes arise in which the rate of change of $f(x)$ with some variable other than x is required, and an alternative notation is then desirable.

In deducing the result (5.2) the change in the variable x was denoted by h and that in y by k. An alternative approach that links the change in the variable with the variable itself is to denote the change in the variable x by Δx and that in y by Δy. Here Δx denotes 'the **change in** x' not 'Δ multiplied by x'. Equation (5.1) for the average rate of change is then replaced by

$$R = \frac{f(x + \Delta x) - f(x)}{\Delta x} = \frac{\Delta y}{\Delta x} \tag{5.3}$$

and equation (5.2) for the instantaneous rate of change of y with respect to x becomes

$$r = \lim_{\Delta x \to 0} \left\{ \frac{f(x + \Delta x) - f(x)}{\Delta x} \right\} = \lim_{\Delta x \to 0} \left(\frac{\Delta y}{\Delta x} \right) = f'(x) . \tag{5.4}$$

It is conventional to write

$$\frac{dy}{dx} \quad \text{in place of} \quad \lim_{\Delta x \to 0} \left(\frac{\Delta y}{\Delta x} \right)$$

to denote the derivative of y with respect to x. This is pronounced 'dee wye by dee ex'. Note that dx does not represent d multiplied by x. dy/dx is sometimes referred to as the **differential coefficient**, and y is then said to be **differentiable** with respect to x. Alternative notations for dy/dx when $y = f(x)$ are

$$f'(x), \quad y', \quad \dot{y}, \quad Dy, \quad \frac{d}{dx} \{f(x)\} .$$

In defining the derivative of $y = f(x)$ it was assumed that y was continuous. A discontinuous function is not differentiable at a point of discontinuity, and a continuous function may not be differentiable at unusual points. For example $y = |x|$ is not differentiable at $x = 0$ since the curve has a corner at that point and a unique tangent cannot be drawn.

Example: Differentiate from first principles

(i) $y = 5x^2 + x$, (ii) $y = x^{\frac{1}{2}}$.

(i) Using the notation of (5.2)

$f(x) = 5x^2 + x$,

$f(x + h) = 5(x + h)^2 + (x + h)$,

$f'(x) = \lim_{h \to 0} \left(\dfrac{5(x + h)^2 + (x + h) - (5x^2 + x)}{h} \right)$

$\qquad = \lim_{h \to 0} \left(\dfrac{5x^2 + 10xh + 5h^2 + x + h - (5x^2 + x)}{h} \right)$

$\qquad = \lim_{h \to 0} \left(\dfrac{10xh + 5h^2 + h}{h} \right)$

$\qquad = \lim_{h \to 0} \{10x + 5h + 1\} = 10x + 1$.

Alternatively, using the notation of (5.4)

$y = 5x^2 + x$,

$y + \Delta y = 5(x + \Delta x)^2 + (x + \Delta x)$,

$\dfrac{\Delta y}{\Delta x} = \dfrac{(y + \Delta y) - y}{\Delta x} = \dfrac{5(x + \Delta x)^2 + (x + \Delta x) - (5x^2 + x)}{\Delta x}$

$\qquad = \dfrac{1}{\Delta x} \{5x^2 + 10x\Delta x + 5(\Delta x)^2 + x + \Delta x - 5x^2 - x\}$

$\qquad = \dfrac{1}{\Delta x} \{10x\Delta x + 5(\Delta x)^2 + \Delta x\}$

$\qquad = 10x + 5\Delta x + 1$,

$\dfrac{dy}{dx} = \lim_{\Delta x \to 0} \left(\dfrac{\Delta y}{\Delta x} \right) = 10x + 1 \quad \text{as before.}$

(ii) $y = x^{\frac{1}{2}}$,

$y + \Delta y = (x + \Delta x)^{\frac{1}{2}}$,

$\dfrac{\Delta y}{\Delta x} = \dfrac{(x + \Delta x)^{\frac{1}{2}} - x^{\frac{1}{2}}}{\Delta x}$.

The resulting limit is of the type encountered in Section 4.5 so that

$\dfrac{\Delta y}{\Delta x} = \dfrac{\{(x + \Delta x)^{\frac{1}{2}} - x^{\frac{1}{2}}\} \{(x + \Delta x)^{\frac{1}{2}} + x^{\frac{1}{2}}\}}{\Delta x\{(x + \Delta x)^{\frac{1}{2}} + x^{\frac{1}{2}}\}}$

$\qquad = \dfrac{(x + \Delta x) - x}{\Delta x\{(x + \Delta x)^{\frac{1}{2}} + x^{\frac{1}{2}}\}} = \dfrac{1}{(x + \Delta x)^{\frac{1}{2}} + x^{\frac{1}{2}}}$,

$\dfrac{dy}{dx} = \lim_{\Delta x \to 0} \left(\dfrac{\Delta y}{\Delta x} \right) = \dfrac{1}{x^{\frac{1}{2}} + x^{\frac{1}{2}}} = \dfrac{1}{2x^{\frac{1}{2}}}$.

In the first of these examples the limiting process is straightforward and involves manipulation of a quadratic expression. In the second example the limiting process is more complicated. These examples illustrate the process of differentiating functions from first principles. They also indicate the need for a more routine process when dealing with commonly occurring functions. Some general results will now be obtained.

5.3 DERIVATIVE OF A CONSTANT

When $y = c$, a constant, the graph is a horizontal straight line, see Fig. 5.3, and the gradient is always zero. Consequently

$$\frac{dy}{dx} = \frac{d(c)}{dx} = 0 \ . \qquad\qquad (5.5)$$

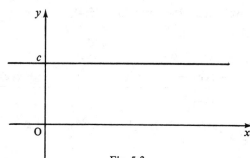

Fig. 5.3

Note that in this case

$$f(x) = f(x + \Delta x) = c$$

and hence $\Delta y = 0$ for all x.

5.4 DERIVATIVE OF x^α

We first consider the case when

$$y = f(x) = x^n \ ,$$

and n is a positive integer. Then

$$\frac{\Delta y}{\Delta x} = \frac{(x + \Delta x)^n - x^n}{\Delta x}$$

and since $x + \Delta x = x \left(1 + \frac{\Delta x}{x}\right)$

$$\frac{\Delta y}{\Delta x} = \frac{x^n \left(1 + \dfrac{\Delta x}{x}\right)^n - x^n}{\Delta x} \ .$$

In this expression x is to remain fixed as $\Delta x \to 0$ in the limiting process so that $(1 + \Delta x/x)^n$ may be expanded by using the binomial theorem, equation (2.15). It is found that

$$\frac{\Delta y}{\Delta x} = \frac{1}{\Delta x} \left\{ x^n \left[1 + n \left(\frac{\Delta x}{x}\right) + \frac{n(n-1)}{1.2} \left(\frac{\Delta x}{x}\right)^2 + \ldots + \left(\frac{\Delta x}{x}\right)^n \right] - x^n \right\}$$

$$= \frac{1}{\Delta x} \left\{ nx^{n-1} \Delta x + \frac{n(n-1)}{1.2} x^{n-2} (\Delta x)^2 + \ldots + (\Delta x)^n \right\}$$

$$= nx^{n-1} + \frac{n(n-1)}{1.2} x^{n-2} \Delta x + \ldots + (\Delta x)^{n-1} .$$

Since all terms beyond the first contain powers of Δx, in the limit

$$\frac{dy}{dx} = \lim_{\Delta x \to 0} \left(\frac{\Delta y}{\Delta x}\right) = nx^{n-1} .$$

It can be shown that this result is valid for any constant real value of α so that if

$$y = x^\alpha$$

then $\dfrac{dy}{dx} = \alpha x^{\alpha-1}$.   (5.6)

This is an important and widely used result.

Application of the binomial theorem for non-integer values of α demonstrates that (5.6) is correct, but a complete proof of the binomial theorem depends on (5.6) and this result will be assumed here. A proof of (5.6), valid for all α, may be obtained by using the logarithmic function discussed in Chapter 12. The validity of (5.6) when $\alpha = \frac{1}{2}$ was shown in part (ii) of the example in Section 5.2.

It should be noted that y and/or dy/dx may not exist for certain values of x. Thus when $\alpha = \frac{1}{2}$ so that

$$y = f(x) = x^{\frac{1}{2}}$$

and $\dfrac{dy}{dx} = \frac{1}{2} x^{-\frac{1}{2}}$

it is clear that the result is not valid for $x < 0$ since it would involve the square root of a negative number. In addition, at the point $x = 0$, y is zero but its derivative becomes infinite.

5.5 DERIVATIVES OF sin x AND cos x

When

$$y = f(x) = \sin x \ ,$$

$$y + \Delta y = f(x + \Delta x) = \sin (x + \Delta x)$$

so that

$$\frac{\Delta y}{\Delta x} = \frac{\sin(x + \Delta x) - \sin x}{\Delta x}$$

$$= \frac{2}{\Delta x} \cdot \sin \left(\frac{x + \Delta x - x}{2} \right) \cos \left(\frac{x + \Delta x + x}{2} \right)$$

$$= \frac{2}{\Delta x} \sin \left(\tfrac{1}{2} \Delta x \right) \cos \left(x + \tfrac{1}{2} \Delta x \right)$$

when the result A25[†] for the difference of two sine functions is used. On setting

$$z = \tfrac{1}{2} \Delta x$$

it is found that

$$\frac{\Delta y}{\Delta x} = \frac{\sin z}{z} \cos(x + z)$$

and since $z \rightarrow 0$ as $\Delta x \rightarrow 0$

$$\lim_{\Delta x \rightarrow 0} \left(\frac{\Delta y}{\Delta x} \right) = \lim_{z \rightarrow 0} \left(\frac{\sin z}{z} \cos(x + z) \right) = \cos x$$

using (4.4). This gives the important result

$$\frac{d}{dx} (\sin x) = \cos x \ . \tag{5.7}$$

By using a similar process it is found that

$$\frac{d}{dx} (\cos x) = -\sin x \ . \tag{5.8}$$

It should be noted that in deriving (5.7) it is necessary to assign a value to $\lim_{z \rightarrow 0} \left(\frac{\sin z}{z} \right)$. The value assigned here, which was derived in Section 4.6 requires the angle to be measured in radians. This requirement persists throughout differential and integral calculus and will apply to all results involving trigonometric functions.

†Appendix A.

5.6 RULES OF DIFFERENTIATION

Many functions encountered in applications of mathematics to practical problems are composed of combinations of elementary functions such as x^n and $\sin x$. It is convenient to have available a set of rules that enable differentiation to be carried out in a straightforward manner for such functions. In the following cases it is assumed that y is the dependent variable, x the independent variable, and u, v and w are variables such that

$$y = f(x), \quad u = u(x), \quad v = v(x), \quad w = w(x) .$$

Rule 1: If y may be expressed as a constant multiple of the variable u so that

$$y = f(x) = cu = cu(x) ,$$

where c is a constant then

$$y + \Delta y = f(x + \Delta x) = c\{u + \Delta u\} = cu(x + \Delta x) ,$$

where Δu is the change in u due to the change Δx in x. Then

$$\frac{\Delta y}{\Delta x} = c \frac{\Delta u}{\Delta x}$$

so that in the limit

$$\frac{dy}{dx} = c \frac{du}{dx} \tag{5.9}$$

or $\qquad f'(x) = cu'(x) .$ $\qquad\qquad\qquad\qquad\qquad$ (5.10)

Rules 2, 3 and 4 are proved by methods similar to those used to prove rule 1, and they will be stated here. Proofs may be found in any standard book on calculus such as Smyrl (1978).

Rule 2: If y may be expressed as the sum (or difference) of the two variables u and v so that

$$y = f(x) = u \pm v = u(x) \pm v(x)$$

then $\qquad \dfrac{dy}{dx} = \dfrac{du}{dx} \pm \dfrac{dv}{dx}$ $\qquad\qquad\qquad\qquad$ (5.11)

or $\qquad f'(x) = u'(x) \pm v'(x) .$ $\qquad\qquad\qquad\qquad$ (5.12)

This rule may be extended to cover the case

$$y = u + v + w$$

when $\qquad \dfrac{dy}{dx} = \dfrac{du}{dx} + \dfrac{dv}{dx} + \dfrac{dw}{dx} .$

Rule 3: *The product rule.* If y may be expressed as the product of the two variables u and v so that

$$y = f(x) = uv = u(x)v(x)$$

then $$\frac{dy}{dx} = u\frac{dv}{dx} + v\frac{du}{dx}$$ (5.13)

or $$f'(x) = u(x)v'(x) + v(x)u'(x) .$$ (5.14)

Rule 4: *The quotient rule.* If y may be expressed as the quotient of the two variables u and v so that

$$y = f(x) = u/v = u(x)/v(x)$$

then $$\frac{dy}{dx} = \frac{v\dfrac{du}{dx} - u\dfrac{dv}{dx}}{v^2}$$ (5.15)

or $$f'(x) = \frac{v(x)u'(x) - u(x)v'(x)}{\{v(x)\}^2} .$$ (5.16)

Note that in rule 2 the plus sign in y is associated with the plus sign in dy/dx or $f'(x)$. In rule 3 it is not significant which function is labelled u and which is labelled v since (5.13) and (5.14) are symmetrical in u and v. In rule 4 it is important to distinguish between u and v since u denotes the numerator and v the denominator.

Example 1: Find dy/dx when (i) $y = 6 \cos x$, (ii) $y = x \cos x$, (iii) $y = x^2 + x \cos x - \sin x$, (iv) $y = \tan x$, (v) $y = \sec x$.

(i) y is expressible as a constant multiple of $u = \cos x$ so that, from (5.8) and (5.9),

$$\frac{dy}{dx} = 6(-\sin x) = -6 \sin x .$$

(ii) Here y is the product of $u = x$ and $v = \cos x$ so that

$$\frac{du}{dx} = 1 \quad \text{and} \quad \frac{dv}{dx} = - \sin x$$

from (5.6) with $n = 1$ and (5.8). The product rule gives

$$\frac{dy}{dx} = x(-\sin x) + 1.\cos x = -x \sin x + \cos x .$$

(iii) In this example y may be expressed as $u + v - w$ where $u = x^2$, $v = x \cos x$, $w = \sin x$. Then

$$\frac{du}{dx} = 2x, \quad \frac{dv}{dx} = -x \sin x + \cos x, \quad \frac{dw}{dx} = \cos x ,$$

where the result in part (ii) has been used to obtain dv/dx. Rule 2 now gives

$$\frac{dy}{dx} = \frac{du}{dx} + \frac{dv}{dx} - \frac{dw}{dx} = 2x - x\sin x + \cos x - \cos x$$

$$= 2x - x\sin x \ .$$

(iv) The variable $y = \tan x = \dfrac{\sin x}{\cos x} = \dfrac{u}{v}$ may be regarded as the quotient of

$u = \sin x$ and $v = \cos x$.

From (5.7) and (5.8)

$$\frac{du}{dx} = \cos x \ , \qquad \frac{dv}{dx} = -\sin x$$

so that the quotient rule gives

$$\frac{dy}{dx} = \frac{(\cos x)(\cos x) - (\sin x)(-\sin x)}{\cos^2 x}$$

$$= \frac{\cos^2 x + \sin^2 x}{\cos^2 x}$$

$$= 1/\cos^2 x$$

$$= \sec^2 x \ .$$

This result ,

$$\frac{d}{dx}(\tan x) = \sec^2 x \ , \qquad\qquad\qquad (5.17)$$

and the result

$$\frac{d}{dx}(\cot x) = -\operatorname{cosec}^2 x \qquad\qquad\qquad (5.18)$$

which may be proved by the same method are important standard results.

(v) $y = \sec x$ may be expressed in the form

$$y = \frac{1}{\cos x} = \frac{u}{v}$$

where $u = 1$, $v = \cos x$ and

$$\frac{du}{dx} = 0, \quad \frac{dv}{dx} = -\sin x \ .$$

The quotient rule gives

$$\frac{dy}{dx} = \frac{(\cos x)(0) - (1)(-\sin x)}{\cos^2 x}$$

$$= \frac{\sin x}{\cos^2 x}$$

$$= \sec x \tan x \ .$$

This result ,

$$\frac{d}{dx}(\sec x) = \sec x \tan x \ , \tag{5.19}$$

and the similar one

$$\frac{d}{dx}(\operatorname{cosec} x) = -\operatorname{cosec} x \cot x \tag{5.20}$$

which may be proved by the same method, are also standard results.

Example 2: Differentiate $f(x) = x^3 - 3x^2 + 2$ and evaluate $f'(x)$ when $x = 0, 1, 2, 3$.

$$f'(x) = 3x^2 - 6x = 3x(x-2) \ .$$

This expression gives a value for $f'(x)$ valid for any value of x. Particular values are obtained by inserting the appropriate values of x. Thus

$$f'(0) = \quad 0 \ ,$$
$$f'(1) = -3 \ ,$$
$$f'(2) = \quad 0 \ ,$$
$$f'(3) = \quad 9 \ .$$

Note that $f'(x)$ may be positive, negative, or zero for certain values of x.

In the discussion following equation (5.2) it was noted that $r = f'(x)$ gives the gradient of the function $f(x)$ at a point of the curve and that this measures the slope of the tangent at that point. The graph of a typical function of x is shown in Fig. 5.4. On sections AB and CD of the curve the tangent makes an acute angle with the x-axis, and its slope, $f'(x)$, is positive. On the section BC the tangent makes an obtuse angle with the x-axis and $f'(x)$ is negative. When $f'(x)$ is zero the tangent is parallel to the x-axis.

Example 3: Determine the equations of the tangents to the curve $y = x^3 - 2x^2 + 1$ at the points $(1, 0)$ and $(-1, -2)$ and find their point of intersection.

Here $y = f(x) = x^3 - 2x^2 + 1 \ ,$

$$\frac{dy}{dx} = f'(x) = 3x^2 - 4x \ .$$

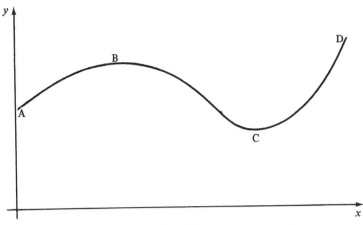

Fig. 5.4

When $x = 1$, $dy/dx = 3 - 4 = -1$ so that the slope of the tangent is -1. Since the tangent passes through $(1, 0)$ it has equation given by

$$y - 0 = -1\,(x - 1)$$

from (3.2) and the final equation of this tangent is

$$y = -x + 1 \ .$$

When $x = -1$, $dy/dx = 3 + 4 = 7$ and using (3.2) the equation of this tangent is

$$y + 2 = 7\,(x + 1) \ ,$$

since $(-1, -2)$ lies on the line. Finally

$$y = 7x + 5$$

gives the equation of the second tangent.

The two tangents intersect when

$$y = -x + 1 = 7x + 5$$

so that $8x = -4$. Hence $x = -1/2$ giving $y = 3/2$. The two tangents intersect in the point $(-1/2, 3/2)$.

In discussing differentiation and its rules it has been assumed that the dependent variable depends on an independent variable labelled x. Clearly the independent variable may be assigned any other letter as a label, and there is no particular significance in the use of x. It is conventional, for example, to use the letter t to denote time in problems involving a rate of change with time when, if y is given as a function of t, dy/dt represents velocity. Results for differentiation with respect to t are identical with those for differentiation with respect to x; the letter t simply replaces the letter x.

Example 4: Protein is synthesised in such a way that the mass m of protein expressed as a function of time t is given by

$$m = 500 + 50t + 4t^2 .$$

Find the reaction rate.

The dependent variable is m and the independent variable is t. From (5.6)

$$\frac{d}{dt}(t) = 1, \quad \frac{d}{dt}(t^2) = 2t ,$$

so that, using rule 2,

$$\frac{dm}{dt} = 50 + 8t .$$

Rule 5: *The chain rule.* By using (5.6) it is possible to differentiate a function such as $x^{\frac{1}{2}}$ but not $(2 + 3x)^{\frac{1}{2}}$. The chain rule enables such functions to be differentiated in a straightforward way. The right-hand side of the equation

$$y = (2 + 3x)^{\frac{1}{2}}$$

assumes a simpler form when the substitution

$$t = 2 + 3x$$

is made since then

$$y = t^{\frac{1}{2}} .$$

Both of these expressions may be differentiated straightforwardly to obtain

$$\frac{dt}{dx} = 3, \quad \frac{dy}{dt} = \tfrac{1}{2}t^{-\frac{1}{2}} .$$

It will be shown that the value of dy/dx is given by the product of these quantities, and in this example

$$\frac{dy}{dx} = \frac{dy}{dt}\frac{dt}{dx} = \frac{3}{2}t^{-\frac{1}{2}} = \frac{3}{2}(2 + 3x)^{-\frac{1}{2}} .$$

In general, when y may be expressed as a function of the variable t which is itself a function of the variable x so that

$$y = g(t) \text{ and } t = h(x) ,$$

the variable y may be regarded as a function of x. It is assumed that $g(t)$ and $h(x)$ are well behaved so that a small change in x produces a small change in t which in turn produces a small change in y. Then

$$\frac{\Delta y}{\Delta x} = \frac{\Delta y}{\Delta t} \cdot \frac{\Delta t}{\Delta x} .$$

In taking the limit as $\Delta x \to 0$ both $\Delta t \to 0$ and $\Delta y \to 0$ so that

$$\frac{dy}{dx} = \frac{dy}{dt}\frac{dt}{dx} . \tag{5.21}$$

This result is known as *the chain rule*.

When making use of the chain rule it is important to be able to visualise the way in which y is constructed since normally $f(x)$ is given and an appropriate form for t must be chosen. The following problems illustrate the approach.

Example 5: Find dy/dx when (i) $y = (3 + 2x)^{\frac{1}{2}}$, (ii) $y = (a + bx)^n$, (iii) $y = (x^2 + 3)^4$, (iv) $y = \sin^4 x$.

(i) The appropriate choice of t for the use of the chain rule is $t = (3 + 2x)$ since then $y = t^{\frac{1}{2}}$ and (5.6) may be used. With

$$y = t^{\frac{1}{2}}, \qquad t = 3 + 2x ,$$

$$\frac{dy}{dt} = \tfrac{1}{2}t^{-\frac{1}{2}}, \qquad \frac{dt}{dx} = 2 ,$$

$$\frac{dy}{dx} = \frac{dy}{dt}\frac{dt}{dx} = (\tfrac{1}{2}t^{-\frac{1}{2}})(2) = (2 + 3x)^{-\frac{1}{2}} .$$

Note that it is normal to express y in terms of the original variable when this is possible.

(ii) Choosing $t = a + bx$ results in

$$y = t^n, \qquad t = a + bx ,$$

$$\frac{dy}{dt} = nt^{n-1}, \qquad \frac{dt}{dx} = b ,$$

so that

$$\frac{dy}{dx} = \frac{dy}{dt}\frac{dt}{dx} = (nt^{n-1})(b) = nb(a + bx)^{n-1} . \tag{5.22}$$

The result (5.22) is a generalisation of the result in part (i).

(iii) In this example write $t = x^2 + 3$ so that

$$y = t^4 , \qquad t = x^2 + 3 ,$$

$$\frac{dy}{dt} = 4t^3 , \qquad \frac{dt}{dx} = 2x ,$$

and

$$\frac{dy}{dx} = (4t^3)(2x) = 8x(x^2 + 3)^3 .$$

(iv) Here $t = \sin x$ results in

$$y = t^4 , \qquad t = \sin x ,$$

$$\frac{dy}{dt} = 4t^3 , \qquad \frac{dt}{dx} = \cos x ,$$

and

$$\frac{dy}{dx} = (4t^3)(\cos x) = 4 \sin^3 x \cos x.$$

Example 6: Differentiate $\sin^3(2x + 3)$.

Let $y = \sin^3(2x + 3)$ and $t = 2x + 3$ so that

$$y = \sin^3 t \ .$$

This is still not in a suitable form but if we set $s = \sin t$ then

$$y = s^3, \ s = \sin t, \ t = 2x + 3 ,$$

$$\frac{dy}{ds} = 3s^2, \ \frac{ds}{dt} = \cos t, \ \frac{dt}{dx} = 2 \ .$$

It is necessary in this example to use the extended chain rule in the form

$$\frac{dy}{dx} = \frac{dy}{ds}\frac{ds}{dt}\frac{dt}{dx} = (3s^2)(\cos t)(2)$$

$$= 6 \sin^2 t \cos t$$

$$= 6 \sin^2(2x + 3) \cos(2x + 3) \ .$$

Example 7: The height h of a plant over the time interval $0 \leqslant t \leqslant 2$ is given as a function of the time t by $h = (4 + 3t + t^2)^{\frac{1}{2}}$. Find the rate at which h is increasing when $t = 1$.

In this problem t is the independent variable. The quantity h may be written

$$h = s^{\frac{1}{2}} \text{ with } s = 4 + 3t + t^2 ,$$

$$\frac{dh}{ds} = \tfrac{1}{2}s^{-\frac{1}{2}} , \ \frac{ds}{dt} = 3 + 2t ,$$

so that the rate of increase of h at time t is given by

$$\frac{dh}{dt} = \frac{dh}{ds}\frac{ds}{dt} = (\tfrac{1}{2}s^{-\frac{1}{2}})(3 + 2t)$$

$$= \frac{3 + 2t}{2(4 + 3t + t^2)^{\frac{1}{2}}} \ .$$

The value of the derivative at time $t = 1$ is denoted by $\left(\dfrac{dh}{dt}\right)_{t\,=\,1}$ or $\dfrac{dh(1)}{dt}$ or $\dfrac{dh}{dt}\Big|_{t\,=\,1}$. Here $\left(\dfrac{dh}{dt}\right)_{t\,=\,1} = \dfrac{5}{2(8)^{\frac{1}{2}}} = \dfrac{5}{4\sqrt{2}}$.

5.7 IMPLICIT DIFFERENTIATION

It is not always possible, or convenient, to express y explicitly as a function of x. For example the equation

$$xy + \sin y = \tan x$$

cannot be solved for y, although the equation implies that y is a function of x. Implicit differentiation permits dy/dx to be calculated in such situations. Every term in the equation may be differentiated with respect to x, by using the basic rules developed in the previous section. In particular we need to be able to differentiate a function of y with respect to x.

In general if w is a function of y then, since y is itself a function of x, from the chain rule

$$\frac{dw}{dx} = \frac{dw}{dy}\frac{dy}{dx}\,.$$

Thus, if $w = y^2 = \{y(x)\}^2$ then

$$\frac{dw}{dy} = 2y \quad \text{and} \quad \frac{dw}{dx} = 2y\frac{dy}{dx} = \frac{d}{dx}(y^2)\,.$$

Note that in this process an explicit form for dy/dx is not known; this is usually the quantity to be determined.

Example 1: If $y = x^{1/3}$ determine dy/dx (i) by differentiating y directly and alternatively (ii) by differentiating the equation $y^3 = x$.

(i) $y = x^{1/3}$, $dy/dx = \dfrac{1}{3}x^{-2/3}$.

(ii) Differentiate both sides of the equation with respect to x to obtain

$$3y^2\frac{dy}{dx} = 1\,,$$

$$\frac{dy}{dx} = \frac{1}{3y^2} = \frac{1}{3x^{2/3}} = \frac{1}{3}x^{-2/3}$$

as before.

Example 2: Find dy/dx when (i) $x^2 + y^2 = 1$, (ii) $y + xy^2 + x^2 = 1$, (iii) $\sin y = 2xy^3$.

(i) Differentiate both sides of the equation with respect to x to obtain

$$2x + 2y\frac{dy}{dx} = 0$$

so that

$$\frac{dy}{dx} = -\frac{x}{y} .$$

(ii) Differentiate both sides of the equation with respect to x. The product rule gives

$$\frac{d}{dx}(xy^2) = y^2 + 2xy\frac{dy}{dx}$$

so that

$$\frac{dy}{dx} + y^2 + 2xy\frac{dy}{dx} + 2x = 0 ,$$

$$\frac{dy}{dx}(1 + 2xy) = -(y^2 + 2x) ,$$

$$\frac{dy}{dx} = -\frac{y^2 + 2x}{1 + 2xy} .$$

(iii) Differentiation with respect to x using

$$\frac{d}{dx}(\sin y) = \cos y\frac{dy}{dx} , \quad \frac{d}{dx}(xy^3) = y^3 + 3xy^2\frac{dy}{dx} ,$$

results in

$$\cos y\frac{dy}{dx} = 2y^3 + 6xy^2\frac{dy}{dx} ,$$

$$\frac{dy}{dx}(\cos y - 6xy^2) = 2y^3 ,$$

$$\frac{dy}{dx} = \frac{2y^3}{\cos y - 6xy^2} .$$

5.8 THE INVERSE RULE

In certain situations it is convenient to express x explicitly in terms of y instead of expressing y in terms of x. It is then useful to write

$$\frac{\Delta y}{\Delta x} = \frac{1}{\Delta x/\Delta y}$$

which in the limit as Δx and Δy tend to zero gives

$$\frac{dy}{dx} = \frac{1}{dx/dy} .$$

This result is useful when differentiating inverse functions, particularly inverse trigonometric functions.

Example: Find dy/dx when (i) $y = \sin^{-1}x$, (ii) $y = \tan^{-1}x$, (iii) $y = \tan^{-1}(x^2)$.

(i) When $y = \sin^{-1}x$, $-1 \leqslant x \leqslant 1$, $-\frac{1}{2}\pi \leqslant y \leqslant \frac{1}{2}\pi$, write

$$x = \sin y$$

so that

$$\frac{dx}{dy} = \cos y = \pm (1 - \sin^2 y)^{\frac{1}{2}} = \pm (1 - x^2)^{\frac{1}{2}} .$$

From the graph of $y = \sin^{-1}x$, Fig. 1.9, it is clear that an increase in the value of x will produce an increase in the value of y so that dy/dx is always positive. Therefore

$$\frac{dx}{dy} = (1 - x^2)^{\frac{1}{2}}$$

and

$$\frac{dy}{dx} = \frac{1}{(1 - x^2)^{\frac{1}{2}}} .$$ (5.23)

(ii) When $y = \tan^{-1}x$, $-\infty < x < \infty$, $-\frac{1}{2}\pi < y < \frac{1}{2}\pi$,

$$x = \tan y$$

$$\frac{dx}{dy} = \sec^2 y = 1 + \tan^2 y = 1 + x^2$$

and

$$\frac{dy}{dx} = \frac{1}{1 + x^2} .$$ (5.24)

The results (5.23) and (5.24) are important standard results. It is also possible to show that if $y = \cos^{-1}x$ then

$$\frac{dy}{dx} = -\frac{1}{(1 - x^2)^{\frac{1}{2}}} ,$$

but this result is less commonly required.

(iii) When $y = \tan^{-1}(x^2)$ we could write $x^2 = \tan y$ and proceed as in part (ii). It is simpler to use the chain rule and (5.24):

$$y = \tan^{-1}(t) , \qquad t = x^2 ,$$

$$\frac{dy}{dt} = \frac{1}{1 + t^2} , \qquad \frac{dt}{dx} = 2x ,$$

$$\frac{dy}{dx} = \frac{dy}{dt}\frac{dt}{dx} = \left(\frac{1}{1 + t^2}\right)(2x) = \frac{2x}{1 + x^4} .$$

5.9 PARAMETRIC DIFFERENTIATION

The chain rule (5.21) has the effect of relating the variables y and x through another variable, or parameter, t. Using the result of Section 5.8 it is possible to write

$$\frac{dy}{dx} = \frac{dy}{dt}\frac{dt}{dx} = \frac{dy/dt}{dx/dt} = \frac{\dot{y}}{\dot{x}} .$$

A practical situation in which this result is useful is that of a point moving in the xy-plane, when the parameter t represents time. Then $\dot{x}$ and $\dot{y}$ represent the components of velocity in the x and y directions.

Example: If $x = \cos^3 t$, $y = \sin^3 t$, find dy/dx.

$$\dot{x} = -3\cos^2 t \sin t, \quad \dot{y} = 3\sin^2 t \cos t,$$

so that

$$\frac{dy}{dx} = \frac{\dot{y}}{\dot{x}} = -\frac{3\sin^2 t \cos t}{3\cos^2 t \sin t} = -\frac{\sin t}{\cos t} = -\tan t .$$

5.10 HIGHER DERIVATIVES

If $y = f(x)$ is given then it is possible to calculate $f'(x) = dy/dx$. As $f'(x)$ is a function of x it may be possible to differentiate it to obtain $f''(x) = \dfrac{d}{dx}(f'(x))$.

This process may usually be continued indefinitely. To denote the connection between $f''(x)$ and y it is conventional to write

$$f''(x) = \frac{d}{dx}\left(\frac{dy}{dx}\right) = \frac{d^2 y}{dx^2}$$

where the 'twos' indicate that this is the second derivative of y with respect to x.

In general the nth derivative of y with respect to x is denoted by

$$\frac{d^n y}{dx^n} = f^{(n)}(x) .$$

Some functions give rise to non-zero derivatives indefinitely; others have higher derivatives which are zero beyond a certain order. When y is given implicitly in terms of x it is possible to calculate second and higher derivatives by using implicit differentiation.

Example: Find the first four derivatives of (i) $y = x^2 - 2x + 4$, (ii) $y = \sin x$.

(i) When $y = x^2 - 2x + 4$,

$$\frac{dy}{dx} = 2x - 2 ,$$

$$\frac{d^2y}{dx^2} = \frac{d}{dx}\left(\frac{dy}{dx}\right) = 2 \ ,$$

$$\frac{d^3y}{dx^3} = \frac{d}{dx}\left(\frac{d^2y}{dx^2}\right) = 0 \ ,$$

$$\frac{d^4y}{dx^4} = 0$$

and all higher derivatives are zero.

(ii) $y = \sin x \ ,$

$$\frac{dy}{dx} = \cos x \ ,$$

$$\frac{d^2y}{dx^2} = -\sin x \ ,$$

$$\frac{d^3y}{dx^3} = -\cos x \ ,$$

$$\frac{d^4y}{dx^4} = \sin x = y \ .$$

In this example all derivatives are non-zero since

$$\frac{d^8y}{dx^8} = \frac{d^4y}{dx^4} = y \ \text{etc.}$$

PROBLEMS

1. The length l, in cm, of a certain snake is given in terms of the time t, in weeks, by the table

t	0	2	4	6	8	10	12	14	16	18	20
l	0	4	10	19	41	60	73	80	86	89	91

Determine, to two decimal places, the average rate of growth of the snake during the time intervals:

(i) 0 to 4, (ii) 4 to 10, (iii) 0 to 10, (iv) 10 to 16, (v) 16 to 20, (vi) 10 to 20, (vii) 0 to 20.

2. The distance s in metres fallen by an object, starting from rest, under the influence of gravity is given approximately by $s = 4.9t^2$ metres where t is measured in seconds. Determine the average speed of the object during each of the first five seconds of its motion and the average speed during the whole of the first five seconds.

3. A lymphocyte cell population reproduces such that its size is given by

    $$N = 1000 \times 2^t ,$$

    where t is measured in hours. Determine the average rate of growth of the population during each of the first four hours and for the whole four-hour period.

4. Hormone is released from a gland in such a way that the mass m, in milligrams, at time t, in seconds, is given by

    $$m = 500 + 50t + 4t^2 .$$

    Determine the average rate of increase in hormone during each of the first three seconds and the average for the whole of the three seconds.

5. In a chemical reaction the mass m, in gm, of protein disintegrates according to the formula

    $$m = \frac{60}{t + 2} ,$$

    where t is the time in seconds. Calculate the rate of increase in m for each of the first four seconds and for the whole of the first four seconds. What is the significance of your negative answer?

6. Differentiate from first principles:

    (i) $x^2 + 2x + 3$,

    (ii) $(x + 1)^{\frac{1}{2}}$,

    (iii) $1/(x - 1)$,

    (iv) $(a - x)^2$ where a is constant,

    (v) $(x + 1)^{-2}$,

    (vi) $x^{-\frac{1}{2}}$,

    (vii) x^3,

    (viii) $x - 2x^2 + 1$,

    (ix) $(x^2 + 1)^2$,

    (x) $\tan x$.

7. Differentiate with respect to x:

    (i) $x^3 + 3x + 5$,

    (ii) $4x^{1.6}$,

    (iii) $x^{\frac{1}{2}} - x^{-\frac{1}{2}}$,

(iv) $(x^2 + 1)^2$,

(v) $(x + 1)^3$,

(vi) $x(2x^2 + 1)$,

(vii) $1/x + 2/x^2$,

(viii) $x + 1/x$,

(ix) $(x + 2)(x - 3)$,

(x) $x^3 - x^{-3}$.

8. Differentiate with respect to x:

(i) $(x^3 + 1)(x^4 - 3x^2 + 2x - 1)$,

(ii) $(x + 1)/(x - 1)$,

(iii) $x^2/(x - 1)$,

(iv) $x(x + 1)^{-1}$,

(v) $1/(1 + x)$,

(vi) $(2x + 1)/(2 - 3x)$,

(vii) $x/(x^2 + 2)$,

(viii) $(x - 1)/(x + 1)^2$,

(ix) $(x + 2)^{-2}$,

(x) $(x^2 - 1)/(x^2 + 1)$.

9. Differentiate with respect to x:

(i) $\sin x \cos x$,

(ii) $x + 2 \sin x$,

(iii) $\cos^2 x$,

(iv) $x \sin x$,

(v) $x^2 \cos x$,

(vi) $x^3 - x \sin x$,

(vii) $\cot x$,

(viii) $\operatorname{cosec} x$,

(ix) $x/\cos x$,

(x) $\sin x/(1 + x)$.

10. Given that

$$m(t) = 500 + 50t + 4t^2 .$$

Calculate $m'(t)$ and evaluate $m'(t)$ when $t = 0, 1, 2, 3$. (Compare with question 4; these values give the instantaneous rate of increase in $m(t)$).

11. Given that

$$m(t) = 60/(t + 2) \, ,$$

calculate $m'(t)$ and evaluate $m'(t)$ when $t = 0,1,2,3,4$. (Compare with question 5).

12. The reaction R of the body to a dose x of a certain drug is given by $R = x(C - x)$, where C is a constant. Calculate the rate of change of R with x (the sensitivity).

13. Protein is synthesised in such a way that the mass m of protein expressed as a function of time t is given by

$$m = 300 + 40t + 3t^2 \, .$$

Find the rates of synthesis when $t = 0, 2, 4$.

14. The rate of net photosynthesis R for given light intensity I is given by

$$R = \frac{I}{a + bI} \, .$$

Show that the photochemical efficiency, dR/dI, is $a(a + bI)^{-2}$. Determine the values of R and dR/dI as I tends to infinity.

15. The pressure p and volume v of a gas are related by $pv = k$, where k is a constant. Prove that $v dp/dv = -p$.

16. The pressure p and volume v of a certain gas are related by $pv^{1.4} = k$, where k is a constant. Show that $v dp/dv = -1.4p$.

17. The relative growth rate of a population of size N is given by $\dfrac{1}{N}\dfrac{dN}{dt}$. If $N = a + bt^2$ show that the relative growth rate is $\dfrac{2bt}{a + bt^2}$.

18. In a forest those cones that fall from pine trees and smash on impact commence development immediately. The distance s fallen by a cone, under the influence of gravity is given approximately by $s = 4.9t^2$ metres where t is measured in seconds. Its velocity is given by $v = ds/dt$ and its acceleration by $f = dv/dt$. Calculate v and f when $t = 2$.

19. The weight W, in kilograms, of a person with sitting height h, in metres, is given approximately by $W = 90h^3$. Show that $h dW/dh = 3W$.

20. In laminar flow of blood through a cylindrical artery the resistance R is inversely proportional to the fourth power of the radius r. Show that $r \mathrm{d}R/\mathrm{d}r + 4R = 0$.

21. Determine the equation of the tangent to the curve $y = x^2 - 3x + 3$ at the point $(1, 1)$.

22. Determine the equations of the tangents at the points $(0, 0)$ and $(1, -1)$ on the curve $y = x^3 - 2x$. Find their point of intersection.

23. Find the points on the curve $y = 2x^3 - 3x^2 - 12x + 2$ where the tangent is parallel to the x-axis.

24. The normal to the curve $y = f(x)$ at a point P is perpendicular to the tangent at P. The slope of the normal is given by $-1/m$ where $m = \mathrm{d}y/\mathrm{d}x$ at P.
     Determine the normal to the curve $y = x^3 - 3x^2 + 1$ at the point $(1, -1)$.

25. Differentiate with respect to x:
     (i) $(2x + 5)^{\frac{1}{2}}$,
     (ii) $(3x^2 + 1)^3$,
     (iii) $(x - 1)^{-3/4}$,
     (iv) $\cos^3 x$,
     (v) $\sin(3x + \pi/4)$,
     (vi) $(1 + \sin x)^3$,
     (vii) $(1 - x)^{\frac{1}{2}}/(1 + x)^{\frac{1}{2}}$,
     (viii) $1/(1 - x^2)^{\frac{1}{2}}$,
     (ix) $\tan^2 x$,
     (x) $\cos x/(1 + \sin x)^2$.

26. The air temperature T, in degrees centigrade, on a certain day is given approximately by
     $$T = 14 + 8 \sin \{\pi(t - 8)/12\} ,$$
     where t is the time in hours measured from midnight. Find the average rate of increase in temperature between 02.00 hours and 14.00 hours. Find the instantaneous rate of change of T at 02.00 hours, 08.00 hours, and 14.00 hours.

27. The height h, in metres, of a deciduous tree over a time period of three years is given as a function of the time t by $h = (9 + 4t + t^2)^{\frac{1}{2}}$. (i) Find the rate of increase of h when $t = 0, 1, 2, 3$. (ii) Calculate the average rate of increase in h between $t = 0$ and 3. Give all answers to two places of decimals.

28. The surface area of a cell during a certain phase is given approximately by

$$A = (a + bt)^3 ,$$

where a, b are constants and t is the time. Show that $dA/dt = 3bA^{2/3}$.

29. In a certain experiment the tension T required to produce an extension ratio of λ in myosin filaments was found to be given approximately by

$$T = \frac{3(\lambda^3 - 1)}{(\lambda^3 + 2)^2}.$$ Find the rate of change of T with λ.

30. Use implicit differentiation to find dy/dx when:

   (i) $x^3 + y^3 = 1$,

   (ii) $xy = 4$,

   (iii) $x^2 - y^2 = 1$,

   (iv) $x^{\frac{1}{2}} + y^{\frac{1}{2}} = 4$,

   (v) $y^2 = x^2 + 1/x^2$,

   (vi) $2 \cos x + \sin y = 1$,

   (vii) $1/y + 1/x = 1$,

   (viii) $\cos(x + y) = \sin(x - y)$.

31. Find the equation of the tangent to the curve $y^2 - x^2 = 24$ at the point $(1,5)$.

32. Differentiate with respect to x:

   (i) $\sin^{-1}(2x)$,

   (ii) $\sin^{-1}(x^2)$,

   (iii) $\sin^{-1}(\cos x)$,

   (iv) $\tan^{-1}(2x)$,

   (v) $x \sin^{-1}x$,

   (vi) $x^2 \tan^{-1}x$.

33. Find dy/dx in terms of t given that:

    (i) $x = \cos t, y = \sin t$,

    (ii) $x = t^2, y = t^3$,

    (iii) $x = \sec t, y = \tan t$,

    (iv) $x = t, y = 1/t$,

    (v) $x = t^2 - t, y = t^2 + t$,

    (vi) $x = \cos^4 t, y = \sin^4 t$,

    (vii) $x = a(t - \sin t), y = a(1 - \cos t)$.

34. A point P on the ellipse $x^2/a^2 + y^2/b^2 = 1$ may be represented in parametric form by $x = a\cos t, y = b\sin t$. Show that $dy/dx = -(b/a)\cot t$ and deduce that the equation of the tangent at P is

    $$bx\cos t + ay\sin t = ab .$$

35. Find the first three derivatives with respect to x of the following:

    (i) $\cos 2x$,

    (ii) $x^2 + 3x - 4$,

    (iii) $2x^3 - 3x^2 + x - 2$,

    (iv) $(1 + x)^{\frac{1}{2}}$,

    (v) $1/(1 - x)$,

    (vi) $x\sin x$,

    (vii) $\sin^2 x$,

    (viii) $(1 + x^2)^{\frac{1}{2}}$.

36. Prove that if

    (i) $y = \sin 2x$ then $\dfrac{d^2y}{dx^2} = -4y$,

    (ii) $y = \tan x$ then $\dfrac{d^2y}{dx^2} = 2y(1 + y^2)$,

    (iii) $xy = \sin x$ then $x\dfrac{d^2y}{dx^2} + 2\dfrac{dy}{dx} + xy = 0$,

    (iv) $y(1 - x^2) = 1$ then $\dfrac{d^2y}{dx^2} = 2y^2(4y - 3)$.

CHAPTER 6

Applications of differentiation

6.1 INTRODUCTION

In Chapter 5 the basic mechanics of differentiation were developed and applications to some simple problems were considered. In this chapter applications to more complicated practical problems will be investigated. The evaluation of maxima and minima will be discussed. This will be followed by some hints on curve sketching. Finally the application of calculus to the evaluation of small errors and limits is considered.

6.2 MAXIMA AND MINIMA

It is assumed here that for a given function $f(x)$ both $f'(x)$ and $f''(x)$ exist. For the curve $y = f(x)$ the value of the function $f'(x) = dy/dx$ measures the slope of the curve at each point (see Section 5.6). Referring to Fig. 6.1, when dy/dx is positive, so that the tangent to the curve makes an acute angle with the positive x-axis, for example on the sections of the curve AB, CD and DE, then y increases as x increases and y is said to be an **increasing function** of x. When dy/dx is negative, on the section BC, y decreases as x increases and y is said to be a **decreasing function** of x. The points B, C and D where dy/dx is zero are called **stationary points**.

On the section AB $dy/dx > 0$, at B dy/dx is zero, and on the section BC $dy/dx < 0$ so that as x increases the slope is successively positive, zero, negative, and B is a **local maximum**. It is termed a local maximum since the function may take a greater value at some point such as E. On the section BC $dy/dx < 0$, at C dy/dx is zero and on CD $dy/dx > 0$. As x increases the slope is successively negative, zero, positive, the conditions for a **local minimum** at C; again local since the value of the function may be less at some other point. At the point D dy/dx is zero and $dy/dx > 0$ on both CD and DE. The point D is called a **point of inflexion**. A more comprehensive definition of a point of inflexion will be given shortly.

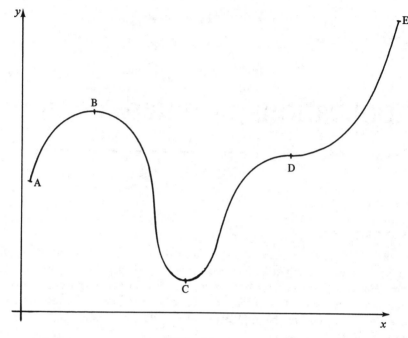

Fig. 6.1

Example 1: Find and identify the stationary points of:

(i) $y = 2x^3 - 3x^2 - 12x + 13$, (ii) $y = x^3 + 6x^2 + 12x - 10$.

(i) $y = 2x^3 - 3x^2 - 12x + 13$,

$$\frac{dy}{dx} = 6x^2 - 6x - 12 = 6(x^2 - x - 2) = 6(x + 1)(x - 2).$$

dy/dx is zero when $x = -1$ and when $x = 2$. When $x < -1, dy/dx > 0$; when $-1 < x < 2, dy/dx < 0$; when $x > 2, dy/dx > 0$. Hence, the point $x = -1, y = 20$ corresponds to a local maximum (dy/dx has the sequence $+, 0, -$). The point $x = 2, y = -7$ is a local minimum (dy/dx has the sequence $-, 0, +$).

(ii) $y = x^3 + 6x^2 + 12x - 10$,

$$\frac{dy}{dx} = 3x^2 + 12x + 12 = 3(x^2 + 4x + 4) = 3(x + 2)^2.$$

dy/dx is zero when $x = -2$ only. Since $dy/dx > 0$ for all other values of x its value does not change sign at $x = -2$, and $(-2, -18)$ is a point of inflexion.

An alternative method of classifying stationary points is to examine the value of the second derivative at the stationary point. If $\dfrac{d^2y}{dx^2} = \dfrac{d}{dx}\left(\dfrac{dy}{dx}\right)$ is positive when $x = a$ say, then dy/dx is an increasing function of x as x increases through $x = a$. In particular, if $dy/dx = 0$ at $x = a$ and dy/dx is an increasing function of x, then the curve is concave up (dish shaped) at $x = a$ and has a **relative minimum**. If $\dfrac{d^2y}{dx^2} = \dfrac{d}{dx}\left(\dfrac{dy}{dx}\right)$ is negative when $x = a$ then dy/dx is a decreasing function of x as x increases through $x = a$. If $dy/dx = 0$ at $x = a$ the curve is concave down (shaped like an inverted dish) at $x = a$ and has a **relative maximum**.

In summary, if $dy/dx = 0$ at $x = a$ and $d^2y/dx^2 > 0$ the curve has a local minimum; if $dy/dx = 0$ at $x = a$ and $d^2y/dx^2 < 0$ the curve has a local maximum. If both d^2y/dx^2 and dy/dx are zero at $x = a$ no information is obtained concerning the stationary point.

If d^2y/dx^2 is zero at $x = a$ then dy/dx has a stationary value. If in addition d^2y/dx^2 changes sign as x increases through $x = a$ then the curve is said to have a point of inflexion at that point. It is important to ensure that both of these requirements are satisfied. It is not necessary for dy/dx to vanish at a point of inflexion.

Example 2: Find and identify the stationary points and points of inflexion of $y = 2x^3 - 3x^2 - 12x + 13$.

$$y = 2x^3 - 3x^2 - 12x + 13 \ ,$$

$$\frac{dy}{dx} = 6x^2 - 6x - 12 = 6(x + 1)(x - 2) \ ,$$

$$\frac{d^2y}{dx^2} = 12x - 6 = 6(2x - 1) \ .$$

dy/dx is zero when $x = -1$ and when $x = 2$.

When $x = -1, y = 20, d^2y/dx^2 = -18$; the curve has a local maximum.

When $x = 2, y = -7, d^2y/dx^2 = 18$; the curve has a local minimum.

These conclusions are in agreement with those reached in example 1(i). $d^2y/dx^2 = 0$ when $x = \frac{1}{2}$ and since d^2y/dx^2 changes sign at this value of x this point corresponds to a point of inflexion.

It should be noted that in the example just considered the turning points constitute a local maximum and a local minimum. If the function is defined for $-3 \leqslant x \leqslant 4$ only then the absolute maximum value of y in this range occurs when $x = 4$ and $y = 45$. The absolute minimum value of y in this range occurs when $x = -3$ and $y = -32$. These absolute values occur at the end points.

Example 3: At time $t = 0$ a population of 1000 bacteria is introduced into a nutrient medium. At time t the size of the population is given by

$$p = \frac{9000(4 + t)}{4(9 + t^2)} .$$

Find the maximum size of the population.

In this example p is the dependent variable ($\equiv y$) and t the independent variable ($\equiv x$).

$$\frac{dp}{dt} = \frac{9000\{(9 + t^2) - 2t(4 + t)\}}{4(9 + t^2)^2}$$

$$= \frac{9000\{9 - 8t - t^2\}}{4(9 + t^2)^2} = \frac{9000(9 + t)(1 - t)}{4(9 + t^2)^2} . \qquad (6.1)$$

dp/dt is zero when $t = -9$ and when $t = 1$. Since p exists for $t \geqslant 0$, only $t = 1$ can give a local maximum. For $0 < t < 1$, $dp/dt > 0$ and for $1 < t$, $dp/dt < 0$ so that $t = 1$ is a local maximum. When $t = 1$, $p = 1125$ and this is an absolute maximum since at the end points $t = 0$, $p = 1000$ and as $t \to \infty$, $p \to 0$.

Note that in this example the change in sign of dp/dt has been used to show that the maximum occurs when $t = 1$. The reason for this is that d^2p/dt^2, obtained by differentiating (6.1), is extremely complicated and it is simpler to use the method used here.

6.3 CURVE SKETCHING

In many practical situations it is important to be able to determine the shape and essential characteristics of a curve without making an accurate plot of the graph.

The first step is to choose scales and axes which allow the important sections of the graph to be shown as large as possible. It is normally relatively easy to find those points in which the curve $y = f(x)$ cuts the y-axis ($x = 0$) and the x-axis ($y = 0$). It is essential to determine the position and nature of all stationary points. The positions of points of inflexion are also useful aids. The behaviour of the curve for large values of $|x|$ and $|y|$ is important, and any points of discontinuity should be noted. In particular if

$$y = f(x) = \frac{g(x)}{h(x)} ,$$

where $g(x)$ and $h(x)$ are polynomials in x, there will be discontinuities in the graph at points $x = a$ where $h(x)$ is zero and $g(x)$ is non-zero. The vertical line $x = a$ is then referred to as a **vertical asymptote** and y becomes infinite for a finite value of x. Horizontal asymptotes also arise when x becomes infinite for finite values of y. These types of asymptote are particular cases of a more general result which will not be considered here (see Smyrl (1978)). The technique of curve sketching is probably best illustrated by means of examples.

Example 1: Sketch the curve $y = x^3 - 3x$.

This curve cuts the y-axis $(x = 0)$ when $y = 0$ and it cuts the x-axis $(y = 0)$ when $x = 0, x = -\sqrt{3}, x = \sqrt{3}$.

$$\frac{dy}{dx} = 3x^2 - 3 = 3(x^2 - 1) = 3(x + 1)(x - 1) ,$$

$$\frac{d^2y}{dx^2} = 6x ,$$

so that there are stationary points when

$$x = -1, \ y = 2, \ d^2y/dx^2 = -6, \text{a maximum};$$

$$x = +1, \ y = -2, \ d^2y/dx^2 = 6, \text{a minimum}.$$

In addition $d^2y/dx^2 = 0$ and changes sign when $x = 0$ so that $x = 0$ is a point of inflexion.

For this equation there are no horizontal or vertical asymptotes. For large values of $|x|$ the term x^3 is the most important on the RHS, so that as $x \to \infty$, $y \to \infty$ and as $x \to -\infty, y \to -\infty$. A sketch of the curve is shown in Fig. 6.2.

Example 2: Sketch the curve $y = \dfrac{1}{1-x}$.

This curve cuts the y-axis when $y = 1$. It does not intersect the x-axis for finite values of x.

$$\frac{dy}{dx} = \frac{1}{(1 - x)^2} ,$$

$$\frac{d^2y}{dx^2} = \frac{2}{(1 - x)^3} ,$$

so that there are no stationary points or points of inflexion. Since $(1 - x)^2 > 0$ for all x other than $x = 1$ the slope is positive for all $x \neq 1$.

When $x = 1, y \to \pm \infty$ so that $x = 1$ is a vertical asymptote.

Since $\lim\limits_{x \to \pm\infty} \dfrac{1}{1 - x} = 0, y = 0$ is a horizontal asymptote.

The information obtained thus far is not adequate to enable the curve to be sketched with certainty. Information on the behaviour of the curve near $x = 1$ may be obtained by writing $x = 1 \pm \epsilon$, where $\epsilon > 0$. For $x = 1 - \epsilon$,

$$y = \frac{1}{1 - (1 - \epsilon)} = \frac{1}{\epsilon}$$

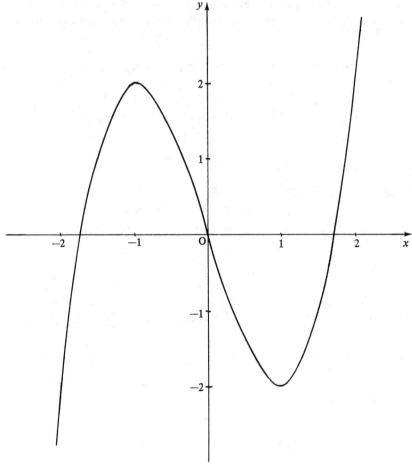

Fig. 6.2

and as $\epsilon \to 0$ (that is, x approaches 1 from below)

$$y \to \lim_{\epsilon \to 0} \left(\frac{1}{\epsilon} \right) = + \infty .$$

Similarly by setting $x = 1 + \epsilon$ it is found that

$$y = \frac{1}{1 - (1 + \epsilon)} = - \frac{1}{\epsilon}$$

and as $\epsilon \to 0$ (x approaches 1 from above)

$$y \to \lim_{\epsilon \to 0} \left(- \frac{1}{\epsilon} \right) = - \infty .$$

Thus the curve approaches $+\infty$ to the left of the line $x = 1$ and approaches $-\infty$ to the right of this line, and there is an infinite discontinuity at $x = 1$. A sketch is shown in Fig. 6.3.

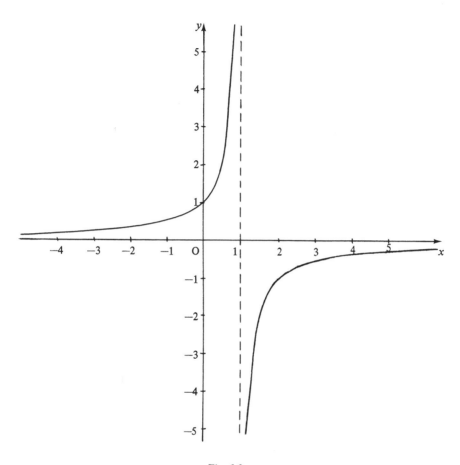

Fig. 6.3

Example 3: The length L of the regenerating posterior end of an earthworm is given approximately by

$$L = 10(6t^2 - 8t^3 + 3t^4)$$

where t is the time. The variable representing time has been scaled so that $t = 0$ corresponds to the start of growth and $t = 1$ corresponds to full maturity. Sketch the graph of L against t.

$$L = 10(6t^2 - 8t^3 + 3t^4) = 10t^2(6 - 8t + 3t^2)$$

so that when $t = 0$, $L = 0$ and when $t = 1$, $L = 10$. When $L = 0$ either $t = 0$ or $3t^2 - 8t + 6 = 0$. An attempt to solve this quadratic equation shows that there are no real roots, so the curve cuts the t-axis in $L = 0$ only. To determine the stationary points, L is differentiated to give

$$\frac{dL}{dt} = 10(12t - 24t^2 + 12t^3) = 120t (1 - t)^2 ,$$

$$\frac{d^2L}{dt^2} = 10(12 - 48t + 36t^2) = 120 (1 - 3t) (1 - t) .$$

Since $dL/dt = 0$ and $d^2L/dt^2 > 0$ when $t = 0$ this is a minimum. Also $dL/dt = 0$ when $t = 1$. Since $d^2L/dt^2 = 0$ when $t = 1$ and changes sign there, the point $t = 1$, $L = 10$ is a (horizontal) point of inflexion.

The point $t = \frac{1}{3}$ is also a point of inflexion. There are no horizontal or vertical asymptotes.

A sketch of the curve for $0 \leqslant t \leqslant 1$ is given in Fig. 6.4. The behaviour of L for $t < 0$ or $t > 1$ is not relevant in this example.

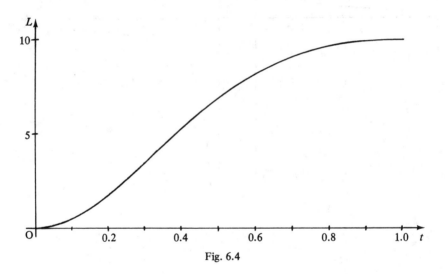

Fig. 6.4

Example 4: The biomass S of an animal is given approximately by

$$S = \tfrac{1}{2}(3t^{\frac{1}{2}} - t^{\frac{3}{2}})$$

where t is the time. The variables have been scaled so that maturity corresponds to $S = 1$, occurs when $t = 1$, and the value of S at $t = 0$ has been ignored.

Sketch S for $0 < t \leqslant 1$.

Here $S = \tfrac{1}{2}(3t^{\frac{1}{2}} - t^{\frac{3}{2}}) = \tfrac{1}{2}t^{\frac{1}{2}}(3 - t)$ so that $S = 0$ when $t = 0$ and $t = 3$ (outside the range of interest). Differentiating to find the stationary points gives

$$\frac{dS}{dt} = \frac{3}{4}t^{-\frac{1}{2}} - \frac{3}{4}t^{\frac{1}{2}} = \frac{3}{4t^{\frac{1}{2}}}(1 - t) ,$$

$$\frac{d^2S}{dt^2} = -\frac{3}{8}t^{-\frac{3}{2}} - \frac{3}{8}t^{-\frac{1}{2}} = -\frac{3}{8t^{\frac{3}{2}}}(1 + t) .$$

Since $dS/dt = 0$ when $t = 1$ only, this corresponds to a maximum.

$d^2S/dt^2 = 0$ when $t = -1$ which is outside the range of interest, so that there are no relevant points of inflexion. Note that $dS/dt \to \infty$ as $t \to 0$.

There are no vertical or horizontal asymptotes.

A sketch is given in Fig. 6.5.

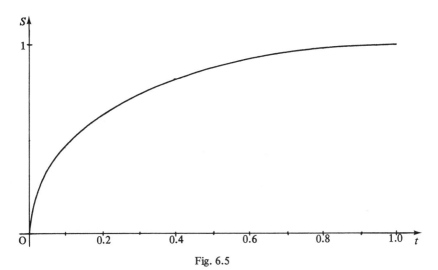

Fig. 6.5

Example 5: Sketch the curve $r = \dfrac{9(4 + t)}{4(9 + t^2)}$.

This example relates to example 3 of Section 6.2 with $r = p/1000$, and results obtained there are used here. A sketch of the graph is given for all t although only that part for which $t \geqslant 0$ relates to the growth of a population.

When $r = 0$, $t = -4$; when $t = 0, r = 1$. Since $9 + t^2$ is never zero there are no vertical asymptotes. A horizontal asymptote is given by $\displaystyle\lim_{t\to\infty} \frac{9(4 + t)}{4(9 + t^2)}$

which is found to be zero by the methods of Chapter 4.

From equation (6.1)

$$\frac{dr}{dt} = \frac{1}{1000}\frac{dp}{dt} = \frac{9(9 + t)(1 - t)}{4(9 + t^2)^2} ,$$

and it is already known that $t = 1$ corresponds to a maximum. dr/dt is zero when $t = -9$, and since the derivative changes sign from negative to positive there, $t = -9$ gives a minimum value. In this particular example the calculation of the second derivative is complicated and not really necessary since an adequate sketch can be obtained without determining possible points of inflexion.

The curve is shown in Fig. 6.6.

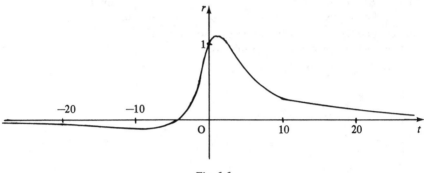

Fig. 6.6

6.4 SMALL CHANGES

In Section 2.6 the binomial theorem was used in order to calculate the effect of a small change in a variable on some given quantity. An alternative approach which is applicable to a more general class of functions than those for which the binomial theorem is valid is based on differential calculus. In effect the method uses the first two terms of the associated Maclaurin series (see Chapter 7).

When y is a given function of x, say $y = f(x)$, the basic definition of the derivative is (see (5.4))

$$\frac{dy}{dx} = \lim_{\Delta x \to 0} \left(\frac{\Delta y}{\Delta x} \right) = \lim_{\Delta x \to 0} \left(\frac{f(x + \Delta x) - f(x)}{\Delta x} \right).$$

In this result Δx is a small change in the variable x, and Δy is the resulting small change in y. Provided that Δx is sufficiently small but not zero (usually $\Delta x / x$ is approximately 5%) then

$$\frac{dy}{dx} \cong \frac{\Delta y}{\Delta x} \quad \text{or} \quad \frac{dy}{dx} \cong \frac{f(x + \Delta x) - f(x)}{\Delta x}. \tag{6.2}$$

The change in y due to a small given change in x is Δy where

$$\Delta y \cong \frac{dy}{dx} \Delta x. \tag{6.3}$$

In equation (6.3) the value of dy/dx at the point x is used. Alternatively, from (6.2)

$$f(x + \Delta x) \cong f(x) + \frac{dy}{dx} \Delta x. \tag{6.4}$$

Equation (6.4) gives the approximate value of the function $f(x)$ at $x + \Delta x$ in terms of its value at the point x and its derivative at that point. Note that Δx may be positive, corresponding to an increase in x, or it may be negative, corresponding to a decrease.

Example 1: At time t the size of a biological population is given by $p = \dfrac{9000(4 + t)}{4(9 + t^2)}$. Calculate the approximate size of the population when $t = 0.1$. (See example 3 of Section 6.2 and example 3 of Section 6.3).
    From (6.1)

$$\frac{dp}{dt} = \frac{9000(9 + t)(1 - t)}{4(9 + t^2)^2} .$$

In this problem the small change in the independent variable t is $\Delta t = 0.1$ and the resulting change in p is Δp where

$$\Delta p \cong \frac{dp}{dt} \Delta t$$

and dp/dt is calculated at $t = 0$. Consequently

$$\Delta p \cong \frac{9000 \times 9}{4 \times 9^2} \times \frac{1}{10} = 25 .$$

The increase in p is approximately 25 so that the size of the population is now approximately $1000 + 25 = 1025$. (The actual value as calculated from the formula for p is 1024).

Example 2: The volume of a spherical cell of radius r is given by $V = \frac{4}{3} \pi r^3$. Show that if r is increased by 1% the resulting increase in V is approximately 3%. (See example 4 of Section 2.5).

Here $$\frac{dV}{dr} = 4\pi r^2$$

and $$\Delta V \cong \frac{dV}{dr} \Delta r = 4\pi r^2 \Delta r .$$

If r is increased by 1% then $\Delta r = r/100$ and

$$\Delta V \cong \frac{4\pi r^3}{100} = \frac{3V}{100} .$$

Hence V increases by approximately 3% when r increases by 1%.

Example 3: Ohm's law states that $I = V/R$, where in an electric circuit I is the current flowing, V is the voltage, and R the resistance. If V remains constant and R is increased by 2% calculate the percentage change in I.

Here
$$\frac{dI}{dR} = -\frac{V}{R^2}$$

and
$$\Delta I \cong \frac{dI}{dR} \Delta R = -\frac{V}{R^2} \Delta R .$$

Since R is increased by 2%, $\Delta R = \frac{2}{100}R$ and

$$\Delta I \cong -\frac{2V}{100R} = -\frac{2}{100}I .$$

The negative sign indicates that I decreases by approximately 2% when R is increased by 2%.

6.5 RELATED RATES OF CHANGE

As an animal grows to maturity so its height, length, breadth, surface area and volume increase, but not all at the same rate. The rate at which the height, length and breadth change with time will be comparable whereas surface area and volume will both increase at much faster rates. The quantities are all functions of time but their rates of change with time are different. These differences arise because the surface area is proportional to the square of the linear dimension and the volume is proportional to the cube of the linear dimension. Since relationships can often be written down relating quantities such as height, surface area and volume it is possible to obtain statements relating the various rates of change. It is not necessary to obtain a specific equation linking say volume and surface area, as is illustrated by the following example.

Example: At time t a spherical cell has radius r, surface area $S = 4\pi r^2$ and volume $V = 4\pi r^3/3$. If the volume increases at a constant rate p find the rates of increase of r and S in terms of p and r.

Differentiating the expression for V gives

$$p = \frac{dV}{dt} = 4\pi r^2 \frac{dr}{dt}$$

so that
$$\frac{dr}{dt} = \frac{p}{4\pi r^2} .$$

Differentiating the expression for S gives

$$\frac{dS}{dt} = 8\pi r \frac{dr}{dt} = \frac{2p}{r} .$$

Note that as $\dot{S} = 2\dot{V}/r$ it is possible to relate $\dot{S}$ and $\dot{V}$ without obtaining an explicit relationship between S and V. The variable r acts as a parameter.

6.6 L'HOSPITAL'S RULE

In Sections 4.5 and 4.6 some methods were discussed that are useful in the evaluation of limits. In certain cases those methods are difficult to employ, and it may then be advantageous to make use of **L'Hospital's rule**. This is stated here without proof.

It can be shown that if $f(a) = g(a) = 0$, so that $f(a)/g(a)$ is indeterminate, and if the limit of $f'(x)/g'(x)$ as x approaches a exists, then

$$\lim_{x \to a} \frac{f(x)}{g(x)} = \lim_{x \to a} \frac{f'(x)}{g'(x)} \ .$$

It is assumed that $f'(x)$ and $g'(x)$ exist.

Example: Evaluate the limits (i) $\lim_{x \to 1} \dfrac{x^2 + x - 2}{x^2 - 1}$, (ii) $\lim_{x \to 0} \dfrac{x}{\tan x}$, (iii) $\lim_{x \to 0} \dfrac{\sin^2 x}{x^2}$.

(i) In this example $f(x) = x^2 + x - 2$, $g(x) = x^2 - 1$, $f'(x) = 2x + 1$, $g'(x) = 2x$ so that

$$\lim_{x \to 1} \frac{x^2 + x - 2}{x^2 - 1} = \lim_{x \to 1} \frac{2x + 1}{2x} = \frac{3}{2} \ .$$

(ii) Here $f(x) = x$, $g(x) = \tan x$, $f'(x) = 1$, $g'(x) = \sec^2 x$ and

$$\lim_{x \to 0} \frac{x}{\tan x} = \lim_{x \to 0} \frac{1}{\sec^2 x} = 1 \ .$$

(iii) Here $f(x) = \sin^2 x$, $g(x) = x^2$, $f'(x) = 2\sin x \cos x$, $g'(x) = 2x$ and

$$\lim_{x \to 0} \frac{\sin^2 x}{x^2} = \lim_{x \to 0} \frac{\sin x \cos x}{x} \ .$$

L'Hospital's rule may now be used again with $F(x) = \sin x \cos x$, $G(x) = x$, $F'(x) = \cos^2 x - \sin^2 x$, $G'(x) = 1$ so that

$$\lim_{x \to 0} \frac{\sin^2 x}{x^2} = \lim_{x \to 0} \frac{\sin x \cos x}{x} = \lim_{x \to 0} \frac{\cos^2 x - \sin^2 x}{1} = 1 \ .$$

PROBLEMS

1. Determine those values of x for which the following functions are (a) increasing, (b) decreasing, and find and identify the stationary points, if any,

    (i) $x^3 + 3x^2 - 9x$,

(ii) $x^3 + 3x^2 + 6x$,

(iii) $x^3 + 3x^2 + 3x$,

(iv) $3/x + x^3$, $x > 0$,

(v) $3/x - x^3$, $x > 0$.

2. Find and identify the stationary points, if any, of the following:

(i) $x^2 + 4x + 3$,

(ii) $(x + 2)^3$,

(iii) $1/(x + 1)$,

(iv) $x/(x - 1)^{\frac{1}{2}}$,

(v) $(3 - x)(1 + x^2)^{\frac{1}{2}}$,

(vi) $x/(x^2 + 1)$.

3. For what values of x is $f(x) = x^2(3 - x)$ (a) negative and (b) positive? Show that the greatest value of $f(x)$ for $x > 0$ is $f(2)$.   (Edinburgh, 1977)

4. When $f(x) = 4x/(1 + x)^2$ evaluate $f(x)$ for $x = 0, \frac{1}{2}, 1, 4, 9$, and use these values to sketch the graph of $f(x)$ for $0 \leqslant x \leqslant 9$. By use of derivatives, prove that the greatest value of $f(x)$ is $f(1)$.   (Edinburgh, 1977)

5. When $y = 4x^3 - x^4$:

(i) Evaluate y for $x = 0, 1, 2, 3, 4$.

(ii) Find the average rate of change of y with respect to x as the value of x increases from 1 to 3.

(iii) Find dy/dx.

(iv) Find the instantaneous rate of change of y when $x = 2$.

(v) For what values of x is y (a) an increasing function and (b) a decreasing function.

(vi) Sketch roughly the graph of y.   (Edinburgh, 1978)

6. The velocity, V, of air flowing through a tube of variable radius r, is given by

$$V = kr^2(R - r),$$

where k and R are positive constants and $0 \leqslant r \leqslant R$. Find the greatest value of V.   (Edinburgh, 1974).

7. The reaction of a person to a drug is measured by the change in body temperature an hour after administration. This is found to depend upon the amount x of the drug administered according to

    $$T = T_0 - 0.25x^2(3 - x) ,$$

    where T_0 is the initial body temperature. For what value of x will the temperature drop be greatest?

8. An object projected vertically upwards at time $t = 0$ with a velocity of 14 metres/second reaches a height, y, in metres, given approximately by $y = 14t - 4.9t^2$, where t is measured in seconds. Determine the maximum height reached by the object. Sketch a graph of y against t.

9. The reaction R of the body to a dose D of a certain drug is such that $R = D^2 \left(\frac{1}{2}C - \frac{1}{3}D\right)$, where C is a constant (the maximum permitted dose). The sensitivity is the rate of change of R with D. For what value of D is the sensitivity a maximum?   (Edinburgh)

10. A population of 1000 bacteria is introduced into a nutrient medium. The population p grows according to $p(t) = \dfrac{15000(1 + t)}{15 + t^2}$ where the time t is measured in hours. Determine the maximum size of the population. Sketch a graph of p with t.

11. A biological cell has the shape of a right circular cylinder. If the volume is fixed, what ratio of length l to radius r will give minimum surface area?

12. Find the volume of the largest right circular cylinder that can be inscribed in a sphere of radius r.

13. One side of a field is bounded by a straight river bank. How should a temporary fence of given length l be used in order to enclose as large a rectangular area as possible?

14. A capsule shaped tablet consists of a right circular cylinder of radius r and length l surmounted at each end by a hemisphere. If the tablet has fixed volume V show that the surface area A is given by $A = \dfrac{4\pi r^2}{3} + \dfrac{2V}{r}$. Deduce that A is a minimum when $r = \{3V/(4\pi)\}^{\frac{1}{3}}$.

15. Sketch graphs of the following:

(i) $y = x^3 + x$,

(ii) $y = x^3 - x$,

(iii) $y = (x^2 - 1)(x^2 + 4)$,

(iv) $y = (x^2 - 1)(x^2 - 4)$,

(v) $y = 1/(x + 1)$,

(vi) $y = (x - 1)/(x + 1)$,

(vii) $y = 1/(x^2 - 1)$,

(viii) $y = 1/(x^3 - 4x)$,

(ix) $y = 3/x - x^3$,

(x) $y = 3/x + x^3$,

(xi) $y = x^2(3 - x)$,

(xii) $y = (x - 1)^3$,

(xiii) $y = x^4 - 4x$,

(xiv) $y = 3x^{\frac{1}{2}} + x^{\frac{3}{2}}$,

(xv) $y^2 = x^3$.

16. When $y = x(x - 1)^2$, find those values of x for which $dy/dx < 0$.
Find the equation of the tangent at the point $(2, 2)$.
Sketch the curve for $0 \leqslant x \leqslant 2$, showing the tangents at the points $(0, 0)$ and $(2, 2)$. (Edinburgh, 1974).

17. The velocity, V, of air flowing through a bronchial tube of variable radius r, is given by

$$V = kr^2(R - r) ,$$

where k and R are positive constants and $0 \leqslant r \leqslant R$. Plot a graph of V against r.

18. When $f(x) = x^2(3C - x)$ show by use of a value of $f'(x)$ that when the value of x is changed from C to $1.01C$, then the value of $f(x)$ is changed by about $0.03C^3$. (Edinburgh, 1977).

19. In a glasshouse the intensity I of light received by a tomato of radius r is given by

$$I = kr^2$$

where k is a constant. If r increases by 2% show that I increases by 4% approximately.

20. The pressure p and volume v of a certain gas are related by $pv^{1.4} = k$, where k is a constant. If the volume is increased by 2% estimate the percentage change in p. Why is your answer negative?

21. The length of the side b of a triangle is calculated by using the formula $b = a \sin B / \sin A$. If the values of a, A and B should be 150, 30° and 45° respectively, find the approximate percentage error in b if a and A are correct but B is 44°.

22. In a certain experiment the tension T required to produce an extension ratio of $\lambda(\geqslant 1)$ in myosin filaments was found to be given approximately by

$$T = \frac{3(\lambda^3 - 1)}{(\lambda^3 + 2)^2}.$$

(i) Find the approximate change in T required to change λ from 1 to 1.05.
(ii) For what value of λ is T a maximum? Plot a graph of T against λ for $\lambda \geqslant 1$.

23. In a dialyser the rate of flow F of blood along a plastic tube of radius r is given approximately by $F = kr^4$ where k is a constant. If r is decreased by 4% find the percentage change in F (i) by calculus, (ii) by the binomial theorem taking the dominant two terms.

24. In a chemical reaction the mass m, in gms, of protein disintegrates according to the formula

$$m = \frac{60}{t + 2}$$

where t is the time in hours. Sketch a graph of m against $t(\geqslant 0)$. Calculate the approximate percentage change in m as t increases from 1 to 1.1.

25. The weight W, in kilograms, of a person with sitting height h, in metres, is given approximately by $W = 90h^3$. Find the percentage increase in W if h increases by 2%.

26. In laminar flow of blood through a cylindrical artery the resistance R is inversely proportional to the fourth power of the radius r. If r is decreased by 2% show that R is increased by 8% approximately.

27. The area of a square is increasing at a rate of 1 cm²/hr. At what rate is the length of each side increasing when the length is (i) 1 cm and (ii) 1 metre?
(Edinburgh).

28. A cube increases in size. If the surface area increases by 25% show that the volume increases by just under 40%.
    When the length of each side of the cube is 10 cm, the rate of increase in surface area is 10 cm^2/min; show that the rate of increase in volume at this instant is 25 cm^3/min.   (Adapted from Edinburgh, 1975)

29. A spherical raindrop accumulates moisture at a rate proportional to its surface area. Show that the radius increases at a constant rate.

30. Liquid flows from a right circular conical funnel (vertex down) at a constant rate. Show that, for any funnel angle, the rate of change of the depth h of the liquid is inversely proportional to h^2.

31. When air expands adiabatically the pressure p and volume v are related by $pv^{1.4} = k$ where k is a constant. Show that $v\dfrac{dp}{dt} = -1.4\,p\dfrac{dv}{dt}$.

32. The shape of a nematode worm can be approximated by a right circular cylinder. If the length is equal to the radius r and the worm grows such that its surface area increases at a constant rate C find the rate of change of r and the volume V at any time t.

33. Use L'Hospital's rule to evaluate:

    (i) $\displaystyle\lim_{x\to 1}\frac{x^4 - 1}{x - 1}$,

    (ii) $\displaystyle\lim_{x\to -1}\frac{x^5 + 1}{x + 1}$,

    (iii) $\displaystyle\lim_{x\to 0}\frac{\sin 4x}{x}$,

    (iv) $\displaystyle\lim_{x\to 0}\frac{1 - \cos x}{x^2}$,

    (v) $\displaystyle\lim_{x\to \frac{\pi}{2}}\frac{x - \pi/2}{\cos x}$,

    (vi) $\displaystyle\lim_{x\to \pi}\frac{\sin x}{\pi - x}$,

    (vii) $\displaystyle\lim_{x\to 0}\frac{x - \sin x}{x^3}$,

    (viii) $\displaystyle\lim_{x\to 0}\frac{x^2}{\tan^2 x}$.

Maclaurin series

7.1 INTRODUCTION

From the discussion of polynomials given in Chapter 3 it is clear that these particular functions are relatively simple to manipulate compared with other types of function. In addition to the properties outlined in Chapter 3 polynomials have the merit of being continuous and differentiable at all points where they are defined. For this reason it is sometimes convenient to be able to approximate complicated functions by a polynomial having similar properties to the given function. This type of approximation is discussed in this chapter.

7.2 MACLAURIN SERIES

To approximate a given function $f(x)$ near $x = 0$ by a polynomial of degree n in x, $p_n(x)$ where

$$p_n(x) = a_0 + a_1x + a_2x^2 + \ldots + a_nx^n \ , \qquad (7.1)$$

it is necessary to formulate a method of calculating the $(n + 1)$ unknown coefficients $a_0, a_1, a_2, \ldots, a_n$. A sufficient number of properties that $f(x)$ and $p_n(x)$ have in common must be stated in order to carry out this calculation. There are a number of ways of imposing a mathematical similarity between $f(x)$ and $p_n(x)$; the one adopted here is to insist that $p_n(0) = f(0)$ and

$$p_n^{(r)}(0) = f^{(r)}(0), \quad r = 1, 2, 3, \ldots, n \ .$$

Thus it is assumed that every derivative of $p_n(x)$ is equal to the corresponding derivative of $f(x)$ at $x = 0$. It is therefore necessary for $f(x)$ to be continuous and differentiable a sufficient number of times.

   The assumption that $p_n(0) = f(0)$ requires $p_n(x)$ to take the same value as $f(x)$ at $x = 0$, but not necessarily elsewhere. The requirement that $p'_n(0) = f'(0)$ states that the first derivative of $p_n(x)$ is equal to the first derivative of $f(x)$ at $x = 0$ but not necessarily elsewhere. A similar interpretation applies to the other derivatives. As a result the shape of $p_n(x)$ will be close to that of $f(x)$ near

$x = 0$, but not necessarily for other values of x. Intuition suggests that it is likely that this correspondence will improve as the value of n is increased.

When $p_n(x)$ is given by (7.1) it is found that

$$p'_n(x) = a_1 + 2a_2x + 3a_3x^2 + \ldots + na_nx^{n-1} , \tag{7.2}$$

$$p''_n(x) = 2a_2 + 2.3a_3x + 3.4a_4x^2 + \ldots + (n-1)na_nx^{n-2} , \tag{7.3}$$

$$p'''_n(x) = 2.3a_3 + 2.3.4a_4x + \ldots + (n-2)(n-1)na_nx^{n-3} , \tag{7.4}$$

$$\cdots\cdots\cdots\cdots\cdots$$

$$p_n^{(n)}(x) = 1.2.3 \ldots (n-2)(n-1)na_n = n!a_n . \tag{7.5}$$

Setting $x = 0$ in (7.1) – (7.5) in turn there results

$$p_n(0) = f(0) = a_0 ,$$
$$p'_n(0) = f'(0) = a_1 ,$$
$$p''_n(0) = f''(0) = 2a_2 ,$$
$$p'''_n(0) = f'''(0) = 2.3a_3 ,$$
$$p_n^{(n)}(0) = f^{(n)}(0) = n!a_n ,$$

so that $\quad a_0 = f(0), \ a_1 = \dfrac{1}{1!}f'(0), \ a_2 = \dfrac{1}{2!}f''(0), \ a_3 = \dfrac{1}{3!}f'''(0),$

$$a_n = \frac{1}{n!}f^{(n)}(0) .$$

Substituting these values of the coefficients into (7.1) gives

$$p_n(x) = f(0) + \frac{f'(0)}{1!}x + \frac{f''(0)}{2!}x^2 + \frac{f'''(0)}{3!}x^3 + \ldots$$

$$\ldots + \frac{f^{(n)}(0)}{n!}x^n . \tag{7.6}$$

The equation (7.6) is called the **Maclaurin expansion of degree** n for $f(x)$.

The polynomial expansion (7.6) may be extended to give an infinite series for $f(x)$, called the **Maclaurin series**, by taking additional terms. It is found that

$$f(x) = f(0) + \frac{f'(0)}{1!}x + \frac{f''(0)}{2!}x^2 + \ldots + \frac{f^{(n)}(0)}{n!}x^n + \ldots$$

$$= \sum_{n=0}^{\infty} \frac{f^{(n)}(0)}{n!}x^n . \tag{7.7}$$

The series representation for $f(x)$ given by (7.7) will give the exact value for $f(x)$ at $x = 0$, and in general it will approximate $f(x)$ near $x = 0$. The Maclaurin series expansion of some functions can be shown to be valid for all values of x, but detailed properties of the series are beyond the scope of this book.

The series representation for $f(x)$ near $x = a$ may be obtained by substituting $z = x - a$. Then $f(x)$ is transformed into a function of z, say $g(z)$, and since $x = a$ is equivalent to $z = 0$ the expansion of $g(z)$ near $z = 0$ is equivalent to an expansion of $f(x)$ near $x = a$. The resulting series in $(x - a)$ is called a **Taylor series** (see Smyrl (1978)).

Example 1: Obtain the Maclaurin series expansion of $\cos x$ up to the term in x^4.

Here
$$f(x) = \cos x, \qquad f(0) = 1 ,$$
$$f'(x) = -\sin x, \qquad f'(0) = 0 ,$$
$$f''(x) = -\cos x, \qquad f''(0) = -1 ,$$
$$f'''(x) = \sin x, \qquad f'''(0) = 0 ,$$
$$f^{iv}(x) = \cos x, \qquad f^{iv}(0) = 1 .$$

Equation (7.7) then gives

$$\cos x = 1 + \frac{0}{1!}x + \frac{(-1)}{2!}x^2 + \frac{0}{3!}x^3 + \frac{1}{4!}x^4 + \ldots$$

$$= 1 - \frac{1}{2!}x^2 + \frac{1}{4!}x^4 - \ldots .$$

The graphs of the three functions

$$f(x) = \cos x ,$$
$$f_1(x) = 1 - \tfrac{1}{2}x^2 ,$$
$$f_2(x) = 1 - \tfrac{1}{2}x^2 + \tfrac{1}{24}x^4 ,$$

for $0 \leqslant x \leqslant \pi$ are shown in Fig. 7.1. These graphs indicate how the Maclaurin series approximates the function $\cos x$. Both $f_1(x)$ and $f_2(x)$ are particularly close to $\cos x$ near $x = 0$. $f_1(x)$ gives a good approximation up to $x = \pi/4$ but is less satisfactory for larger values of x. $f_2(x)$ is a closer approximation than $f_1(x)$ beyond $x = \pi/4$.

Example 2: Obtain the first five terms of the Maclaurin series expansion of $\dfrac{1}{1-x}$.

Here
$$f(x) = \frac{1}{1 - x} , \qquad\qquad f(0) = 1 ,$$

$$f'(x) = \frac{1}{(1 - x)^2} , \qquad\qquad f'(0) = 1 ,$$

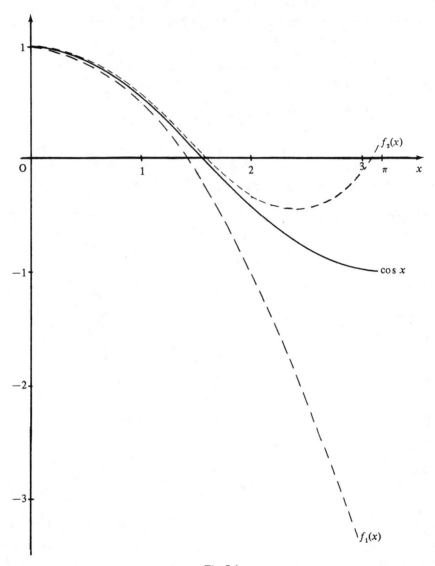

Fig. 7.1

$$f''(x) = \frac{2}{(1-x)^3} \; , \qquad f''(0) = 2! \; ,$$

$$f'''(x) = \frac{2.3}{(1-x)^4} \; , \qquad f'''(0) = 3! \; ,$$

$$f^{iv}(x) = \frac{2.3.4}{(1-x)^5} \; , \qquad f^{iv}(0) = 4! \; ,$$

so that from (7.7)

$$\frac{1}{1-x} = 1 + x + x^2 + x^3 + x^4 + \ldots . \qquad (7.8)$$

Equation (7.8) is the binomial expansion of $(1-x)^{-1}$, see equation (2.12).

The binomial series expansion of $(1+x)^\alpha$, where α is a real number, may similarly be proved by using the Maclaurin series expansion.

Example 3: The height h of a plant over the time interval $0 \leqslant t \leqslant 2$ is given as a function of the time t by $h = (4 + 3t + t^2)^{\frac{1}{2}}$. Obtain the first three terms of the Maclaurin series expansion for h and use it to calculate h when $t = 0.2$ to four places of decimals.

Here $$h = f(t) = (4 + 3t + t^2)^{\frac{1}{2}} \; , \qquad (7.9)$$

$$\frac{dh}{dt} = f'(t) = \tfrac{1}{2}(3 + 2t)(4 + 3t + t^2)^{-\frac{1}{2}} \; ,$$

$$\frac{d^2h}{dt^2} = f''(t) = (4 + 3t + t^2)^{-\frac{1}{2}} - \tfrac{1}{4}(3 + 2t)^2 (4 + 3t + t^2)^{-\frac{3}{2}} \; ,$$

and when $t = 0$

$$f(0) = 2, \; f'(0) = \frac{3}{4}, \; f''(0) = \frac{1}{2} - \frac{9}{32} = \frac{7}{32} \; ,$$

so that $$h = f(t) = 2 + \frac{3}{4}t + \frac{7}{64}t^2 + \ldots .$$

When $t = 0.2$ this equation gives

$$h \cong 2 + 0.15 + 0.004375 = 2.154375 \cong 2.1544 \; .$$

The value given by (7.9) is $h = (4.64)^{\frac{1}{2}} \cong 2.1541$ so that for this value of t the Maclaurin series gives a very good estimate.

The graphs of the three functions

$$h = (4 + 3t + t^2)^{\frac{1}{2}} \; ,$$

$$h_1 = 2 + \frac{3}{4}t \; ,$$

$$h_2 = 2 + \frac{3}{4}t + \frac{7}{64}t^2 \ ,$$

for $0 \leqslant t \leqslant 2$ are shown in Fig. 7.2. These curves give an indication of how the Maclaurin series expansion approximates the true function.

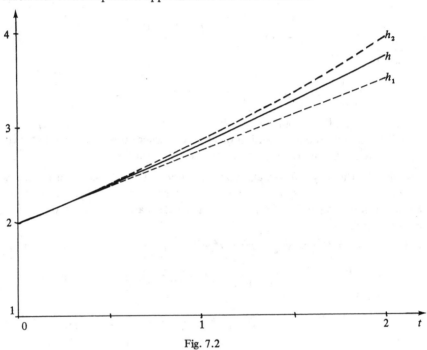

Fig. 7.2

7.3 SUMMARY

The Maclaurin series discussed in this chapter give a relatively simple approximation to a given function. Normally the series is terminated to form a polynomial approximation to the series, and it is then important to be able to estimate the error involved. The Taylor series expansion about the point $x = a$ instead of about $x = 0$ is also important in certain circumstances, particularly when numerical techniques are needed. Further discussion of these and other aspects of series expansions is to be found in Smyrl (1978).

PROBLEMS

1. Obtain the first three non-zero terms in the Maclaurin series expansions of:

    (i) $\sin x$,

    (ii) $1/(1 + x)$,

(iii) $\tan x$,

(iv) $(1 + x)^{\frac{1}{2}}$,

(v) $(1 - x)^{\frac{1}{2}}$,

(vi) $(1 + x)^{-\frac{1}{2}}$,

(vii) $\tan^{-1}x$,

(viii) $1/(1 - x^2)$.

2. The percentage efficiency of a herbicide of concentration m deteriorates with time according to the formula

$$m = 180/(t + 2) ,$$

where t is the time in days. Obtain the first three terms in the Maclaurin series expansion for m and use it to calculate m at $t = 0.2$.

3. A population of 1000 bacteria is introduced into a nutrient medium at time $t = 0$. The size of the population is given by

$$p(t) = \frac{15000(1 + t)}{15 + t^2}$$

where t is measured in hours. Determine the first three terms in the Maclaurin series expansion for $p(t)$. Sketch graphs comparing the series with the correct value for $p(t)$.

4. The influence of photoperiodicity on the activity of an insect, y_1, and its predator, y_2, is described by the equations

$$y_1 = 2a + a\cos(\pi t/3) ,$$

$$y_2 = 2a - a\cos\{\pi(t - 1)/5\} ,$$

where t is the time. Obtain the first three terms in the Maclaurin series expansions of y_1 and y_2.

5. The height h of a plant over the time interval $0 \leqslant t \leqslant 3$ is given by $h = (9 + 4t + t^2)^{\frac{1}{2}}$. Obtain the first three terms in the Maclaurin series expansion of h. Sketch graphs of this approximation to h and the true function h for $0 \leqslant t \leqslant 3$.

CHAPTER 8

The solution of equations

8.1 INTRODUCTION

In many practical problems it is necessary to solve an equation of the form

$$f(x) = 0$$

where $f(x)$ is a function of x. Some discussion of the available techniques was given in Chapter 3 when $f(x)$ is a polynomial. The methods used there are now applied to the solution of more general equations. The techniques used were factorisation, approximate solution by graphical methods, or numerical approximation.

This chapter commences with a discussion of the solution of trigonometric equations and concludes with an account of Newton's method of solving equations. Further discussion of numerical methods is given in Chapter 17.

8.2 TRIGONOMETRIC EQUATIONS

The principle objective when solving a trigonometric equation (that is one containing trigonometric functions only) is to obtain factors which when set equal to zero can be written in the form of a trigonometric function of x equalling a constant. Solution values of x may be obtained by using the results of Section 1.15. The given trigonometric equation will not normally factorise on sight, and it is usually necessary to make use of the results listed in Appendix A in order to obtain a suitable equation. It should be noted that there are frequently several suitable ways of solving a given trigonometric equation. It should also be recognised that each trigonometric factor will give rise to an infinite number of solutions owing to the periodic property of these functions discussed in Section 1.15. For example $\sin x = 0$ has solutions $x = 0, \pm \pi, \pm 2\pi, \ldots$. Values of x may be required for a restricted range such as $0 \leqslant x \leqslant \pi$ and sometimes for the whole range. If the equation does not factorise then use must be made of graphical or numerical methods.

The method of solution by factorisation is illustrated in the following examples.

Example 1: Solve the equation

$$2 \sin^2 x + 3 \cos x = 3 \quad \text{for } 0 \leqslant x \leqslant 2\pi .$$

The equation contains terms in $\sin x$ and in $\cos x$. To obtain a solution it is necessary to express the equation in terms of one of these functions only. Using A 18[†] there results

$$2(1 - \cos^2 x) + 3 \cos x = 3$$

so that $2 \cos^2 x - 3 \cos x + 1 = 0 .$

Factorising gives

$$(2 \cos x - 1)(\cos x - 1) = 0$$

so that $\cos x = \frac{1}{2}$ or $\cos x = 1 .$

When $\cos x = \frac{1}{2}$ the principal value is $\pi/3$ and the other solution between 0 and 2π is $2\pi - \pi/3 = 5\pi/3$. When $\cos x = 1$ the principal value is 0 and the other solution is 2π.

Hence the solutions of the equation in the appropriate range are 0, $\pi/3$, $5\pi/3$, 2π.

Example 2: Solve the equation $\sin 2\theta = 3 \sin \theta$ for $0 \leqslant \theta \leqslant 2\pi$.

The simplest approach is to make use of A 14[†] to express $\sin 2\theta$ in terms of $\sin \theta$ and $\cos \theta$. The equation then becomes

$$2 \sin \theta \cos \theta - 3 \sin \theta = 0 , \tag{8.1}$$

$$\sin \theta (2 \cos \theta - 3) = 0$$

so that $\sin \theta = 0$ or $\cos \theta = 3/2 .$

Since $\cos \theta \neq 3/2$ for real values of θ the only solutions are given by $\sin \theta = 0$ and are $\theta = 0, 2\pi$.

Note that it is important not to cancel the $\sin \theta$ term in equation (8.1) or solutions will be lost.

8.3 GRAPHICAL METHODS

Any equation for x that can be expressed in the form

$$F(x) = G(x) \tag{8.2}$$

may be solved approximately by plotting the curves $y = F(x)$ and $y = G(x)$. The points of intersection of the two curves correspond to the roots of (8.2) since at these points $F(x)$ and $G(x)$ take the same value. This method was mentioned in Chapter 3 as a method of solving polynomial equations, but it

†Appendix A.

may be applied generally, particularly to equations that do not factorise. It has the merit of being relatively simple to use, but against this it does not give very accurate values. The best approach is to sketch rough graphs of $F(x)$ and $G(x)$ and then make a more accurate plot in the region of an expected root; the final accuracy of the method depends on the accuracy of the graphs. The method gives an approximate value to a root quickly, and this value can usually be made more accurate by using a numerical method such as Newton's method which is discussed in the next section.

Example 1: Find the solution of $2 \sin x = x$ for $x > 0$.

The most suitable form for the equation is

$$\sin x = \tfrac{1}{2} x$$

rather than

$$2 \sin x = x$$

since $\sin x$ is a standard tabulated function of x, and $y = \tfrac{1}{2} x$ gives a straight line graph. A rough sketch indicates that there is a root at $x = 0$ and a root between $x = \tfrac{1}{2}\pi$ and $x = \pi$. There can be no root beyond $x = 2$ since for $x > 2, \tfrac{1}{2} x > 1$ and $\sin x$ is always less than or equal to 1. Consequently the two curves cannot intersect again. Graphs of $y = \sin x$ and $y = \tfrac{1}{2} x$ are given in Fig. 8.1 for $0 \leqslant x \leqslant \pi$.

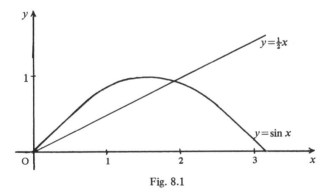

Fig. 8.1

The curves intersect near the point where $x = 1.9$ so that this value of x gives an approximate value for the root.

Example 2: The height h_1 of a plant is given by $h_1 = (4 + 2t + t^2)^{\frac{1}{2}}$ for $t \geqslant 0$. The height h_2 of a similar plant which is planted one unit of time later and given a different nutrient treatment is given by $h_2 = \{4 + 4(t - 1) + 3(t - 1)^2\}^{\frac{1}{2}}$ for $t \geqslant 1$. Find the time at which the two plants are the same height.

The two plants will be the same height when $h_1 = h_2$ so that, if graphs of the two functions are plotted, the point of intersection will give the appropriate time. Graphs of the two expressions are shown in Fig. 8.2 from which it is seen that $h_1 = h_2$ when $t = 2.2$ approximately.

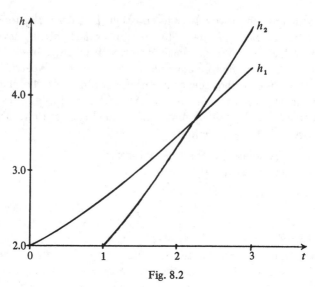

Fig. 8.2

8.4 NEWTON'S METHOD

The graphical method considered in Section 8.3. gives an approximate value for the root of an equation which is not usually very accurate. When an accurate value for the root is required it is usual to use a numerical method to improve the approximation. With the aid of computers, numerical methods are relatively easy to apply.

Suppose that an approximate root, $x = x_1$, of the equation

$$f(x) = F(x) - G(x) = 0 \tag{8.3}$$

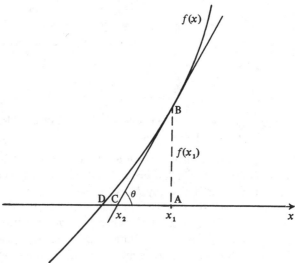

Fig. 8.3

is known, either by plotting $F(x)$ and $G(x)$, or by guesswork. The graph of $y = f(x)$ cuts the x-axis at D near the point A(x_1, 0) (see Fig. 8.3). The point on the curve corresponding to $x = x_1$, is B($x_1, f(x_1)$). The tangent to the curve at B cuts the x-axis at the point C where $x = x_2$ and makes an angle θ with the x-axis. The slope of this tangent may be calculated in two ways and is given by

$$\tan \theta = \frac{f(x_1)}{x_1 - x_2} = f'(x_1) .$$

This equation may be solved for the unknown quantity x_2 to give

$$x_2 = x_1 - \frac{f(x_1)}{f'(x_1)} . \tag{8.4}$$

The result (8.4) is the basis of **Newton's method.** The value of x_2 calculated from (8.4) will normally give a better approximation to the root of (8.3), that is, C is closer to D than A. The procedure may now be continued by replacing x_1 by x_2 in (8.4) to give x_3 where

$$x_3 = x_2 - \frac{f(x_2)}{f'(x_2)} ,$$

then by replacing x_2 by x_3 and so on. The general formula is that, after the nth calculation, x_{n+1} is given in terms of the preceding value x_n by

$$x_{n+1} = x_n - \frac{f(x_n)}{f'(x_n)} , \quad n \geqslant 1 . \tag{8.5}$$

The process is terminated when two successive values are the same, to the required accuracy.

The curve in Fig. 8.3 is shown concave upwards. When the curve is concave downwards as in Fig. 8.4 the point D where the curve cuts the x-axis will lie

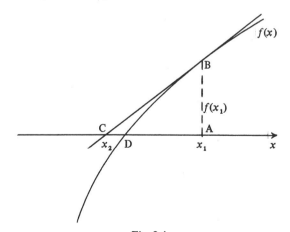

Fig. 8.4

between C and A. No modification of the method is required in this case, and the process will usually converge.

Newton's method is usually a satisfactory method of solving equations provided that the given function may be differentiated. Difficulties may arise when the derivative is very small or when two roots lie close together. In such cases the method is usually still successful if a better first approximation to the root is chosen.

Example 1: Determine, to three places of decimals, the root of $x^3 + x^2 - 12x + 8 = 0$ that lies between $x = 0$ and $x = 1$. (See example 2 of Section 3.5).

$$\text{Here } f(x) = x^3 + x^2 - 12x + 8 \ ,$$
$$f'(x) = 3x^2 + 2x - 12 \ .$$

With the starting value $x_1 = 0$

$$f(x_1) = 8.0 \ ,$$
$$f'(x_1) = -12.0 \ .$$

Applying Newton's method

$$x_2 = 0 + \frac{8.0}{12.0} \cong 0.6667 \ ,$$

$$f(x_2) = (0.6667)^2 + (0.6667)^2 - 12(0.6667) + 8 \cong 0.7408 \ ,$$
$$f'(x_2) = 3(0.6667)^2 + 2(0.6667) - 12 \cong -9.333 \ ,$$

$$x_3 = 0.6667 + \frac{0.7408}{9.333} \cong 0.7461 \ ,$$

$$f(x_3) \cong 0.0189, \ f'(x_3) \cong -8.838, \ x_4 \cong 0.7482 \ ,$$
$$f(x_4) \cong 0.0002, \ f'(x_4) \cong -8.824, \ x_5 \cong 0.7482 \ .$$

To three places of decimals the value of the root is 0.748. The other two roots of the equation may also be calculated, using as starting values the results from Section 3.5.

Example 2: Find the solution of $2 \sin x = x$, for $x > 0$, to three places of decimals. (See example 1 of Section 8.3).

The solution is known to be close to 1.9 from the graphical solution of Section 8.3.

The equation is written in the form $f(x) = 0$, and to illustrate the power of Newton's method the starting value $x_1 = 2.0$ chosen. Then

$$f(x) = \sin x - \tfrac{1}{2} x \ ,$$
$$f'(x) = \cos x - \tfrac{1}{2} \ ,$$

so that $f(x_1) = \sin (2.0) - 1.0 \cong -0.0907 \ ,$
$$f'(x_1) = \cos (2.0) - 0.5 \cong -0.9292 \ .$$

Applying the formula

$$x_2 = 2.0 - \frac{0.0907}{0.9292} \cong 1.9024 ,$$

$$f(x_2) = \sin(1.9024) - 0.9512 \cong -0.0057 ,$$

$$f'(x_2) = \cos(1.9024) - 0.5 \cong -0.8255 ,$$

$$x_3 = 1.9024 - \frac{0.0057}{0.8255} \cong 1.8955 ,$$

$$f(x_3) = \sin(1.8955) - 0.9478 \cong -0.0001 ,$$

$$f'(x_3) = \cos(1.8955) - 0.5 \cong -0.8190 ,$$

$$x_4 = 1.8955 - \frac{0.0001}{0.8190} \cong 1.8954 .$$

After this calculation x_3 and x_4 differ by 1 in the fourth place of decimals so that to three places of decimals the root is 1.895.

8.5 SUMMARY

The methods discussed in this chapter for the solution of various types of equation represent a brief introduction only. In particular, a number of other techniques for numerical solution are available, see for example Hamming (1971) and also Chapter 17.

PROBLEMS

1. Solve the following equations for x, $0 \leqslant x \leqslant 2\pi$:

    (i) $4 \cos^2 x = 1$,

    (ii) $4 \sin x \cos x = 1$,

    (iii) $\sin^2 x = \cos x$,

    (iv) $\sin 2x = \sin x$,

    (v) $\tan x \tan 2x = 1$,

    (vi) $2 \sin x \sin 2x = 1 - \cos 2x$,

    (vii) $\cos x + \cos 2x = \sin x \sin 2x$,

    (viii) $\cos x \sin 2x = \sin x (\sin x + 1)$.

2. The time T, in hours, an insect spends foraging each day is related to the mean daily temperature θ. If the relationship is of the form

    $$T = 14 + 8 \sin \{\pi(\theta - 8)/12\} ,$$

    at what temperature is T

    (i) a maximum,

    (ii) a minimum,

    (iii) equal to 18?

3.   In the blood stream, the biorhythmical appearance of gametocytes from two different forms of malaria, y_1 and y_2, is described by the equations

$$y_1 = 2a + a \cos(\pi t/3),$$
$$y_2 = 2a - a \cos\{\pi(t - 1)/15\}.$$

Find the first positive value of t when

    (i) y_1 and y_2 take the same value,

    (ii) $y_1 + y_2 = 4a$.

4.   Solve, to two decimal places, (a) graphically and (b) by Newton's method the equations:

    (i) $4x^3 - 4x^2 - 5x + 3 = 0,$ $0 < x < 1,$

    (ii) $2x^3 + x^2 + 4x - 15 = 0,$

    (iii) $x^3 + x - 1 = 0,$

    (iv) $x \tan x = 1,$ $0 < x < 1,$

    (v) $x \tan x = 2,$ $0 < x < 1.5,$

    (vi) $2x \cot x = 1,$ $0 < x < 1.5,$

    (vii) $2x \cot x + 1 = 0,$ $1 < x < 2,$

    (viii) $x \cot x + 2 = 0,$ $2 < x < 3.$

5.   The velocity, V, with which an electrical charge travels along a nerve axon of variable radius r, is given by $V = kr^2(R - r)$, where k and R are positive constants and $0 < r < R$. Find, to two places of decimals, the values of r for which $V = kR^3/10$. Use both a graphical method and Newton's method.

6.   In a certain experiment the tension T required to produce an extension ratio of $\lambda(\geqslant 1)$ in myosin filaments was found to be given approximately by

$$T = \frac{3(\lambda^3 - 1)}{(\lambda^3 + 2)^2}.$$

Find, to two decimal places, the values of λ for which $T = 1/5$ by (i) graphical methods, (ii) Newton's method, and (iii) by solving a quadratic in λ^3.

CHAPTER 9

Functions of several variables

9.1 INTRODUCTION

In Chapters 5 and 6 attention was focussed on a dependent variable which was a function of a single independent variable. The results obtained there are sufficient to explain many physical processes, and a number of practical examples were given. However, there are many practical situations in which a dependent variable is a function of two or more independent variables. For example the volume V of a right circular cylinder of height h and radius r is $\pi r^2 h$ and depends on r and h independently.

The number N of people to die from an infectious disease depends on the number of susceptibles s and the number of infectives r present in the population. N is said to be a function of r and s, written $N = f(r, s)$. Note the insertion of a comma between r and s to indicate that they are independent variables.

The speed S at which a bird flies depends on its weight w and its wing-span x. S is a function of w and x, written $S = f(w, x)$.

The concentration C of a solute introduced into a solvent in a tube will depend on the point P at which the measurement is taken and the time t at which the measurement is taken. The quantity C depends on x, the distance of P from some reference point and t, so that $C = f(x, t)$.

The flow of blood, the behaviour of skin and tendon, and the transmission of nerve impulses are some of the many examples in which quantities depend on more than one variable. A complete discussion of the mathematics relevant to these problems is beyond the scope of this book, and only the basic concepts will be touched on here.

9.2 DIFFERENTIATION OF A FUNCTION OF TWO VARIABLES

For simplicity attention is restricted initially to a variable w which is a **function of two independent variables** x and y so that $w = f(x, y)$. It is assumed that this function is continuous in both x and y and is defined within some region R of the xy-plane, see Fig. 9.1. For every point P of the region R with coordinates x, y there is a corresponding value of $w = f(x, y)$.

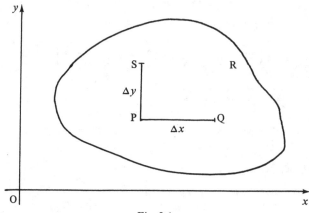

Fig. 9.1

If y is held constant while x changes to $x + \Delta x$ then we move along a line parallel to the x-axis to the point Q with coordinates $x + \Delta x$, y. The corresponding value of w is $f(x + \Delta x, y)$. The change in w resulting from this change in the value of x is Δw where

$$\Delta w = f(x + \Delta x, y) - f(x, y) ,$$

and the average rate of change of w between P and Q is given by

$$\frac{\Delta w}{\Delta x} = \frac{f(x + \Delta x, y) - f(x, y)}{\Delta x} .$$

The **partial derivative** of w with respect to x is denoted by $\partial w/\partial x$ and is defined by

$$\frac{\partial w}{\partial x} = \lim_{\Delta x \to 0} \left(\frac{\Delta w}{\Delta x} \right) = \lim_{\Delta x \to 0} \left(\frac{f(x + \Delta x, y) - f(x, y)}{\Delta x} \right). \qquad (9.1)$$

This partial derivative measures the rate of change of w with respect to x when the other variable y is held constant. In computing $\partial w/\partial x$ for a given function the mechanics of the process are the same as those used in calculating an ordinary derivative with respect to x since y does not change as x changes. An alternative notation for $\partial w/\partial x$ is w_x.

If the variable x is held constant while y changes to $y + \Delta y$ so that we move from P to S with coordinates x, $y + \Delta y$ then the corresponding value of w is $f(x, y + \Delta y)$. The average rate of change of w between P and S is

$$\frac{f(x, y + \Delta y) - f(x, y)}{\Delta y}$$

and the partial derivative of w with respect to y is given by

$$\frac{\partial w}{\partial y} = w_y = \lim_{\Delta y \to 0} \left\{ \frac{f(x, y + \Delta y) - f(x, y)}{\Delta y} \right\}. \tag{9.2}$$

The quantity $\partial w/\partial y$ measures the rate of change of w with respect to y when x is kept constant.

Note that it is customary to use the symbol ∂ instead of d to denote the partial derivative. This convention should be followed since the use of ∂ indicates that the function being differentiated depends on more than one variable.

The rules of differentiation for the sum, product, and quotient of two functions apply to functions of two variables without modification. For a function such as

$$w = \tan^{-1}(x/y) ,$$

so that w depends on the single variable t where

$$w = \tan^{-1}(t) \quad \text{and} \quad t = x/y$$

the chain rule, in the form

$$\frac{\partial w}{\partial x} = \frac{dw}{dt} \frac{\partial t}{\partial x} , \tag{9.3}$$

still holds.

Example 1: If $w = 2x^2y + x^3 - 3y^2x$ find $\dfrac{\partial w}{\partial x}$ and $\dfrac{\partial w}{\partial y}$ and show that $x \dfrac{\partial w}{\partial x} + y \dfrac{\partial w}{\partial y} = 3w$.

Differentiate w with respect to x, keeping y constant, to obtain

$$\frac{\partial w}{\partial x} = 4xy + 3x^2 - 3y^2 .$$

Similarly $\dfrac{\partial w}{\partial y} = 2x^2 - 6yx$.

$$x \frac{\partial w}{\partial x} + y \frac{\partial w}{\partial y} = (4x^2y + 3x^3 - 3xy^2) + (2x^2y - 6y^2x)$$

$$= 6x^2y + 3x^3 - 9y^2x = 3w .$$

Example 2: If $w = y \tan^{-1}(x/y)$ find $\partial w/\partial x$ and $\partial w/\partial y$.

Here $\quad w = y \tan^{-1}(t)$ where $t = x/y$.

From (9.3) $\dfrac{\partial w}{\partial x} = \dfrac{y}{1 + t^2} \dfrac{\partial t}{\partial x} = \dfrac{y}{1 + \dfrac{x^2}{y^2}} \left(\dfrac{1}{y} \right) = \dfrac{y^2}{x^2 + y^2} ,$

$$\frac{\partial w}{\partial y} = \tan^{-1}\left(\frac{x}{y}\right) + \frac{y}{1 + t^2} \frac{\partial t}{\partial y}$$

$$= \tan^{-1}\left(\frac{x}{y}\right) + \frac{y}{1 + \dfrac{x^2}{y^2}} \left(-\frac{x}{y^2}\right)$$

$$= \tan^{-1}\left(\frac{x}{y}\right) - \frac{xy}{x^2 + y^2} .$$

Example 3: The temperature distribution at the point with coordinate x at time t is given by

$$T = C(x^4 + 12x^2t + 12t^2)$$

where C is a constant. Find the rate of change of T with x and with t.

$$\frac{\partial T}{\partial x} = C(4x^3 + 24xt) ,$$

$$\frac{\partial T}{\partial t} = C(12x^2 + 24t) .$$

9.3 HIGHER DERIVATIVES

When discussing a function of a single variable it was found to be useful and convenient to introduce second and higher derivatives. In the case of a function of two variables it is similarly useful to define second and higher derivatives. For $w = f(x, y)$ we define **second derivatives** with respect to x and y by

$$\frac{\partial^2 w}{\partial x^2} = \frac{\partial}{\partial x}\left(\frac{\partial w}{\partial x}\right) , \quad \frac{\partial^2 w}{\partial y^2} = \frac{\partial}{\partial y}\left(\frac{\partial w}{\partial y}\right) . \qquad (9.4)$$

In addition to (9.4) it is possible to define the **mixed derivatives**

$$\frac{\partial^2 w}{\partial x \partial y} = \frac{\partial}{\partial x}\left(\frac{\partial w}{\partial y}\right) \text{ and } \frac{\partial^2 w}{\partial y \partial x} = \frac{\partial}{\partial y}\left(\frac{\partial w}{\partial x}\right) .$$

It can be shown that for the majority of commonly occurring functions these two quantities are equal, so that the order in which the differentiation is carried out for a mixed derivative is not important. This holds for all examples included in this book, so that it may be assumed that

$$\frac{\partial^2 w}{\partial x \partial y} = \frac{\partial^2 w}{\partial y \partial x} . \qquad (9.5)$$

Derivatives higher than the second may be defined by a simple extension of (9.4) and (9.5). For example

$$\frac{\partial^3 w}{\partial x^3} = \frac{\partial}{\partial x}\left(\frac{\partial^2 w}{\partial x^2}\right), \quad \frac{\partial^3 w}{\partial x \partial y^2} = \frac{\partial}{\partial x}\left(\frac{\partial^2 w}{\partial y^2}\right) = \frac{\partial}{\partial y}\left(\frac{\partial^2 w}{\partial x \partial y}\right).$$

Example 1: The temperature distribution across a tissue section is given by $T = C(x^5 + 20x^3 t + 60xt^2)$ where C is a constant. Show that T satisfies the heat conduction equation $\dfrac{\partial^2 T}{\partial x^2} = \dfrac{\partial T}{\partial t}$. Verify that $\dfrac{\partial}{\partial x}\left(\dfrac{\partial T}{\partial t}\right) = \dfrac{\partial}{\partial t}\left(\dfrac{\partial T}{\partial x}\right)$.

$$\frac{\partial T}{\partial x} = C(5x^4 + 60x^2 t + 60t^2) ,$$

$$\frac{\partial^2 T}{\partial x^2} = C(20x^3 + 120xt) ,$$

$$\frac{\partial T}{\partial t} = C(20x^3 + 120xt) ,$$

and clearly $\dfrac{\partial^2 T}{\partial x^2} = \dfrac{\partial T}{\partial t}$.

$$\frac{\partial}{\partial x}\left(\frac{\partial T}{\partial t}\right) = C\frac{\partial}{\partial x}(20x^3 + 120xt) = C(60x^2 + 120t) ,$$

$$\frac{\partial}{\partial t}\left(\frac{\partial T}{\partial x}\right) = C\frac{\partial}{\partial t}(5x^4 + 60x^2 t + 60t^2) = C(60x^2 + 120t) ,$$

and these two results are identical.

Example 2: Show that $w = (A \cos x + B \sin x)(C \cos t + D \sin t)$ where A, B, C, D are constants, satisfies the equation $\dfrac{\partial^2 w}{\partial x^2} = \dfrac{\partial^2 w}{\partial t^2}$.

Here
$$\frac{\partial w}{\partial x} = (-A \sin x + B \cos x)(C \cos t + D \sin t) ,$$

$$\frac{\partial^2 w}{\partial x^2} = (-A \cos x - B \sin x)(C \cos t + D \sin t) = - w ,$$

$$\frac{\partial w}{\partial t} = (A \cos x + B \sin x)(-C \sin t + D \cos t) ,$$

$$\frac{\partial^2 w}{\partial t^2} = (A \cos x + B \sin x)(-C \cos t - D \sin t) = - w ,$$

and it is seen that the given equation is satisfied.

There are a number of situations in the biological sciences where a process is governed by a relationship between partial differential coefficients of the type stated in these two examples. Such a relationship is called a **partial differential equation**. Diffusion processes as well as heat conduction are governed by the equation

$$\frac{\partial^2 T}{\partial x^2} = \frac{\partial T}{\partial t} .$$

The equation

$$\frac{\partial^2 w}{\partial x^2} = \frac{\partial^2 w}{\partial t^2}$$

relates to the propagation of disturbances and has applications in the propagation of nerve impulses. A discussion of the methods of solving equations such as these is beyond the scope of this book. For further discussion see Hildebrand (1948).

9.4 MORE THAN TWO VARIABLES

The extension of the results of Sections 9.2 and 9.3 to functions that depend on more than two variables is straightforward. However, a pictorial representation becomes difficult as the number of variables increases. For example if the variable w is a function of three variables x, y and z so that $w = f(x, y, z)$ then $\partial w/\partial x$ is defined by (9.1) and now both y and z remain constant during the calculation. The process of calculating $\partial w/\partial x$ for a particular function is essentially the same as that used in Section 9.2. Second and higher derivatives may also be calculated by using the methods of Section 9.3.

Example: If $w = \dfrac{x}{y} + \dfrac{y}{z} + \dfrac{z}{x}$ show that $x \dfrac{\partial w}{\partial x} + y \dfrac{\partial w}{\partial y} + z \dfrac{\partial w}{\partial z} = 0.$

$$\frac{\partial w}{\partial x} = \frac{1}{y} - \frac{z}{x^2} ,$$

$$\frac{\partial w}{\partial y} = -\frac{x}{y^2} + \frac{1}{z} ,$$

$$\frac{\partial w}{\partial z} = -\frac{y}{z^2} + \frac{1}{x} ,$$

$$x \frac{\partial w}{\partial x} + y \frac{\partial w}{\partial y} + z \frac{\partial w}{\partial z} = x\left(\frac{1}{y} - \frac{z}{x^2}\right) + y\left(-\frac{x}{y^2} + \frac{1}{z}\right) + z\left(-\frac{y}{z^2} + \frac{1}{x}\right)$$

$$= \left(\frac{x}{y} - \frac{z}{x}\right) + \left(-\frac{x}{y} + \frac{y}{z}\right) + \left(-\frac{y}{z} + \frac{z}{x}\right) = 0 .$$

9.5 SUMMARY

The behaviour of functions of several variables has only been touched upon in this chapter. It is possible to consider maxima and minima of such functions and to compute small changes by using methods close to those used in Chapter 6 for a function of a single variable. A necessary condition for $w(x, y)$ to have a maximum or minimum is that both $\partial w/\partial x$ and $\partial w/\partial y$ are zero. There are, in addition, many other applications, particularly through partial differential equations. Further details are to be found in Smyrl (1978), Hildebrand (1948).

PROBLEMS

1. Find $\partial w/\partial x$ and $\partial w/\partial y$ when:

   (i) $w = ax^2 + bxy + cy^2 - 2$,

   (ii) $w = \cos(x + y) + \sin(x - y)$,

   (iii) $w = \sin(x + y) - \cos(x - y)$,

   (iv) $w = x \cos y - y \sin x$,

   (v) $w = \sin^{-1}(x/y)$,

   (vi) $w = (x^2 + y^2)^{\frac{1}{2}} + (x^2 + y^2)^{-\frac{1}{2}}$,

   (vii) $w = \tan^{-1}(y/x)$,

   (viii) $w = 2x^2y^3 - 3x^3y$.

2. Verify that $\partial^2 w/\partial x \partial y = \partial^2 w/\partial y \partial x$ when:

   (i) $w = 2x^2y^3 - 3x^3y$,

   (ii) $w = \cos(x + y) + \sin(x - y)$,

   (iii) $w = \sin^{-1}(x/y)$,

   (iv) $w = \tan^{-1}(y/x)$.

3. If $w = x^2 + 3xy - 6y^2$ show that $x\partial w/\partial x + y\partial w/\partial y = 2w$.

4. If $w = (x^2 + y^2)^{-\frac{1}{2}}$ show that $\dfrac{\partial^2 w}{\partial x^2} + \dfrac{\partial^2 w}{\partial y^2} = w^3$.

5. If $w = x^3 - 3xy^2 + 3x^2y - y^3$ show that $\dfrac{\partial^2 w}{\partial x^2} + \dfrac{\partial^2 w}{\partial y^2} = 0$.

6. Show that $w = \{A \cos nx + B \sin nx\}\ \{C \cos mt + D \sin mt\}$, where n and m are constants, satisfies $m^2 \dfrac{\partial^2 w}{\partial x^2} = n^2 \dfrac{\partial^2 w}{\partial t^2}$.

7. The temperature distribution across a tissue section is given by $T = C(x^4 + 12x^2t + 12t^2)$, where C is a constant. Show that $\dfrac{\partial^2 T}{\partial x^2} = \dfrac{\partial T}{\partial t}$.

8. The temperature distribution across a tissue section is given by $T = C(x^6 + 30x^4t + 180x^2t^2 + 120t^3)$, where C is a constant. Show that $\dfrac{\partial^2 T}{\partial x^2} = \dfrac{\partial T}{\partial t}$.

9. A nerve impulse is represented by $w = A\cos(x - ct)$, where A and c are constants. Show that $\dfrac{\partial^2 w}{\partial x^2} = \dfrac{1}{c^2}\dfrac{\partial^2 w}{\partial t^2}$.

10. In a predator-prey system the size of the prey population oscillates according to $w = A + B\sin(x - ct)$, where A, B and c are constants. Show that $\dfrac{\partial^2 w}{\partial x^2} = \dfrac{1}{c^2}\dfrac{\partial^2 w}{\partial t^2}$.

CHAPTER 10

Integration I.
The indefinite integral

10.1 INTRODUCTION

In Chapter 5, the process of differentiation was developed. Given a function of x, its derivative with respect to x can be found by using established results and rules. In this chapter the process of integration is developed by regarding it as the reverse of differentiation. An alternative view of integration is presented in the next chapter.

Integration can be used to find the relationship between two quantities when the rate of change of one quantity with respect to the other is known. This procedure is called solving a differential equation and is covered more fully in Chapter 15. Another use of integration is in the calculation of areas and volumes. These applications, and others, are covered in the succeeding chapters.

In order to carry out the integration of a general function it is again desirable to establish a table of standard results and rules. The first results and rules of integration will come from a second look at some of the properties of differentiation. Throughout this chapter a lower-case letter will be used to denote the derivative of a function which is itself denoted by the same upper-case letter, thus $f(x) = dF/dx$.

10.2 INTEGRATION

Integration of a function is a process which produces a second function having the property that its derivative is the original function. Thus the integration of $f(x)$ produces a function, $F(x)$ say, sometimes called the **antiderivative function** of $f(x)$, having the property that $dF/dx = f(x)$. Examination of the integration process begins by considering the derivative of some common functions.

The simplest function is the constant function $F(x) \equiv C$, where C is a constant. Differentiating gives

$$f(x) = \frac{dF}{dx} = \frac{dC}{dx} = 0 \ .$$

So the integration of $f(x) \equiv 0$ must produce $F(x) \equiv C$, a constant. However, the function $F_1(x) \equiv C_1$ gives $f_1(x) = dF_1/dx = 0$ as well. Thus integration of 0 (zero) could give rise to any constant. This result, giving as it does a constant which is unknown, may appear to be disadvantageous, but it will turn out to be very important in applications involving integration.

Because integration of zero gives an arbitrary constant, the integration of a general function $f(x)$, with respect to x, results in an expression $F(x) + C$. Here the function $F(x)$ is any function satisfying $dF/dx = f(x)$, and C is a constant called the **arbitrary constant of integration**. The expression $F(x) + C$ is known as the value of the **indefinite integral** of $f(x)$, with respect to x, and includes all possible integrals of $f(x)$. The integration of $f(x)$, the **integrand,** with respect to the **variable of integration** x, is denoted by

$$\int f(x)\,dx = F(x) + C . \tag{10.1}$$

Whenever an integration is performed it is essential that the arbitrary constant be introduced. Notice that as the integration process is reversible, through differentiation, the results can easily be checked, using

$$\frac{d}{dx}(F(x) + C) = \frac{dF}{dx} + \frac{dC}{dx} = f(x) + 0 = f(x) .$$

From the two differentiation rules

$$\frac{d}{dx}[kF(x)] = kf(x), \, k \text{ a constant,}$$

and $$\frac{d}{dx}[F(x) + G(x)] = f(x) + g(x)$$

come the integration rules

$$\int kf(x)\,dx = k\int f(x)\,dx = kF(x) + C$$

and $$\int [f(x) + g(x)]\,dx = \int f(x)\,dx + \int g(x)\,dx = F(x) + G(x) + C.$$

By considering the derivative of the standard functions obtained in Chapter 5 some standard integrals will now be obtained. A list is included as Appendix B.

10.3 STANDARD INTEGRALS

The integral of x^α, where α is a constant not equal to minus one, arises from the differentiation of the function $F(x) = \dfrac{x^{\alpha+1}}{\alpha + 1}$. Differentiating, $f(x) = dF/dx = x^\alpha$, so the indefinite integral, with respect to x, of x^α is from (10.1),

$$\int x^\alpha \, dx = \frac{x^{\alpha+1}}{\alpha + 1} + C \qquad (\alpha \neq -1) .$$

The case when $\alpha = -1$ is not given by this result (since it would lead to a division by zero) and will be examined separately in Chapter 12.

Example 1: Integrate with respect to x:

$$\text{(i) } x^{3/2}, \text{ (ii) } 2x^3 + 3x^{-3/2}, \text{ (iii) } 3x^2 - 5x^{2/3} + x^{-1/2}, \text{ (iv) } \left(x + \frac{1}{x}\right)^2 .$$

(i) $\int x^{3/2} dx = \dfrac{x^{5/2}}{5/2} + C = \dfrac{2}{5} x^{5/2} + C$.

(ii) $\int (2x^3 + 3x^{-3/2}) dx = 2\int x^3 dx + 3\int x^{-3/2} dx$

$$= 2\frac{x^4}{4} + 3\frac{x^{-1/2}}{(-1/2)} + C = \frac{x^4}{2} - 6x^{-1/2} + C .$$

(iii) $\int (3x^2 - 5x^{2/3} + x^{-1/2}) dx = 3\int x^2 dx - 5\int x^{2/3} dx + \int x^{-1/2} dx$

$$= 3\frac{x^3}{3} - 5\frac{x^{5/3}}{5/3} + \frac{x^{1/2}}{1/2} + C$$

$$= x^3 - 3x^{5/3} + 2x^{1/2} + C .$$

(iv) $\displaystyle\int \left(x + \frac{1}{x}\right)^2 dx = \int \left(x^2 + 2 + \frac{1}{x^2}\right) dx = \int (x^2 + 2x^0 + x^{-2}) dx$

$$= \frac{x^3}{3} + 2x - x^{-1} + C .$$

The integrals of $\cos(ax)$ and $\sin(ax)$ where a is a non-zero constant can be easily found. Taking $F(x) = \dfrac{1}{a} \sin ax, f(x) = dF/dx = \cos ax$, so

$$\int \cos ax \, dx = \frac{1}{a} \sin ax + C .$$

Similarly, if $F(x) = -\dfrac{1}{a} \cos ax$ then $f(x) = dF/dx = \sin ax$, so

$$\int \sin ax \, dx = -\frac{1}{a} \cos ax + C .$$

Example 2: Find $\int (2 \cos 4x + 3 \sin 3x) dx$.

$$\int (2 \cos 4x + 3 \sin 3x) dx = 2\int \cos 4x \, dx + 3\int \sin 3x \, dx$$

$$= 2 \left(\frac{1}{4} \sin 4x\right) + 3 \left(-\frac{1}{3} \cos 3x\right) + C$$

$$= \tfrac{1}{2} \sin 4x - \cos 3x + C .$$

Sometimes an integrand involving trigonometric functions needs to be rearranged, using the trigonometric identities given in Appendix A, before it can be integrated, as in the following example.

Example 3: Integrate (i) $\sin^2 x$ and (ii) $\cos x \cos 2x$.

(i) As $\sin^2 x = \frac{1}{2}(1 - \cos 2x)$, using A16,

$$\int \sin^2 x \, dx = \frac{1}{2}\int(1 - \cos 2x)dx = \frac{1}{2}(x - \frac{1}{2}\sin 2x) + C \ .$$

(ii) $\int \cos x \cos 2x \, dx = \frac{1}{2}\int(\cos 3x + \cos x)dx$, using A29 with $A = 1, B = 2$

$$= \frac{1}{2}\left(\frac{1}{3}\sin 3x + \sin x\right) + C \ .$$

Other integrals involving trigonometric functions that can be obtained easily from the results in Chapter 5 are:

$$\int \sec^2 x \, dx = \tan x + C \qquad\qquad \text{taking } F(x) = \tan x \ .$$
$$\int \csc^2 x \, dx = -\cot x + C \qquad\qquad \text{taking } F(x) = -\cot x \ .$$
$$\int \sec x \tan x \, dx = \sec x + C \qquad\qquad \text{taking } F(x) = \sec x \ .$$
$$\int \csc x \cot x \, dx = -\csc x + C \qquad\qquad \text{taking } F(x) = -\csc x \ .$$

Example 4: Integrate (i) $\sec^2 x + \csc^2 x$ and (ii) $\tan^2 x$.

(i) $\int(\sec^2 x + \csc^2 x)dx = \tan x - \cot x + C \ .$

(ii) $\int \tan^2 x \, dx = \int(\sec^2 x - 1)dx \qquad\qquad \text{using A19}$
$$= \tan x - x + C \ .$$

The final two standard integrals obtainable by direct reversal of a standard differentiation result involve the inverse trigonometric functions $\sin^{-1}x$ and $\tan^{-1}x$.

First let $y = F(x) = \sin^{-1}\dfrac{x}{a}$ where $|x| \leqslant a$. To differentiate y the chain rule is used with $t = \dfrac{x}{a}$. Thus $y = \sin^{-1}t$ and

$$f(x) = \frac{dF}{dx} = \frac{dy}{dx} = \frac{dy}{dt}\frac{dt}{dx} = \frac{1}{\sqrt{(1-t^2)}} \cdot \frac{1}{a} = \frac{1}{\sqrt{(1-(x/a)^2)}} \cdot \frac{1}{a}$$

$$= \frac{1}{\sqrt{(a^2-x^2)}} \ .$$

Hence $\displaystyle\int \frac{1}{\sqrt{(a^2-x^2)}} \, dx = \sin^{-1}\frac{x}{a} + C \ .$

Note that a is taken to be positive.

Example 5: Integrate $(4 - x^2)^{-\frac{1}{2}}$ with respect to x.

Here $a = 2$, so that

$$\int \frac{1}{\sqrt{(4 - x^2)}}\, dx = \sin^{-1}\frac{x}{2} + C .$$

Next let $y = F(x) = \dfrac{1}{a}\tan^{-1}\dfrac{x}{a}$. Again let $t = \dfrac{x}{a}$ so that $y = \dfrac{1}{a}\tan^{-1}t$ and

$$f(x) = \frac{dF}{dx} = \frac{dy}{dx} = \frac{dy}{dt}\frac{dt}{dx} = \frac{1}{a}\cdot\frac{1}{1 + t^2}\cdot\frac{1}{a} = \frac{1}{a\left(1 + \left(\dfrac{x}{a}\right)^2\right)a}$$

$$= \frac{1}{a^2 + x^2} .$$

Hence $\quad\displaystyle\int \frac{dx}{a^2 + x^2} = \frac{1}{a}\tan^{-1}\frac{x}{a} + C .$ (10.2)

Example 6: Find $\displaystyle\int \frac{dx}{9 + 4x^2}$.

This is not in the form required by (10.2). However,

$$9 + 4x^2 = 4\left(\frac{9}{4} + x^2\right) = 4\left(\left(\frac{3}{2}\right)^2 + x^2\right), \text{ so}$$

$$\int \frac{dx}{9 + 4x^2} = \frac{1}{4}\int \frac{dx}{(\frac{3}{2})^2 + x^2} = \frac{1}{4}\cdot\frac{2}{3}\tan^{-1}\frac{2x}{3} + C = \frac{1}{6}\tan^{-1}\frac{2x}{3} + C .$$

10.4 INTEGRATING A RATE OF CHANGE

Problems arise in the biological sciences which require the relationship between two variables to be found when information on the rate of change of one variable with respect to the other is given. If the two variables are denoted by y and x then the mathematical problem is to find y, given dy/dx, as a function of x. The solution to this kind of problem is obtained by integration.

If $dy/dx = f(x)$ then $y = F(x) + C$ where

$$F(x) + C = \int f(x)dx = \int \frac{dy}{dx}\, dx = y(x) .$$ (10.3)

Notice that y is not given uniquely as a function of x by this procedure. In order that the value of the arbitrary constant of integration C may be determined more information is required. Usually this additional information specifies the value taken by y at a particular value of x.

Formulations of practical problems frequently result in equations containing derivatives. Such an equation is known as a **differential equation** and requires the use of integration to find its solution. Further types of differential equation are covered in Chapter 15.

Example: A species of migratory bird is observed to gather at its summer territory at a rate given approximately by $12t - 3t^2$ where t is measured in months from the beginning of April. If the first bird arrives during April find an expression for the subsequent size of the mature bird population. Assume no mature birds die. When is the number of mature birds greatest? When have all the mature birds departed? Note that birds born during the summer are not counted.

Let $p(t)$ denote the size of the mature bird population, then

$$\frac{dp}{dt} = 12t - 3t^2 . \tag{10.4}$$

Hence

$$p(t) = \int \frac{dp}{dt}\, dt = \int (12t - 3t^2)\, dt = 6t^2 - t^3 + C .$$

When $t = 0$ (the beginning of April) no birds have arrived, so $p(0) = 0$. Therefore $C = 0$, and $p(t) = 6t^2 - t^3$ will give the number of mature birds in the colony t months from the beginning of April (see Fig. 10.1).

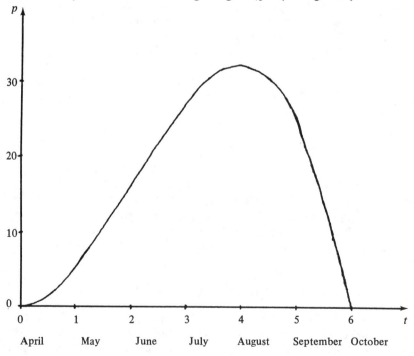

Fig. 10.1

The greatest number of mature birds occurs when $dp/dt = 0$. From (10.4) this gives $3t(4 - t) = 0$, so $t = 0$ or $t = 4$. At $t = 0$ there are no birds, so this is a minimum. When $t = 4$, $d^2p/dt^2 = 12 - 6t = -12$, so the population is greatest at the beginning of August and numbers 32 birds.

All the birds are away when $p(t) = t^2(6 - t) = 0$; that is, when $t = 0$ or $t = 6$. Departures will only occur after $t = 0$, so the birds will have migrated by the beginning of October ($t = 6$). Note that the function $p(t)$ is not applicable for t outside the interval from 0 to 6.

This example illustrates how the value to be assigned to the arbitrary constant of integration may, in a given situation, be determined. In general an arbitrary constant C must be introduced whenever $dy/dx = f(x)$ is integrated to give $y = F(x) + C$. The significance of C will now be examined.

The equation $y = F(x)$ represents some curve in the xy-plane, (see Fig. 10.2). $y = F(x) + C$ represents a similarly shaped curve, but for each value of x the y value is changed by the same constant amount C, each value of C giving a different curve. If $y = k$ when $x = 0$ then the curve $y = F(x) - k$ is the only curve in the family of curves $y = F(x) + C$ which passes through the origin.

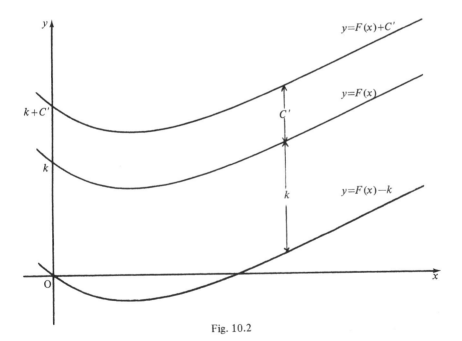

Fig. 10.2

In a given physical problem only one particular curve from the family of curves is required. Thus, in the example, the population size is given by $p(t) = 6t^2 - t^3 + C$ in general. If, when $t = 0$, one bird has just arrived, $p(0) = 1$, so the constant $C = 1$. Whereas, if the number of birds is not counted until the beginning of May, when $t = 1$, and is then found to number eight, $p(1) = 5 + C = 8$, so $C = 3$.

10.5 CHANGE OF VARIABLE

The chain rule, given in Chapter 5, enables the differentiation of certain complicated functions to be carried out easily. This was achieved by introducing a new variable chosen to bring about a simplification in the function to be differentiated. In this section it is shown how it is possible to introduce a new variable into an integral, replacing the original variable of integration. By a suitable choice of this new variable it may be possible to simplify the expression being integrated, so enabling the original integral to be found.

Consider $y(x) = \int f(x)dx$. Introducing a new variable t related to x (that is x is some function of t, or t is some function of x) the integrand $f(x)$ becomes the function $g(t)$, say, in the variable t. So if $x = x(t)$ then $f(x) = f(x(t)) = g(t)$. By the chain rule of differentiation

$$f(x) = \frac{dy}{dx} = \frac{dy}{dt}\frac{dt}{dx} = g(t)$$

so

$$\frac{dy}{dt} = g(t) \left/ \frac{dt}{dx} \right. = g(t)\frac{dx}{dt} .$$

Integrating this with respect to t gives

$$y(x(t)) = \int g(t)\frac{dx(t)}{dt}\,dt = H(t) + C \text{ say} .$$

But $y(x) = \int f(x)dx$, therefore,

$$\int f(x)dx = \int f(x(t))\frac{dx(t)}{dt}\,dt .\qquad(10.5)$$

In (10.5) the integral on the left is in terms of the variable x while that on the right is in terms of the variable t.

In summary, to change the variable of integration x in $\int f(x)dx$ to another variable, t, related by $x = x(t)$, replace $f(x)$ by the equivalent function $g(t) = f(x(t))$ and replace dx by $\frac{dx(t)}{dt}\,dt$ to obtain the integral $\int g(t)\frac{dx(t)}{dt}\,dt$.

When the transformed integral $\int g(t)\frac{dx}{dt}\,dt$ has been evaluated to give a function of t, substitute for t, using the relationship between x and t, in order to return to the original variable x.

To use the result (10.5) to evaluate a complicated integral a function of t must be chosen so that the new integrand $f(x(t)) \dfrac{dx(t)}{dt}$ can be integrated easily. Note that there are often several choices for the function of t that will bring about a simplification. The choice of function is most clearly illustrated through examples.

Example 1: Evaluate $\int(2x + 1)^2 dx$.

Here the integrand can be rewritten so that

$$\int (2x + 1)^2 dx = \int [4x^2 + 4x + 1] dx = \frac{4x^3}{3} + 2x^2 + x + C .$$

However, if the expression $2x + 1$ were raised to a high power, or to a fractional power, it would not be practical to expand the integrand before performing the integration. For an alternative approach replace $2x + 1$ by t. Then

$$f(x) = (2x + 1)^2 = t^2 = g(t) .$$

Here $t = 2x + 1$ or $x = \frac{1}{2}(t - 1)$ so $\dfrac{dx}{dt} = \frac{1}{2}$, and

$$\int (2x + 1)^2 dx = \int t^2 \cdot \frac{1}{2} \, dt = \frac{1}{2} \cdot \frac{t^3}{3} + K = \frac{(2x + 1)^3}{6} + K$$

where K is the constant of integration. Hence

$$\int (2x + 1)^2 dx = \frac{4}{3} x^3 + 2x^2 + x + \left(\frac{1}{6} + K \right)$$

on expanding $\dfrac{(2x + 1)^3}{6}$.

The two approaches give the same results for the dependence on x, and differ only in the form of the constant term, which is arbitrary in any event.

The integral in example 1 is a particular case of the frequently occurring integral

$$\int f(ax + b) dx$$

where a and b are constants with a non-zero. To evaluate such integrals set

$$t = ax + b, \text{ or } x = \frac{t - b}{a}, \text{ so that } \frac{dx}{dt} = \frac{1}{a}.$$

Then $$\int f(ax + b) dx = \int f(t) \frac{1}{a} \, dt = \frac{1}{a} F(t) + C = \frac{1}{a} F(ax + b) + C .$$

Example 2: Integrate (i) $(3x + 2)^{3/2}$, (ii) $\cos(3x + 2)$ and (iii) $\operatorname{cosec}^2 3x$.

(i) $\int (3x + 2)^{3/2} dx = \dfrac{1}{3} \int t^{3/2} dt \quad$ where $t = 3x + 2$

$$= \frac{1}{3} \cdot \frac{t^{5/2}}{5/2} + C = \frac{2}{15} (3x + 2)^{5/2} + C \;.$$

(ii) $\int \cos(3x + 2) dx = \dfrac{1}{3} \int \cos t \, dt \quad$ where $t = 3x + 2$

$$= \frac{1}{3} \sin t + C = \frac{1}{3} \sin(3x + 2) + C \;.$$

(iii) $\int \operatorname{cosec}^2 3x \, dx = \dfrac{1}{3} \int \operatorname{cosec}^2 t \, dt \quad$ where $t = 3x$

$$= \frac{1}{3} (-\cot t) + C = -\frac{1}{3} \cot 3x + C \;.$$

Example 3: Evaluate $\displaystyle\int \frac{x}{(1 - x)^{\frac{1}{2}}} \, dx$.

The denominator is simplified by setting $t = 1 - x$, then $x = 1 - t$ so $\dfrac{dx}{dt} = -1$ and

$$\int \frac{x}{(1 - x)^{\frac{1}{2}}} \, dx = \int \frac{1 - t}{t^{\frac{1}{2}}} (-1) dt = \int (t^{\frac{1}{2}} - t^{-\frac{1}{2}}) dt$$

$$= \frac{2}{3} t^{3/2} - 2t^{1/2} + C$$

$$= \frac{2}{3} (1 - x)^{3/2} - 2(1 - x)^{1/2} + C \;.$$

In applying the rule for a change of variable to some integration problems the quantity dx/dt and possibly $f(x)$ may, in their simplest form, still contain the variable x explicitly. This often arises when it is more convenient to express t as a function of x, that is $t = t(x)$. In such a case the inverse rule of differentiation is used to obtain dx/dt $\left(\text{since } \dfrac{dx}{dt} = 1 \left/ \dfrac{dt}{dx} \right. \right)$ and results in dx/dt being a function of x. Instead of expressing dx/dt and $f(x)$ in terms of t separately it is advisable to form the product $f(x) \, dx/dt$ first, for often terms in x will cancel.

Example 4: Evaluate (i) $\int x(3x^2 - 1)^{1/4} \, dx$ and (ii) $\int \sin^2 x \cos x \, dx$.

(i) To find $\int x(3x^2 - 1)^{1/4} \, dx$ the replacement of $3x^2 - 1$ by t brings about a simplification in the integrand. Taking $t = 3x^2 - 1$ gives

$$\frac{dx}{dt} = 1 \bigg/ \frac{dt}{dx} = \frac{1}{6x} \quad \text{and} \quad f(x)\frac{dx}{dt} = x(3x^2 - 1)^{1/4}\frac{1}{6x} = \frac{t^{1/4}}{6}. \text{ So}$$

$$\int x(3x^2 - 1)^{1/4} dx = \frac{1}{6}\int t^{1/4} \, dt = \frac{1}{6}\cdot\frac{t^{5/4}}{5/4} + C = \frac{2}{15}(3x^2 - 1)^{5/4} + C \ .$$

(ii) $\int \sin^2 x \cos x \, dx$. In this integral substitute $t = \sin x$.

Then
$$\frac{dx}{dt} = 1 \bigg/ \frac{dt}{dx} = \frac{1}{\cos x}$$

and
$$f(x)\frac{dx}{dt} = \sin^2 x \cos x \frac{1}{\cos x} = \sin^2 x = t^2$$

so
$$\int \sin^2 x \cos x \, dx = \int t^2 \, dt = \frac{t^3}{3} + C = \frac{1}{3}\sin^3 x + C \ .$$

For integrals of the form $\int \sin^\alpha x \cos x \, dx$, where α is a non-zero constant, the substitution $t = \sin x$ will, as in the previous example, simplify the integration. To evaluate the similar integral, $\int \cos^\alpha x \sin x \, dx$, the substitution $t = \cos x$ can be used.

Example 5: Find $\int \dfrac{\sin x}{\cos^3 x} \, dx = \int(\cos x)^{-3} \sin x \, dx.$

Here $\alpha = -3$, the power of $\cos x$, so let $t = \cos x$, then $\dfrac{dx}{dt} = \dfrac{1}{-\sin x}$ and

$$f(x)\frac{dx}{dt} = (\cos x)^{-3} \sin x \left(\frac{-1}{\sin x}\right) = -t^{-3}.$$

So
$$\int\frac{\sin x}{\cos^3 x} \, dx = -\int t^{-3} dt = \tfrac{1}{2}t^{-2} + C = \tfrac{1}{2}(\cos x)^{-2} + C$$

$$= \tfrac{1}{2}\sec^2 x + C \ .$$

Example 6: Evaluate $\int \cos^3 x \, dx$.

This integral is not of the type just considered but, since $\cos^2 x = 1 - \sin^2 x$, it may be rewritten.

$$\int\cos^3 x \, dx = \int\cos x \cdot \cos^2 x \, dx = \int\cos x (1 - \sin^2 x)dx$$

$$= \int\cos x \, dx - \int\sin^2 x \cos x \, dx$$

$$= \sin x - \frac{1}{3}\sin^3 x + C \ ,$$

using the result of example 4(ii) to evaluate the second integral.

Example 7: Evaluate $\int\sqrt{(1-x^2)}\,dx$.

The substitution used to evaluate this integral is not immediately obvious. However, a hint comes from the standard integral

$$\int \frac{1}{\sqrt{(1-x^2)}}\,dx = \sin^{-1}x + C.$$

Let $x = \sin t$, then $dx/dt = \cos t$, so

$$\begin{aligned}
\int\sqrt{(1-x^2)}dx &= \int\sqrt{(1-\sin^2 t)}\,\cos t\,dt = \int\cos^2 t\,dt \\
&= \tfrac{1}{2}\int(1+\cos 2t)dt = \tfrac{1}{2}(t+\tfrac{1}{2}\sin 2t) + C \\
&= \tfrac{1}{2}t+\tfrac{1}{2}\sin t\cos t + C = \tfrac{1}{2}\sin^{-1}x + \tfrac{1}{2}x\sqrt{(1-x^2)} + C.
\end{aligned}$$

10.6 INTEGRATION BY PARTS

The product rule of differentiation forms the basis of the technique for integrating a product known as **integration by parts**. Integrating both sides of the identity

$$\frac{d}{dx}(uv) = u\frac{dv}{dx} + v\frac{du}{dx}$$

with respect to x gives

$$uv = \int\frac{d}{dx}(uv)\,dx = \int\left(u\frac{dv}{dx}+v\frac{du}{dx}\right)dx = \int u\frac{dv}{dx}\,dx + \int v\frac{du}{dx}\,dx$$

or
$$\int u\frac{dv}{dx}\,dx = uv - \int v\frac{du}{dx}\,dx . \tag{10.6}$$

The name integration by parts arises because only part of the original integrand $u\dfrac{dv}{dx}$ is integrated. No arbitrary constant is included when dv/dx is integrated to obtain v as one introduced at this stage would always combine with the arbitrary constant which must be introduced when the final integration is completed.

The integrand, which is a product, should first be examined for a possible change of the variable, and failing that integration by parts can be tried. To use this technique it is necessary to decide which part of the integrand will be differentiated, that is called u, and which part will be integrated, that is denoted dv/dx. In choosing the functions denoted by u and dv/dx certain points need to be borne in mind. The main requirement is that dv/dx can be easily integrated to give v. Secondly, it is desirable to obtain a simple integral on the right-hand side of (10.6). The following examples illustrate the use of this procedure.

Example 1: Evaluate $\int x \cos x \, dx$.

In this example both x and $\cos x$ can be differentiated and integrated. Taking $u = x$ and $dv/dx = \cos x$, so that $v = \sin x$,

$$\int x \cos x \, dx = x \sin x - \int \sin x \cdot 1 \cdot dx = x \sin x + \cos x + C.$$

If instead we had taken $u = \cos x$ and $dv/dx = x$, then $v = x^2/2$ and

$$\int x \cos x \, dx = \cos x \cdot \frac{x^2}{2} - \int \frac{x^2}{2}(-\sin x)dx = \frac{x^2}{2}\cos x + \int \frac{x^2}{2}\sin x \, dx.$$

The new integral is harder than the original! Notice that by differentiating x it is eliminated from the integrand, thus the first choice for u is better.

Example 2: Integrate $x^2 \sin x$ with respect to x.

In $\int x^2 \sin x \, dx$ the power of x will be reduced by differentiation, so take $u = x^2$ and $dv/dx = \sin x$. Then $du/dx = 2x$ and $v = -\cos x$ so that

$$\int x^2 \sin x \, dx = x^2(-\cos x) - \int(-\cos x)2x \, dx = -x^2 \cos x + 2\int x \cos x \, dx$$
$$= -x^2 \cos x + 2(x \sin x + \cos x) + C,$$

using example 1.

Even when there is just one function present in the integrand, integration by parts can still be used, if the function is differentiable, by taking $dv/dx = 1$.

Example 3: Find $\int \sin^{-1}x \, dx$.

Write the integral as $\int \sin^{-1}x \cdot 1 \, dx$ and take $u = \sin^{-1}x$, $dv/dx = 1$ so that

$$\frac{du}{dx} = \frac{1}{\sqrt{(1-x^2)}}, v = x. \text{ Then}$$

$$\int \sin^{-1}x \, dx = \sin^{-1}x \cdot x - \int x \frac{1}{\sqrt{(1-x^2)}} \, dx = \sin^{-1}x - \int \frac{x}{\sqrt{(1-x^2)}} \, dx.$$

The integral $\int \dfrac{x \, dx}{\sqrt{(1-x^2)}}$ can be evaluated by using the substitution $t = 1-x^2$.

$$\frac{dx}{dt} = 1 \bigg/ \frac{dt}{dx} = \frac{1}{-2x}$$

and $f(x)\dfrac{dx}{dt} = \dfrac{x}{\sqrt{(1-x^2)}} \cdot \dfrac{-1}{2x} = \dfrac{-t^{-\frac{1}{2}}}{2}$

so $\int \dfrac{x}{\sqrt{(1-x^2)}} \, dx = -\tfrac{1}{2}\int t^{-\frac{1}{2}}dt = -t^{\frac{1}{2}} + C = -\sqrt{(1-x^2)} + C.$

Therefore, $\int \sin^{-1}x \, dx = x \sin^{-1}x + \sqrt{(1-x^2)} + C.$

Example 4: Show, by integrating by parts, that, for any positive integer n greater than two,

$$\int \sin^n x \, dx = \frac{-\sin^{n-1}x \cos x}{n} + \frac{n-1}{n} \int \sin^{n-2}x \, dx \ .$$

Hence evaluate $\int \sin^6 x \, dx$.

From the form of the answer it can be seen that the index n has to be reduced. This can be achieved through differentiation when the integration by parts procedure is carried out.

The decomposition of the integrand as $u = \sin^n x$, $\dfrac{dv}{dx} = 1$ leads to

$$\frac{du}{dx} = n \sin^{n-1}x \cos x \text{ and } v = x$$

so that $\quad \int \sin^n x \, dx = \sin^n x . x - \int n \sin^{n-1}x \cos x . x \, dx \ .$

The expression on the right now contains an x which is not present in the required answer, and so this choice of u and dv/dx is not suitable.

The introduction of x can be avoided by choosing a different decomposition of the integrand. As $n > 2$, take $u = \sin^{n-1}x$ and $\dfrac{dv}{dx} = \sin x$

so that $\quad \dfrac{du}{dx} = (n-1)\sin^{n-2}x \cos x \text{ and } v = -\cos x \ .$

Then $\quad \int \sin^n x \, dx = -\sin^{n-1}x \cos x - \int (n-1) \sin^{n-2}x \cos x \, (-\cos x) \, dx$

$$= -\sin^{n-1}x \cos x + (n-1) \int \sin^{n-2}x \cos^2 x \, dx$$

$$= -\sin^{n-1}x \cos x + (n-1) \int \sin^{n-2}x \, (1 - \sin^2 x) \, dx$$

$$= -\sin^{n-1}x \cos x + (n-1) \int \sin^{n-2}x \, dx$$

$$- (n-1) \int \sin^n x \, dx.$$

Therefore

$$[1 + (n-1)] \int \sin^n x \, dx = -\sin^{n-1}x \cos x + (n-1) \int \sin^{n-2}x \, dx$$

so that $\quad \int \sin^n x \, dx = \dfrac{-\sin^{n-1}x \cos x}{n} + \dfrac{n-1}{n} \int \sin^{n-2}x \, dx \ .$

This expression for $\int \sin^n x \, dx$, involving as it does $\int \sin^{n-2}x \, dx$, is an example of a **reduction formula**. A reduction formula is a relationship between an integral containing a parameter (here n) and a similar integral having the value of the parameter reduced. By successive applications of the reduction formula the original integral can be reduced to an expression containing an integral that can be easily found.

When $n = 6$ the reduction formula gives

$$\int \sin^6 x \, dx = \frac{-\sin^5 x \cos x}{6} + \frac{5}{6} \int \sin^4 x \, dx \ .$$

For $n = 4$,

$$\int \sin^4 x \, dx = \frac{-\sin^3 x \cos x}{4} + \frac{3}{4} \int \sin^2 x \, dx \ .$$

From example 3(i) of Section 10.3,

$$\int \sin^2 x \, dx = \frac{1}{2} x - \frac{1}{4} \sin 2x + C \ .$$

Thus
$$\int \sin^4 x \, dx = \frac{-\sin^3 x \cos x}{4} + \frac{3}{8} x - \frac{3}{16} \sin 2x + C_1, \qquad \left(C_1 = \frac{3}{4} C \right)$$

and
$$\int \sin^6 x \, dx = -\frac{1}{6} \sin^5 x \cos x - \frac{5}{24} \sin^3 x \cos x + \frac{5}{16} x$$
$$- \frac{5}{32} \sin 2x + C_2 \ .$$

Here $C_2 = \frac{5}{6} C_1$, and is chosen as a more convenient arbitrary constant.

One further integration technique, employing partial fractions, is post-poned until Chapter 13. It is applicable when the integrand is a rational poly-nomial.

PROBLEMS

1. Integrate the following expressions with respect to x:

    (a) $x^{2/5}$,

    (b) $(x^3 - 1)/x^2$,

    (c) $\left(x - \dfrac{1}{x} \right)^2$,

    (d) $2x - 3x^{1/2} - x^{-4/3}$,

    (e) $(1 + x^2)/\sqrt{x}$,

    (f) $x^2(1 - \sqrt{x})^2$.

2. Integrate the following trigonometric expressions with respect to x:

    (a) $2 \sin 3x + 3 \cos 4x$,

    (b) $3 \sin 2x - 4 \cos 3x$,

    (c) $(1 + \sin x)/\cos^2 x$,

    (d) $(1 + \cos x)/\sin^2 x$.

3. Rewrite, using trigonometric identities, and then integrate with respect to x, the following expressions:

(a) $\cos^2 3x$,

(b) $\sin x \cos x$,

(c) $\sin x \sin 2x$,

(d) $\sin x \cos 2x$,

(e) $\cot^2 x$,

(f) $\sin^4 x$.

4. Integrate the following expressions with respect to x:

(a) $\dfrac{1}{\sqrt{(9 - x^2)}}$,

(b) $\dfrac{1}{\sqrt{(1 - 9x^2)}}$,

(c) $(4 - 9x^2)^{-\frac{1}{2}}$,

(d) $(4 + 9x^2)^{-1}$,

(e) $\dfrac{x^2}{x^2 + 4}$,

(f) $\dfrac{x^2}{9x^2 + 1}$.

5. An insect population, initially numbering 100, increases to a population of size $p(t)$ after a time t (measured in days). If the growth rate at time t is given by $2t + 3t^2$ determine the size of the insect population after (a) 1 day, (b) 10 days.

6. Over twelve hours of daylight the net rate of production of oxygen during photosynthesis in a pond, is given by $\dfrac{1664 - 144t - t^3}{52(144 + t^2)}$ where t is measured in hours from daybreak. During the night oxygen is consumed at a constant rate. Show that the maximum concentration of oxygen in the pond occurs eight hours after daybreak. If the minimum oxygen concentration in the pond is 1, in some units, obtain expressions giving the oxygen concentration throughout a 24-hour period.

7. After hatching from their eggs the speed at which miracidia move radially outwards in search of a snail host is given by

$$\frac{dr}{dt} = \frac{32 - 2t}{5} \text{ cm/hr}, \quad 0 \leqslant t \leqslant 16 \text{ hrs.}$$

If the miracidia only live for sixteen hours what is the maximum area of discovery they will cover?

8. Integrate, by changing the variable, the following expressions:

(a) $(3x + 5)^{\frac{1}{2}}$,

(b) $(2 - 3x)^{-3}$,

(c) $(2x - 3)^{49}$,

(d) $\cos\left(2x + \dfrac{\pi}{3}\right)$,

(e) $\sin\left(3x - \dfrac{\pi}{4}\right)$,

(f) $\sec 2x \tan 2x$.

9. By making a suitable change of variable, integrate the following expressions:

   (a) $x/(x + 2)^3$,   (f) $\sin^3 x \cos x$,

   (b) $x(x - 2)^{15}$,   (g) $\cos^3 x \sin x$,

   (c) $x\sqrt{(2x + 1)}$,   (h) $\sin^3 x$,

   (d) $x(1 - x^2)^{\frac{1}{2}}$,   (i) $\sec^2 x \sqrt{(1 + \tan x)}$.

   (e) $x(1 - x^2)^{-\frac{1}{2}}$,

10. Integrate the following expressions by using the given substitution:

   (a) $x^3(x^4 + 1)^{1/3}$, $t = x^4 + 1$;

   (b) $x^3(x^2 + 2)^{3/2}$, $t = x^2 + 2$;

   (c) $\dfrac{x^2}{\sqrt{(1 - x^2)}}$, $x = \sin t$;

   (d) $\dfrac{1}{x\sqrt{(x^2 - 1)}}$, $t = \dfrac{1}{x}$;

   (e) $\sqrt{(1 - \cos x)}$, $x = 2t$.

11. The rate of change of the temperature inside a cold frame is given by

$$\frac{dT}{dt} = \frac{\pi}{2} \sin\left(\frac{\pi t}{12} - \frac{\pi}{4}\right)$$

   where T is the internal temperature (Centigrade) and t is the time in hours measured from midnight. Find an expression giving the internal temperature during the day if the minimum temperature inside the cold frame is 4°C.

12. Use integration by parts to integrate, with respect to x, the following expressions:

   (a) $x(x - 1)^{5/2}$,   (d) $x^2 \cos x$,

   (b) $x \sin x$,   (e) $x \tan^{-1} x$,

   (c) $x \cos 2x$,   (f) $x \sec^2 x \tan x$.

13. Use integration by parts to verify the following reduction formulae (n is an integer greater than one). Hence find the given integral.

   (a) $\displaystyle\int (1 + x^2)^{-n} dx = \frac{1}{(2n - 2)} \cdot \frac{x}{(1 + x^2)^{n-1}} + \frac{2n - 3}{2n - 2} \int (1 + x^2)^{1-n} dx$;

   $\int (1 + x^2)^{-2} dx$,

(b) $\int x^n \sin x \, dx = -x^n \cos x + nx^{n-1} \sin x - n(n-1) \int x^{n-2} \sin x \, dx;$

$\int x^4 \sin x \, dx,$

(c) $\int \sec^n x \, dx = \dfrac{\tan x \sec^{n-2}x}{n-1} + \dfrac{n-2}{n-1} \int \sec^{n-2} x \, dx;$

$\int \sec^4 x \, dx.$

14.   The rate at which weight is lost by an animal suffering from a viral infection is observed to satisfy the relationship $\dfrac{dL}{dt} = \dfrac{t}{72}(8-t)^{1/3}$ for $0 \leqslant t \leqslant 8$.

Here L is the weight loss, as a percentage of the animal's original weight, t weeks from the time of infection. Find the function giving the percentage weight loss L at time t, and determine the percentage weight loss when the animal dies eight weeks later.

15.   The rate at which air is drawn into the lungs, during a respiratory cycle, may be approximated by the equation

$$\frac{dV}{dt} = t(2t-5)(t-5)/20 \text{ litres/sec}, \quad 0 \leqslant t \leqslant 5 \text{ secs.}$$

If, at the start of the respiratory cycle $(t = 0)$ the volume of air in the lungs is 0.1 litres, show that the maximum volume of air contained in the lungs is, approximately, 1.077 litres.

CHAPTER 11

Integration II.
The definite integral

11.1 INTRODUCTION

The chapter starts with the determination of the area under a curve in terms of an integral. This leads to the definition of a definite integral followed by its properties. The effect of changing the variable of integration in a definite integral is also discussed. Finally the representation of a definite integral by an infinite sum is derived; this equivalence is used in many applications involving integration, see Chapter 14, and in the numerical approximation of an integral, see Chapter 17.

11.2 AREA UNDER A CURVE

The integral calculus can be used to obtain an expression for the area under a curve. Consider the region between the x-axis and a continuous smooth curve lying above the x-axis given by the equation $y = f(x)$, see Fig. 11.1. On the curve take two adjacent points P and R, with coordinates (x,y) and $(x + \Delta x, y + \Delta y)$ respectively, and draw lines parallel to the axes through these two points. Denote the area bounded by TU, UR, the curve PR, and TP by ΔA. (This quantity is the increase in the area A under the curve due to an increase of Δx in the base length). If the curve has positive slope between P and R

$$\text{area TUQP} < \Delta A < \text{area TURS} \ .$$

That is $\quad y \, \Delta x < \Delta A < (y + \Delta y)\Delta x$

so $\qquad y < \dfrac{\Delta A}{\Delta x} < y + \Delta y \ .$ $\hfill (11.1)$

In the limit as $\Delta x \to 0$, $\Delta y \to 0$ (since the curve is continuous and smooth) and $\dfrac{\Delta A}{\Delta x} \to \dfrac{\mathrm{d}A}{\mathrm{d}x}$, so (11.1) gives

$$\mathrm{d}A/\mathrm{d}x = y \ .$$

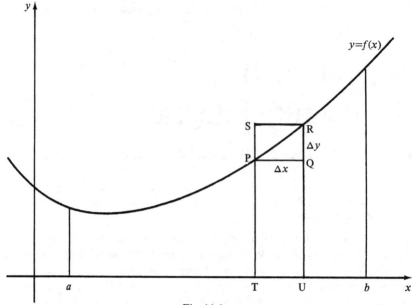

Fig. 11.1

This relationship can be shown to hold whatever the slope of the curve between P and R. Since $dA/dx = y$,

$$A(x) = \int \frac{dA}{dx}\, dx = \int y\, dx = \int f(x)\, dx = F(x) + C$$

where $dF/dx = f(x)$.

If the area under the curve is measured from the line $x = a$ then the area will be zero when $x = a$. Hence

$$A(a) = F(a) + C = 0, \text{ so } C = -F(a) \text{ and}$$

$$A(x) = \int f(x)\, dx = F(x) - F(a) .$$

When $x = b$ $(b > a)$, $A(b) = F(b) - F(a)$ will be the total area bounded by the curve $y = f(x)$ $(f(x) \geqslant 0$ for $a \leqslant x \leqslant b)$, the x-axis and the lines $x = a$ and $x = b$.

It is conventional to write

$$F(b) - F(a) = \left[F(x) \right]_a^b = \int_{x=a}^{x=b} f(x)\, dx = \int_a^b f(x)\, dx$$

where $dF/dx = f(x)$ and $\displaystyle\int_a^b f(x)\, dx$ denotes the **definite integral,** from a to b, of

$f(x)$. The quantities a and b are called the **limits of integration** and prescribe the range of the variable of integration x. Thus

$$\int_a^b f(x)\,dx = \left[F(x)\right]_a^b = F(b) - F(a) \tag{11.2}$$

gives the area bounded by the curve $y = f(x)$ ($f(x) \geqslant 0$ for $a \leqslant x \leqslant b$), the lines $x = a$, $x = b$ and the x-axis. Notice that the definite integral does not give rise to a function of x but a real number. Furthermore, there is no constant of integration present. (If one were introduced in the usual way it would cancel, since

$$\int_a^b f(x)\,dx = \left[F(x) + C\right]_a^b = (F(b) + C) - (F(a) + C)$$

$$= F(b) - F(a).)$$

Example 1: Find the area bounded by $y = mx$, the x-axis, $x = 0$ and $x = l$ ($l > 0$).

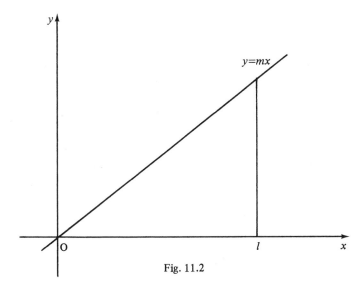

Fig. 11.2

Here $f(x) = mx$, so the area of the triangular region, see Fig. 11.2, is given by

$$\int_0^l mx\,dx = \left[\tfrac{1}{2}mx^2\right]_0^l = \tfrac{1}{2}ml^2 - 0 = \tfrac{1}{2}ml^2 .$$

This agrees with the well known result for the area of a triangle of base l and height ml.

Example 2: Find the area bounded by $y = \sqrt{x}$, $x = 1$, $x = 4$ and the x-axis (see Fig. 11.3).

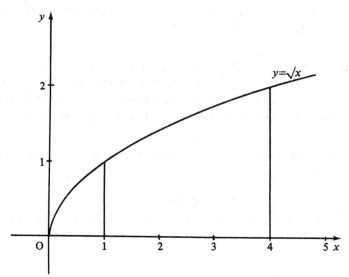

Fig. 11.3

$$\text{Area} = \int_1^4 x^{\frac{1}{2}} \, dx = \left[\frac{2}{3}x^{3/2}\right]_1^4 = \frac{2}{3}(8 - 1) = \frac{14}{3}.$$

11.3 THE DEFINITE INTEGRAL

The concept of a definite integral has been introduced by discussing the area under a curve. Applications involving definite integrals also arise in situations not related to the area under a curve, and some will be discussed in Chapter 14.

The steps leading to the definition of the definite integral given in Section 11.2 require the integrand to be a positive continuous smooth function between the limits of integration. These requirements can be relaxed by using the following properties of the definite integral.

(i) $\displaystyle\int_a^b k \, f(x) \, dx = k \int_a^b f(x) \, dx$ where k is a constant.

Proof. $\displaystyle\int_a^b k \, f(x) \, dx = \left[k \, F(x)\right]_a^b = kF(b) - kF(a) = k(F(b) - F(a))$

$$= k\left[F(x)\right]_a^b = k \int_a^b f(x) \, dx.$$

In particular, this result enables the integration of functions taking negative values to be carried out. Notice that now the integral need no longer give the area under the curve since a negative value for an area is not defined.

(ii) $\displaystyle\int_a^a f(x)\,dx = 0$.

Proof. $\displaystyle\int_a^a f(x)\,dx = F(a) - F(a) = 0$.

(iii) Interchanging the limits of integration changes the sign of the integral, that is

$$\int_a^b f(x)\,dx = -\int_b^a f(x)\,dx .$$

Proof. $\displaystyle\int_a^b f(x)\,dx = F(b) - F(a) = -(F(a) - F(b)) = -\int_b^a f(x)\,dx$.

(iv) If c satisfies $a < c < b$ then

$$\int_a^b f(x)\,dx = \int_a^c f(x)\,dx + \int_c^b f(x)\,dx .$$

Proof. $\displaystyle\int_a^b f(x)\,dx = F(b) - F(a) = (F(c) - F(a)) + (F(b) - F(c))$

$$= \int_a^c f(x)\,dx + \int_c^b f(x)\,dx .$$

This result can be used to integrate certain discontinuous functions by choosing c to be the value of x at which the discontinuity occurs.

Example 1: Evaluate the following definite integrals:

(i) $\displaystyle\int_0^{\pi/2} \cos x\,dx$, (ii) $\displaystyle\int_0^1 \frac{dx}{1+x^2}$, (iii) $\displaystyle\int_{-1}^1 x\,dx$, and (iv) $\displaystyle\int_0^{\pi/2} x \cos x\,dx$.

(i) $\displaystyle\int_0^{\pi/2} \cos x\,dx = \Big[\sin x\Big]_0^{\pi/2} = \sin\frac{\pi}{2} - \sin 0 = 1 - 0 = 1$.

(ii) $\displaystyle\int_0^1 \frac{dx}{1+x^2} = \left[\tan^{-1} x\right]_0^1 = \tan^{-1} 1 - \tan^{-1} 0 = \frac{\pi}{4} - 0 = \frac{\pi}{4}.$

(iii) $\displaystyle\int_{-1}^1 x\,dx = \left[\frac{x^2}{2}\right]_{-1}^1 = \frac{1}{2} - \frac{1}{2} = 0.$

(iv) $\displaystyle\int_0^{\pi/2} x \cos x\,dx = \left[x \sin x\right]_0^{\pi/2} - \int_0^{\pi/2} 1 . \sin x\,dx$

$$= \left(\frac{\pi}{2} - 0\right) + \left[\cos x\right]_0^{\pi/2} = \frac{\pi}{2} + (0 - 1) = \frac{\pi}{2} - 1.$$

Notice that the integrated part of the expression obtained by integrating by parts is evaluated like a separate definite integral.

Example 2: Using the calculus find the area of the triangle with vertices at $(0, 0)$, $(1, 1)$ and $(3, 0)$.

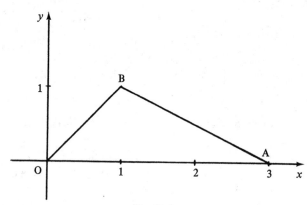

Fig. 11.4

(From the formula $\frac{1}{2}$ base $\times$ height the area is $3/2$).

The curve giving the upper boundary to the area inside the triangle, see Fig. 11.4, has the equation

$$y = \begin{cases} x & \text{for} \quad 0 \leqslant x \leqslant 1 \quad \text{(OB)} \\[2mm] \dfrac{3}{2} - \dfrac{1}{2}x & \text{for} \quad 1 \leqslant x \leqslant 3 \quad \text{(BA)} \end{cases}$$

with a corner at $x = 1$.

The area is formed by two triangular regions and is given by

$$\int_0^3 y \, dx = \int_0^1 x \, dx + \int_1^3 \left(\frac{3}{2} - \frac{1}{2}x\right) dx = \left[\frac{x^2}{2}\right]_0^1 + \left[\frac{3}{2}x - \frac{x^2}{4}\right]_1^3$$

$$= \left(\frac{1}{2} - 0\right) + \left(\frac{9}{4} - \frac{5}{4}\right) = \frac{3}{2} .$$

In some problems the definite integral may be such that it is not possible to arrive at a meaningful value by simply substituting the limits of integration directly into the function $F(x)$. The use of a limiting process may enable a value to be obtained, otherwise the integral is said not to exist. This limiting approach must always be used whenever an infinity is present.

Example 3: Evaluate $\int_0^\infty \dfrac{dx}{(1+x)^2}$.

$$\int_0^\infty \frac{dx}{(1+x)^2} = \left[\frac{-1}{1+x}\right]_0^\infty = \lim_{x \to \infty} \left(\frac{-1}{1+x}\right) - \left(\frac{-1}{1+0}\right) = 0 + 1 = 1.$$

11.4 CHANGE OF VARIABLE IN A DEFINITE INTEGRAL

The limits of integration in the definite integral $\int_a^b f(x) \, dx$ mean that the integration, with respect to x, is taken from $x = a$ to $x = b$. If a change in the variable x is made to a new variable t then the integrand becomes $f(x(t)) \dfrac{dx(t)}{dt}$. Suppose the value of t when $x = a$ is c, and the value of t when $x = b$ is d, then

$$\int_{x=a}^{x=b} f(x) \, dx = \int_{t=c}^{t=d} f(x(t)) \frac{dx(t)}{dt} \, dt . \tag{11.3}$$

It is essential to change the limits of integration when a change of variable is made. Notice that the lower limits a and c correspond, as do the upper limits b and d.

Since the replacement of x by t, through $x = t$, results in the statement

$$\int_a^b f(x) \, dx = \int_a^b f(t) \, dt = F(b) - F(a) ,$$

the letter denoting the variable of integration in a definite integral is unimportant.

Example 1: Evaluate $\displaystyle\int_0^1 \frac{x}{(x+1)^3}\,dx$.

Let $t = x + 1$, then $dx/dt = 1$ and when $x = 0$, $t = 1$; and when $x = 1$, $t = 2$. Thus

$$\int_0^1 \frac{x}{(x+1)^3}\,dx = \int_1^2 \frac{t-1}{t^3}\,dt = \int_1^2 (t^{-2} - t^{-3})dt$$

$$= \left[-t^{-1} + \frac{t^{-2}}{2}\right]_1^2 \tag{11.4}$$

$$= \left(-\frac{1}{2} + \frac{1}{8}\right) - \left(-1 + \frac{1}{2}\right) = \frac{1}{8}.$$

Alternatively, once the integration with respect to t has been performed the original variable x could be reintroduced together with the appropriate limits of integration for x. Then, continuing from line (11.4),

$$\int_0^1 \frac{x}{(x+1)^3}\,dx = \left[-\frac{1}{x+1} + \frac{1}{2(x+1)^2}\right]_0^1$$

$$= \left(-\frac{1}{2} + \frac{1}{8}\right) - \left(-1 + \frac{1}{2}\right) = \frac{1}{8}.$$

Example 2: Evaluate $\displaystyle\int_0^{\pi/2} \sin^2 x \cos^3 x \,dx$.

$$\int_0^{\pi/2} \sin^2 x \cos^3 x \,dx = \int_0^{\pi/2} \sin^2 x \cos x (1 - \sin^2 x)dx$$

$$= \int_0^{\pi/2} (\sin^2 x - \sin^4 x)\cos x \,dx .$$

Put $t = \sin x$, then $\dfrac{dx}{dt} = \dfrac{1}{\cos x}$ and when $x = 0$, $t = 0$; when $x = \dfrac{\pi}{2}$, $t = 1$. Therefore

$$\int_0^{\pi/2} \sin^2 x \cos^3 x \,dx = \int_0^1 (t^2 - t^4)dt = \left[\frac{t^3}{3} - \frac{t^5}{5}\right]_0^1 = \frac{1}{3} - \frac{1}{5} = \frac{2}{15} .$$

Example 3: Find the area bounded by $y = \sqrt{(a^2 - x^2)}$ and the x-axis.
    As the equation $y = \sqrt{(a^2 - x^2)}$ is equivalent on squaring, to $x^2 + y^2 = a^2$

with $y \geqslant 0$, the curve is a semicircle of radius a with centre at the coordinate origin, see Fig. 11.5.

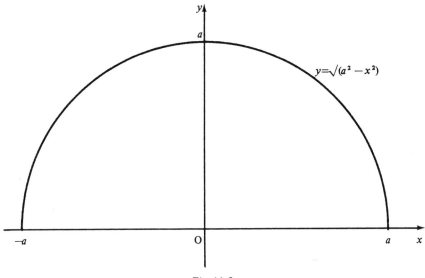

Fig. 11.5

The area, A, is given by

$$A = \int_{-a}^{a} \sqrt{(a^2 - x^2)} \, dx \ .$$

Let $x = a \sin t$ then $dx/dt = a \cos t$.

When $x = a$, $\sin t = 1$ so $t = \sin^{-1} 1 = \pi/2$.

When $x = -a$, $\sin t = -1$ so $t = \sin^{-1}(-1) = -\pi/2$.

Hence
$$A = \int_{-\pi/2}^{\pi/2} (a^2 - a^2 \sin^2 t)^{\frac{1}{2}} a \cos t \, dt = a^2 \int_{-\pi/2}^{\pi/2} \cos^2 t \, dt$$

$$= \frac{a^2}{2} \int_{-\pi/2}^{\pi/2} (1 + \cos 2t) dt = \frac{a^2}{2} \left[t + \frac{1}{2} \sin 2t \right]_{-\pi/2}^{\pi/2}$$

$$= \frac{a^2}{2} \left[\left(\frac{\pi}{2} + 0 \right) - \left(-\frac{\pi}{2} + 0 \right) \right] = \frac{\pi}{2} a^2 \ ,$$

half the area of a circle of radius a.

In this example, returning to the original variable before applying the limits of integration would not be so simple.

For definite integrals in which the integrand becomes infinite at the end of the interval of integration the theory presented in Section 11.2 cannot be used. However, by making an appropriate change in the variable the original integral can, sometimes, be replaced by one which can be evaluated.

Example 4: Show that $\displaystyle\int_0^1 \frac{dx}{\sqrt{x}} = 2$.

In this definite integral the integrand becomes infinite as x approaches zero, the lower limit of integration. By letting $x = t^2$, so that $dx/dt = 2t$,

$$\int_0^1 \frac{dx}{\sqrt{x}} = \int_0^1 \frac{1}{t} 2t \, dt = \int_0^1 2 \, dt \ .$$

The last integral presents no problems and can be evaluated with ease since

$$\int_0^1 2 \, dt = \left[2t \right]_0^1 = 2 - 0 = 2.$$

11.5 DIFFERENTIATION WITH RESPECT TO A LIMIT OF INTEGRATION

In the definite integral given by

$$\int_a^x f(t) \, dt = \left[F(t) \right]_a^x = F(x) - F(a) \tag{11.5}$$

the upper limit of integration is not fixed but is a variable. The right-hand side of (11.5) is a function of the variable x and is defined whenever $F(x) - F(a)$ exists. Differentiating (11.5), with respect to x, gives

$$\frac{d}{dx}\left(\int_a^x f(t) \, dt \right) = \frac{d}{dx}\left(F(x) - F(a) \right) = \frac{dF}{dx} = f(x) \ . \tag{11.6}$$

11.6 THE DEFINITE INTEGRAL AS AN INFINITE SUM

There is an alternative derivation of the area bounded by the x-axis, the lines $x = a$, $x = b$, $(a < b)$, and the curve $y = f(x)$, where $f(x)$ is continuous and positive for $a \leqslant x \leqslant b$, and by its use the definite integral $\int_a^b f(x) \, dx$ can be obtained. The approach is based on the concept of dividing the area into small strips and then summing the areas of these strips.

Consider the area divided into n strips, each of width $\Delta x = (b - a)/n$, by lines drawn parallel to the y-axis, see Fig. 11.6. Let the lines defining the strips

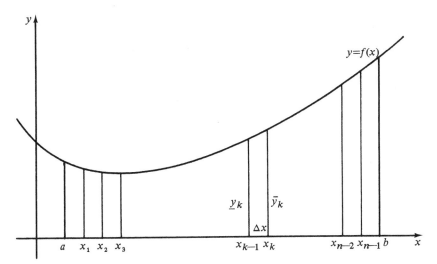

Fig. 11.6

intersect the x-axis at $x_0, x_1, x_2, \ldots, x_n$ with $x_0 = a, x_n = b$. The kth strip is then bounded by $x = x_{k-1}$ and $x = x_k$ with $x_k - x_{k-1} = \Delta x$. Denote the maximum and minimum values of $f(x)$ in the kth strip by $\bar{y}_k$ and $\underline{y}_k$ respectively, and let the area of the kth strip under the curve be A_k. Then

$$\underline{A}_k = \underline{y}_k \, \Delta x \leqslant A_k \leqslant \bar{y}_k \, \Delta x = \bar{A}_k \ .$$

Adding all the n strips (elements of area) together gives

$$\underline{A} = \sum_{k=1}^{n} \underline{A}_k \leqslant A = \sum_{k=1}^{n} A_k \leqslant \bar{A} = \sum_{k=1}^{n} \bar{A}_k \ .$$

The approximations to the area under the curve contained in this statement can be improved by taking narrower, and thus more, strips. In the limit as n becomes infinite both $\underline{A}$ and $\bar{A}$ tend to the true area A. Furthermore, it can be shown that any value of $f(x)$ in the kth strip, say Y_k, can be taken instead of $\bar{y}_k$ or $\underline{y}_k$. Thus the area A can be obtained from the result

$$A = \lim_{n \to \infty} \left(\sum_{k=1}^{n} Y_k \, \Delta x \right) . \tag{11.7}$$

Hence a definite integral may be thought of as the value of an infinite sum, or vice-versa, with

$$\int_{a}^{b} f(x) \, dx = \lim_{n \to \infty} \left(\sum_{k=1}^{n} Y_k \, \Delta x \right) \tag{11.8}$$

where $\Delta x = (b - a)/n$ and Y_k is any value taken by $f(x)$ for x between $a + (k-1)\Delta x$ and $a + k \, \Delta x$, inclusive.

Notice that this derivation of the definite integral does not require deriva-tives and so provides an alternative starting point for integration without using the idea of antiderivatives. It can be shown that the two approaches give the same results.

For example, in example 1 of Section 11.2 the area bounded by the curve $y = mx$, the x-axis, the lines $x = 0$ and $x = l$ was found to be $\frac{1}{2} ml^2$. To find the area using (11.8) divide the base into n segments of length $\Delta x = (l-0)/n = l/n$.

Y_k is some value of mx for x satisfying $(k-1)\dfrac{l}{n} \leqslant x \leqslant k\dfrac{l}{n}$.

Take $Y_k = m \cdot \dfrac{kl}{n}$ then

$$\sum_{k=1}^{n} Y_k \, \Delta x = \sum_{k=1}^{n} m\frac{kl}{n} \cdot \frac{l}{n} = \frac{ml^2}{n^2} \sum_{k=1}^{n} k$$

$$= \frac{ml^2}{n^2} \cdot \frac{1}{2} n(n+1) \qquad \text{using the result obtained in}$$
$$\text{example 1(b) of Section 2.3.}$$

$$= \frac{ml^2}{2}\left(1 + \frac{1}{n}\right) \rightarrow \frac{1}{2} ml^2 \quad \text{as } n \rightarrow \infty .$$

The area given by this approach is in agreement with that obtained pre-viously.

One disadvantage of this second approach to integration is the need to sum a series. This can only be done easily in certain cases. However, the result given in (11.8) has important consequences in the approximate evaluation of definite integrals numerically, see Chapter 17.

PROBLEMS

1. Find the area bounded by the x-axis, the curve $y = f(x)$, and the vertical lines $x = a, x = b$ when:

    (a) $f(x) = 9 - 2x$, $a = 1, b = 3$;

    (b) $f(x) = (2x + 1)^{\frac{1}{2}}$, $a = 0, b = 4$;

    (c) $f(x) = (2x + 1)^{-\frac{1}{2}}$, $a = 0, b = 4$;

    (d) $f(x) = x(2x^2 + 1)^{\frac{1}{2}}$, $a = 0, b = 2$;

    (e) $f(x) = x(2x^2 + 1)^{-\frac{1}{2}}$, $a = 0, b = 2$;

    (f) $f(x) = 5x - \frac{1}{2}x^2$, $a = 2, b = 8$.

2. Find the area contained between the x-axis and one arch of the curve $y = \cos 3x$.

3. Verify the following definite integrals:

(a) $\displaystyle\int_{-1}^{2} x(x + 1)dx = \frac{9}{2}$,

(b) $\displaystyle\int_{0}^{\pi/4} \sec^2 x \, dx = 1$,

(c) $\displaystyle\int_{0}^{3} \frac{dx}{9 + x^2} = \frac{\pi}{12}$,

(d) $\displaystyle\int_{0}^{1/3} (1 - 4x^2)^{-\frac{1}{2}} dx = \frac{1}{2} \sin^{-1}\left(\frac{2}{3}\right)$,

(e) $\displaystyle\int_{0}^{\pi/8} \tan^2(2x)dx = \frac{1}{2} - \frac{\pi}{8}$,

(f) $\displaystyle\int_{0}^{3} \frac{dx}{\sqrt{(4 - x)}} = 2$,

(g) $\displaystyle\int_{0}^{\pi/2} \cos x \, \sin^{\frac{1}{2}} x \, dx = \frac{2}{3}$,

(h) $\displaystyle\int_{0}^{\frac{1}{2}} \sin^{-1}(2x)dx = \frac{\pi - 2}{4}$,

(i) $\displaystyle\int_{0}^{\pi} x^2 \cos\left(\frac{x}{2}\right) dx = 2\pi^2 - 16$,

(j) $\displaystyle\int_{-\pi}^{\pi} (1 + \sin x)^2 dx = 3\pi$,

(k) $\displaystyle\int_{0}^{\infty} \frac{dx}{4 + x^2} = \frac{\pi}{4}$,

(l) $\displaystyle\int_{-\infty}^{\infty} \frac{dx}{1 + x^2} = \pi$.

4. Use the given substitution to evaluate the following definite integrals:

(a) $\displaystyle\int_0^1 \frac{x}{(x+1)^4}\,dx,$ $\qquad t = x + 1\ ;$

(b) $\displaystyle\int_0^{\pi/4} \frac{\cos x}{1 + \sin^2 x}\,dx,$ $\qquad t = \sin x\ ;$

(c) $\displaystyle\int_0^1 \left(\frac{x}{1-x}\right)^{\frac{1}{2}}\,dx,$ $\qquad x = \sin^2 t\ ;$

(d) $\displaystyle\int_{\sqrt{2}}^2 \frac{dx}{x\sqrt{(x^2 - 2)}},$ $\qquad t^2 = x^2 - 2\ ;$

(e) $\displaystyle\int_1^2 \frac{dx}{x\sqrt{(x^2 - 1)}},$ $\qquad x = \operatorname{cosec} t\ ;$

(f) $\displaystyle\int_0^1 \frac{x}{\sqrt{(x^2 + 1)}}\,dx,$ $\qquad t = x^2 + 1\ .$

5. Verify the following reduction formulae, in which n is an integer greater than one, and hence evaluate the given integral:

(a) $\displaystyle\int_0^{\pi/2} \cos^n x\,dx = \frac{n-1}{n}\int_0^{\pi/2} \cos^{n-2} x\,dx,\qquad \int_0^{\pi/2} \cos^5 x\,dx\ ;$

(b) $\displaystyle\int_0^{\pi/2} x^n \cos x\,dx = \left(\frac{\pi}{2}\right)^n + (n - n^2)\int_0^{\pi/2} x^{n-2}\cos x\,dx,\qquad \int_0^{\pi/2} x^4 \cos x\,dx;$

(c) $\displaystyle\int_0^{\pi/4} \tan^n x\,dx = \frac{1}{n-1} - \int_0^{\pi/4} \tan^{n-2} dx\,,\qquad \int_0^{\pi/4} \tan^4 x\,dx.$

6. Approximate the following definite integrals by a finite sum with n terms and hence evaluate them. Check your answer by direct integration.

(a) $\displaystyle\int_0^1 x^3\,dx,$

(b) $\displaystyle\int_1^3 x^2\,dx,$

(c) $\int_0^1 \cos x \, dx.$

Note that $\sum_{k=1}^{n} k = \frac{1}{2}n(n+1)$, $\sum_{k=1}^{n} k^2 = \frac{1}{6}n(n+1)(2n+1)$,

$\sum_{k=1}^{n} k^3 = \frac{1}{4}n^2(n+1)^2$, $\sum_{k=1}^{n} \cos k\theta = \frac{\cos[(n+1)\theta/2]\sin(n\theta/2)}{\sin(\theta/2)}$.

7. The amount of embryonic development that takes place during a day is proportional to the area under the temperature curve for that day. If the temperature curve is given as a function of time t, measured in hours, by

(i) $T = 20$,

(ii) $T = \begin{cases} 10 + t & 0 < t < 16 \\ 58 - 2t & 16 < t < 24 \end{cases}$,

(iii) $T = 18 - 8 \sin\left(\frac{\pi t}{12} + \frac{\pi}{4}\right)$,

find the amount of development that takes place during a day in each case. Take the constant of proportionality to be 10^{-3}.

8. The shape of a leaf is given, approximately, by the curve

$$y^2 = x(5 - x)^2$$

for $0 \leqslant x \leqslant 5$ cm. Show that the total surface area of the leaf is $\frac{80\sqrt{5}}{3}$ cm^2.

CHAPTER 12

Logarithmic and exponential functions

12.1 INTRODUCTION

In this chapter two functions which are basic to mathematics and its applications to problems in the biological sciences are introduced. They are the logarithmic and exponential functions.

A function, which is defined in terms of a definite integral, is shown to have properties similar to a logarithm. This function is called the logarithmic function. A number of standard results for integrals are also obtained.

The exponential function is defined as the solution to a differential equation and is then shown to be the inverse of the logarithmic function. Properties of the exponential function are exhibited and an application to the quantitative description of radioactive decay processes is given.

The functions frequently arise in solutions to practical problems described by differential equations, see Chapter 15, and also in statistics, see Chapter 22.

12.2 THE LOGARITHMIC FUNCTION

The integration of a power of x, given in Chapter 10, by the formula

$$\int x^{\alpha} dx = \frac{x^{\alpha+1}}{\alpha + 1} + C$$

is incomplete since this result is not valid when $\alpha = -1$. None of the elementary functions already encountered give rise to $1/x$, when differentiated. As expressions taking the form of a reciprocal occur frequently when applying mathematics to physical situations; for example, whenever one quantity varies in inverse proportion to another, it is natural to examine the integral of $1/x$.

The definite integral of a positive function was shown in Chapter 11 to give the area bounded by the function and the axis of the variable of integration between the limits of integration. The function $y = 1/t$ is defined for all t,

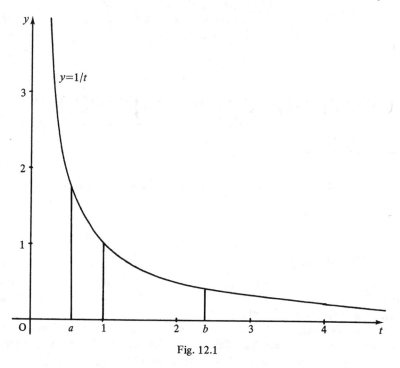

Fig. 12.1

except $t = 0$, and is positive if t is positive. Fig. 12.1 portrays the positive portion of the graph of $y = 1/t$ and the area indicated would be given by

$$\int_a^b \frac{1}{t} \, dt.$$

Fixing a to equal one and allowing b to vary, say $b = x$, the area, as given by $\int_1^x \frac{1}{t} \, dt$ for $x > 1$, depends on x. The function, giving the value of the area, defined by $\int_1^x \frac{1}{t} \, dt$ is called the **logarithmic function** and is denoted by $\ln x$.

Thus for $x > 1$, $\ln x$ is the area enclosed by $t = 1, t = x$, the curve $y = 1/t$ and the t-axis, that is,

$$\ln x = \int_1^x \frac{1}{t} \, dt \ .$$

The definition of the logarithmic function can be extended to include values of x in the range $0 < x < 1$. Since interchanging the limits of integration leads to

$$\int_1^x \frac{1}{t} \, dt = -\int_x^1 \frac{1}{t} \, dt \ ,$$

ln x, for $0 < x < 1$, is minus the area enclosed by the curve $y = 1/t$ and the
t-axis between $t = x$ and $t = 1$. When $x = 1$ the area reduces to zero, so ln $1 = 0$.
Thus ln x is defined by

$$\ln x = \int_1^x \frac{1}{t}\, dt \tag{12.1}$$

for all $x > 0$. Note that the logarithmic function is not defined for values of the
argument less than or equal to zero.

Since

$$\frac{d}{dx} \int_1^x f(t)\, dt = f(x) \,,$$

$$\frac{d}{dx} \ln x = \frac{d}{dx} \int_1^x \frac{1}{t}\, dt = \frac{1}{x} \,. \tag{12.2}$$

The derivative of ln x, given in (12.2), is a standard result.

Example: Differentiate with respect to x:

(i) $\ln(4 - 3x)$, (ii) $\ln(x^2 + 5)$, (iii) $\ln(\sin x)$.

(i) Let $y = \ln(4 - 3x)$. Using the chain rule with $t = 4 - 3x$, $y = \ln t$ and

$$\frac{dy}{dx} = \frac{dy}{dt}\frac{dt}{dx} = \frac{1}{t} \cdot (-3) = \frac{-3}{4 - 3x} = \frac{3}{3x - 4} \,.$$

(ii) If $y = \ln(x^2 + 5)$ then $y = \ln t$ with $t = x^2 + 5$ so, by the chain rule,

$$\frac{dy}{dx} = \frac{1}{t} \cdot 2x = \frac{2x}{x^2 + 5} \,.$$

(iii) If $y = \ln(\sin x)$ then $y = \ln t$ with $t = \sin x$ so

$$\frac{dy}{dx} = \frac{1}{t} \cdot \cos x = \frac{\cos x}{\sin x} = \cot x \,.$$

12.3 PROPERTIES OF ln x

If x and a are both positive

$$\frac{d}{dx}\ln(ax) = \frac{1}{ax} \cdot a = \frac{1}{x} = \frac{d}{dx}\ln x \,.$$

Integrating with respect to x the first and last terms in this expression gives

$$\ln(ax) = \ln x + C \qquad \text{for some constant } C \,.$$

This equation must hold for all positive values of x. On setting $x = 1$ the appropriate value of C is found to be given by $C = \ln a$, since $\ln 1 = 0$. Hence $\ln(ax) = \ln x + \ln a$. Replacing x by b then gives

$$\ln(ab) = \ln a + \ln b .$$ (12.3)

This result indicates why the name logarithmic function is appropriate; it behaves like the ordinary logarithm studied in Chapter 1.

Putting $b = 1/a$ in (12.3) gives

$$\ln a + \ln \frac{1}{a} = \ln 1 = 0$$

so $$\ln \frac{1}{a} = -\ln a .$$

Furthermore,

$$\ln \frac{a}{b} = \ln a + \ln \frac{1}{b} = \ln a - \ln b .$$

The final law of logarithms, namely that

$$\ln(a^r) = r \ln a$$

where r is a rational number, also holds.

When $r = n$, a positive integer, the result follows from

$$\ln a^n = \ln(\underbrace{a \times a \times \ldots \times a}_{n \text{ times}}) = \underbrace{\ln a + \ln a + \ldots + \ln a}_{n \text{ times}} = n \ln a .$$

The index law can also be extended to cover irrational indices.

The value of x for which $\ln x = 1$ is denoted by e, e $= 2.171828 \ldots$, and is an irrational number. Another notation for the logarithmic function is to write $\ln x = \log_e x$. The logarithmic function is also called the **logarithm to base** e or **natural logarithm**. A table giving values of $\ln x$ for x between 1 and 10 is included as Appendix C. Alternatively the relationship established in Section 1.11 viz.,

$$\ln x = \ln 10 . \log_{10} x$$

can be used, enabling calculations to be made using logarithms to base 10.

Example 1: Evaluate $\ln 98.7$.

The table of values of $\ln x$ only gives values for x lying between 1 and 10, so write 98.7 as 9.87×10. Then

$$\ln 98.7 = \ln(9.87 \times 10) = \ln 9.87 + \ln 10$$
$$= 2.2895 + 2.3026$$
$$= 4.5921 .$$

Alternatively, $\ln 98.7 = \ln 10.\log_{10} 98.7$

$$= 2.3026 \times 1.9943 = 4.5921 \ .$$

The properties of the logarithmic function enable certain expressions to be differentiated more easily, as the following example shows.

Example 2: Differentiate (i) $\ln x^4$, (ii) $\ln \left(\dfrac{x-1}{x+1}\right)^{\frac{1}{2}}$, (iii) $(x-1)^2(x+2)^3$.

(i) $y = \ln x^4 = 4 \ln x$ so $\dfrac{dy}{dx} = \dfrac{4}{x}$.

(ii) $y = \ln \left(\dfrac{x-1}{x+1}\right)^{\frac{1}{2}} = \dfrac{1}{2} \ln \left(\dfrac{x-1}{x+1}\right) = \dfrac{1}{2}\ln (x-1) - \dfrac{1}{2}\ln (x+1).$

Therefore $\dfrac{dy}{dx} = \dfrac{1}{2}\cdot\dfrac{1}{x-1} - \dfrac{1}{2}\cdot\dfrac{1}{x+1} = \dfrac{1}{x^2-1}$.

(iii) $y = (x-1)^2(x+2)^3$. Instead of using the product rule take the natural logarithm of the expression. Thus

$$\ln y = \ln[(x-1)^2(x+2)^3] = \ln(x-1)^2 + \ln(x+2)^3$$
$$= 2\ln(x-1) + 3\ln(x+2) \ .$$

Differentiating with respect to x, using implicit differentiation for $\ln y$, gives

$$\frac{1}{y}\frac{dy}{dx} = 2\cdot\frac{1}{x-1} + 3\cdot\frac{1}{x+2} = \frac{5x+1}{(x-1)(x+2)} \ .$$

So $\dfrac{dy}{dx} = \dfrac{y(5x+1)}{(x-1)(x+2)} = (5x+1)(x-1)(x+2)^2 \ .$

Example 2(iii) illustrates a technique known as **logarithmic differentiation**. The natural logarithm of the equation is taken, and, after using properties of logarithms to write it in terms of simpler functions, it is then differentiated.

By using logarithmic differentiation the result that

$$\frac{d}{dx}(x^\alpha) = \alpha x^{\alpha-1} \tag{12.4}$$

where α is any real number can be proved. Let $y = x^\alpha$ then $\ln y = \ln(x^\alpha) = \alpha \ln x$, so

$$\frac{1}{y}\frac{dy}{dx} = \alpha\frac{1}{x} \ .$$

Therefore $\dfrac{dy}{dx} = \alpha\dfrac{y}{x} = \alpha\dfrac{x^\alpha}{x} = \alpha x^{\alpha-1} \ .$

The logarithmic function, $y = \ln x$, is defined only for $x > 0$ and has derivative $dy/dx = 1/x$. For $x > 0$, $dy/dx = 1/x > 0$ so the function is always increasing, although less steeply as x increases, and has no stationary points. The function is zero when $x = 1$, and $\ln x > 0$ if $x > 1$, whilst $\ln x < 0$ if $0 < x < 1$. Setting $x = 2^n$, $\ln x = \ln 2^n = n \ln 2$ which becomes infinite as $n \to \infty$ and tends to minus infinity as $n \to -\infty$. Hence

$$\ln x \to \infty \text{ as } x \to \infty \ ,$$

and $\ln x \to -\infty \text{ as } x \to 0$.

Figure 12.2 can be compared with the graph given in Fig. 1.3 of the common logarithm.

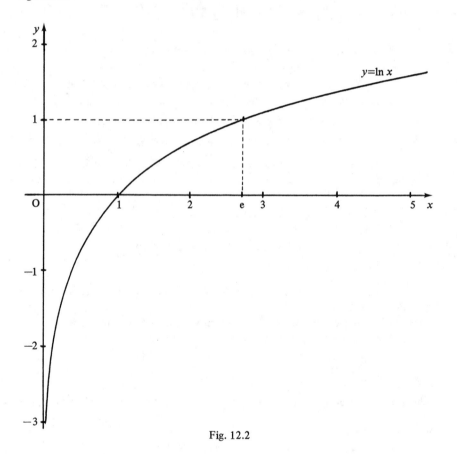

Fig. 12.2

As $\ln x$ is not defined for $x = 0$ a Maclaurin series expansion for $\ln x$ cannot be obtained. Instead the standard expansion is that obtained from the function $\ln(1 + x)$.

$$f(x) = \ln(1 + x) \qquad f(0) = \ln 1 = 0 ,$$

$$f'(x) = \frac{1}{1 + x} \qquad f'(0) = 1 ,$$

$$f''(x) = \frac{-1}{(1 + x)^2} \qquad f''(0) = -1 ,$$

$$f'''(x) = \frac{2}{(1 + x)^3} \qquad f'''(0) = 2 ,$$

$$f^{iv}(x) = \frac{-3!}{(1 + x)^4} \qquad f^{iv}(0) = -3!$$

So

$$\ln(1 + x) = 0 + 1.x - \frac{1}{2!}x^2 + \frac{2!}{3!}x^3 - \frac{3!}{4!}x^4 + \ldots$$

$$= x - \frac{1}{2}x^2 + \frac{1}{3}x^3 - \frac{1}{4}x^4 + \ldots$$

$$= \sum_{n=1}^{\infty} \frac{(-1)^{n+1}x^n}{n} . \tag{12.5}$$

The expansion is valid for x satisfying $-1 < x \leqslant 1$.

The logarithmic function is defined only for positive arguments but integrals such as

$$\int_{-3}^{-2} \frac{1}{x} \, dx$$

do occur. The interval between the limits of integration does not include the point $x = 0$ where the function is discontinuous (see Section 11.3), so by changing to the new variable t, given by $t = -x$, the integral becomes

$$\int_{3}^{2} \frac{1}{t} \, dt = \left[\ln t \right]_{3}^{2} = \ln 2 - \ln 3 = \ln \frac{2}{3} .$$

Here $\ln t$ exists, since $t > 0$.

Hence, if $x < 0$, set $t = -x$ so that

$$\int \frac{1}{x} \, dx = \int \frac{1}{t} \, dt = \ln t + C_1 = \ln(-x) + C_1 .$$

But

$$\int \frac{1}{x} \, dx = \ln x + C_2 \quad \text{if } x > 0 .$$

In order not to have to distinguish between the cases when x is positive or negative we make use of the modulus sign introduced in Chapter 1, to write

$$\int \frac{1}{x}\,dx = \ln|x| + C\,, \qquad x \neq 0\,. \tag{12.6}$$

In practical applications involving integrals resulting in the logarithmic function it is often convenient to take the constant of integration C to be in the form of a natural logarithm. For example, take $C = \ln K$ where K is a positive arbitrary constant. (Notice that the constant C will be negative when K is less than one.) Then

$$\int \frac{1}{x}\,dx = \ln|x| + \ln K = \ln|Kx|\,.$$

A frequently occurring problem in integration requires the evaluation of $\int \frac{1}{ax+b}\,dx$. The integration is performed by changing to a new variable $t = ax + b$. Then $\dfrac{dx}{dt} = \dfrac{1}{a}$ and

$$\int \frac{1}{ax+b}\,dx = \int \frac{1}{t}\cdot\frac{1}{a}\,dt = \frac{1}{a}\ln|t| + C = \frac{1}{a}\ln|ax+b| + C.$$

Example 3: Integrate (i) $\dfrac{1}{2x-1}$, (ii) $\dfrac{1}{1-2x}$, (iii) $\dfrac{x}{x-1}$.

(i) $\displaystyle\int \frac{dx}{2x-1} = \frac{1}{2}\ln|2x-1| + C\,.$

(ii) $\displaystyle\int \frac{dx}{1-2x} = -\frac{1}{2}\ln|1-2x| + C\,.$ Note that (i) and (ii) are related.

(iii) $\displaystyle\int \frac{x\,dx}{x-1} = \int \frac{(x-1)+1}{x-1}\,dx = \int\left(1 + \frac{1}{x-1}\right)dx$

$$= x + \ln|x-1| + C\,.$$

The expression for the integral of $1/(ax+b)$ with respect to x is really a special case of the more general result given by

$$\int \frac{g'(x)}{g(x)}\,dx = \ln|g(x)| + C\,. \tag{12.7}$$

The integrand in (12.7) has as numerator the derivative of the denominator. The result is obtained by using the change of variable $t = g(x)$, so that

$$\int \frac{g'(x)}{g(x)} dx = \int \frac{g'(x)}{t} \frac{1}{g'(x)} dt = \int \frac{1}{t} dt = \ln|t| + C = \ln|g(x)| + C .$$

Example 4: Integrate $\dfrac{x}{x^2 + a}$ where a is a constant.

Let $t = x^2 + a$, then $\dfrac{dx}{dt} = 1 \left/ \dfrac{dt}{dx} = \dfrac{1}{2x} \right.$ and

$$\int \frac{x}{x^2 + a} dx = \int \frac{x}{t} \cdot \frac{1}{2x} dt = \frac{1}{2} \int \frac{1}{t} dt = \frac{1}{2} \ln|t| + C .$$

So $$\int \frac{x}{x^2 + a} dx = \frac{1}{2} \ln|x^2 + a| + C .$$

Example 5: Integrate (i) $\cot x$, (ii) $\tan x$.

(i) $\displaystyle\int \cot x \, dx = \int \frac{\cos x}{\sin x} dx$. Since $\cos x = \dfrac{d}{dx} \sin x$,

$$\int \cot x \, dx = \ln|\sin x| + C .$$

(ii) $\displaystyle\int \tan x \, dx = \int \frac{\sin x}{\cos x} dx$. $\dfrac{d}{dx} \cos x = -\sin x$ so

$$\int \tan x \, dx = -\ln|\cos x| + C = \ln|\sec x| + C .$$

These results provide two more standard integrals.

Example 6: Evaluate (i) $\displaystyle\int_1^5 \frac{dx}{x + 3}$, (ii) $\displaystyle\int_0^1 \frac{x^2}{x^3 + 2} dx$, (iii) $\displaystyle\int_0^{\pi/2} \frac{\cos x}{1 + 2\sin x} dx$.

(i) $\displaystyle\int_1^5 \frac{dx}{x + 3} = \Big[\ln|x + 3|\Big]_1^5 = \ln 8 - \ln 4 = \ln \frac{8}{4} = \ln 2 .$

(ii) $\displaystyle\int_0^1 \frac{x^2}{x^3 + 2} dx$. Let $t = x^3 + 2$, then $\dfrac{dx}{dt} = \dfrac{1}{3x^2}$.

When $x = 0$, $t = 2$, and when $x = 1$, $t = 3$. Therefore

$$\int_0^1 \frac{x^2}{x^3 + 2} dx = \int_2^3 \frac{1}{3t} dt = \frac{1}{3} \Big[\ln|t|\Big]_2^3 = \frac{1}{3}(\ln 3 - \ln 2) = \frac{1}{3} \ln \frac{3}{2} .$$

(iii) $\displaystyle\int_0^{\pi/2} \frac{\cos x}{1 + 2\sin x}\,dx$. Let $t = 1 + 2\sin x$, then $\dfrac{dx}{dt} = \dfrac{1}{2\cos x}$.

When $x = 0$, $t = 1$; when $x = \dfrac{\pi}{2}$, $t = 3$, so

$$\int_0^{\pi/2} \frac{\cos x}{1 + 2\sin x}\,dx = \int_1^3 \frac{1}{2t}\,dt = \frac{1}{2}\left[\ln|t|\right]_1^3 = \frac{1}{2}(\ln 3 - \ln 1) = \frac{1}{2}\ln 3 \ .$$

The integral of $\ln|x|$ is obtained by using integration by parts. In the formula (10.6) take $u = \ln|x|$, $\dfrac{dv}{dx} = 1$ then $\dfrac{du}{dx} = \dfrac{1}{x}$ and $v = x$, so

$$\int \ln|x|\,dx = x\,\ln|x| - \int x \cdot \frac{1}{x}\,dx = x\,\ln|x| - x + C \ .$$

Similarly, integration by parts may be used to evaluate $\int x^\alpha \ln x\,dx$ for any real number α.

The final two integrals in this section are standard results; however, their derivation involves observing that the integrand can be rewritten. Thus

$$\int \sec x\,dx = \int \frac{\sec x\,(\sec x + \tan x)}{\sec x + \tan x}\,dx \ .$$

Now $\dfrac{d}{dx}(\sec x + \tan x) = \sec x \tan x + \sec^2 x$, the numerator, so

$$\int \sec x\,dx = \ln|\sec x + \tan x| + C \ .$$

Similarly,

$$\int \operatorname{cosec} x\,dx = \int \frac{\operatorname{cosec} x\,(\operatorname{cosec} x + \cot x)}{\operatorname{cosec} x + \cot x}\,dx$$

$$= -\ln|\operatorname{cosec} x + \cot x| + C \ .$$

12.4 THE EXPONENTIAL FUNCTION

The function considered in this section is perhaps the most important of all the basic functions used in the biological sciences. This stems from its use in the analysis of processes involving absorption, decay, excretion and growth. The function also arises in the theory of probability.

Consider $y = a^x$, $a > 0$.

Then by taking logarithms

$$\ln y = \ln(a^x) = x \ln a \ .$$

On differentiating this with respect to x,

$$\frac{1}{y}\frac{dy}{dx} = \ln a$$

so
$$\frac{dy}{dx} = y \ln a = a^x \ln a .$$

As $\ln e = 1$, this equation takes on a particularly simple form when $a = e$. Thus $y = e^x$ has derivative $dy/dx = e^x$, that is

$$\frac{d}{dx}(e^x) = e^x.$$

This function e^x, having the property that it is its own derivative, is called the **exponential function**, also written $\exp(x)$. Usually the index form is used, but when the argument is complicated it is more convenient to use the 'exp' form. From its derivation the exponential function is always positive, must obey the rules of indices and is the inverse function of the logarithmic function, see Section 1.11.

Example 1: Simplify $e^{\ln x}$.
   Taking the natural logarithm of $y = e^{\ln x}$,

$$\ln y = \ln(e^{\ln x}) = \ln x . \ln e = \ln x, \text{ since } \ln e = 1.$$

So $y = x$, that is, $x = e^{\ln x}$.

The graph of the curve $y = e^x$ is shown in Fig. 12.3.

The main features of the curve $y = e^x$ are

   (i) it is always positive,

   (ii) e^x is an increasing function ($e > 1$) so the curve increases indefinitely as x increases and goes to zero as x approaches minus infinity,

   (iii) it has no stationary points,

   (iv) it cuts the y-axis at $y = 1$.

Figure 12.4 shows the graph of the related function $y = e^{-x} = \frac{1}{e^x}$. A table giving values of e^x and e^{-x} is included as Appendix D.

   The Maclaurin series expansion of e^x is

$$e^x = 1 + x + \frac{x^2}{2!} + \frac{x^3}{3!} + \frac{x^4}{4!} + \ldots = \sum_{n=0}^{\infty} \frac{x^n}{n!} \qquad (12.8)$$

and holds for all x.

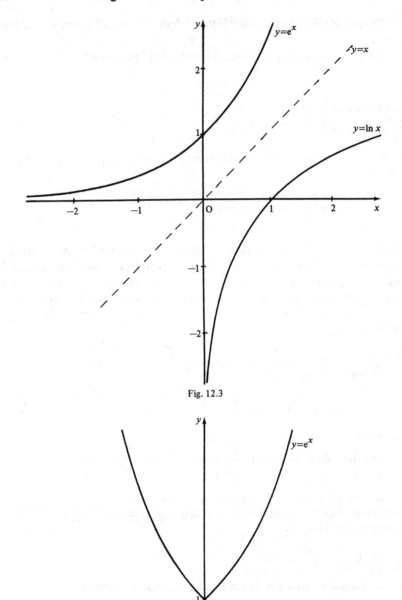

Fig. 12.3

Fig. 12.4

This result follows because $f(x) = f^{(n)}(x) = e^x$ for all n, and e^x equals one when x is zero.

An important limit involving the exponential function is

$$\lim_{x \to \infty} x^\alpha \, e^{-x} = 0 \tag{12.9}$$

for any real constant α.

Proof. For $x > 0, x^\alpha \, e^{-x} > 0$ and

$$(x^\alpha \, e^{-x})^{-1} = \frac{e^x}{x^\alpha} = \sum_{n=0}^{\infty} \frac{x^{n-\alpha}}{n!} > \frac{x^{p-\alpha}}{p!}$$

where p is any fixed positive integer greater than α.

Taking the reciprocal gives

$$0 < x^\alpha \, e^{-x} < p! \, x^{\alpha-p} \ .$$

As $x \to \infty, x^{\alpha-p} \to 0$ since $\alpha < p$, so $x^\alpha \, e^{-x}$ must tend to zero as well.

Thus, as x becomes infinite, the term e^{-x} goes to zero faster than any power of x can increase.

Putting $e^{-x} = y$ and $\alpha = 1$ in (12.9) we obtain the result that

$$\lim_{y \to 0} y \ln y = 0 \ .$$

Example 2: Find (i) the derivative, and (ii) the integral, of e^{ax}.

Let $y = e^{ax}$, then putting $t = ax$, $y = e^t$.

(i) $\dfrac{d}{dx}(e^{ax}) = \dfrac{dy}{dx} = \dfrac{dy}{dt}\dfrac{dt}{dx} = e^t \cdot a = ae^{ax}.$

(ii) $\displaystyle\int e^{ax} \, dx = \int y \, dx = \int e^t \frac{dx}{dt} \, dt = \frac{1}{a} \int e^t \, dt = \frac{1}{a} e^t + C = \frac{1}{a} e^{ax} + C \ .$

In order to use the established properties of the exponential function, expressions having an exponent are often written in terms of exponential functions. The identity

$$a^b = e^{kb} \ ,$$

where $k = \ln a$, makes this possible.

By using this result the derivative of $y = a^x$ can be found.

$$y = a^x = e^{kx} \qquad \text{where } k = \ln a,$$

so, $\qquad \dfrac{dy}{dx} = ke^{kx} \qquad \text{using example 2(i)}$

$$= \ln a \cdot a^x \ .$$

This result completes the list of standard results given in Appendix B.

An important application of the exponential function occurs in the study of **radioactive decay**. In this process a radioactive atom changes its atomic structure by emitting a particle from its nucleus. Owing to the large number of atoms involved the decay process can be quantified accurately by the equation

$$M = M_0 \, e^{-\lambda t} \, ,$$
(12.10)

see Chapter 15. Here M is the mass of the unchanged radioactive atoms present at time t. M_0 is the original value of M (when t is zero) and λ is a parameter governing the rate of decay. In place of the mass M other measures of the amount of radioactive substance present can be used, such as the density or the number of atoms remaining unchanged.

Associated with each radioactive substance is a quantity called its **half-life** which characterises the element's decay. The half-life is the time taken for the amount of the radioactive element present at any time T to be reduced by one half.

Denoting the half-life by τ,

$$\frac{1}{2} M_0 \, e^{-\lambda T} = M_0 \, e^{-\lambda (T+\tau)} \quad \text{when } t = T+\tau \, .$$

Cancelling $M_0 \, e^{-\lambda T}$ and taking logarithms

$$\ln \frac{1}{2} = -\lambda \tau$$

so $\qquad \lambda = \dfrac{\ln 2}{\tau} \, .$

Notice that this result does not depend on the amount of radioactive substance present nor on the time T.

One practical use of equation (12.10) is in **radiocarbon dating**. In this technique the age of an object containing dead organic matter can be determined approximately by measuring the relative proportions of the carbon isotopes C^{12} and C^{14}. C^{14} decays to C^{12} with a half-life τ of 5730 years. On death the intake of carbon by the organic matter will cease and the isotope C^{14} absorbed by the organic matter at that time will subsequently decay into C^{12}. If it is assumed that the naturally occurring proportion of the two isotopes has not changed over the years, measuring the relative proportions of carbon in the object and similar organic matter, whose age is known, enables the time of death to be established.

Let Φ and Φ_0 be the proportions of C^{14} to C^{12} in the specimen to be dated and the specimen whose age is known, respectively. The age, T, of the object is then given by,

$$\Phi = \Phi_0 \, \exp\!\left(-\ln 2 \cdot \frac{T}{\tau}\right) \, ,$$

so $\qquad T = \dfrac{5730}{\ln 2} \ln\!\left(\dfrac{\Phi_0}{\Phi}\right)$ years.

As a numerical illustration suppose a wooden spoon is discovered during an archaeological dig. If the spoon's carbon content contains 10% of C^{14} while wood from a growing tree contains 13%, how old is the spoon?

$$\frac{\Phi_o}{\Phi} = \frac{13}{10} = 1.3, \text{ so the age of the spoon is approximately}$$

$$T = \frac{5730}{\ln 2} \ln 1.3 \cong 2169 \text{ years.}$$

Example 3: Differentiate (i) e^{x^2+1}, (ii) $e^{\sin x}$, (iii) xe^x, (iv) $e^x \cos x$.
The first two require the use of the chain rule, the last two the product rule.

(i) Let $y = e^{x^2+1} = e^t$, where $t = x^2 + 1$, then

$$\frac{dy}{dx} = e^t \frac{dt}{dx} = e^{x^2+1} \cdot 2x = 2xe^{x^2+1}.$$

(ii) Let $y = e^{\sin x} = e^t$, where $t = \sin x$, then $\dfrac{dy}{dx} = e^t \dfrac{dt}{dx} = e^{\sin x} \cdot \cos x$.

(iii) Let $y = xe^x$, $\dfrac{dy}{dx} = 1 \cdot e^x + x \cdot e^x = (1 + x) e^x$.

(iv) Let $y = e^x \cos x$, $\dfrac{dy}{dx} = e^x \cos x + e^x(-\sin x) = (\cos x - \sin x) e^x$.

Example 4: Integrate (i) e^{2x-3}, (ii) xe^{x^2}, (iii) xe^{-ax}.

(i) $\displaystyle\int e^{2x-3} dx = \int e^{2x} e^{-3} dx = e^{-3} \int e^{2x} dx = e^{-3} \cdot \frac{1}{2} e^{2x} + C = \frac{1}{2} e^{2x-3} + C.$

Alternatively, let $t = 2x - 3$, then $dx/dt = \frac{1}{2}$ and

$$\int e^{2x-3} dx = \int e^t \frac{1}{2} dt = \frac{1}{2} e^t + C = \frac{1}{2} e^{2x-3} + C .$$

(ii) $\displaystyle\int xe^{x^2} dx$. Let $t = x^2$, then $\dfrac{dx}{dt} = \dfrac{1}{2x}$ and

$$x e^{x^2} \frac{dx}{dt} = x e^{x^2} \frac{1}{2x} = \frac{1}{2} e^t, \text{ so}$$

$$\int x e^{x^2} dx = \frac{1}{2} \int e^t dt = \frac{1}{2} e^t + C = \frac{1}{2} e^{x^2} + C .$$

(iii) $\int x\,e^{-ax}\,dx$. Integration by parts is used to evaluate this integral.

Take $u = x$, $\dfrac{dv}{dx} = e^{-ax}$ then $\dfrac{du}{dx} = 1$, $v = -\dfrac{1}{a}e^{-ax}$, so

$$\int x\,e^{-ax}\,dx = -\frac{xe^{-ax}}{a} - \int 1\left(\frac{-e^{-ax}}{a}\right)dx = -\frac{xe^{-ax}}{a} - \frac{e^{-ax}}{a^2} + C\ .$$

Example 5: Show that, if $a > 0$,

(i) $\displaystyle\int_0^\infty x\,e^{-ax}\,dx = \frac{1}{a^2}$, (ii) $\displaystyle\int_0^\infty x^2\,e^{-ax}\,dx = \frac{2}{a^3}$.

(i) Using example 4(iii), $\displaystyle\int_0^\infty x\,e^{-ax}\,dx = \left[-\frac{xe^{-ax}}{a} - \frac{e^{-ax}}{a^2}\right]_0^\infty$

$$= \lim_{x\to\infty}\left(-\frac{xe^{-ax}}{a} - \frac{e^{-ax}}{a^2}\right) - \left(0 - \frac{1}{a^2}\right)\ .$$

To evaluate the limit let $t = ax$, then $t \to \infty$ as $x \to \infty$, so

$$\lim_{x\to\infty}\left(-\frac{xe^{-ax}}{a} - \frac{e^{-ax}}{a^2}\right) = \lim_{t\to\infty}\left(-\frac{te^{-t}}{a^2} - \frac{e^{-t}}{a^2}\right) = 0 - 0 = 0$$

using (12.9).

So $\displaystyle\int_0^\infty x\,e^{-ax}\,dx = \frac{1}{a^2}$.

(ii) $\displaystyle\int_0^\infty x^2\,e^{-ax}\,dx = \left[-\frac{x^2 e^{-ax}}{a}\right]_0^\infty + \int_0^\infty \frac{2x\,e^{-ax}}{a}\,dx$

$$= \left[\lim_{x\to\infty}\left(-\frac{x^2 e^{-ax}}{a}\right) - 0\right] + \frac{2}{a}\cdot\frac{1}{a^2}\quad\text{using (i)}$$

$$= \lim_{t\to\infty}\left(-\frac{t^2 e^{-t}}{a^3}\right) + \frac{2}{a^3}\quad\text{setting } t = ax$$

$$= \frac{2}{a^3}\quad\text{using (12.9).}$$

PROBLEMS

1. Find the value of $\ln x$, using the table in Appendix C, when x is:

   (a) 1.234, (b) 12.34, (c) 123.4, (d) 0.1234.

2. Differentiate with respect to x the following expressions:

    (a) $\ln(2x - 3)$,

    (b) $\ln(3 - 2x)$,

    (c) $x \ln x - x$,

    (d) $x(\ln x)^2$,

    (e) $\ln\left(\tan \dfrac{x}{2}\right)$,

    (f) $\ln(\sec x + \tan x)$.

3. Using properties of a logarithmic function simplify and then differentiate the following expressions:

    (a) $\ln\sqrt{(x^2 - 1)}$,

    (b) $\ln\left(\dfrac{1 + x}{1 - x}\right)^{1/3}$,

    (c) $\ln\sqrt{\left(\dfrac{x^2 - 1}{x^2 + 1}\right)}$,

    (d) $\ln\left(\dfrac{1 + \sin x}{1 - \sin x}\right)^{\frac{1}{2}}$,

    (e) $\log_{10} x$,

    (f) $\ln\left(\dfrac{x^2 + 2x - 3}{x + 1}\right)$.

4. Use logarithmic differentiation to differentiate the following expressions with respect to x:

    (a) $\dfrac{(x - 1)^2 (x - 2)^3}{x + 1}$,

    (b) $\dfrac{(x + 1)^2 (x - 1)^3}{\sqrt{x}}$,

    (c) $\dfrac{\sqrt{x}(x - 1)^8}{(x + 1)^8}$,

    (d) $(x - 1)^2 (x + 1)^{1/3} \cos x$,

    (e) x^x,

    (f) $x^{\cos x}$.

5. If $x = y \ln(x^2 y)$ show that $\dfrac{dy}{dx} = \dfrac{xy - 2y^2}{xy + x^2}$.

6. Use the first three terms of the Maclaurin series expansion of $\ln(1 + x)$ to calculate, correct to six decimal places,

    (a) $\ln(1.01)$,

    (b) $\ln(0.99)$.

7. Find the first three non-zero terms in a power series expansion of $\ln(1 + \sin x)$.

8. Find the Maclaurin series expansion of $\ln(\cos x)$ as far as the term in x^4.

    By taking $x = \dfrac{\pi}{4}$ deduce that $\ln 2 \cong \dfrac{\pi^2}{16}\left(1 + \dfrac{\pi^2}{96}\right)$.

9. Integrate the following expressions with respect to x:

    (a) $\dfrac{1}{3x + 2}$,

    (c) $\dfrac{x + 1}{2x + 1}$,

    (e) $\dfrac{x^3}{2 - x^4}$,

    (b) $\dfrac{1}{2 - 3x}$,

    (d) $\dfrac{x + 1}{x^2 + 4}$,

    (f) $\dfrac{\sin x}{2 + \cos x}$.

10. Evaluate the following integrals:

    (a) $\displaystyle\int_0^1 \dfrac{dx}{2x + 3}$,

    (f) $\displaystyle\int_0^{\pi/4} \dfrac{\sec^2 x}{1 + \tan x}\, dx$,

    (b) $\displaystyle\int_{-1}^1 \dfrac{dx}{4 - 3x}$,

    (g) $\displaystyle\int_0^{\pi/2} \dfrac{\sin 2x}{1 + \cos^2 x}\, dx$,

    (c) $\displaystyle\int_0^2 \dfrac{x}{4x^2 + 9}\, dx$,

    (h) $\displaystyle\int_0^{\pi/8} \tan 2x\, dx$,

    (d) $\displaystyle\int_0^1 \dfrac{x}{2 - x^2}\, dx$,

    (i) $\displaystyle\int_{\pi/4}^{\pi/2} \dfrac{\sin x - \cos x}{\sin x + \cos x}\, dx$.

    (e) $\displaystyle\int_{-1}^0 \dfrac{x^2}{x^3 - 1}\, dx$,

11. Use integration by parts to show that $\displaystyle\int_0^{\pi/4} x \sec^2 x\, dx = \dfrac{\pi}{4} - \dfrac{1}{2}\ln 2$.

12. Integrate by parts to find the following indefinite integrals:

    (a) $\displaystyle\int x \ln x\, dx$,

    (d) $\displaystyle\int \dfrac{1}{x}\ln x\, dx$,

    (b) $\displaystyle\int x^2 \ln x\, dx$,

    (e) $\displaystyle\int x \ln(1 + x)\, dx$,

    (c) $\displaystyle\int \dfrac{1}{x^2}\ln x\, dx$,

    (f) $\displaystyle\int \ln(1 + x^2)\, dx$.

13. If α is a real number and n is an integer verify that

$$\int x^\alpha (\ln x)^n dx = \frac{x^{\alpha+1}}{\alpha+1}(\ln x)^n - \frac{n}{\alpha+1}\int x^\alpha (\ln x)^{n-1} dx, \qquad \text{if } \alpha \neq -1,$$

$$= \frac{(\ln x)^{n+1}}{n+1} + C, \qquad \text{if } \alpha = -1 \text{ and } n \neq -1,$$

$$= \ln|\ln x| + C, \qquad \text{if } \alpha = -1 \text{ and } n = -1.$$

14. At the start of an epidemic the fraction of the population that are carriers of the disease is a. The fraction of the population contracting the disease during the epidemic, s, is given by

$$s = \int_0^a (-\ln x) dx.$$

Calculate s when a is (i) 1/5 and (ii) 1/2.

15. The proportion of an area of farmland which is snail habitat is b. If N cattle graze on this land the number which become infected by a parasite transmitted by the snail is given by

$$-N\int_0^b 4x \ln x \, dx \ .$$

Find the number of cattle which become infected when b is (i) 1/5 and (ii) 1/2. Take N to be fifty.

16. Express the following quantities in the form e^{kx} for some constant k:

(a) 2^x,   (c) 3^x,

(b) $(\tfrac{1}{2})^x$,   (d) 10^x.

17. Differentiate, with respect to x, the following expressions:

(a) $x^2 e^x$,   (d) $e^{\cos x}$,

(b) $\exp(-x^2)$,   (e) $e^{2x} \tan 3x$,

(c) $e^x \ln x$,   (f) $(e^x - 1)/(e^{-x} + 1)$.

18. Find $\dfrac{dy}{dx}$ if $y^3 + (y+1)e^{2x} = x^3$.

19. For $y = c[e^{-at} - e^{-bt}]$, $b > a > 0$, $c > 0$, $t \geqslant 0$ show that

    (i) $y > 0$ when $t > 0$,

    (ii) $y = 0$ when $t = 0$,

    (iii) $y \rightarrow 0$ as $t \rightarrow \infty$,

    (iv) y is greatest when $t = \dfrac{\ln(b/a)}{b - a}$.

The function given by y is often used to describe the variation, with time, in the concentration of a drug injected into the blood stream.

20. Show that, in the following, y satisfies the given differential equation.

    (a) $y = x^2 e^{2x}$, $\dfrac{d^2 y}{dx^2} - 4\dfrac{dy}{dx} + 4y = 2e^{2x}$;

    (b) $y = \exp(x^2)$, $y\dfrac{d^2 y}{dx^2} = 2y^2 + \left(\dfrac{dy}{dx}\right)^2$;

    (c) $y = \exp(\tan^{-1}x)$, $(1 + x^2)\dfrac{d^2 y}{dx^2} + (2x - 1)\dfrac{dy}{dx} = 0$;

    (d) $y = \exp(\sin^{-1}x)$, $(1 - x^2)\dfrac{d^2 y}{dx^2} - x\dfrac{dy}{dx} - y = 0$.

21. Verify that the maximum value taken by the following expressions is as shown:

    (a) $2 + 2x - e^x$, $2 \ln 2$;

    (b) $1 - x \ln x$, $(e + 1)/e$;

    (c) xe^{-x}, $1/e$;

    (d) $e^{-x}\sin x$, $\dfrac{1}{\sqrt{2}}e^{-\pi/4}$.

22. Show that $y = xe^{-2x}$

    (i) satisfies the differential equation $\dfrac{d^2 y}{dx^2} + \dfrac{dy}{dx} - 2y + 3e^{-2x} = 0$,

    (ii) has a maximum turning point at $\left(\dfrac{1}{2}, \dfrac{1}{2e}\right)$,

    (iii) has a point of inflexion at $(1, 1/e^2)$.

23. By using Maclaurin series expansions show that

$$e^{i\theta} = \cos \theta + i \sin \theta$$

where i is the complex number representing $\sqrt{-1}$ and θ is any real number. Hence deduce that

$$(\cos \theta + i \sin \theta)^{\alpha} = \cos \alpha\theta + i \sin \alpha\theta$$

for any real number α.

24. Use the Maclaurin series expansions of $\sin x$ and e^x to obtain a power series, as far as the term in x^4, for:

(i) $e^x \sin x$,   (ii) $\sin(e^x - 1)$,   (iii) $\exp(\sin x)$.

By taking a suitable value of x in the result obtained for (iii), show that

$$e \cong 1 + \frac{\pi}{2} + \frac{\pi^2}{8} - \frac{\pi^4}{128} .$$

25. Integrate, with respect to x, the following expressions:

(a) e^{2-3x},   (d) $x \exp(-x^2)$,

(b) $x^2 \exp(x^3)$,   (e) xe^{-x},

(c) $\cos x \, e^{\sin x}$,   (f) 2^x.

26. Evaluate the following definite integrals:

(a) $\displaystyle\int_0^1 xe^{-2x}dx$,   (c) $\displaystyle\int_{-1}^1 \frac{e^x \, dx}{1 + e^x}$,

(b) $\displaystyle\int_0^1 x^2 \exp(-x^3)dx$,   (d) $\displaystyle\int_{-\infty}^{\infty} \frac{dx}{e^x + e^{-x}}$ (set $t = e^x$).

27. Use integration by parts to show that

$$\int e^{ax} \sin bx \, dx = \frac{(a \sin bx - b \cos bx) e^{ax}}{a^2 + b^2} + C .$$

28. Sketch, using the same axes, curves corresponding to the functions e^x, e^{-x} and $\frac{1}{2}(e^x + e^{-x})$.

If $y = \frac{1}{2}(e^x + e^{-x})$ show that

$$y^2 - \left(\frac{dy}{dx}\right)^2 = 1 .$$

Find the area enclosed by each of the curves, the x-axis and the lines $x = 0, x = 1$.

29. The reaction to a given dose of a drug t hours after administration is given by a function $r(t)$ (measured in appropriate units). Find the total reaction to the dose of the drug, as given by $\int_{0}^{\infty} r(t)\mathrm{d}t$, when $r(t)$ is

   (i) $(1 + t^2)^{-1}$,   (ii) $te^{-t/2}$,   (iii) $t\exp(-t^2)$.

30. The rate of change in the concentration of a particular radioactive tracer is given by 2^{-t} where the time t is measured in hours from the time of administration. If the initial concentration is 1 microgram per litre, determine the concentration at subsequent times.

31. A radioactive isotope loses a tenth of its radiation in ten years. Show that the isotope has a half-life of about 66 years.

32. The bones from a skeleton unearthed on a building site contain one third of the amount of carbon 14 which living bones contain. Given that the half-life of C^{14} is 5730 years find the age of the skeleton.

33. The intensity of sunlight, I, in a pond at a depth of d metres is given by
$$I = I_{o}\, e^{-1.5d} ,$$
   where I_o is the intensity of sunlight at the surface. At what depths are (a) 50% and (b) 90% of the sun's light absorbed?

34. A radioactive tracer having a half-life of 12 hours was injected into some tissue cells and after one hour the amount of radioactive tracer present in the tissue was measured and found to be 80% of its original value. How much of the radioactive tracer administered has been removed from the tissue during this time?

35. When cancerous cells are irradiated with x-rays the fraction surviving such a treatment, for varying doses of radiation, falls exponentially. Thus the fraction F surviving after a dose d is given by $F = e^{-kd}$ where k is a positive constant. What dose should be given if 90% of the cells are to be destroyed when it is known that a dose of 3 units results in 30% being destroyed?

CHAPTER 13

The integration of rational polynomials

13.1 INTRODUCTION

In this chapter methods for evaluating integrals in which the integrand is a rational polynomial will be given. That is, the integral is of the form $\int \dfrac{p(x)}{q(x)} \, dx$ where $p(x)$ and $q(x)$ are polynomials. The procedures for completing the square of a quadratic expression, and writing a rational polynomial in terms of partial fractions, are discussed. These enable the integral to be rewritten in a form which can be easily integrated by using standard results.

The integrals considered in this chapter mostly have integrands which are proper rational polynomials, that is the degree of the numerator, $p(x)$, is less than the degree of the denominator, $q(x)$. If this is not the case then the integrand must first be divided, using the technique discussed in Section 3.7. The subsequent treatment depends on the form of the denominator $q(x)$. Integrals in which $q(x)$ is linear have been dealt with in Chapter 12.

13.2 DENOMINATOR QUADRATIC

The treatment of integrals of the form

$$\int \frac{dx + e}{ax^2 + bx + c} \, dx \ ,$$

where a, b, c, d and e are constants, depends on the quadratic expression $ax^2 + bx + c$.

Three cases can occur:

(a) the quadratic has no real factors,
(b) the quadratic is a perfect square,
(c) the quadratic has distinct real factors.

(a) *No Real Factors*

In this case the procedure is to first complete the square in the quadratic terms and then make a change of variable.

13.2.1 Completing the Square

A given quadratic expression $ax^2 + bx + c$ may be rewritten in several alternative forms. One way may be to write it as the product of two real linear factors, as in Chapter 3, another is to **complete the square** in x. This last procedure consists of rewriting $ax^2 + bx$ as the difference of a squared quantity involving x and a constant. Thus

$$ax^2 + bx = a\left(x + \frac{b}{2a}\right)^2 - \frac{b^2}{4a} .$$

(Note that this operation is always possible).

Hence $ax^2 + bx + c = a\left(x + \frac{b}{2a}\right)^2 + c - \frac{b^2}{4a}$ (13.1)

where the right-hand side of (13.1) gives the completion of the square of $ax^2 + bx + c$. Notice that if $c = b^2/(4a)$ the quadratic can be written as a perfect square. From (13.1) it is possible to derive the formula stated in Section 3.4 for solving the quadratic equation $ax^2 + bx + c = 0$, since

$$a\left(x + \frac{b}{2a}\right)^2 + c - \frac{b^2}{4a} = 0 ,$$

$$a\left(x + \frac{b}{2a}\right)^2 = \frac{b^2}{4a} - c = \frac{b^2 - 4ac}{4a} ,$$

$$x + \frac{b}{2a} = \frac{\pm\sqrt{(b^2 - 4ac)}}{2a} ,$$

$$x = \frac{-b \pm \sqrt{(b^2 - 4ac)}}{2a} .$$

Example 1: Complete the square on (i) $x^2 - 2x + 2$, (ii) $2x^2 + 2x + 3$.

(i) $x^2 - 2x + 2$. Here $a = 1, b = -2, c = 2$, so
$$x^2 - 2x + 2 = (x - 1)^2 + 2 - 1 = (x - 1)^2 + 1 .$$

(ii) $2x^2 + 2x + 3$. In this example $a = 2, b = 2, c = 3$, so
$$2x^2 + 2x + 3 = 2(x + \tfrac{1}{2})^2 + 3 - \tfrac{1}{2} = 2(x + \tfrac{1}{2})^2 + \tfrac{5}{2} .$$

Example 2: Evaluate (i) $\displaystyle\int \frac{dx}{x^2 - 2x + 2}$, (ii) $\displaystyle\int \frac{2x + 5}{2x^2 + 2x + 3}dx$.

(i) $\displaystyle\int \frac{dx}{x^2 - 2x + 2} = \int \frac{dx}{(x-1)^2 + 1}$ using example 1(i)

$\displaystyle = \int \frac{dt}{t^2 + 1}$ putting $t = x - 1$

$= \tan^{-1} t + C = \tan^{-1}(x - 1) + C$.

(ii) $\displaystyle\int \frac{2x + 5}{2x^2 + 2x + 3}\, dx = \int \frac{2x + 5}{2(x + \frac{1}{2})^2 + \frac{5}{2}}\, dx$ using example 1(ii)

$\displaystyle = \int \frac{2(t - \frac{1}{2}) + 5}{2t^2 + \frac{5}{2}}\, dt$ putting $t = x + \frac{1}{2}$

$\displaystyle = \int \frac{t + 2}{t^2 + \frac{5}{4}}\, dt = \int \frac{t}{t^2 + \frac{5}{4}}\, dt + \int \frac{2}{t^2 + \frac{5}{4}}\, dt$

$\displaystyle = \frac{1}{2}\ln(t^2 + \frac{5}{4}) + 2\,\frac{2}{\sqrt{5}}\tan^{-1}\left(\frac{2t}{\sqrt{5}}\right) + C$

using the result of example 4 in Section 12.3 with $a = 5/4$ to evaluate the first integral. Thus

$\displaystyle\int \frac{2x + 5}{2x^2 + 2x + 3}\, dx = \frac{1}{2}\ln\left((x + \frac{1}{2})^2 + \frac{5}{4}\right) + \frac{4}{\sqrt{5}}\tan^{-1}\left(\frac{2x + 1}{\sqrt{5}}\right) + C$.

(b) *Perfect Square*

Integrals in which the denominator is a perfect square are evaluated by using a change of variable as shown in the following example.

Example 3: Evaluate $\displaystyle\int \frac{x}{x^2 + 2x + 1}\, dx$.

$\displaystyle\int \frac{x}{x^2 + 2x + 1}\, dx = \int \frac{x}{(x + 1)^2}\, dx$

$\displaystyle = \int \frac{t - 1}{t^2}\, dt$ putting $t = x + 1$

$\displaystyle = \int \left(\frac{1}{t} - \frac{1}{t^2}\right) dt$

$\displaystyle = \ln|t| + \frac{1}{t} + C = \ln|x + 1| + \frac{1}{x + 1} + C$.

(c) *Distinct Real Factors*

When the quadratic expression in the denominator has distinct real factors the integrand is expressed in terms of these factors by using the method of partial fractions.

13.2.2 Partial Fractions

The expression $\dfrac{1}{2x+1} + \dfrac{2}{x-3}$ can be written as a single fraction by introducing the common denominator $(2x+1)(x-3)$. Thus

$$\frac{1}{2x+1} + \frac{2}{x-3} = \frac{(x-3)+2(2x+1)}{(2x+1)(x-3)} = \frac{5x-1}{(2x+1)(x-3)}.$$

The method of **partial fractions** provides a means of performing the reverse operation. That is, starting with $\dfrac{5x-1}{(2x+1)(x-3)}$ the technique enables the fractions $\dfrac{1}{2x+1} + \dfrac{2}{x-3}$ to be obtained, and so a complicated rational polynomial may be written in terms of simpler components.

The decomposition of an expression of the form $\dfrac{ax+b}{(cx+d)(ex+f)}$, where the values of a, b, c, d, e and f are known and $cf \neq de$, is governed by the rule:

To every distinct linear factor $(\alpha x + \beta)$ of the denominator there corresponds a partial fraction of the form $\dfrac{\gamma}{\alpha x + \beta}$.

Thus $$\frac{ax+b}{(cx+d)(ex+f)} = \frac{A}{cx+d} + \frac{B}{ex+f}$$

where the constants A and B are to be determined. To find the values of A and B multiply through by $(cx+d)(ex+f)$ to obtain

$$ax+b = A(ex+f) + B(cx+d).$$

As this equation is to hold for all values of x it will hold, in particular, for $x = -f/e$. This choice of x makes the term involving A disappear, so enabling B to be found. Similarly, taking $x = -d/c$, the term in B disappears, leaving a simple calculation to obtain A.

Example 4: Express in terms of partial fractions (i) $\dfrac{1}{x^2-1}$ and (ii) $\dfrac{2-x}{x^2+x-2}$.

(i) $\dfrac{1}{x^2-1} = \dfrac{1}{(x-1)(x+1)} = \dfrac{A}{x-1} + \dfrac{B}{x+1}$.

Multiplying by $(x - 1)(x + 1)$ gives

$$1 = A(x + 1) + B(x - 1) .$$

When $x = -1$, $\qquad 1 = -2B \quad$ so $B = -\frac{1}{2}$.

When $x = 1$, $\qquad 1 = 2A \quad$ so $A = \frac{1}{2}$.

Therefore $\quad \dfrac{1}{x^2 - 1} = \dfrac{\frac{1}{2}}{x - 1} - \dfrac{\frac{1}{2}}{x + 1}$.

(ii) $\dfrac{2 - x}{x^2 + x - 2} = \dfrac{2 - x}{(x - 1)(x + 2)} = \dfrac{A}{x - 1} + \dfrac{B}{x + 2}$.

Therefore $2 - x = A(x + 2) + B(x - 1)$.

Putting $x = -2$ gives $4 = -3B$ so $B = -\frac{4}{3}$,

and $\quad x = 1$ gives $1 = 3A$ so $A = \frac{1}{3}$.

Hence $\quad \dfrac{2 - x}{x^2 + x - 2} = \dfrac{\frac{1}{3}}{x - 1} - \dfrac{\frac{4}{3}}{x + 2}$.

Example 5: Evaluate (i) $\displaystyle\int \dfrac{dx}{x^2 - 1}$, (ii) $\displaystyle\int \dfrac{x^2}{x^2 + x - 2} \, dx$.

(i) $\displaystyle\int \dfrac{dx}{x^2 - 1} = \int \left(\dfrac{\frac{1}{2}}{x - 1} - \dfrac{\frac{1}{2}}{x + 1} \right) dx \qquad$ using example 4(i).

$$= \tfrac{1}{2} \ln|x - 1| - \tfrac{1}{2} \ln|x + 1| + C = \tfrac{1}{2} \ln \left| \dfrac{x - 1}{x + 1} \right| + C .$$

(ii) The integrand is not a proper rational polynomial. Dividing out

$$\dfrac{x^2}{x^2 + x - 2} = 1 + \dfrac{2 - x}{x^2 + x - 2}$$

$$= 1 + \dfrac{\frac{1}{3}}{x - 1} - \dfrac{\frac{4}{3}}{x + 2} \qquad \text{using example 4(ii)}$$

Therefore

$$\int \dfrac{x^2}{x^2 + x - 2} \, dx = \int \left(1 + \dfrac{\frac{1}{3}}{x - 1} - \dfrac{\frac{4}{3}}{x + 2} \right) dx$$

$$= x + \dfrac{1}{3} \ln|x - 1| - \dfrac{4}{3} \ln|x + 2| + C .$$

Example 6: Evaluate $\displaystyle\int_1^2 \dfrac{2x + 1}{x(x + 1)} \, dx$.

Introducing partial fractions $\dfrac{2x+1}{x(x+1)} = \dfrac{A}{x} + \dfrac{B}{x+1}$,

so $2x + 1 = A(x+1) + Bx$.

Putting $x = 0$ gives $1 = A$,

and $x = -1$ gives $-1 = -B$.

Therefore

$$\int_1^2 \frac{2x+1}{x(x+1)}\,dx = \int_1^2 \left(\frac{1}{x} + \frac{1}{x+1}\right)dx = \left[\ln|x| + \ln|x+1|\right]_1^2$$

$$= (\ln 2 + \ln 3) - (\ln 1 + \ln 2) = \ln 3 .$$

13.3 DENOMINATOR CUBIC

A cubic polynomial may factorise into (a) three distinct linear factors, (b) three identical linear factors, (c) a linear factor and a perfect square, or (d) a linear factor and a non-factorising quadratic factor.

The rule for obtaining partial fractions given in Section 13.2.2 is not restricted to problems having just two linear factors but applies whenever the denominator contains distinct linear factors. Hence this rule is applicable to case (a).

In cases (b) and (c) there are repeated factors. These are handled by the rule:

To every repeated linear factor $(\alpha x + \beta)^r$ in the denominator there corresponds the partial fractions

$$\frac{A_1}{\alpha x + \beta} + \frac{A_2}{(\alpha x + \beta)^2} + \ldots + \frac{A_r}{(\alpha x + \beta)^r} .$$

This rule is valid for all r and in particular for $r = 2$ and $r = 3$.

The non-factorising quadratic term in case (d) gives rise to a term determined by the rule:

To every non-factorising quadratic term $Q(x)$ in the denominator there corresponds a partial fraction of the form $\dfrac{Ax + B}{Q(x)}$.

The method used to compute values for the constants $A_1, A_2, \ldots, A_r, A$ and B is essentially the same as that for the case of distinct linear factors, that is, cross-multiply and substitute values for x.

Example 1: Evaluate $\displaystyle\int \frac{x}{(x-1)(x+1)^2}\,dx.$

The integrand can be written, using partial fractions, as

$$\frac{x}{(x-1)(x+1)^2} = \frac{A}{x-1} + \frac{B}{x+1} + \frac{C}{(x+1)^2} .$$

Therefore $x = A(x + 1)^2 + B(x - 1)(x + 1) + C(x - 1)$.

$\quad\quad x = \quad 1 \quad$ gives $\quad 1 = \quad 4A \quad\quad$ so $A = \quad \frac{1}{4}$,

$\quad\quad x = -1 \quad$ gives $-1 = -2C \quad\quad$ so $C = \quad \frac{1}{2}$,

$\quad\quad x = \quad 0 \quad$ gives $\quad 0 = A - B - C \quad$ so $B = -\frac{1}{4}$.

Hence

$$\int \frac{x}{(x - 1)(x + 1)^2}\, dx = \int \left(\frac{\frac{1}{4}}{x - 1} - \frac{\frac{1}{4}}{x + 1} + \frac{\frac{1}{2}}{(x + 1)^2} \right) dx$$

$$= \frac{1}{4}\ln|x - 1| - \frac{1}{4}\ln|x + 1| - \frac{1}{2(x + 1)} + C .$$

Example 2: Evaluate $\displaystyle\int \frac{x}{(x - 1)(x^2 + 1)}\, dx.$

Here $\quad \dfrac{x}{(x - 1)(x^2 + 1)} = \dfrac{A}{x - 1} + \dfrac{Bx + C}{x^2 + 1}$.

Therefore $x = A(x^2 + 1) + (Bx + C)(x - 1)$.

$\quad\quad x = \quad 1 \quad$ gives $\quad 1 = 2A \quad\quad\quad\quad$ so $A = \quad \frac{1}{2}$,

$\quad\quad x = \quad 0 \quad$ gives $\quad 0 = A - C \quad\quad\quad$ so $C = \quad \frac{1}{2}$,

$\quad\quad x = -1 \quad$ gives $-1 = 2A + 2(B - C) \;$ so $B = -\frac{1}{2}$.

Hence

$$\int \frac{x}{(x - 1)(x^2 + 1)}\, dx = \int \left(\frac{\frac{1}{2}}{x - 1} + \frac{-\frac{1}{2}x + \frac{1}{2}}{x^2 + 1} \right) dx$$

$$= \frac{1}{2}\int \left(\frac{1}{x - 1} - \frac{x}{x^2 + 1} + \frac{1}{x^2 + 1} \right) dx$$

$$= \frac{1}{2}\left(\ln|x - 1| - \frac{1}{2}\ln(x^2 + 1) + \tan^{-1}x \right) + C .$$

PROBLEMS

1. Integrate with respect to x the following expressions:

(a) $\dfrac{1}{x^2 + 4x + 5}$,

(d) $\dfrac{1}{x^2 + 4x + 4}$,

(b) $\dfrac{x}{x^2 + 4x + 8}$,

(e) $\dfrac{x + 1}{4x^2 - 4x + 1}$,

(c) $\dfrac{x^2}{x^2 - 2x + 5}$,

(f) $\dfrac{x^2}{x^2 + 2x + 1}$.

2. Evaluate:

(a) $\displaystyle\int_{-1}^{1} \frac{dx}{x^2 + 6x + 10}$,

(d) $\displaystyle\int_{0}^{1} \frac{x - 1}{x^2 + 2x + 1}\, dx$,

(b) $\displaystyle\int_{1}^{2} \frac{x\, dx}{x^2 - 2x + 2}$,

(e) $\displaystyle\int_{0}^{1} \frac{dx}{x^2 - 4x + 4}$,

(c) $\displaystyle\int_{0}^{2} \frac{dx}{x^2 + 2x + 1}$,

(f) $\displaystyle\int_{0}^{1} \frac{x + 1}{4 - 2x - x^2}\, dx$.

3. Express in terms of partial fractions and then integrate the following:

(a) $\displaystyle\frac{x + 1}{(x - 3)(x - 4)}$,

(d) $\displaystyle\frac{x + 7}{2x^2 + 3x - 2}$,

(b) $\displaystyle\frac{x - 13}{x^2 - x - 6}$,

(e) $\displaystyle\frac{x^2}{1 - x^2}$,

(c) $\displaystyle\frac{8 - x}{2 + x - x^2}$,

(f) $\displaystyle\frac{x + 10}{x^2 - x - 12}$.

4. Evaluate:

(a) $\displaystyle\int_{1}^{2} \frac{x + 3}{x(x + 2)}\, dx$,

(d) $\displaystyle\int_{4}^{6} \frac{dx}{x^2 - x - 6}$,

(b) $\displaystyle\int_{0}^{2} \frac{dx}{9 - x^2}$,

(e) $\displaystyle\int_{2}^{3} \frac{x}{x^2 - 4x - 5}\, dx$,

(c) $\displaystyle\int_{2}^{4} \frac{2x - 1}{x^2 + x - 2}\, dx$,

(f) $\displaystyle\int_{4}^{5} \frac{x + 2}{x^2 - 2x - 3}\, dx$.

5. Use the given substitution to show that

(a) $\displaystyle\int_{\pi/3}^{\pi/2} \frac{\sin\theta}{\cos^2\theta + \cos\theta - 2}\, d\theta = \frac{1}{3}\ln\frac{2}{5}$, $t = \cos\theta$;

(b) $\displaystyle\int_{-1}^{1} \frac{dx}{1 + e^x} = 1$, $t = e^x$.

6. Express in terms of partial fractions and then integrate the following:

(a) $\dfrac{x - 1}{(x + 1)(x^2 + 1)}$,

(d) $\dfrac{10x^2 + 9x - 7}{x^2(x + 2)}$,

(b) $\dfrac{1}{x^3 - x}$,

(e) $\dfrac{3 - x}{(x + 1)^3}$,

(c) $\dfrac{2x + 1}{(x - 1)(x^2 + x + 1)}$,

(f) $\dfrac{x + 4}{x(x + 1)(x + 2)}$.

7. Evaluate:

(a) $\displaystyle\int_2^3 \dfrac{x + 1}{(x - 1)(x^2 + 1)}\, dx$,

(c) $\displaystyle\int_2^3 \dfrac{9}{(x - 1)(x + 2)^2}\, dx$,

(b) $\displaystyle\int_0^{\frac{1}{2}} \dfrac{x}{(2x + 1)(1 - x)^2}\, dx$,

(d) $\displaystyle\int_1^3 \dfrac{x + 3}{x(x + 1)(x + 2)}\, dx$.

8. Show that $\displaystyle\int_0^3 \dfrac{2x}{(1 + x^2)(3 + x^2)}\, dx = \dfrac{1}{2}\ln\dfrac{5}{2}$.

9. In a period of t days the number of bacteria in a culture increases from N_0 to N according to the relationship

$$t = \int_{N_0}^{N} \dfrac{B\, dn}{n(A - n)} ,$$

where A and B are positive constants, with $A > N_0$. Find N as a function of t.

10. In t hours a chemical reaction producing an enzyme results in a concentration q of that enzyme given by the relationship

$$t = \int \dfrac{dq}{(a - q)(b + q)} ,$$

in appropriate units, where a and b are positive constants. Find an equation relating q and t.

CHAPTER 14

Applications of integration

14.1 INTRODUCTION

In this chapter applications requiring the evaluation of a definite integral are presented. These include procedures for finding areas, volumes, lengths of lines, surface areas and mean values. The integral formulation for each of these quantities is obtained by first approximating the required quantity by a finite sum of similar small component parts. Then by allowing the number of such parts to become large the sum can be represented by an integral as outlined in Section 11.6.

14.2 AREA BETWEEN TWO CURVES

In example 2 of Section 11.2 the region bounded by $x = 1, x = 4, y = \sqrt{x}$ and $y = 0$ was found to have an area of 14/3. Consider now the region bounded by $x = 1$, $x = 4$, $y = \sqrt{x}$ and $y = 1$ as shown in Fig. 14.1. The area of this new

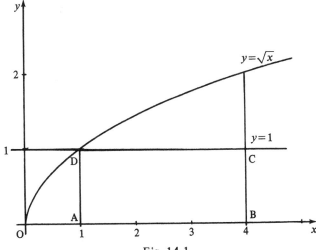

Fig. 14.1

region can be found by subtracting the area of the rectangle ABCD, which is 3, from the original area of 14/3 to give

$$14/3 - 3 = 5/3 \ .$$

This problem indicates how the area between two curves can be found. The area between each curve and the x-axis is computed and their difference gives the required area.

Let the two curves be given by $y = f_1(x)$ and $y = f_2(x)$ with $f_1(x) \geqslant f_2(x) \geqslant 0$ for $a \leqslant x \leqslant b$ (see Fig. 14.2).

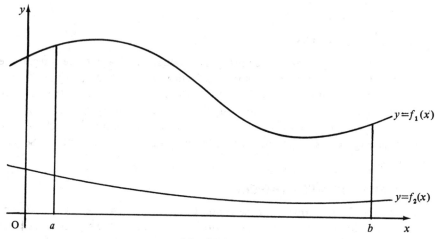

Fig. 14.2

From Chapter 11, the area A_1 enclosed by $y = f_1(x)$, $x = a$, $x = b$ and $y = 0$ is given by

$$A_1 = \int_a^b f_1(x) \ dx \ .$$

Similarly, the area A_2 enclosed by $y = f_2(x)$, $x = a$, $x = b$ and $y = 0$ is

$$A_2 = \int_a^b f_2(x) \ dx \ .$$

Hence the area enclosed by the two curves $y = f_1(x)$ and $y = f_2(x)$, between $x = a$ and $x = b$ is

$$A = A_1 - A_2 = \int_a^b [f_1(x) - f_2(x)] \ dx \ . \tag{14.1}$$

As areas are positive in value it is essential that the upper curve, $y = f_1(x)$, be taken first in the integrand.

The result given in (14.1) can also be used to calculate areas when the curves are below the x-axis. If the two curves cross, say when $x = c$, $a < c < b$, the total area is given by the sum of the individual areas between $x = a$ and $x = c$, and between $x = c$ and $x = b$, etc. The individual areas must be expressed in terms of integrals with integrands given by the functions corresponding to the upper curve minus the lower curve.

Example 1: Find the area between $y = \sqrt{x}, y = 1, x = 1$ and $x = 4$.

As $\sqrt{x} \geqslant 1$ for $1 \leqslant x \leqslant 4$,

$$y = \sqrt{x} = f_1(x), y = 1 = f_2(x) \text{ and the area is given by}$$

$$\int_1^4 (\sqrt{x} - 1)\, dx = \left[\frac{2}{3}x^{3/2} - x\right]_1^4 = \left(\frac{2}{3}8 - 4\right) - \left(\frac{2}{3} - 1\right)$$

$$= \frac{5}{3}.$$

Example 2: Find the total area enclosed by the curves $y = x^3$ and $y = x$.

The two curves cross when $x^3 = x$, that is when $x = -1, 0$ and 1.

From Fig. 14.3 the two areas are given by

$$A_1 = \int_0^1 (x - x^3)\, dx = \left[\frac{x^2}{2} - \frac{x^4}{4}\right]_0^1 = \left(\frac{1}{2} - \frac{1}{4}\right) - 0 = \frac{1}{4},$$

$$A_2 = \int_{-1}^0 (x^3 - x)\, dx = \left[\frac{x^4}{4} - \frac{x^2}{2}\right]_{-1}^0 = 0 - \left(\frac{1}{4} - \frac{1}{2}\right) = \frac{1}{4}.$$

So the total area required is $A = A_1 + A_2 = \frac{1}{4} + \frac{1}{4} = \frac{1}{2}$.

(Notice that the integral

$$\int_{-1}^1 (x^3 - x)\, dx = \left[\frac{x^4}{4} - \frac{x^2}{2}\right]_{-1}^1 = \left(\frac{1}{4} - \frac{1}{2}\right) - \left(\frac{1}{4} - \frac{1}{2}\right) = 0 \neq A).$$

An alternative approach to equation (14.1) is via summation. The area between the two curves can be divided into n strips, parallel to the y-axis, each of width $\Delta x = (b - a)/n$. A typical strip, the kth say, can be approximated by a rectangle, so the area A_k of the strip (Fig. 14.4) is given approximately by

$$A_k \cong [f_1(x_k) - f_2(x_k)]\, \Delta x$$

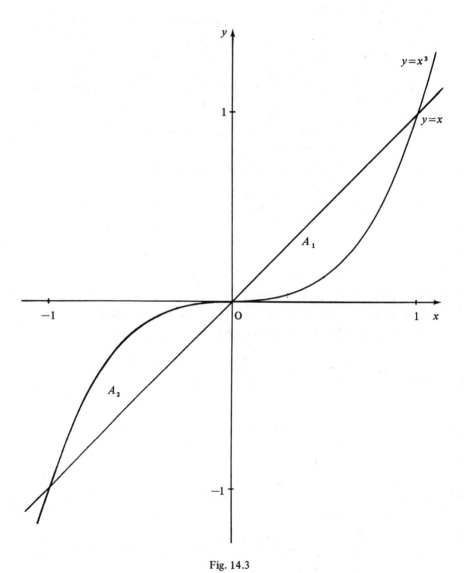

Fig. 14.3

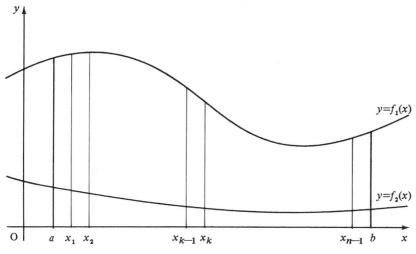

Fig. 14.4

where $x_k = a + k\,\Delta x$. The total area A between the two curves is then given, approximately, by

$$A = \sum_{k=1}^{n} A_k \cong \sum_{k=1}^{n} [f_1(x_k) - f_2(x_k)]\,\Delta x .$$

As $n \to \infty$, $\Delta x \to 0$ and the summation can be replaced by an integral (see Section 11.6). Thus

$$A = \int_{a}^{b} [f_1(x) - f_2(x)]\,dx .$$

In some problems the boundary of the region whose area is to be found is described in terms of functions of y, such as $x = g(y)$. As the next example shows it is not necessary to rewrite the curves as functions of x in order to find the area.

Example 3: Find the area bounded by $x = 0$, $x = 2y - y^2$, $y = 0$ and $y = 2$.
    The region is shown in Fig. 14.5.

$$\text{Area} = \int_{0}^{2} g(y)\,dy = \int_{0}^{2} (2y - y^2)\,dy = \left[y^2 - \frac{y^3}{3} \right]_{0}^{2} = 4 - \frac{8}{3} = \frac{4}{3} .$$

Treating the region in terms of functions of x the equation $x = 2y - y^2$ is given by the two curves

$$y = f_1(x) = 1 + \sqrt{(1-x)} \quad \text{and} \quad y = f_2(x) = 1 - \sqrt{(1-x)} .$$

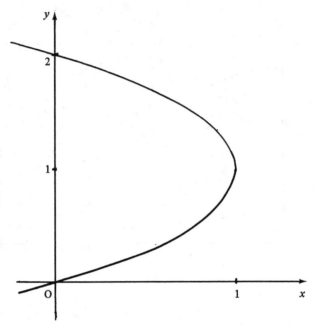

Fig. 14.5

Hence the area is obtained from

$$\int_0^1 [f_1(x) - f_2(x)] \, dx = \int_0^1 \{[1 + \sqrt{(1-x)}] - [1 - \sqrt{(1-x)}]\} \, dx$$

$$= 2 \int_0^1 \sqrt{(1-x)} \, dx = 2 \left[-\frac{2}{3}(1-x)^{3/2} \right]_0^1$$

$$= 2 \left[0 - \left(-\frac{2}{3} \right) \right] = \frac{4}{3} \, , \qquad \text{as above .}$$

If the region is described in terms of a parameter t the area is found by the rule for changing the variable of integration. Thus

$$\text{Area} = \int_a^b y(x) \, dx = \int_c^d y(x(t)) \frac{dx(t)}{dt} \, dt \ .$$

Example 4: Find the area of the region given by the equations $x = 2 \cos t$, $y = \sin t$ for $0 \leqslant t \leqslant \pi$.

The region is shown in Fig. 14.6.

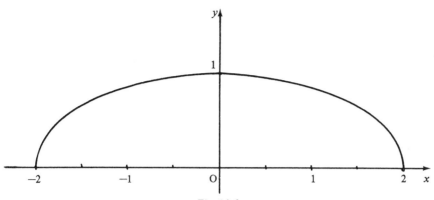

Fig. 14.6

$$\text{Area} = \int_{-2}^{2} y(x)\, dx = \int_{\pi}^{0} \sin t\,(-2\sin t)\, dt$$

$$= \int_{0}^{\pi} 2\sin^2 t\, dt = \int_{0}^{\pi} [1 - \cos 2t]\, dt$$

$$= [t - \tfrac{1}{2}\sin 2t]_{0}^{\pi} = \pi .$$

14.3 VOLUMES

A solid bounded by planes perpendicular to the x-axis through the points $x = a$, $x = b$ may be thought of as being built up from n thin plates each of thickness Δx $(\Delta x = (b-a)/n)$. See Fig. 14.7.

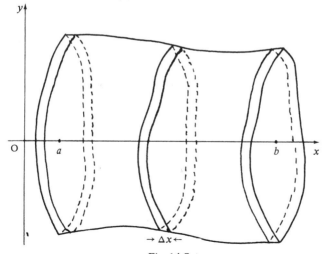

Fig. 14.7

The volume V_k of a typical plate, the kth, is given, approximately, by

$$V_k \cong A_k \, \Delta x \ ,$$

where A_k is the area of the plate. Hence the total volume V of the solid is given, approximately, by

$$V = \sum_{k=1}^{n} V_k \cong \sum_{k=1}^{n} A_k \, \Delta x$$

In the limit as n increases to infinity the value of this expression gives the true volume. Thus

$$V = \lim_{n \to \infty} \left(\sum_{k=1}^{n} A_k \, \Delta x \right) = \int_a^b A(x) \, dx \qquad (14.2)$$

where $A(x)$ is the area of the cross-section of the solid perpendicular to the x-axis at the point x.

The volumes of solids having the x-axis as an axis of symmetry can be easily found by using (14.2). Suppose a **volume of revolution** is formed by rotating, through 2π radians (or $360°$), about the x-axis, the curve $y = f(x)$ between $x = a$ and $x = b$. As the cross-section of the solid at the point x is circular, with radius $y = f(x)$, $A(x) = \pi y^2 = \pi [f(x)]^2$. Thus the volume of revolution is given by

$$V = \pi \int_a^b [f(x)]^2 \, dx = \pi \int_a^b y^2 \, dx \ .$$

Example 1: Find the volume of a right circular cone of height H and base radius R.

The cone may be thought of as being formed by rotating the line $y = Rx/H$ about the x-axis from $x = 0$ to $x = H$, see Fig. 14.8. Therefore

$$V = \pi \int_0^H \left(\frac{R}{H} x \right)^2 dx = \frac{\pi R^2}{H^2} \int_0^H x^2 \, dx = \frac{\pi R^2}{H^2} \left[\frac{x^3}{3} \right]_0^H = \frac{1}{3} \pi R^2 H \ .$$

Example 2: Find the volume of a sphere of radius a.

The volume required may be obtained by rotating the curve $y = \sqrt{(a^2 - x^2)}$, which describes a semi-circle of radius a, about the x-axis, see Fig. 14.9.

Thus $\quad V = \pi \int_{-a}^{a} (a^2 - x^2) \, dx = \pi \left[a^2 x - \frac{x^3}{3} \right]_{-a}^{a} = \pi \left[\frac{2a^3}{3} - \left(-\frac{2a^3}{3} \right) \right]$

$$= \frac{4}{3} \pi a^3 \ .$$

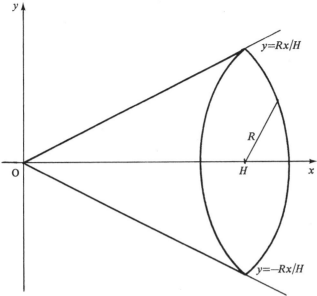

Fig. 14.8

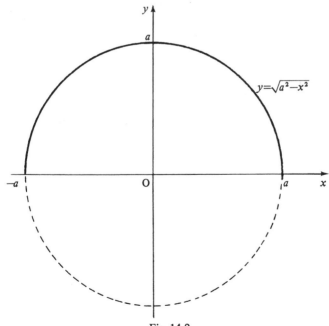

Fig. 14.9

14.4 LENGTH OF ARC OF A PLANE CURVE

In this section a formula is derived for the length of a continuous curve. Suppose the curve is given by $y = f(x)$ and the required arc length, s, is from A to B, having $x = a$ and $x = b$ respectively, see Fig. 14.10.

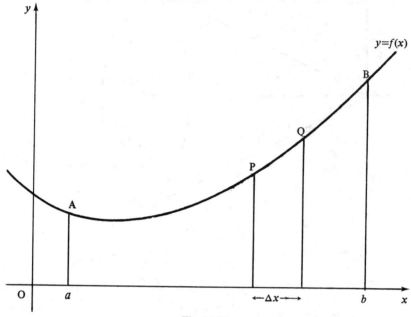

Fig. 14.10

Divide the x-axis between $x = a$ and $x = b$ into n equal parts of length $\Delta x = (b - a)/n$. Let that part of the arc AB between $x = x_{k-1}$ and $x = x_k$, PQ say, be of length Δs_k. This length is approximately the same (Fig. 14.11) as that of the chord PQ which, by Pythagoras' Theorem, has length

$$[(\Delta x)^2 + (\Delta y_k)^2]^{\frac{1}{2}} = \left[1 + \left(\frac{\Delta y_k}{\Delta x}\right)^2\right]^{\frac{1}{2}} \Delta x \ .$$

Fig. 14.11

So the length of arc s, from A to B, is given, approximately, by

$$s = \sum_{k=1}^{n} \Delta s_k \cong \sum_{k=1}^{n} \left[1 + \left(\frac{\Delta y_k}{\Delta x} \right)^2 \right]^{\frac{1}{2}} \Delta x .$$

In the limit as n becomes infinite this approximation becomes exact, so

$$s = \lim_{n \to \infty} \left\{ \sum_{k=1}^{n} \left[1 + \left(\frac{\Delta y_k}{\Delta x} \right)^2 \right]^{\frac{1}{2}} \Delta x \right\} .$$

As $n \to \infty$, $\Delta x \to 0$ and $\dfrac{\Delta y_k}{\Delta x} \to \dfrac{dy}{dx} = f'(x_k)$, therefore, from (11.8), the arc length is given by

$$s = \int_{a}^{b} [1 + [f'(x)]^2]^{\frac{1}{2}} \, dx = \int_{a}^{b} \left[1 + \left(\frac{dy}{dx} \right)^2 \right]^{\frac{1}{2}} \, dx .$$

Example: Find the length of the curve $y = x^{3/2}$ between $x = 0$ and $x = 1$ (Fig. 14.12).

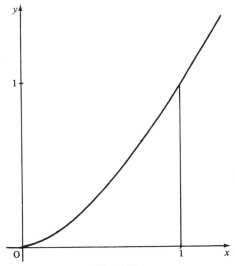

Fig. 14.12

Here $y = x^{3/2}$ so $\dfrac{dy}{dx} = \dfrac{3}{2} x^{\frac{1}{2}}$ and $1 + \left(\dfrac{dy}{dx} \right)^2 = 1 + \dfrac{9x}{4}$.

Hence $s = \displaystyle\int_{0}^{1} \left(1 + \frac{9x}{4} \right)^{\frac{1}{2}} dx = \left[\frac{2}{3} \cdot \frac{4}{9} \left(1 + \frac{9x}{4} \right)^{3/2} \right]_{0}^{1}$

$$= \frac{8}{27} \left[\left(\frac{13}{4} \right)^{3/2} - 1 \right] = \frac{13^{3/2} - 8}{27} .$$

14.5 AREA OF SURFACE OF REVOLUTION

The area of the surface of revolution formed by rotating, through 2π radians ($360°$), about the x-axis the curve $y = f(x)$ between $x = a$ and $x = b$ is given by

$$S = 2\pi \int_a^b f(x) \, [1 + [f'(x)]^2]^{\frac{1}{2}} \, dx = 2\pi \int_a^b y \left[1 + \left(\frac{dy}{dx} \right)^2 \right]^{\frac{1}{2}} dx \ .$$

(14.3)

This result is obtained by considering the corresponding volume of revolution to be made up of n thin discs of thickness Δx. The curved surface area of a disc, the kth say, is then, approximately, the circumference, $2\pi y_k$, of the disc times the portion of arc length forming the edge, Δs_k. That is, for the kth disc,

$$S_k \cong 2\pi y_k \, \Delta s_k \cong 2\pi y_k \left[1 + \left(\frac{\Delta y_k}{\Delta x} \right)^2 \right]^{\frac{1}{2}} \Delta x$$

so the total curved surface area will be

$$S = \sum_{k=1}^{n} S_k \ .$$

Again the approximation improves as the number of discs increases until, in the limit of an infinite number, the total curved surface area of revolution can be represented by the definite integral (14.3).

Example: Find the surface area of a sphere of radius a.
     This surface is formed by rotating the curve $y = (a^2 - x^2)^{\frac{1}{2}}$ about the x-axis, see Fig. 14.9. Now

$$y' = -x \, (a^2 - x^2)^{-\frac{1}{2}} \quad \text{and} \quad 1 + (y')^2 = 1 + \frac{x^2}{a^2 - x^2} = \frac{a^2}{a^2 - x^2} \ ,$$

therefore,

$$S = 2\pi \int_{-a}^{a} (a^2 - x^2)^{\frac{1}{2}} \cdot \frac{a}{(a^2 - x^2)^{\frac{1}{2}}} \, dx = 2\pi a \int_{-a}^{a} dx = 2\pi a \left[x \right]_{-a}^{a}$$

$$= 2\pi a [a - (-a)] = 4\pi a^2 \ .$$

14.6 MEAN VALUES

The **arithmetic mean**, or **average value**, of the n numbers $a_1, a_2, \ldots, a_n$ is given by

$$\bar{a} = (a_1 + a_2 + \ldots + a_n)/n = \frac{1}{n} \sum_{k=1}^{n} a_k \ .$$

Taking the n numbers a_k, $k = 1, 2, \ldots, n$, to be the values of the continuous function $f(x)$ at the points $x_k = a + k\,\Delta x$, where $\Delta x = (b-a)/n$, that is $a_k = f(x_k)$, then

$$\bar{a} = \frac{1}{n} \sum_{k=1}^{n} a_k = \frac{1}{b-a} \sum_{k=1}^{n} f(x) \cdot \frac{b-a}{n} = \frac{1}{b-a} \sum_{k=1}^{n} f(x)\,\Delta x \ .$$

$$(14.4)$$

As n increases, (14.4) gives the average value of many more points on the curve $y = f(x)$ between $x = a$ and $x = b$. In the limit as $n \to \infty$ this amounts to taking all values of $f(x)$ in the interval $a \leqslant x \leqslant b$. The mean value, $\bar{f}$, of $f(x)$ for x between a and b is then given by

$$\bar{f} = \frac{1}{b-a} \int_a^b f(x)\,dx \ .$$

Example: The number of migratory birds present on a nature reserve during the months from April to October was found in the example of Section 10.4 to be given by $p(t) = 6t^2 - t^3$ for $0 \leqslant t \leqslant 6$. Find the average number of birds present during this period.

$$\bar{p} = \frac{1}{6-0} \int_0^6 [6t^2 - t^3]\,dt$$

$$= \frac{1}{6} \left[6\frac{t^3}{3} - \frac{t^4}{4} \right]_0^6 = 18 \ .$$

The average number of birds present is 18, which can be compared with the maximum number of 32 during August.

PROBLEMS

1. Find the finite area bounded by the following curves:
    (a) $y = x^2$ and $y = x^5$,
    (b) $y = x^2$ and $y^2 = x$,
    (c) $y = \cos x$ and $y = \sin x$ $\left(\dfrac{\pi}{4} \leqslant x \leqslant \dfrac{5\pi}{4} \right)$,
    (d) $y = e^x, y = e^{-x}$ and $x = 2$,
    (e) $x = 1$ and $x = y^2 - 2y - 2$.

2. Find the total finite area enclosed by the following curves:
    (a) $y = x^3 - x^2 + x - 1$ and $y = 2x^2 - x - 1$,
    (b) $y = (x-1)(x-2)(x-3)$ and $y = 3(x-2)$,
    (c) $y = \sin x$ and $y = 0$, for $0 \leqslant x \leqslant 2\pi$,
    (d) $y = x \sin 2x$ and $y = x$, for $0 \leqslant x \leqslant \dfrac{5\pi}{4}$.

3. Sketch the region bounded by the line $2y = x$, the circle $x^2 + y^2 = 5$, and the positive portion of the x-axis. Find its area by integration.

4. Find the area between the x-axis and the curve given by the parametric equations

$$x = 1 - \cos^2\left(\frac{t}{2}\right), \quad y = \sin^2 t$$

for $0 \leqslant t \leqslant \pi$.

5. Show that the area between the curve described in polar coordinates by $r = f(\theta)$, the two lines $\theta = \alpha$ and $\theta = \beta$, and the origin O is given by

$$\int_\alpha^\beta \frac{r^2}{2} \, d\theta .$$

[Consider the total area to be made up from n elementary sectors each having internal angle $\Delta\theta = (\beta - \alpha)/n$. Find an approximation to the area of a typical sector; sum the sectors and then take the limit as n becomes infinite.]
Hence find the area of

(a) the cardioid $r = 1 - \sin \theta$,

(b) a leaf described by $r = 3 \sin 2\theta$, for $0 \leqslant \theta \leqslant \frac{\pi}{2}$.

6. The area covered by leucocytes migrating from a capillary tube is to be found. In t hours leucocytes migrating in a direction making an angle θ with some fixed line, move radially a distance r mm from the tube where r is given by

$$r = \frac{t}{9}(2 + \cos \theta) .$$

Find the area covered by leucocytes in 12 hours.
[Use the formula given in question 5 for the area inside the curve $r = f(\theta)$ where r, θ are polar coordinates.]

7. Find the volume generated when the region, bounded by the given curves, is rotated through $360°$ about the x-axis.

(a) $y = 2x - x^2$, $\quad x = 0, x = 2$,

(b) $y = e^x$, $\quad x = 0, x = 1$,

(c) $y = \sin x$, $\quad x = 0, x = \pi$.

8. Sketch the curves given by $y = 2x^2 + 3$ and $x + y = 4$ using the same axes. Find

    (i) the area enclosed by the curves,

    (ii) the volume of the solid formed by rotating the area in (i) through $360°$ about the x-axis.

9. The shape of a rodent's body can be approximated by an ellipsoid. Hence the volume of the rodent's body is approximately the same as that generated by rotating the area inside the ellipse $b^2x^2 + a^2y^2 = a^2b^2$ about the x-axis. Show that the volume of the ellipsoid is $\frac{4}{3}\pi ab^2$.

10. Sketch the curves represented by the equations $y = 2x^2$ and $y = 3 - x^2$, and find the volume generated by the revolution of the enclosed area about the x-axis.

11. A hemispherical bowl of internal radius 9 cm contains water to a depth of 3 cm. Show that the volume of water in the bowl is 72π cm^3.

12. Sketch the curve represented by the equation $y = x - 1/x$ for $x > 0$. Find the area enclosed by this curve, the x-axis and the lines $x = 1, x = 2$ and determine the volume of the body formed when this area is rotated through $360°$ about (a) the x-axis, (b) the y-axis.

13. Use the formula for the length of arc of a curve to verify that the circumference of a circle of radius a, given by $x^2 + y^2 = a^2$, is $2\pi a$.

14. Show that the length of the curve given by $y = \dfrac{x^4 + 3}{6x}$, between the points $\left(1, \dfrac{2}{3}\right)$ and $\left(2, \dfrac{19}{12}\right)$, can be obtained by evaluating

$$\int_1^2 \frac{x^4 + 1}{2x^2} \, dx \ .$$

    Hence find the length of this curve.

15. Estimate the perimeter of a leaf whose shape is given by the loop of the curve

$$y^2 = x\left(1 - \frac{x}{3}\right)^2 .$$

16. Find the area enclosed by the curve represented by the equation $y = 2x^{2/3}$, the x-axis and the line $x = 1$. Determine the volume and the curved surface area of the body formed when this area is rotated through $360°$ about the x-axis. (Substitute $t^2 = 9\,x^{2/3} + 16$ to evaluate the integral giving the surface area).

17. For the portion of the curve $y = \frac{1}{2}(e^x + e^{-x})$ between $x = 0$ and $x = 1$ find

   (i) the area under the curve,

   (ii) the volume of revolution generated when this area is rotated about the x-axis,

   (iii) the length of the curve,

   (iv) the curved surface area of revolution.

18. Find the mean value of the following expressions in the given interval:

   (a) x^2, $0 \leqslant x \leqslant 2$,

   (b) $(4 + x^2)^{-1}$, $0 \leqslant x \leqslant 2$,

   (c) $\sin^2 x$, $0 \leqslant x \leqslant \pi$,

   (d) $\dfrac{x + 10}{x^2 - x - 12}$, $-2 \leqslant x \leqslant 2$.

19. One thousand cells are spread uniformly over a 1 cm diameter disc. How many cells are there in each square millimetre? What is the average distance of a cell from the centre of the disc?
   [Consider the disc to be made up from n rings each of width Δr. Obtain an approximation to the number of cells in the ring bounded by circles of radius r and $r + \Delta r$. The cells in this ring are at a distance r, approximately, from the centre. Thus the average distance, in millimetres, of a cell from the centre of the disc is given by $\displaystyle\int_0^5 Cr^2 dr$, for some constant C.]

20. A swarm of wasps is clustered tightly round the queen to form a sphere of radius 10 cm. Estimate the average distance of a wasp from the queen.
   [Follow a similar procedure to that outlined in question 19, but this time take the sphere to be made up from concentric shells each of thickness Δr].

CHAPTER 15

Differential equations

15.1 INTRODUCTION

Many problems that arise in real life involve the concept of rate of change. As rates of change are expressed mathematically by derivatives, differential equations; that is, equations involving unknown functions, their derivatives and the variables upon which they depend; naturally play an important role in the application of mathematics to practical problems. If the unknown functions are functions of one variable (the independent variable) then the derivatives involved are ordinary derivatives and the equation is called an **ordinary differential equation.** When the unknown functions depend on more than one variable, the derivatives that occur will be partial derivatives and the equation is called a **partial differential equation.** Partial differential equations will not be considered in this book.

Ordinary differential equations involving one unknown function y and the independent variable x are equations containing x, y, dy/dx, d^2y/dx^2, etc. If $d^n y/dx^n$ is the highest derivative occurring in this relationship then the ordinary differential equation is said to be of the nth **order.** A **linear** n^{th} order differential equation is one in which y and its derivatives appear linearly (that is, do not occur as powers or products); hence it can be written in the form

$$a_n(x)\frac{d^n y}{dx^n} + a_{n-1}(x)\frac{d^{n-1}y}{dx^{n-1}} + \ldots + a_1(x)\frac{dy}{dx} + a_0(x)y = R(x)$$

where $a_0(x)$, $a_1(x)$, $\ldots$, $a_n(x)$ and $R(x)$ are known functions of x. Otherwise the equation is said to be **nonlinear.** A differential equation is **homogeneous** if, when y and its derivatives are set equal to zero, the equation is satisfied; otherwise it is said to be **inhomogeneous.**

Examples of ordinary differential equations are given below

(i) $\dfrac{dy}{dx} + \dfrac{y}{x} = 0$ order 1, linear, homogeneous,

(ii) $\dfrac{d^2y}{dx^2} - y = 2x$ order 2, linear, inhomogeneous,

(iii) $\dfrac{d^2y}{dx^2} = \sec^2 y \tan y$ order 2, nonlinear, homogeneous.

A **solution** of an ordinary differential equation is any relationship between x and y which results in the differential equation being satisfied. For the differential equations above

(i) $y = \dfrac{C}{x}$ is a solution of $\dfrac{dy}{dx} + \dfrac{y}{x} = 0$, for any constant C, since

$$\frac{dy}{dx} = -\frac{C}{x^2} \text{ and } \frac{dy}{dx} + \frac{y}{x} = -\frac{C}{x^2} + \frac{C}{x} \cdot \frac{1}{x} = 0 \ .$$

(ii) $y = Ae^{-x} + Be^x - 2x$ is a solution of $y'' - y = 2x$, for any constants A and B. Here $y' = -Ae^{-x} + Be^x - 2$ and $y'' = Ae^{-x} + Be^x$, so

$$y'' - y = Ae^{-x} + Be^x - [Ae^{-x} + Be^x - 2x] = 2x \ .$$

(iii) $\sin y = De^{-x}$ is a solution of $y'' = \sec^2 y \tan y$, for any constant D. Differentiating implicitly gives $\cos y \, y' = -De^{-x}$ or

$$y' = -\frac{De^{-x}}{\cos y} = -\frac{\sin y}{\cos y} = -\tan y \ .$$

Differentiating again gives $y'' = -\sec^2 y \, y' = \sec^2 y \tan y$, which is the original differential equation, so verifying that $\sin y = De^{-x}$ is a solution. Notice that the solution to a differential equation may be an implicit relationship.

It is a good practice to check any solution you obtain by substitution in the original differential equation.

A solution of an nth order ordinary differential equation involves at most n independent arbitrary constants. The solution which contains exactly n independent arbitrary constants is called the **general solution**. The solutions given for equations (i) and (ii) are the general solutions to the differential equations since the number of arbitrary constants equals the order of the differential equation in each case. This is not so for the solution given for equation (iii) which is of order 2 yet the solution contains only one arbitrary constant D. Any solution obtained by assigning specific values to each of the arbitrary constants in the general solution is called a **particular solution**. A particular solution of equation (ii) is $y = e^x - 2x$. Here the arbitrary constants A and B in the general solution have been given the values 0 and 1, respectively.

The solution of a differential equation which is describing mathematically a real-life process has often to satisfy certain specified conditions corresponding to the values taken by physical quantities at particular instants. If the conditions are all given at the same value of the independent variable, for example the position and velocity of a body at the start of the process, they are called **initial conditions**. The ordinary differential equation and initial conditions together constitute an **initial value problem**. If the conditions are given at two values of the independent variable, for example the concentration of a dissolved chemical at two different times, they are known as **boundary conditions**. With the ordinary differential equation this results in a **two-point boundary value problem**. In other problems the form of the solution may be specified in a given region, for example the behaviour of the solution as the independent variable becomes large. The n arbitrary constants in the general solution of an nth order ordinary differential equation can be determined uniquely by specifying n conditions.

This chapter only deals with ordinary differential equations having solutions which can be expressed in terms of a finite combination of standard functions. It should be emphasised that in practice differential equations often arise which have no such solution and have instead to be solved by using numerical methods. Some numerical techniques for solving differential equations are considered in Chapter 17.

15.2 FIRST ORDER DIFFERENTIAL EQUATIONS

Two types of first order differential equation, known as variables separable and linear, are considered in this section. By way of illustration an introduction to the behaviour of population size and epidemics is included.

15.2.1 Variables Separable Differential Equations

The type of differential equation considered here arises frequently in the study of practical problems. The differential equation takes the form

$$\frac{dy}{dx} = f(x)g(y) \tag{15.1}$$

where $f(x)$ is a function of the independent variable x alone, while $g(y)$ is a function of the dependent variable y alone. Equation (15.1) can be rewritten as

$$f(x) = \frac{1}{g(y)} \frac{dy}{dx} \, ,$$

so integrating this with respect to x gives

$$\int f(x)dx = \int \frac{1}{g(y)} \frac{dy}{dx} dx = \int \frac{1}{g(y)} dy \, ,$$

on using the rule for a change of variable (see Section 10.5). Thus the problem reduces to two integrations having respectively x and y as their variables of integration. Notice that although there are two separate integrations involved there will only be one independent arbitrary constant in the solution. This is because the two constants of integration can be combined into a single constant, see example 1.

Example 1: Solve $\dfrac{dy}{dx} = \dfrac{x}{y}$.

Here $f(x) = x, g(y) = \dfrac{1}{y}$ so

$$\int x \, dx = \int y \, dy \ ,$$

that is $\qquad \dfrac{x^2}{2} + C_1 = \dfrac{y^2}{2} + C_2 \ ,$

or $\qquad y^2 = x^2 + C$

where $C = 2(C_1 - C_2)$. This is the general solution of the first order equation $\dfrac{dy}{dx} = \dfrac{x}{y}$ and contains one arbitrary constant.

Differentiating $y^2 = x^2 + C$ with respect to x gives

$$2y \frac{dy}{dx} = 2x$$

or $\qquad \dfrac{dy}{dx} = \dfrac{x}{y}$.

This verifies that $y^2 = x^2 + C$ is a solution of the differential equation.

Example 2: Find the solution of the differential equation

$$\frac{dy}{dx} = ky^2$$

where k is a constant, having $y = 2$ when $x = 0$ and $y = 4$ when $x = 1$.

$$\frac{dy}{dx} = ky^2$$

so $\qquad \displaystyle\int k \, dx = \int \frac{1}{y^2} \, dy$.

Therefore,

$$kx = -\frac{1}{y} + C \ .$$

Using the specified values of x and y the constants C and k can be found. As $y = 2$ when $x = 0$, $0 = -\frac{1}{2} + C$, so $C = \frac{1}{2}$.

Thus $\qquad kx = -\dfrac{1}{y} + \dfrac{1}{2}$.

When $x = 1$, $y = 4$ so $k = -\dfrac{1}{4} + \dfrac{1}{2} = \dfrac{1}{4}$.

Therefore,

$$\frac{1}{4}x = -\frac{1}{y} + \frac{1}{2}$$

or $\qquad y = \dfrac{4}{2-x}$.

This solution may be checked by differentiating since

$$\frac{dy}{dx} = \frac{4}{(2-x)^2} = \frac{1}{4}y^2 \quad \text{and} \quad k = \frac{1}{4} \ .$$

Also $y = 4/(2-x)$ satisfies the boundary conditions $y = 2$ when $x = 0$, and $y = 4$ when $x = 1$.

Example 3: In the study of the radioactive decay of elements the rate at which an element decays into another substance is proportional to the amount of unchanged material present. Thus, if m denotes the mass of the radioactive element at time t,

$$\frac{dm}{dt} = -km$$

where k is a positive constant. Find m as a function of t.

Separating the variables and integrating

$$\int \frac{dm}{m} = \int -k \, dt \ ,$$

so $\qquad \ln m = -kt + C$,

or $\qquad m = e^{C-kt} = Ae^{-kt} \qquad$ if $A = e^C$.

If initially $m = m_0$, $A = m_0$ $\quad$ so

$$m = m_0 e^{-kt} \ .$$

This relationship was the one used in Section 12.4.

Example 4: An object in a steady stream of gas, or liquid, is found to cool at a rate proportional to the difference between its temperature and the temperature of the surrounding gas, or liquid. This is **Newton's law of cooling**. For an object having a temperature T at time t in surroundings maintained at a constant temperature S ($<T$) the rate at which the object cools is given by

$$\frac{dT}{dt} = -k(T - S)$$

where the constant of proportionality k is positive and the negative sign indicates that the temperature T is decreasing. Find the general solution to the differential equation. Hence obtain the particular solution which gives the temperature of a body having an initial temperature of 110°C if the body is observed to cool down to 80°C after 5 minutes. The body is in an air stream maintained at a temperature of 20°C. How hot will this body be when 10 minutes have elapsed?

Rearranging the differential equation and integrating

$$\int \frac{dT}{T - S} = - \int k \, dt \ ,$$

so $\ln|T - S| = -kt + C$,

or $T - S = e^{C-kt} = e^{C}e^{-kt} = Ae^{-kt}$

on writing $A = e^{C}$. Therefore,

$$T = S + Ae^{-kt} \ .$$

This is the general solution and A is the arbitrary constant.

The particular solution satisfying the information supplied can now be found. Since the temperature of the surroundings is 20°C, S = 20, and the initial temperature of the body is 110°C, so that when $t = 0$, $T = 110$. Thus

$$110 = 20 + A$$

so $A = 90$.

Therefore, $T = 20 + 90e^{-kt}$.

When $t = 5$ the body's temperature $T = 80$ so

$$80 = 20 + 90e^{-k5} \ .$$

Therefore, $e^{-5k} = 2/3$, so that $e^{-k} = (2/3)^{1/5}$.

Whence $T = 20 + 90e^{-kt} = 20 + 90(2/3)^{t/5}$.

Notice that it is not necessary to find the numerical value of k.
The temperature of the body at time $t = 10$ is then given by

$$T = 20 + 90(2/3)^{10/5} = 20 + 90(2/3)^2 = 60°C \ .$$

Example 5: When heat is lost from an object through natural convection it is sometimes assumed that the rate at which the temperature of the object falls is given by the differential equation

$$\frac{dT}{dt} = -c(T - S)^{5/4} \ .$$

Here T is the object's temperature at time t, S is the temperature of the surroundings, assumed constant, and c is a positive constant. Find the solution of the differential equation, which satisfies the same numerical values as those used in example 4, and determine the object's temperature after ten minutes.

Rearranging and then integrating this equation gives

$$\int \frac{\mathrm{d}T}{(T-S)^{5/4}} = -\int c\,\mathrm{d}t \ .$$

Therefore, $-4(T-S)^{-1/4} = -ct - b$, on taking the arbitrary constant of integration to be $-b$. Rearranging this gives

$$T = S + \left(\frac{4}{b+ct}\right)^4 \ .$$

With the same values as in example 4, that is, $S = 20$, $T = 110$ when $t = 0$, and $T = 80$ when $t = 5$, the temperature T, at time t, is given by

$$T = 20 + \frac{90}{\left\{1 + \left[(3/2)^{1/4} - 1\right]\dfrac{t}{5}\right\}^4}\,^\circ\mathrm{C} \ .$$

When $t = 10$,

$$T = 20 + \frac{90}{[(24)^{1/4} - 1]^4} \cong 61.52\,^\circ\mathrm{C} \ .$$

As may be expected, the body has not cooled down as much through natural convection as it did when placed in an air stream.

15.2.2 Population models

In the construction of mathematical models for studying the size of a population, differential equations are often employed. This is particularly so for a large population as its size may be regarded as changing continuously with the passage of time. This is because an individual's birth or death does not create a significant change in the population size. Such models are useful in predicting trends in population size. Other models, which take into consideration the discrete nature of the problem, can be constructed by using recurrence equations, see Chapter 16.

If the population numbers N at time t and $N + \Delta N$ at a subsequent time $t + \Delta t$ then the change in the size of the population, ΔN, can be expected to be proportional to the time Δt that has elapsed. The simplest model of population growth assumes that the change ΔN is also proportional to the size of the population N, so that $\Delta N = kN\Delta t$ or $\Delta N/\Delta t = kN$, where k is the constant of proportionality. As this relationship is assumed to be valid at all times, in the limit as $\Delta t \to 0$ the equation becomes

$$\frac{\mathrm{d}N}{\mathrm{d}t} = kN \ . \tag{15.2}$$

The constant of proportionality k is the **relative rate at which the population changes,** that is

$$k = \frac{1}{N}\frac{dN}{dt} .$$

If k is positive the population increases, if k is negative it declines.
    The solution of (15.2) is found by integrating the equation

$$\int \frac{dN}{N} = \int k\, dt .$$

Therefore, $\ln N = kt + C$

or $N = e^{kt + C} .$

If N_0 denotes the size of the population when $t = 0$ this can be written as

$$N = N_0 e^{kt} \tag{15.3}$$

since $e^C = N_0$.

    Because of the form of the solution, when k is positive, the model is said to represent **exponential growth** of a population. This model of population growth is unrealistic for large values of t as any natural population is restricted in its maximum size by the environment in which it lives.

    The next model goes some way towards meeting the restriction on the maximum size of the population. The assumptions behind this second model are that (i) the environment can only support a maximum population size K, known as the **carrying capacity of the environment,** and (ii) the relative rate of change in the population is proportional to $K - N$, the number by which the population can still grow. Thus as the population approaches its maximum value K the relative rate of growth falls linearly to zero. With these assumptions the governing differential equation is

$$\frac{dN}{dt} = cN(K - N)$$

where c is the constant of proportionality. This equation is separable, so

$$\int c\, dt = \int \frac{dN}{N(K - N)}$$

$$= \frac{1}{K}\int \left(\frac{1}{N} + \frac{1}{K - N} \right) dN \qquad \text{using partial fractions.}$$

Integrating,

$$ct + A = \frac{1}{K}(\ln N - \ln|K - N|) = \frac{1}{K}\ln\left| \frac{N}{K - N} \right| .$$

If $N = N_0$ when $t = 0$,

$$A = \frac{1}{K} \ln \left| \frac{N_0}{K - N_0} \right| .$$

Therefore

$$Kct = \ln \left| \frac{N(K - N_0)}{N_0(K - N)} \right| , \qquad (15.4)$$

or rearranging

$$N = \frac{KN_0}{N_0 + (K - N_0)e^{-Kct}} . \qquad (15.5)$$

This is the equation of a **logistic curve**, see Fig. 15.1, and is widely used to describe the size of insect, fish, and other populations.

There is a period of rapid (exponential) growth due to the cNK term in the differential equation and then a levelling off as K is approached.

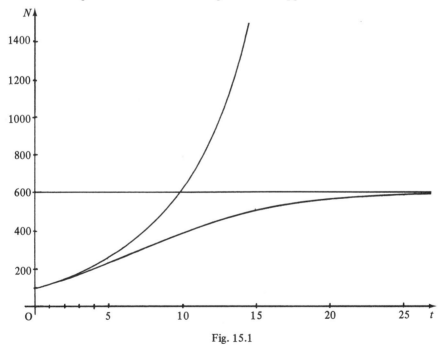

Fig. 15.1

Example 6: A colony of bacteria is to be cultured in a dish on which an environment capable of supporting 600 bacteria is artificially established. Initially 100 bacteria are introduced and one hour later the number of bacteria is observed to be 120. Find the size of the bacterial population at subsequent times.

Here $K = 600$ and $N_0 = 100$.

When $t = 1, N = 120$ so, from (15.4),

$$600c = \ln \frac{120(600 - 100)}{100(600 - 120)} = \ln \frac{5}{4}.$$

Therefore, at time t, the population size, as given by (15.5), is

$$N = \frac{60000}{100 + 500 \exp\left[-\left(\ln \frac{5}{4}\right)t\right]} = \frac{600}{1 + 5\left(\frac{4}{5}\right)^t}.$$

If the simpler model of exponential growth is used the population size would be given by

$$N = 100\left(\frac{6}{5}\right)^t,$$

from (15.3) with $k = \ln \frac{6}{5}$.

Table 15.1 gives the size of the population as predicted by the two models.

Table 15.1

| | Size of Population | |
| | Restricted maximum size | Unrestricted maximum size |
time t	$N = \dfrac{KN_0}{N_0 + (K - N_0)e^{-Kct}}$	$N = N_0 e^{kt}$
0	100	100
1	120	120
2	143	144
3	169	173
4	199	207
5	227	248
10	390	619
15	510	1541
20	567	3834
25	589	9540

These values are the ones used to plot the curves in Fig. 15.1. In the early stages there is little difference between the two predicted values, but later on the difference is dramatic.

15.2.3 Epidemic Models

Differential equations also arise in the study of epidemics. In the case of a 'simple' epidemic it is assumed that one infectious person is introduced into a population of n people all of whom can, with equal chance, contract the disease. Suppose that at a later time t there are s people who have not become infected, called **susceptibles**, then there will be $(n + 1 - s)$ people who are infected, called **infectives**. The rate at which susceptibles become infected in a 'simple' epidemic is then given by

$$\frac{ds}{dt} = -cs(n + 1 - s)$$

where c is the rate of contact between individuals.

To solve this equation, separating the variables and integrating gives

$$\int -c \, dt = \int \frac{ds}{s(n + 1 - s)}$$

$$= \frac{1}{n + 1} \int \left(\frac{1}{s} + \frac{1}{n + 1 - s} \right) ds \, , \quad \text{using partial fractions,}$$

so $\qquad A - ct = \frac{1}{n + 1} [\ln s - \ln(n + 1 - s)] = \frac{1}{n + 1} \ln \left(\frac{s}{n + 1 - s} \right).$

When $t = 0$, $s = n$, so $A = \dfrac{1}{n + 1} \ln n.$

Therefore $\qquad \dfrac{s}{n(n + 1 - s)} = e^{-(n+1)ct}.$

Rearranging,

$$s = \frac{n(n + 1)}{n + \exp[(n + 1)ct]} \, .$$

In practice people that are infected with a disease may be removed from the population by death, isolation, or immunity after recovery. Introducing this feature into the model we have the situation known as a **'general' epidemic**.

Denote the number of infectives $n + 1 - s$ by u, the total number removed from the population by r, and the rate of removal by b; then it can be shown that the following equations hold

$$\frac{ds}{dt} = -csu, \quad \frac{dr}{dt} = bu, \quad \frac{du}{dt} = csu - bu \, .$$

By using the first two equations u and t can be eliminated, thus

$$\frac{ds}{dt} \bigg/ \frac{dr}{dt} = -csu/(bu) \ ,$$

that is, $\dfrac{ds}{dr} = -\dfrac{cs}{b}$, a separable equation in s and r.

Therefore, $\displaystyle\int \frac{ds}{s} = \int -\frac{c}{b}\, dr \ ,$

that is $\quad \ln s = D - \dfrac{cr}{b} \ .$

When $t = 0, s = n, r = 0$ so $D = \ln n$.

Hence $\quad s = n \exp\left(\dfrac{-cr}{b}\right) \ .$

A comprehensive and interesting treatment of epidemic models can be found in Bailey (1957).

15.2.4 Linear Differential Equations

A first order linear differential equation in $y(x)$ has the form

$$\frac{dy}{dx} + p(x)y = f(x) \tag{15.6}$$

where $p(x)$ and $f(x)$ are known functions of x.

To solve this equation a function of x, called an **integrating factor**, is introduced in order that the left-hand side of the equation can be integrated easily. Denoting the integrating factor by $\mu(x)$, the equation, (15.6), is multiplied by $\mu(x)$ to give

$$\mu(x)\frac{dy}{dx} + \mu(x)p(x)y = \mu(x)f(x) \ . \tag{15.7}$$

The left-hand side of (15.7) will be easily integrated if $\mu(x)$ is chosen so that

$$\mu(x)\frac{dy}{dx} + \mu(x)p(x)y = \frac{d}{dx}\Big(\mu(x)y\Big) = \mu(x)\frac{dy}{dx} + \frac{d\mu}{dx}y \ .$$

Thus $\mu(x)$ is required to satisfy the differential equation

$$\frac{d\mu}{dx} = \mu(x)p(x) \ .$$

Solving this separable equation gives

$$\mu(x) = A e^{\int p(x)\,dx} \tag{15.8}$$

where A is an arbitrary constant.

With this choice of $\mu(x)$, (15.7) can be written as

$$\frac{d}{dx}\left(\mu(x)y\right) = \mu(x)f(x)$$

which, on integrating, gives

$$\mu(x)y = \int \mu(x)f(x)\,dx + C , \qquad (15.9)$$

C being the arbitrary constant of integration.

As the arbitrary constant A in the integrating factor $\mu(x)$ given by (15.8) may be cancelled it is common practice to set it equal to one from the outset. Thus the integrating factor is taken to be

$$\mu(x) = \exp\left[\int p(x)\,dx\right] . \qquad (15.10)$$

In the particular case when $p(x) = a$, a constant, (15.6) becomes the linear constant coefficient equation

$$\frac{dy}{dx} + ay = f(x) .$$

The integrating factor is now

$$\mu(x) = \exp\left[\int p(x)\,dx\right] = \exp\left[\int a\,dx\right] = e^{ax} ,$$

and hence the solution of the differential equation, from (15.9), is given by

$$e^{ax}y = \int e^{ax} f(x)\,dx + C ,$$

so
$$y = Ce^{-ax} + e^{-ax}\int e^{ax} f(x)\,dx . \qquad (15.11)$$

Example 7: Find the solution of the linear differential equation

$$\frac{dy}{dx} - \frac{y}{x} = -xe^{-x}$$

which remains finite for all positive values of x.

In this example $p(x) = -\dfrac{1}{x}$ and $f(x) = -xe^{-x}$. Thus

$$\mu(x) = \exp\left[\int p(x)\,dx\right] = \exp\left[\int\left(-\frac{1}{x}\right)dx\right] = \exp[-\ln|x|]$$

$$= \exp\left[\ln\left|\frac{1}{x}\right|\right] = \frac{1}{x} .$$

So
$$\frac{1}{x}\frac{dy}{dx} - \frac{y}{x^2} = \frac{d}{dx}\left(\frac{y}{x}\right) = -\frac{1}{x}xe^{-x} = -e^{-x} .$$

Integrating,

$$\frac{y}{x} = \int -e^{-x} dx = e^{-x} + C ,$$

therefore $y = xe^{-x} + Cx$.

For y to remain finite as $x \to \infty$, the constant of integration C must be taken to be zero. Note that $xe^{-x} \to 0$ as $x \to \infty$, see equation (12.9). Therefore

$$y = xe^{-x}$$

is the desired solution.

Example 8: The elimination of a drug present in the bloodstream, through absorption by the various organs of the body, in particular the kidneys, can be modelled by a differential equation.

The concentration of a drug in the bloodstream can be monitored, and the rate at which the concentration decreases is found to be proportional to the current concentration. Determine the concentration as a function of time.

If Q is the concentration of the drug and t is time then

$$\frac{dQ}{dt} = -kQ ,$$

where k is a positive constant of proportionality. The minus sign indicates that the concentration falls with increasing time.

This equation can be solved as follows. Separating the variables and integrating,

$$\int \frac{dQ}{Q} = \int -k \, dt ,$$

that is, $\ln Q = -kt + A$, where A is the constant of integration.

So $\quad Q = e^{-kt+A} = e^{A} e^{-kt} = Be^{-kt}$

on writing $B = e^{A}$.

If the concentration has a value Q_0 when $t = 0, B = Q_0$, so $Q = Q_0 e^{-kt}$. Thus the concentration falls exponentially with time.

Example 9: In the treatment of certain diseases a patient is given a drug intravenously. Let the rate of infusion be $r(t)$ and the concentration of the drug in the blood be Q. The drug will be removed from the bloodstream by various organs at a rate proportional to the concentration. Therefore the concentration of the drug satisfies the differential equation.

$$\frac{dQ}{dt} = r(t) - kQ ,$$

where k is a positive constant of proportionality. Find the concentration as a function of time.

The equation can be written in the standard form for a linear equation in $Q(t)$, namely,

$$\frac{dQ}{dt} + kQ = r(t) .$$

Here $p(t) = k$, a constant, and $f(t) = r(t)$. The integrating factor is e^{kt} so

$$e^{kt} \frac{dQ}{dt} + k e^{kt} Q = \frac{d}{dt}(e^{kt}Q) = r(t)e^{kt} .$$

Integrating

$$e^{kt}Q = \int^t r(t)e^{kt}dt + C ,$$

where C is the constant of integration. If the rate of infusion of the drug is constant, r, the integral appearing in this expression can be immediately evaluated to give

$$e^{kt}Q = \frac{r}{k} e^{kt} + C .$$

Therefore $Q = \frac{r}{k} + Ce^{-kt}$.

If the concentration is Q_0 when $t = 0$,

$$C = Q_0 - \frac{r}{k}$$

so $Q = \frac{r}{k} + \left(Q_0 - \frac{r}{k} \right) e^{-kt}$.

In the case of $r = 0$ this solution reduces to the one given in example 8 when there was no infusion.

From (15.9) the general solution of the first order linear equation

$$\frac{dy}{dx} + p(x)y = f(x)$$

is $y = \dfrac{C}{\mu(x)} + \dfrac{1}{\mu(x)} \displaystyle\int^x \mu(x)f(x)dx$ (15.12) where $\mu(x) = \exp[\int p(x)dx]$ and C is an arbitrary constant. Equation (15.12) is made up of two parts. The first part, which contains the arbitrary constant, is called the **complementary function** and depends only on the function $p(x)$, which multiplies y, through the integrating factor $\mu(x)$. In fact, as may be verified by substitution, the complementary function $C/\mu(x)$ is a solution of the homogeneous linear equation

$$\frac{dy}{dx} + p(x)y = 0 \ .$$

The second part, known as the **particular integral**, is a solution of the full inhomogeneous linear equation

$$\frac{dy}{dx} + p(x)y = f(x) \ ,$$

and does not contain any arbitrary constants. Thus in example 7, the general solution, before the arbitrary constant was found to be zero, was

$$y = Cx + xe^{-x} \ .$$

Cx is the complementary function and xe^{-x} the particular integral.

15.3 LINEAR CONSTANT COEFFICIENT DIFFERENTIAL EQUATIONS

The solution of higher order linear constant coefficient differential equations will now be discussed. In particular we shall concentrate on the second order equation

$$a\frac{d^2y}{dx^2} + b\frac{dy}{dx} + cy = f(x) \ , \tag{15.13}$$

where a, b and c are constants with a non-zero, and $f(x)$ is a given function, since this will illustrate the features that can occur in higher order equations.

The solution of (15.13) also consists of two parts, a complementary function which is the general solution of

$$a\frac{d^2y}{dx^2} + b\frac{dy}{dx} + cy = 0 \ , \tag{15.14}$$

and a particular integral. This time the complementary function consists of a linear combination of two functions, as (15.13) is second order. The complementary function is found by looking for solutions of (15.14) having the form $y = Ce^{mx}$ where C and m are constants. (This form is suggested by the solution of the first order linear constant coefficient equation $dy/dx + ay = 0$ which, by (15.11), is $y = Ce^{-ax}$). C can be taken to be non-zero since $C = 0$ corresponds to the trivial solution $y \equiv 0$ which is of no interest. Substituting $y = Ce^{mx}$ into (15.14) gives

$$a\frac{d^2y}{dx^2} + b\frac{dy}{dx} + cy = aCm^2e^{mx} + bCme^{mx} + cCe^{mx}$$

$$= Ce^{mx}[am^2 + bm + c]$$

$$= 0$$

for a solution. As Ce^{mx} is non-zero, m must satisfy the quadratic equation

$$am^2 + bm + c = 0 . \tag{15.15}$$

This equation is called the **auxiliary equation** of (15.14). The roots of the auxiliary equation (15.15) are given by

$$m_1 = \frac{-b + \sqrt{(b^2 - 4ac)}}{2a} \text{ and } m_2 = \frac{-b - \sqrt{(b^2 - 4ac)}}{2a} .$$

Three possibilities arise, either $b^2 - 4ac$ is positive, zero, or negative.

(i) If $b^2 - 4ac > 0$, the roots of the auxiliary equation are different real numbers. In this case the general solution of (15.14), and hence the complementary function of (15.13), is given by

$$y = Ae^{m_1x} + Be^{m_2x}$$

where A and B are arbitrary constants.

(ii) If $b^2 - 4ac = 0$, the two roots of the auxiliary equation are the same. The general solution of (15.14) must now have one term modified by a factor x. Thus the complementary function will be

$$y = (A + Bx)e^{px}$$

where $p = -b/(2a)$ is the repeated root and A, B are arbitrary constants.

(iii) When $b^2 - 4ac < 0$, the roots of the auxiliary equation are complex numbers. By making use of the polar form of a complex number, see Section 3.4.1, the general solution of (15.14) now consists of exponential and trigonometric functions in the form

$$y = e^{px} (A \cos qx + B \sin qx)$$

where $p = -\dfrac{b}{2a}, q = \dfrac{\sqrt{(4ac - b^2)}}{2a}$ and A, B are the arbitrary constants.

These solutions can all be verified by substitution in the homogeneous differential equation (15.14).

The determination of a particular integral can only be done easily for certain functions $f(x)$ appearing on the right-hand side of the inhomogeneous equation (15.13). Equations for which a particular integral can be found easily have $f(x)$ composed of functions which can occur in a complementary function, that is, exponentials, polynomials, cosines and sines. This is because differentiation will result in another function having these elements. Therefore the particular integral may be obtained by trying a form of function as suggested in Table 15.2.

Table 15.2

Term in $f(x)$	Appropriate trial functions to take in particular integral
$e^{\alpha x}$	$c_0 e^{\alpha x}$
x^n	$c_0 + c_1 x + c_2 x^2 + \ldots + c_n x^n$
$\cos \alpha x$ or $\sin \alpha x$	$c_1 \cos ax + c_2 \sin ax$

$\alpha, c_0, c_1, c_2, \ldots, c_n$ are constants, n a positive integer.

Combinations of these functions appearing in $f(x)$ can be handled by taking a similar combination of trial functions for the particular integral. If a term in the form suggested for the particular integral already appears in the complementary function then the whole form, as given in the table, needs to be multiplied by x. This is repeated until no terms are common to both the complementary function and the trial form of the particular integral.

The values of the constants $c_0, c_1, \ldots, c_n$ are determined by requiring the form given by Table 15.2 to satisfy the full inhomogeneous differential equation (15.13).

Once the general solution, the sum of the complementary function and particular integral, is found the two arbitrary constants in the complementary function can be determined by using, if available, known values of y at two values of x (boundary conditions) or the value of y and y' at the same point x (initial conditions).

Example 1: Find the general solution of

$$y'' - 5y' + 6y = e^x \ .$$

The homogeneous equation $y'' - 5y' + 6y = 0$ has the auxiliary equation

$$m^2 - 5m + 6 = 0.$$

This has roots $m = 2$ and $m = 3$, therefore the complementary function is

$$y = Ae^{3x} + Be^{2x} .$$

To find the particular integral, Table 15.2 suggests that corresponding to e^x the form ce^x should be tried. Taking

$$y = ce^x ,$$

$$y' = y'' = ce^x ,$$

so substituting

$$y'' - 5y' + 6y = ce^x - 5ce^x + 6ce^x = 2ce^x .$$

This must equal e^x, so c requires to be $\frac{1}{2}$.

Therefore the general solution, complementary function plus particular integral, is

$$y = Ae^{3x} + Be^{2x} + \tfrac{1}{2}e^x .$$

Example 2: Find the general solution of

$$y'' - 2y' - 3y = 4e^{3x} - 3x .$$

To obtain the complementary function consider

$$y'' - 2y' - 3y = 0 .$$

The auxiliary equation is

$$m^2 - 2m - 3 = 0$$

with roots $m = 3$ and $m = -1$.

Therefore, the complementary function is

$$y = Ae^{3x} + Be^{-x} .$$

For the particular integral Table 15.2 suggests that corresponding to the $4e^{3x}$ term the form ce^{3x} should be considered. However, a term of this form viz. Ae^{3x}, already appears in the complementary function, so the modified form cxe^{3x} must be taken. Corresponding to the $-3x$ term Table 15.2 suggests the form $c_0 + c_1 x$ be taken. The total form of the particular integral is then

$$y = cxe^{3x} + c_0 + c_1 x .$$

Hence $y' = c(1 + 3x)e^{3x} + c_1$

and $y'' = c(6 + 9x)e^{3x} .$

Substituting

$$y'' - 2y' - 3y = c(6 + 9x)e^{3x} - 2[c(1 + 3x)e^{3x} + c_1]$$
$$- 3[cxe^{3x} + c_0 + c_1x]$$
$$= 4ce^{3x} - 2c_1 - 3c_0 - 3c_1x .$$

This is required to be equivalent to $4e^{3x} - 3x$,

that is, $4ce^{3x} - 2c_1 - 3c_0 - 3c_1x \equiv 4e^{3x} - 3x .$

To satisfy this equation for all values of x the coefficients of the corresponding terms are equated. Thus

$$4c = 4, \ -3c_1 = -3 \text{ and } -2c_1 - 3c_0 = 0 .$$

Solving these equations, $c = 1, c_1 = 1$ and $c_0 = -2/3$.

Therefore the particular integral is

$$y = xe^{3x} + x - 2/3 .$$

Hence the general solution is

$$y = (A + x)e^{3x} + Be^{-x} + x - 2/3 .$$

Example 3: Find the solution of

$$y'' + 2y' + y = 2 \sin x$$

having $y = 0$ and $y' = 0$ when $x = 0$.

The auxiliary equation is

$$m^2 + 2m + 1 = 0$$

which has two equal roots $m = -1$. Therefore the complementary function is

$$y = (A + Bx)e^{-x} .$$

For the particular integral, since $2\sin x$ is the inhomogeneous part of the equation, try

$$y = a \cos x + b \sin x .$$

Then $y' = -a \sin x + b \cos x ,$

           $y'' = -a \cos x - b \sin x .$

Substituting into the original differential equation

$$y'' + 2y' + y = -a \cos x - b \sin x + 2[-a \sin x + b \cos x]$$
$$+ a \cos x + b \sin x$$
$$= 2b \cos x - 2a \sin x .$$

Therefore $2b \cos x - 2a \sin x \equiv 2\sin x .$

Equating coefficients of the sine and cosine terms gives

$$a = -1 \text{ and } b = 0 .$$

So the particular integral is
$$y = -\cos x$$
and the general solution is
$$y = (A + Bx)e^{-x} - \cos x \ .$$
When $x = 0$, $y = 0$ so $0 = A - 1$, therefore $A = 1$.

Now $\quad y' = (B - A - Bx)e^{-x} + \sin x$

and when $x = 0$, $y' = 0$ so $0 = B - A$, therefore $B = 1$.

The desired solution is thus
$$y = (1 + x)e^{-x} - \cos x \ .$$

Example 4: Find the general solution of
$$y'' - 2y' + 5y = 0 \ .$$
The auxiliary equation $m^2 - 2m + 5 = 0$ has complex roots since
$$b^2 - 4ac = (-2)^2 - 4.15 = -16 < 0 \ .$$

Here $\quad p = \dfrac{-b}{2a} = \dfrac{-(-2)}{2.1} = 1$

and $\quad q = \dfrac{\sqrt{(4ac - b^2)}}{2a} = \dfrac{\sqrt{16}}{2.1} = 2 \ .$

Thus the complementary function and also, since the equation is homogeneous, the general solution, is
$$y = e^x \left[A \cos 2x + B \sin 2x \right] \ .$$

15.4 SYSTEMS OF FIRST ORDER DIFFERENTIAL EQUATIONS

An important field of study in the biological sciences is that of ecology. Here the interaction of many factors within an environment is studied. As a simple example consider two species coexisting in an environment. If these two species compete for the same resources the rate of growth of the population of one species will be affected by the number of the other species present and the resources available. Mathematically this may be represented by the pair of equations

$$\dot{x} = a_{11}x + a_{12}y + f(t) \tag{15.16}$$

and $\quad \dot{y} = a_{21}x + a_{22}y + g(t) \ , \tag{15.17}$

where x and y denote the sizes of the two populations at time t, $a_{11}, a_{12}, a_{21},$ a_{22} are constants, and $f(t)$, $g(t)$ are specified functions of t. Equation (15.16) shows that the rate of change of the population with size x, depends upon three terms. The first, $a_{11}x$, is the rate of change for the population x, in the absence of any other influence, and is here assumed to be proportional to the current size of the population. The second term, $a_{12}y$, represents the effect of the second population on the first, again assumed proportional to the current size of the second population. If the species compete as in the **predator-prey** situation, with the population of size x being the prey, the coefficient a_{12} would be negative. Similarly, if the species with population size x parasitizes the other species, as in the **host-parasite** situation, a_{12} would be negative. However, if the species benefit each other, as in a **symbiotic relationship**, a_{12} would be positive. The third term present in the equation, $f(t)$, could represent the rate at which resources are available within the environment. Similar interpretations apply to equation (15.17).

If either $a_{12} = 0$ or $a_{21} = 0$ then one of the equations is a first order linear equation involving one unknown which can be solved by the technique given in Section 15.2.4. Substituting the solution obtained into the other equation will result in another first order linear equation for the remaining unknown function which can then be solved.

The solution of the pair of equations is more interesting when both a_{12} and a_{21} are non-zero. The solution is obtained by eliminating one of the unknown functions, y say, leaving x to be found by solving a linear second order equation. This is achieved by differentiating (15.16) with respect to t to obtain an equation containing $\ddot{x}$, $\dot{x}$ and $\dot{y}$. Using (15.17), $\dot{y}$ is eliminated and then (15.16) is used again to replace y. This results, finally, in a second order linear constant coefficient equation which can be solved for x by the method given in Section 15.3. Once x has been determined y can easily be found from (15.16).

Example: The equations

$$\dot{x} = 2x - 2y ,$$
$$\dot{y} = -x + y ,$$

describe the influence on the growth rate of two competing species with population sizes x and y. If the initial population sizes are $x(0) = 250$ and $y(0) = 100$ find the population sizes of both species at future times.

$$\dot{x} = 2x - 2y , \qquad\qquad (15.18)$$
$$\dot{y} = -x + y . \qquad\qquad (15.19)$$

Differentiating (15.18)

$$\ddot{x} = 2\dot{x} - 2\dot{y}$$
$$= 2\dot{x} - 2(-x + y) \qquad \text{using (15.19)}$$
$$= 2\dot{x} + 2x + (\dot{x} - 2x) \qquad \text{using (15.18)} ,$$

therefore $\ddot{x} - 3\dot{x} = 0$.

The auxiliary equation $m^2 - 3m = 0$ has roots $m = 0$ and $m = 3$, so

$$x(t) = Ae^{0t} + Be^{3t} = A + Be^{3t} .$$

From (15.18),

$$y(t) = \frac{1}{2}(2x - \dot{x}) = A + Be^{3t} - \frac{1}{2}3Be^{3t}$$

$$= A - \frac{B}{2}e^{3t} .$$

The initial conditions $x(0) = 250$ and $y(0) = 100$ result in the simultaneous equations

$$A + B = 250 ,$$

$$A - \frac{B}{2} = 100 ,$$

for the constants A and B. Thus $A = 150$, $B = 100$, and the population sizes are then given by

$$x(t) = 150 + 100e^{3t} \tag{15.20}$$

and $$y(t) = 150 - 50e^{3t} .$$

The population of the second species, given by y, becomes extinct when $y = 0$, that is when $150 - 50e^{3t} = 0$. This happens when $e^{3t} = 3$ or $t = \frac{1}{3}\ln 3$. For t greater than $\frac{1}{3}\ln 3$ the second species is extinct and the differential equations given originally are no longer appropriate. Suppose that for $t \geqslant \frac{1}{3}\ln 3$, the growth rate of the population of the first species is given by

$$\dot{x} = x .$$

The general solution of this equation is

$$x(t) = Ce^t .$$

The size of this species' population when $t = \frac{1}{3}\ln 3$ is, from (15.20),

$$x(\tfrac{1}{3}\ln 3) = 150 + 100 \exp(\ln 3) = 450 .$$

Using this value the constant C can be found. Since $450 = C3^{1/3}$,

$$C = \frac{450}{3^{1/3}} = 3^{2/3}.150 .$$

Thus the population sizes of the species are given by

$$x(t) = \begin{cases} 150 + 100e^{3t} & 0 \leqslant t \leqslant \dfrac{1}{3}\ln 3 \\[2em] 3^{2/3}.150e^t & t \geqslant \dfrac{1}{3}\ln 3 \end{cases}$$

and
$$y(t) = \begin{cases} 150 - 50e^{3t} & 0 \leqslant t \leqslant \frac{1}{3}\ln 3 \\ \\ 0 & t \geqslant \frac{1}{3}\ln 3 . \end{cases}$$

Graphs of these functions are shown in Fig. 15.2.

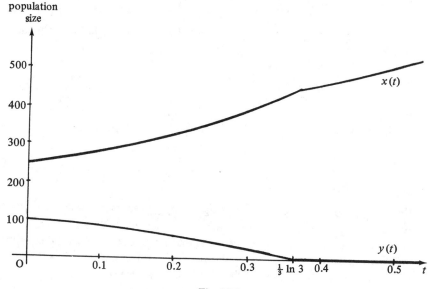

Fig. 15.2

More complicated interrelationships can be studied, although their solution is not at all easy and generally requires numerical techniques, see Section 17.5. A nonlinear system that arises in the study of predator-prey and host-parasite interrelationships is

$$\dot{x} = x(a_{11}x + a_{12}y + b_1)$$
$$\dot{y} = y(a_{21}x + a_{22}y + b_2)$$

where $a_{11}, a_{12}, a_{21}, a_{22}, b_1, b_2$ are constants and x and y give the population sizes. This system was studied by Lotka and Volterra for the special case $a_{11} = a_{22} = 0, a_{12} < 0, a_{21} > 0, b_1 > 0, b_2 < 0$, see Lotka (1956).

To overcome some deficiencies in the Lotka–Volterra model, Leslie (1948) proposed the pair of nonlinear equations

$$\dot{x} = x(b_1 - a_1x - a_2y)$$
$$\dot{y} = y\left(b_2 - a_0\frac{y}{x}\right)$$

where a_0, a_1, a_2, b_1 and b_2 are all positive constants, to represent the predator-prey case. Here x is the size of the prey population and y is the size of the predator population. Interesting reading on predator-prey and host-parasite systems can be found in Pielou (1969) and May (1973).

PROBLEMS

1. Obtain general solutions to the following differential equations:

(a) $\dfrac{dy}{dx} = \dfrac{\sin x}{\cos y}$,

(e) $\dfrac{dy}{dx} + \dfrac{y^3}{x+1} = 0$,

(b) $\dfrac{dy}{dx} = \tan x \tan y$,

(f) $\dfrac{dy}{dx} = \dfrac{y}{x(4-x)}$,

(c) $\dfrac{dy}{dx} = \dfrac{y^2}{x^2+1}$,

(g) $\dfrac{dy}{dx} = \dfrac{4y}{x(y-3)}$,

(d) $\dfrac{dy}{dx} + \dfrac{x^3}{(1+y)^2} = 0$,

(h) $\dfrac{dy}{dx} = \dfrac{x(1-y)}{x^2-1}$.

2. Solve the following initial value problems:

(a) $\dfrac{dy}{dx} = 2x(y^2+1)$, $y(0) = 0$;

(b) $\dfrac{dy}{dx} = \dfrac{1}{y(1-x^2)^{\frac{1}{2}}}$, $y(0) = 2$;

(c) $\dfrac{dy}{dx} = x^2(3-2y)$, $y(0) = 1$;

(d) $\dfrac{dy}{dx} = \dfrac{1-y}{1+x}$, $y(0) = 0$.

3. The growth of a cell depends on the flow of nutrients through its surface. Let $W(t)$ be the weight of the cell at time t. Assume that for a limited time the growth rate dW/dt is proportional to $W^{2/3}$. (If the density remains constant then W is proportional to D^3, where D is the diameter of the cell, and the surface area is proportional to D^2, or equivalently $W^{2/3}$). Hence $\dfrac{dW}{dt} = kW^{2/3}$ for some positive constant k.

Find the general solution to this differential equation.

(Edinburgh 1977, amended)

4. The *Arrhenius* equation

$$\frac{d(\ln k)}{dT} = \frac{E}{RT^2}$$

describes the variation of k (the velocity coefficient) with T (the temperature) in a certain chemical reaction where E and R are constants.

(i) Show that $k = A \exp\left(\frac{-E}{RT}\right)$, where A is a constant.

(ii) Express $\log_{10} k$ in terms of $\ln k$ and hence show theoretically that a plot of values of $\log_{10} k$ against corresponding values of $1/T$ is a straight line. Find the slope of this line. (Edinburgh, 1975)

5. Let x be a measurement of part of a body (x could be a length, a cross-sectional area, a perimeter, a weight, etc.); similarly let y denote a measurement of another part elsewhere in the body. As the body grows the values of x and y will increase. If increments Δx and Δy are observed in the values of x and y between the times t and $t + \Delta t$, the two average relative growth rates are often found to be proportional, that is,

$$\frac{1}{y}\frac{\Delta y}{\Delta t} = k\frac{1}{x}\frac{\Delta x}{\Delta t}$$

or $\frac{\Delta y}{\Delta x} = k\frac{y}{x}$ where k is the constant of proportionality. In the limit as $\Delta t \to 0$ this leads to the differential equation $\frac{dy}{dx} = \frac{ky}{x}$. Show that the general solution to this differential equation is $y = Ax^k$ for some arbitrary constant A. An equation of this type relating x and y is known as an *allometric law*.

Determine the allometric law relating the following measurements

(i) Weight and height of a man: given that for a height of 1.90 m the weight is 120 kg, whilst when the height is 1.40 m the weight is 60 kg.
(ii) Height and spread of a bush: given that when the bush is 0.4 m high the spread is 0.5 m, whilst when the height is 1.6 m the spread is 1.5 m.

6. In certain chemical reactions the concentration of a substance, at any instant, reduces at a rate proportional to the nth power of its concentration at that instant. Determine the concentration Q in terms of the time t for (i) $n = 3/2$, (ii) $n = 2$, (iii) $n = 3$, given that, when $t = 0$, the concentration has the value Q_0 while its rate of reduction is then R_0. Find, in each case, the time at which the concentration is half its initial value.

7. In a study of fasting, the weight of a volunteer decreased from 80 kg to 60 kg in 30 days. During this period it was observed that the weight loss per day was proportional to the current weight of the volunteer. What is the differential equation that gives the weight of the volunteer as a function of time? Determine the weight of the volunteer after 15 days. How many days will it take before the volunteer weighs 40 kg? Comment on the validity of this model.

8. The rate at which a liquid decreases in volume due to evaporation is proportional to the free surface area of the liquid. If the liquid adopts a cylindrical shape 0.5 cm high when lying on a smooth horizontal surface show that the rate at which the radius r of the disc changes is given by $dr/dt = -k(r + 1)$ where k is a positive constant. If the radius of a disc is observed to decrease from 4 cm to 3 cm in 10 minutes determine the time it takes for all the liquid in a disc of radius 3 cm to disappear.

9. The rate at which a body cools is proportional to the difference between the temperature of the body and that of its surroundings. If the surroundings are maintained at $20°C$ and the body cools from $80°C$ to $60°C$ in 10 minutes find how long it would take to cool to $40°C$.

10. A body having a temperature of $50°C$ is immersed in a flow of water maintained at $25°C$, and after 2 minutes the temperature of the body is observed to be $34°C$. Assuming Newton's law of cooling applies to the temperature of the body find the temperature of the body (i) 3 minutes, (ii) 5 minutes, after immersion.

11. At any time t the amount of active ferment in a culture of yeast is increasing at a rate which is directly proportional to the amount of actively fermenting yeast already in the culture. Given that the amount doubles in the first hour, find the time it takes for the amount to reach five times its original value.

12. The rate of destruction of spores exposed to a high temperature is given by

$$\frac{dN}{dt} = -kNe^{cT} ,$$

where N is the number of spores, T the temperature, t the time, and c and k are constants.

If initially there are N_0 spores determine the number of undamaged spores at time $t = 1$ in the three cases

(i) $T = T_0$,

(ii) $T = T_0 + \dfrac{\alpha}{c} t$,

(iii) $T = T_0 - \dfrac{1}{c} \ln \left(\dfrac{t^2}{2} - t + 1 \right)$.

T_0 and α are constants.

13. A population of mice, initially numbering 30, is kept in conditions that can support a population size of 120. The rate of increase in the population size N is assumed to be given by the equation

$$\frac{dN}{dt} = cN(K - N) ,$$

where c and K are positive constants. Find the formula giving the number of mice t months later, if, after one month, the population numbers 80.

14. A curve that is sometimes used in studying the growth of a population is the *Gompertz growth curve* $N = A \exp[-be^{-kt}]$. Here N is the population size at time t, A and k are positive constants, and $b = \ln (A/N_0)$ where N_0 is the size of the population when t is zero.

   Deduce that the Gompertz growth curve is the solution to each of the following differential equations

   (i) $\dfrac{dN}{dt} = kbNe^{-kt}$,

   (ii) $\dfrac{dN}{dt} = kN(\ln A - \ln N)$.

   [In (ii) use the substitution $u = \ln N - \ln A$].

   What happens to the population size as t becomes large?

15. The rate at which a population of animals numbering N changes during the course of a year is given by $\dfrac{dN}{dt} = CN \cos \left(\dfrac{\pi t}{6} + \pi \right)$ where t is the time in months from the start of the year and C is a positive constant. Find the equation for the population size N in terms of t if, at the beginning of January, there are N_0 animals alive. For $C = \frac{1}{3}$, show that the population size oscillates between $0.53 N_0$ and $1.89 N_0$.

16. A modification to the logistic equation which accounts for the influence of the resources available on the growth of the population results in the equation

$$\frac{dN}{dt} = c\frac{(K - N)N}{K + bN}$$

where b, c and K are positive constants. Determine the solution of this equation, given that $N = N_0$ when $t = 0$. Show that the population size N tends to the carrying capacity of the environment K as t increases.

17. In a 'simple' epidemic the rate at which the number of susceptibles, s in a population of total size $n + 1$, become infected is given by

$$\frac{ds}{dt} = -cs(n + 1 - s)$$

where c is a positive constant.

   Show that, if an epidemic is 'simple', the number of new cases of the disease recorded per unit time will be given by

$$\frac{cn(n + 1)^2\exp[(n + 1)ct]}{(n + \exp[(n + 1)ct])^2} \ .$$

18. The rate at which the concentration of a chemical decreases inside a cell due to diffusion through its membrane is proportional to the difference between the concentration in the cell and the concentration in the outside medium. The initial concentration within the cell is 100 gm/ml and after 30 minutes it has dropped to 80 gm/ml. If the concentration outside the cell is maintained at 30 gm/ml how long does it take for the concentration inside to drop to 40 gm/ml?

19. A gene with two alleles A and a occurs in a population in the proportions $p(t)$ and $q(t) = 1 - p(t)$, respectively, at time t. Suppose that the allele A mutates to the allele a at the relative rate μ per unit of time. This means that $dp/dt = -\mu p$. The constant μ is called the mutation rate. As mutations between the alleles A and a may occur in both directions μ is also called the forward mutation rate, and the backward mutation rate is denoted by ν. So, more generally,

$$dp/dt = -\mu p + \nu q = -\mu p + \nu(1 - p) = \nu - (\mu + \nu)p.$$

(a) Determine $p(t)$ (and $q(t)$) in terms of $p(0)$, $(q(0))$, μ and ν.

(b) Prove that for large values of t the proportions become $\dfrac{v}{\mu + v}$ and $\dfrac{\mu}{\mu + v}$. (These are the equilibrium gene frequencies).

(c) If $p(0) = 0.5$ determine, in terms of μ and v, the length of time for $p(t)$ to decrease to 0.3.

20. The rate at which enzyme A is converted to enzyme B is governed by the differential equation $dq/dt = k(a - q)(b + q)$ where q denotes the concentration of enzyme B produced after t hours. Here a and b are the initial concentrations of enzyme A and enzyme B, respectively, while k is a positive constant.

    If the initial concentrations, a and b, respectively, are 100 and 5 moles/litre find the equation giving the concentration of enzyme B produced in the reaction after t hours when the concentration of enzyme B produced in the first hour is 10 moles/litre.

21. The second order differential equation $d^2x/dt^2 = -k^2x$ models the motion of a body on which a force proportional to the distance, x, from a fixed point $(x = 0)$ and always directed towards the fixed point, acts. k is a constant. Since $d^2x/dt^2 = v\,dv/dx$, where v is the velocity of the body, obtain a relationship between x and v. Hence, by integrating again, show that

$$x = a \sin(kt + \alpha)$$

    for some arbitrary constants a and α. This particular periodic motion is known as *Simple Harmonic Motion* (SHM). The period is $2\pi/k$.

22. Find the general solution of the following linear differential equations:

    (a) $y' + 2y = e^{-x}\cos x$,   (d) $x(1 + x)y' - y = 3x^4$,

    (b) $(x + 1)y' + y = x^3 + x^2$,   (e) $(\sin x)y' - 2y\cos x = \cos x$.

    (c) $y' + 2xy = x$,

23. The population of a country changes at a rate proportional to the current size of the population and the prevailing economic situation. If the economy has a ten year cycle resulting in periods of immigration or emigration

$$\frac{dN}{dt} = kN + c \sin\left(\frac{\pi t}{5} + a\right)$$

    where N is the size of the population at time t, measured in years, and k, a and c are constants. Find an expression for the size of the population at any time.

24.   DDT absorbed in silt at the bottom of a pond is slowly degraded by
      bacterial action into harmless products, the rate of degradation being
      proportional to the amount of DDT present. Write down an equation that
      corresponds to this statement.

      If 20% of the initial amount of DDT is degraded in 10 years, show that
      the half-life, τ, of DDT is about 31 years. After how many years will 75%
      be degraded?

      If fresh DDT is still being absorbed by the silt at a rate proportional to
      e^{-ct}, where $c = \dfrac{\ln 2}{2\tau}$, write down the differential equation corresponding
      to this situation. Show that the solution to this differential equation has
      the form $Q = A\, e^{-ct} + B\, e^{-2ct}$ where Q is the amount of DDT absorbed
      in the silt, t is the time in years, and A and B are constants.

                                                     (Edinburgh 1977, amended)

25.   Verify that the differential equation for logistic growth $dN/dt = cN(K-N)$
      can be converted into the linear differential equation $dx/dt + cKx = c$

      by replacing N by $1/x$ $\left(\text{so that } x = 1/N \text{ and } \dfrac{dx}{dt} = \dfrac{-1}{N^2}\dfrac{dN}{dt}\right)$.

      Obtain the equation of logistic growth by first solving the differential
      equation for x as a function of t and then reintroducing N.

26.   When the rate at which the size of a population changes due to seasonal
      factors (such as temperature or food availability) the differential equation
      for logistic growth may be modified by replacing the constants by func-
      tions of the time. The differential equation can now no longer be solved
      by separating the variables. Instead the equation is converted into a
      linear equation by the method outlined in question 25.

      Suppose a population, initially numbering N_0 individuals, has a growth
      rate, over the course of a year, given by

      $$\frac{dN}{dt} = \frac{\pi}{12}\cos\left(\frac{\pi t}{12}\right)N - \frac{\pi}{96N_0}\sin\left(\frac{\pi t}{6}\right)N^2$$

      where N is the size of the population after t months. Show that this model
      predicts that the population size will rise to a maximum of $\dfrac{4e}{5}N_0$ individ-
      uals at the end of the sixth month and that after one year it will again
      number N_0.

      (Note that $\sin\dfrac{\pi t}{6} = 2\sin\dfrac{\pi t}{12}\cos\dfrac{\pi t}{12}$ and $\displaystyle\int \cos at\, e^{b\sin at}\, dt = \dfrac{e^{b\sin at}}{ab} + C$).

27. In a certain epidemic, healthy susceptible people are infected at a rate a per unit of the healthy population per unit time, and infected people die or recover at the rates b and c respectively. People who recover from the disease have immunity from further infection. Using the notation s, u,d and i to denote the number of healthy susceptibles, infectives, dead and immune people in the population at time t from the start of the outbreak, show that $du/dt = as - (b + c)u$.

   Write down the other three differential equations governing the course of the epidemic and hence find successively, s, u, d and i as functions of t.

   Show that when the epidemic is over $b/(b + c)$ of the population will have died, and $c/(b + c)$ of the population will be immune.

28. Find general solutions to the following second order differential equations:

    (a) $y'' + y' - 6y = e^{2x}$, (d) $y'' + 6y' + 8y = 4e^{-2x}$,

    (b) $y'' + 4y' + 5y = 5x$, (e) $y'' - y' - 2y = 2x$,

    (c) $y'' + 2y' + 2y = 2$, (f) $y'' + y = \cos x$.

29. Solve the following pairs of simultaneous linear differential equations:

    (a) $\dot{x} = 2x - y$
    $\dot{y} = x$,

    (b) $\dot{x} = y - x$
    $\dot{y} = x - y + t$,

    (c) $\dot{x} = x - y$
    $\dot{y} = 3x - y$.

30. The equations

    $$\dot{x} = \frac{1}{3}(x + 2y - 1) ,$$

    $$\dot{y} = \frac{1}{3}(4x - y - 4) ,$$

    describe a situation in which two species, with populations of size x and y at time t, complement one another. If initially $x(0) = 1$ and $y(0) = 3$ find expressions giving the size of each population at subsequent times.

31. In a particular environment the effect of competition for resources between two species, with populations of size x and y, is described by the
equations

$$\dot{x} = x - y + 30 + 5e^{-t}$$

and $\dot{y} = 2x - y$.

If initially the populations number 50 and 65 for x and y respectively
find the equations giving the size of the populations at time t. Verify that
neither species becomes extinct (population size is zero).

32. The number of predators, y, and prey, x, in an environment may be
modelled by the equations

$$\dot{x} = x(a - by) ,$$

$$\dot{y} = y(-c + dx) ,$$

where a, b, c and d are positive constants. Obtain, by using the result
that $\dfrac{dx}{dy} = \dfrac{\dot{x}}{\dot{y}}$, a variables separable differential equation in x and y. Hence
find a relationship between the population sizes of the predator and its
prey.

CHAPTER 16

Recurrence equations

16.1 INTRODUCTION

In many situations that arise in the biological sciences information is only available at certain fixed intervals of time or distance. For example, the size of a population may be recorded daily, monthly, or yearly; the number of seeds shed by a plant and lying within a radius of 10, 20, 30, . . . cm of the plant may be counted. Although one could assume that the quantity being observed varied continuously between the points of observation it is sometimes necessary, or more convenient, to work with quantities that are only given values at fixed points in time or space. For example, the growth in the number of cells present in a culture may be studied by observing the change occurring each day. Owing to a finite, as opposed to an infinitesimal, time interval being involved this would lead to a recurrence equation, rather than a differential equation, arising in the mathematical model.

The information relating to a particular quantity will often be in the form of a collection of real numbers ordered by the time of observation. Let the values of the dependent variable at time t be denoted by u_t. For ease of computation the values of t will be chosen to be equally spaced and furthermore take the values 0, 1, 2, ... Note that the value $t = 0$ may not always be included.

Thus the collection of real numbers

$$u_0, u_1, u_2, \ldots, u_t, \ldots$$

form the real sequence $\{u_t\}$, see Section 2.2. The terms in a sequence may be given explicitly, but sometimes they are obtained from preceding terms. For example, the general term of the sequence

	0, 1, 4, 9, 16, . . .	is	$u_t = t^2$,
for	1, 2, 4, 8, 16, . . .	it is	$u_t = 2^t$,
and for	0, 1, 3, 6, 10, . . .	it is given by	$u_t = u_{t-1} + t, u_0 = 0$.

The relationship giving the terms in the last sequence is called a **recurrence equation**.

In many problems the dependent variable u_t is not given directly by a formula involving t but by an equation which relates values of u_t for different values of t. Our task is to solve the resulting recurrence equation in order that the value of u_t can be found directly for each value of the discrete variable t.

The general form for a recurrence equation, also called a **difference equation,** consists of an equation involving the quantities t, u_t, u_{t+1}, ..., u_{t+n}, for some $n \geqslant 1$. If the largest difference in the subscripts in such an equation is n (from the terms u_{t+n} and u_t) the equation is said to be of **order** n. Like differential equations, with which recurrence equations have many features in common, recurrence equations are said to be **linear** if the dependent variable appears only in a linear combination, that is the equation contains no power or product of the dependent variable; otherwise they are said to be **nonlinear.** A recurrence equation is **homogeneous** if, when the dependent variables are set equal to zero, the equation is satisfied; otherwise it is **inhomogeneous.**

Examples:

$$tu_{t+1} - u_t = 0 \qquad \text{order 1, linear, homogeneous.}$$

$$u_{t+2} - u_t = 2t + 2 \qquad \text{order 2, linear, inhomogeneous.}$$

$$u_{t+2} u_t = u_{t+1} \qquad \text{order 2, nonlinear, homogeneous.}$$

16.2 FIRST ORDER RECURRENCE EQUATIONS

A typical first order equation has the form

$$u_{t+1} = f(u_t, t) \qquad \text{for } t = 0, 1, 2, \ldots$$

From this relationship u_t could be found in terms of u_0, for any t, by successive substitution and often lengthy algebra. For example, if

$$u_{t+1} = \frac{2u_t}{1 - (u_t)^2} \qquad \text{and } u_0 = 2$$

then
$$u_1 = \frac{2u_0}{1 - (u_0)^2} = \frac{2[2]}{1 - [2]^2} = -\frac{4}{3},$$

$$u_2 = \frac{2u_1}{1 - (u_1)^2} = \frac{2[-4/3]}{1 - [-4/3]^2} = \frac{24}{7},$$

$$u_3 = \frac{2u_2}{1 - (u_2)^2} = \frac{2[24/7]}{1 - [24/7]^2} = -\frac{336}{527},$$

etc.

No obvious relationship between the value of t and u_t is appearing. In fact the solution of this recurrence equation can be written compactly as

$$u_t = \tan(2^t A)$$

where $A = \tan^{-1}u_0$. That this is the solution may be verified by substitution. Thus

$$\frac{2u_t}{1 - (u_t)^2} = \frac{2\tan(2^tA)}{1 - [\tan(2^tA)]^2} = \tan[2(2^tA)] \quad \text{using A 23}^\dagger$$

$$= \tan(2^{t+1}A) = u_{t+1} .$$

Also $u_0 = \tan(2^0A) = \tan A$.

Example 1: The number of cells present in a culture is observed to double each day. Measuring t in days and denoting the number of cells present at the end of the tth day by n_t find an expression for n_t valid for all t.

The equation giving the number of cells is

$$n_{t+1} = 2n_t .$$

Thus
$$n_1 = 2n_0 ,$$
$$n_2 = 2n_1 = 2.2n_0 = 2^2n_0 ,$$
$$n_3 = 2n_2 = 2.2^2n_0 = 2^3n_0 ,$$
$$n_4 = 2n_3 = 2.2^3n_0 = 2^4n_0 ,$$
$$\text{etc.}$$

The pattern emerging suggests that the solution is

$$n_t = 2^tn_0 \qquad \text{for } t = 0, 1, 2, \ldots$$

This may be checked by substituting into the original equation. Thus

$$2n_t = 2.2^tn_0 = 2^{t+1}n_0 = n_{t+1} .$$

Notice that the solution contains one arbitrary constant n_0. If at the start of the experiment there are 5 cells present then $n_0 = 5$. In this case the solution will be given by

$$n_t = 5.2^t .$$

The **linear first order homogeneous recurrence equation**

$$u_{t+1} = a(t)u_t ,$$

where $a(t)$ is a given function of t, can be solved easily by successive substitutions. Here

$$u_1 = a(0)u_0 ,$$
$$u_2 = a(1)u_1 = a(1)\,[a(0)u_0] ,$$
$$u_3 = a(2)u_2 = a(2)\,[a(1)a(0)u_0] ,$$

and so on. The pattern emerging suggests that

$$u_t = a(t-1)a(t-2) \ldots a(1)a(0)u_0$$

is the general solution. This can be shown to be the case by substitution.

$\dagger$Appendix A.

When $a(t) = a$, a constant for all t, the solution will have the form

$$u_t = a^t u_0 .$$

This agrees with the solution found for example 1 where $a = 2$.

Example 2: Find the solution of

$$u_{t+1} = 3tu_t , \qquad\qquad (16.1)$$

where t can take the values $1, 2, 3, \ldots$.

Here the independent variable t starts with the value of one and $a(t) = 3t$.

Now $u_2 = 3.1u_1$,

$u_3 = 3.2u_2 = 3.2[3.1u_1] = 3^2 2.1u_1$,

$u_4 = 3.3u_3 = 3.3[3^2.2.1u_1] = 3^3 3!u_1$,

$u_5 = 3.4u_4 = 3.4[3^3 3!u_1] = 3^4 4!u_1$.

This suggests that

$$u_t = 3^{t-1}(t-1)!u_1$$

is the solution to (16.1). Substituting this into the right-hand side of (16.1),

$$3tu_t = 3t.3^{t-1}(t-1)!u_1 = 3^t t!u_1 = u_{t+1} ,$$

the left hand side, so verifying the solution.

When the recurrence equation is linear but inhomogeneous, that is

$$u_{t+1} = a(t)u_t + b(t) , \qquad\qquad (16.2)$$

where $a(t)$ and $b(t)$ are given functions of t, the solution can still be found by successive substitution. The general solution is now

$$u_t = a(t-1)a(t-2)\ldots a(1)a(0)u_0 + a(t-1)a(t-2)\ldots a(1)b(0)$$
$$+ a(t-1)a(t-2)\ldots a(2)b(1) + \ldots + a(t-1)b(t-2) + b(t-1).$$
$$(16.3)$$

Example 3: Solve the recurrence equation

$$u_{t+1} = (t+1)u_t + 2 \qquad \text{for } t = 0, 1, 2, \ldots$$

with $u_0 = 1$.

From (16.3), with $a(t) = t + 1$ and $b(t) = 2$, the general solution is

$$u_t = t(t-1)\ldots 2.1u_0 + t(t-1)\ldots 2.2 + t(t-1)\ldots 3.2 + \ldots$$
$$+ t.2 + 2$$

$$= u_0 t! + 2.t! \left[1 + \frac{1}{2!} + \frac{1}{3!} + \ldots + \frac{1}{(t-1)!} + \frac{1}{t!} \right]$$

$$= u_0 t! + 2.t! \sum_{r=1}^{t} \frac{1}{r!} .$$

Substituting this into the right-hand side of the original equation,

$$(t + 1)u_t + 2 = (t + 1)\left[u_0 t! + 2.t! \sum_{r=1}^{t} \frac{1}{r!}\right] + 2$$

$$= u_0(t + 1)! + 2\left[(t + 1)!\left(\sum_{r=1}^{t} \frac{1}{r!}\right) + 1\right]$$

$$= u_0(t + 1)! + 2(t + 1)! \sum_{r=1}^{t+1} \frac{1}{r!}$$

$$= u_{t+1} ,$$

so verifying the solution.

The **general solution** of the linear first order inhomogeneous equation given in (16.3) consists of a term multiplied by u_0, called the **complementary function**, and the remaining part, known as the **particular function**, is a prescribed function of t. This is similar to the solution of a linear differential equation, see Section 15.2.4. The complementary function is the solution of the associated homogeneous equation

$$u_{t+1} = a(t)u_t ,$$

whilst the particular function is a solution of the full inhomogeneous equation (16.2).

When $a(t) = a$, a constant, the recurrence equation is said to be a **linear first order constant coefficient recurrence equation**. The solution of such an equation can often be found easily by methods analogous to those used in the solution of linear constant coefficient differential equations.

16.3 LINEAR CONSTANT COEFFICIENT RECURRENCE EQUATIONS

A linear constant coefficient recurrence equation of order n is of the form

$$a_n u_{t+n} + a_{n-1}u_{t+n-1} + \ldots + a_1 u_{t+1} + a_0 u_t = b(t) \quad (16.4)$$

where $a_0, a_1, \ldots, a_n$ are constants, with a_n non-zero, and $b(t)$ is a known function of t. As noted in Section 16.2, many similarities exist between the solution of such a recurrence equation and the solution of a linear constant coefficient differential equation of order n.

The *general solution* of the recurrence equation (16.4) will contain n independent arbitrary constants and consists of two parts: a *complementary function*, containing the arbitrary constants, which satisfies the associated homogeneous recurrence equation

$$a_n u_{t+n} + a_{n-1} u_{t+n-1} + \ldots + a_1 u_{t+1} + a_0 u_t = 0 , \quad (16.5)$$

and a *particular function* which satisfies the full equation (16.4).

The complementary function is found by looking for solutions of (16.5) having the form $u_t = C\lambda^t$ where C and λ are constants. (This form is suggested by the solution to the first order linear constant coefficient equation $u_{t+1} = au_t$ which, by (16.3), is $u_t = u_0 a^t$). Substituting $u_t = C\lambda^t$ into (16.5) gives

$$a_n C\lambda^{t+n} + a_{n-1} C\lambda^{t+n-1} + \ldots + a_1 C\lambda^{t+1} + a_0 C\lambda^t = 0 \; ,$$

or $\qquad C\lambda^t [a_n \lambda^n + a_{n-1}\lambda^{n-1} + \ldots + a_1 \lambda + a_0] = 0 \; .$

If $u_t \neq 0$, both C and λ must be non-zero, so λ must satisfy the **auxiliary equation**.

$$a_n \lambda^n + a_{n-1}\lambda^{n-1} + \ldots + a_1 \lambda + a_0 = 0 \; .$$

The auxiliary equation for the recurrence equation (16.4) is a polynomial of degree n in λ and so has n roots which may be real or complex numbers. As with differential equations, repeated roots of the auxiliary equation introduce solutions modified by a power of t. The complementary function, v_t, is then formed by taking a linear combination of the n independent solutions given by the roots of the auxiliary equation.

The particular function can only be found easily when the function $b(t)$ in the homogeneous equation (16.4) is composed of functions which can occur in a complementary function. As with differential equations, the general form of the particular function, w_t, can then be deduced, and the values of the constants appearing therein can be determined exactly by substitution. Table 16.1, gives the appropriate expression required in the particular function w_t corresponding to the term that appears in $b(t)$.

Table 16.1

Term in $b(t)$	Appropriate trial function in particular function
α^t	$c\alpha^t$
α	c
t^m	$c_0 + c_1 t + c_2 t^2 + \ldots + c_m t^m$
$\cos \alpha t$ or $\sin \alpha t$	$c_1 \cos \alpha t + c_2 \sin \alpha t$

$\alpha, c, c_0, c_1, c_2, \ldots, c_m$ are constants, m a positive integer.

Recurrence equations in which combinations of these functions appear in $b(t)$ can be solved by taking a similar combination of trial functions for the particular function.

If a term of the form suggested for the particular function already appears in the complementary function then the whole form, as given in the table, needs to be multiplied by t. This is repeated until no terms are common to both the complementary function and the trial form of the particular function.

Once the general solution, the sum of the complementary function, v_t, and particular function w_t, is found the n arbitrary constants in the complementary function can be determined by using, if available, the values of the first n terms $u_0, u_1, \ldots, u_{n-1}$.

Second order recurrence equations illustrate the three types of complementary function that can arise, so let us consider the equation

$$au_{t+2} + bu_{t+1} + cu_t = R(t) .$$

The auxiliary equation associated with this recurrence equation is the quadratic equation in λ,

$$a\lambda^2 + b\lambda + c = 0$$

with roots

$$\lambda = \frac{-b \pm \sqrt{(b^2 - 4ac)}}{2a} .$$

(i) $b^2 - 4ac > 0$, roots are real and distinct.

The roots are

$$\lambda_1 = \frac{-b + \sqrt{(b^2 - 4ac)}}{2a} \quad \text{and} \quad \lambda_2 = \frac{-b - \sqrt{(b^2 - 4ac)}}{2a} .$$

The complementary function is then

$$v_t = A\lambda_1^t + B\lambda_2^t$$

where A and B are arbitrary constants.

Example 1: Find the solution of

$$u_{t+2} - u_{t+1} - 2u_t = 4t$$

with $u_0 = 0, u_1 = 2$.

The auxiliary equation is $\lambda^2 - \lambda - 2 = 0$ with roots $\lambda = 2$ and $\lambda = -1$. Therefore the complementary function is

$$v_t = A2^t + B(-1)^t .$$

To find the particular function, since $R(t) = 4t$, Table 16.1 suggests the function

$$w_t = a + bt$$

is tried. Substituting this expression for the dependent variable u_t in the full inhomogeneous equation gives

$$[a + b(t + 2)] - [a + b(t + 1)] - 2[a + bt] = 4t ,$$

or $-2bt + b - 2a = 4t .$

The particular function is determined by the values of a and b which make this equation hold identically.

Equating coefficients of powers of t,

$$t: \quad -2b = 4 \ ,$$

$$\text{constant:} \quad b - 2a = 0 \ .$$

So $b = -2$ and $a = -1$. Therefore the particular function is

$$w_t = -1 - 2t \ .$$

The general solution is then

$$u_t = v_t + w_t = A2^t + B(-1)^t - 1 - 2t \ .$$

When $t = 0$, $u_0 = A + B - 1 = 0$.

When $t = 1$, $u_1 = 2A - B - 1 - 2 = 2$.

Solving this pair of equations gives $A = 2$ and $B = -1$.

Hence the desired solution is

$$u_t = 2^{t+1} + (-1)^{t+1} - 1 - 2t \ .$$

(ii) $b^2 - 4ac = 0$, roots real and equal.

The roots are $\lambda_1 = \lambda_2 = \dfrac{-b}{2a}$.

One solution of the homogeneous equation is λ_1^t whilst the second distinct solution is found to be $t\lambda_1^t$ (as may be verified by substitution). Hence the complementary function is

$$v_t = (A + Bt)\lambda_1^t$$

where A and B are arbitrary constants.

Example 2: Find the general solution of the equation

$$u_{t+2} - 4u_{t+1} + 4u_t = 2t \ .$$

The auxiliary equation $\lambda^2 - 4\lambda + 4 = 0$ has both roots equal to 2. The complementary function is therefore

$$v_t = (A + Bt)2^t \ .$$

For the particular function the table suggests a function of the form $c2^t$ ($\alpha = 2$). However, the complementary function contains a term of the same form, viz. $A2^t$, so the form must be modified by t to become $ct2^t$. As a term of this form also appears in the complementary function we must multiply again by t to obtain $ct^2 2^t$. This term is different from any that appear in the complementary function, so $w_t = ct^2 2^t$ is the form to take for the particular function. Substituting into the inhomogeneous equation gives

$$c(t + 2)^2 \, 2^{t+2} - 4c(t + 1)^2 \, 2^{t+1} + 4ct^2 a^t = 2^t \ ,$$

that is $4c[t^2 + 4t + 4]2^t - 8c[t^2 + 2t + 1]2^t + 4ct^2 2^t = 2^t$,

which simplifies to $8c = 1$.

Therefore $w_t = \frac{1}{8} t^2 2^t$.

The general solution is thus

$$u_t = [A + Bt + \tfrac{1}{8} t^2] \, 2^t .$$

(iii) $b^2 - 4ac < 0$, roots are complex numbers.

The roots are $\lambda_1 = p + iq$ and $\lambda_2 = p - iq$

where

$$p = \frac{-b}{2a}, \quad q = \frac{\sqrt{(4ac - b^2)}}{2a} .$$

The complementary function is now

$$v_t = A_1 (p + iq)^t + A_2 (p - iq)^t$$

where A_1 and A_2 are constants, or equivalently, by introducing the polar forms, see Section 3.4.1, the complementary function can be written in the more convenient form

$$v_t = r^t [A \cos \theta t + B \sin \theta t]$$

where

$$r = \sqrt{(c/a)}, \quad \theta = \tan^{-1} \left[\frac{-\sqrt{(4ac - b^2)}}{b} \right], \, (0 < \theta < \pi),$$

and A, B are arbitrary constants.

Example 3: Find the general solution of

$$u_{t+2} + 9u_t = 0 .$$

The auxiliary equation $\lambda^2 + 9 = 0$ has roots $\lambda = \pm 3i$.

So $r = \sqrt{(9/1)} = 3$ and $\theta = \tan^{-1}(3/0) = \tan^{-1}(\infty) = \dfrac{\pi}{2}.$

The original recurrence equation is homogeneous, so the complementary function is the general solution in this case. Therefore

$$u_t = 3^t \left[A \cos \left(\frac{\pi}{2} t \right) + B \sin \left(\frac{\pi}{2} t \right) \right] .$$

Example 4: A sequence that has its origins in nature is the **Fibonacci sequence**. This sequence arose from a study, by Fibonacci in the early thirteenth century, on the problem of the size of a rabbit colony. It was assumed that each pair of adult rabbits produced a pair of young rabbits every month; the newborn rabbits taking two months to become adults and produce their first pair of offspring. Starting with one pair of adult rabbits, determine the subsequent size of the rabbit colony. Assume that no rabbits die.

Let t be the month and a_t be the number of adult pairs at time t. Then the number of adult pairs alive at time $t + 2$ is given by

$$a_{t+2} = a_{t+1} + a_t , \tag{16.6}$$

since the pairs born at time t, a_t in number, are just becoming adults at time $t + 2$. The total size of the colony at time $t + 2$, denoted c_{t+2}, consists of the number of adult pairs alive, a_{t+2}, and the number of pairs born at times $t + 2$ and $t + 1$. The number of newly born pairs equals the number of adult pairs alive at that moment of time, so

$$\begin{aligned} c_{t+2} &= a_{t+2} + (a_{t+2} + a_{t+1}) = 2a_{t+2} + a_{t+1} \tag{16.7} \\ &= 2(a_{t+1} + a_t) + a_{t+1} \qquad \text{using (16.6)} \\ &= 3a_{t+1} + 2a_t . \end{aligned}$$

Now (16.7) holds for any time t, so

$$c_{t+1} = 2a_{t+1} + a_t \text{ and } c_t = 2a_t + a_{t-1} .$$

Therefore $\begin{aligned}[t] c_{t+2} &= c_{t+1} + a_{t+1} + a_t = c_{t+1} + (a_t + a_{t-1}) + a_t \\ &= c_{t+1} + 2a_t + a_{t-1} \\ &= c_{t+1} + c_t . \end{aligned}$ $\tag{16.8}$

The two recurrence equations (16.6) and (16.8) are formally the same, the value of successive terms in these sequences being found from the sum of the last two preceding terms. Thus, in general,

$$u_{t+2} = u_{t+1} + u_t ,$$

or $\qquad u_{t+2} - u_{t+1} - u_t = 0 ,$

a linear second order constant coefficient recurrence equation for u_t.

The associated auxiliary equation $\lambda^2 - \lambda - 1 = 0$ has roots given by

$$\lambda = \frac{1 \pm \sqrt{(1+4)}}{2} = \frac{1 \pm \sqrt{5}}{2} .$$

Therefore the general solution of the recurrence equation is

$$u_t = A\left(\frac{1 + \sqrt{5}}{2}\right)^t + B\left(\frac{1 - \sqrt{5}}{2}\right)^t .$$

When $u_0 = 0$ and $u_1 = 1$, which are the appropriate starting values for the number of adult pairs a_t, the constants A and B are $1/\sqrt{5}$ and $-1/\sqrt{5}$ respectively, so that

$$a_t = u_t = \frac{1}{\sqrt{5}}\left(\frac{1 + \sqrt{5}}{2}\right)^t - \frac{1}{\sqrt{5}}\left(\frac{1 - \sqrt{5}}{2}\right)^t .$$

The total number of rabbit pairs in the colony at time t would be given by $c_t = u_{t+1}$ for $t = 1, 2, 3, \ldots$.

The early terms in the Fibonacci sequence starting with $u_0 = 0$ are

$$0, \ 1, \ 1, \ 2, \ 3, \ 5, \ 8, \ 13, \ 21, \ 34, \ 55, \ \ldots$$

The numbers in the Fibonacci sequence are known as **Fibonacci numbers**. These frequently arise in botanical studies; for example, the number of petals in a flower head and the arrangement of leaves on stems of plants usually involve such numbers.

As with differential equations it is sometimes necessary to consider systems of linear recurrence equations involving several dependent variables. By elimination of all but one dependent variable a linear recurrence equation can be obtained for the remaining variable.

Example 5: Solve the simultaneous equations

$$u_{t+1} = u_t - 2v_t , \tag{16.9}$$

$$2v_{t+1} = 3u_t + 2v_t . \tag{16.10}$$

From (16.9),

$$
\begin{aligned}
u_{t+2} &= u_{t+1} - 2v_{t+1} \\
&= u_{t+1} - (3u_t + 2v_t) && \text{using (16.10)} \\
&= u_{t+1} - 3u_t + u_{t+1} - u_t && \text{using (16.9).}
\end{aligned}
$$

So $\qquad u_{t+2} - 2u_{t+1} + 4u_t = 0 .$

The auxiliary equation associated with this recurrence equation is $\lambda^2 - 2\lambda + 4 = 0$ with roots $\lambda = 1 \pm i\sqrt{3}$. Therefore

$$u_t = 2^t \left[A \cos\left(\frac{\pi}{6}t\right) + B \sin\left(\frac{\pi}{6}t\right) \right] .$$

From (16.9),

$$v_t = \tfrac{1}{2}(u_t - u_{t+1})$$

$$= \tfrac{1}{2}\left[2^t\left(A \cos\left(\frac{\pi}{6}t\right) + B \sin\left(\frac{\pi}{6}t\right) \right) \right.$$

$$\left. - 2^{t+1}\left(A \cos\left(\frac{\pi}{6}(t+1)\right) + B \sin\left(\frac{\pi}{6}(t+1)\right) \right) \right]$$

$$= 2^{t-1}\left[A \cos\left(\frac{\pi}{6}t\right) + B \sin\left(\frac{\pi}{6}t\right) - 2A\left\{\cos\left(\frac{\pi}{6}t\right) \cos\frac{\pi}{6} \right.\right.$$

$$\left.\left. - \sin\left(\frac{\pi}{6}t\right) \sin\frac{\pi}{6} \right\} - 2B\left\{\sin\left(\frac{\pi}{6}t\right) \cos\frac{\pi}{6} + \cos\left(\frac{\pi}{6}t\right) \sin\frac{\pi}{6} \right\} \right] ,$$

Therefore

$$v_t = 2^{t-1}\left[(A(1 - \sqrt{3}) - B) \cos\left(\frac{\pi}{6}t\right) \right.$$

$$\left. + (A + B(1 - \sqrt{3})) \sin\left(\frac{\pi}{6}t\right) \right] .$$

PROBLEMS

1. Solve the following first order recurrence equations by recognising the pattern that emerges using successive substitutions:

    (a) $2u_{t+1} + u_t = 0$ $u_0 = 3$

    (b) $u_{t+1} = 3^t u_t$ $u_0 = 2$

    (c) $2t\,u_{t+1} = u_t$ $u_1 = 2$

    (d) $2u_{t+1} + t^2 u_t = 0$ $u_1 = 1$.

2. The relative change in the size of a population numbering N_t at time t and N_{t+1} at the time $t + 1$ is observed to be proportional to $1/(t+1)$, so that

    $$\frac{N_{t+1} - N_t}{N_t} = \frac{k}{t + 1} ,$$

    where k is the constant of proportionality.
    Find the formula giving the population size N_t in terms of t if, when $t = 0$, the population numbers 10, whilst when $t = 1$ it contains 20 individuals.

3. Find the general solution of the following linear constant coefficient recurrence equations:

    (a) $u_{t+1} + 2u_t = 10$,   (c) $2u_{t+1} - u_t = t$,

    (b) $15u_{t+1} - 10u_t = 3$,   (d) $3u_{t+1} - 2u_t = 2^t$.

4. Find the solution to the following recurrence equations:

    (a) $u_{t+1} = u_t$ $u_0 = 5$

    (b) $3u_{t+1} + u_t = 0$ $u_0 = 2$

    (c) $u_{t+1} - u_t = 10$ $u_0 = 2$

    (d) $2u_{t+1} + 6u_t = t^2$ $u_0 = 0$

    (e) $2u_{t+1} - u_t = 1$ $u_0 = 1$

    (f) $u_{t+1} - 2u_t = 2\cos\left(\dfrac{\pi t}{3}\right)$ $u_0 = 0$.

5. The derivative dx/dt is sometimes approximated by the difference quotient $\dfrac{x(t + \Delta t) - x(t)}{\Delta t}$. Use this result with $\Delta t = 1$ to express the following differential equations as first order recurrence equations:

    (a) $\dfrac{dN}{dt} = kN$,   (b) $\dfrac{dN}{dt} = cN(K - N)$.

By considering the solutions to the above differential equations, that is

$N = N_0 e^{kt}$ and $N = \dfrac{KN_0}{N_0 + (K - N_0)e^{-Kct}}$ respectively, obtain first order

recurrence equations for N_{t+1} in terms of N_t.

Compare the form just obtained with the corresponding one found in the first part of the question.

6. An isolated group of people is exposed to a disease to which they have no immunity. Let H_t denote the number of healthy people in the population that are susceptible to the disease at time t, S_t be the number that are sick, and I_t be the number that have gained immunity after their illness or have died. If a week is taken as the unit of time then the number of individuals that are ill with the disease is observed to equal a constant fraction, λ, of the healthy susceptible population one week earlier. Within a week a person with the disease will either recover or die.

Set up three recurrence equations for H, S and I, and hence show that

$$I_t = [1 - (1 - \lambda)^t] \frac{N_0}{1 - \lambda} \qquad t = 0, 1, 2, \ldots$$

where N_0 is the size of the population initially.

7. Find the general solution of the following recurrence equations:

(a) $2u_{t+2} - 5u_{t+1} + 2u_t = 1$

(b) $u_{t+2} - 3u_{t+1} + 2u_t = 3^t - t$

(c) $u_{t+2} + 2u_{t+1} + u_t = 4t$

(d) $u_{t+2} - 2u_{t+1} + 2u_t = 2^t$

(e) $u_{t+2} - 6u_{t+1} + 5u_t = 16t$

(f) $u_{t+2} - 7u_{t+1} + 10u_t = t3^t$.

8. Find the solution to the following recurrence equations:

(a) $u_{t+2} - 5u_{t+1} + 6u_t = 2$ $u_0 = 1$, $u_1 = 2$

(b) $u_{t+2} - 3u_{t+1} + 2u_t = 1$ $u_0 = 0$, $u_1 = 3$

(c) $u_{t+2} - 4u_{t+1} + 4u_t = 3^t$ $u_0 = 3$, $u_1 = 4$

(d) $u_{t+2} + 2u_t = 0$ $u_0 = 1$, $u_1 = \sqrt{2}$

(e) $u_{t+2} - 6u_{t+1} + 9u_t = 4$ $u_0 = 2$, $u_1 = 1$

(f) $u_{t+2} + u_{t+1} + u_t = t$ $u_0 = 1$, $u_1 = 0$.

9.   Find the general solution of

$$u_{t+2} - (a + b)u_{t+1} + abu_t = a^t \qquad (ab \neq 0)$$

Consider the two cases $a \neq b$ and $a = b$.

10. In a population of worms whose life cycle takes three years, the total number N present will depend on the number of viable eggs laid two years ago and the number surviving the larval stage. This may be represented by the second order recurrence equation

$$N_{t+2} = aN_{t+1} + bN_t \qquad\qquad t = 0, 1, 2, \ldots$$

where a and b are positive constants.

Find the solution to this equation, given that $N_0 = 8$ and $N_1 = 10$, in the three cases

(i) $a = 1/2, b = 15/16$,

(ii) $a = 1/2, b = 3/16$,

(iii) $a = 0, \quad b = 1$.

Examine the behaviour of the population in each case after a long period.

11. In a study of the spread of infectious diseases, a record is kept of the outbreaks of a certain disease. It is estimated that the probability p_n of at least one new case in the nth week after an outbreak starts satisfies the recurrence equation

$$p_{n+2} = p_{n+1} - \tfrac{1}{6} p_n$$

for $n = 0, 1, 2, \ldots$. If $p_0 = 0$ and $p_1 = 1$, find p_n.

What happens to p_n as n becomes large? (For details of probability see Chapter 21).   (Edinburgh, 1977)

12. Find the solution of the following simultaneous recurrence equations:

(a) $u_{t+1} = u_t + \tfrac{1}{2} v_t$ (b) $u_{t+1} = u_t - v_t + t$

$\ v_{t+1} = \tfrac{1}{8} u_t + v_t$, $\ v_{t+1} = -u_t + v_t$.

13. The recurrence equations

$$9u_{t+1} = 28u_t + 2v_t \ ,$$
$$9v_{t+1} = -5u_t + 17v_t \ ,$$

describe the interrelationship in an environment containing two competing species with population sizes u_t and v_t at time t. Find the population sizes at subsequent times if initially $u_0 = 10$ and $v_0 = 40$. When does one of the species become extinct?

14. The recurrence equations

$$u_{t+1} = 2u_t - 5v_t + 80 ,$$
$$v_{t+1} = u_t - 2v_t + 40 ,$$

describe a situation in which two species, having populations numbering u_t and v_t at time t, coexist in harmony. If $u_0 = 30$ and $v_0 = 20$ find the expressions giving the population sizes at subsequent times.

15. Find the solution of the following linear constant coefficient recurrence equations:

(a) $u_{t+3} - 3u_{t+1} - 2u_t = 0$ $u_0 = 1, u_1 = 3, u_2 = 2$

(b) $u_{t+3} - 6u_{t+2} + 11u_{t+1} - 6u_t = 0$ $u_0 = 1, u_1 = 1, u_2 = -1$

(c) $u_{t+3} - u_{t+2} + u_{t+1} - u_t = 0$

(d) $u_{t+4} + 4u_{t+3} + 6u_{t+2} + 4u_{t+1} + u_t = 0.$

CHAPTER 17

Numerical methods

17.1 INTRODUCTION

The application of mathematical ideas to real-life problems often leads to an equation to be solved or an expression to be evaluated. In some cases it may be possible to find the solution, or value, required by using the procedures and techniques already discussed. In other situations these methods cannot provide the answer, however, it is sometimes possible to find an approximate answer directly by employing a numerical method. As an illustration, recall the use of Newton's method for finding the root of an equation given in Chapter 8.

Earlier methods often give the answer to a problem via a formula. Such a formula is often described as being the **analytical solution** to the problem. Numerical methods do not involve the analytical solution but go directly to the final numerical answer. As a consequence numerical methods do not produce general results but ones that are associated with the particular problem under consideration. Because the method attempts to go straight to the numerical answer it is important to know that an answer does exist before a numerical method is applied to a problem, otherwise spurious results may be generated. It is also helpful to have an idea of the answer to be expected so that any breakdown in the numerical procedure can be spotted. Often this is not possible, and so a second different numerical method should be tried wherever possible and the results of the two methods compared.

Following a discussion of errors in numerical work, numerical methods are presented for finding the roots of an equation, the evaluation of definite integrals and the solution of differential equations. The numerical determination of maxima and minima will not be considered here. The **algorithms,** that is numerical procedures, given in this chapter are ones that are suitable for use with hand-held calculators or mathematical tables. Computer programs, which are available in most computer libraries, can be used to solve the problems discussed here with more accuracy and speed.

17.2 ERRORS

In any numerical work the effect of errors will lead to an inaccurate result. Ideally the result of a calculation should be exact, but this is not always possible in practice. For example, if $x^2 = 1$ then $x = \pm 1$ but if $x^2 = 2$, $x = \pm\sqrt{2}$, and $\sqrt{2}$ can only be given approximately by a decimal number, see Section 1.2.

Errors in numerical work may be due to:

(i) **Human errors** – such as mistakes in arithmetic, miscopying or the incorrect recall of basic values or formulae.

(ii) **Inherent errors in the formula being used** – often the formula is only an approximation to the correct relationship between the variables.

(iii) **Truncation errors** – these occur when a formula is approximated by taking the first few dominant terms from a series expansion. For example, suppose a quantity p can be expressed as a power series in h, that is

$$p = p_0 + hf_1(p_0) + h^2 f_2(p_0) + h^3 f_3(p_0) + \ldots$$

where p_0 is the value of p when $h = 0$ and $f_i(p_0)$, $i = 1, 2, \ldots$, are known numbers depending on p_0. See the Maclaurin and Taylor series expansions of functions given in Section 7.2. The approximation of p by $p_0 + hf_1(p_0)$ is said to be of **order** h^2 since the terms omitted from the full expansion contain powers of h from h^2 upwards. Similarly $p_0 + hf_1(p_0) + h^2 f_2(p_0)$ will be an approximation to p of order h^3. The truncation errors in these approximations to p are said to be of order h^2 and h^3 respectively. To emphasise the terms that are ignored in obtaining the approximation to p the symbol $0(h^n)$ is introduced, where n is some positive integer. Thus

$$p = p_0 + hf_1(p_0) + 0(h^2)$$

signifies that the terms involving powers of h greater than or equal to 2 are being ignored.

(iv) **Errors due to inaccurate data** – in many practical applications the values of quantities appearing in a formula may have been obtained from observations or measurements.

(v) **Propagation errors** – these arise in a calculation through the use of values obtained from previous calculations which contain errors.

(vi) **Round-off errors** – these arise in most numerical work owing to the practical requirement that numbers with only a limited number of significant figures, SF, or decimal places, D, can be used in the computations (see Section 1.4).

To minimise the risk of errors due to human factors it is important to set out the calculations in a logical and tidy way. This can often be achieved by using a tabular form for the working. It will be assumed that there are no errors in the numerical procedure due to human factors and that truncation, propagation and round-off errors will be the most influential.

The effect of round-off errors can be minimised by organising the calculation appropriately, as the following examples show.

Example 1: Working to 1D find $\dfrac{ab}{c}$ when $a = 5.1, b = 3.5, c = 2.1$.

The calculation can be performed in several ways:

$$(ab)/c = (5.1 \times 3.5)/2.1 = 17.8/2.1 = 8.5 \ ,$$
$$(a/c)b = (5.1/2.1) \times 3.5 = (2.4) \times 3.5 = 8.4 \ ,$$
$$a(b/c) = 5.1 \times (3.5/2.1) = 5.1 \times (1.7) = 8.7 \ ,$$

each giving a different answer.

The true value of $\dfrac{ab}{c}$ is 8.5

$$\left(\frac{ab}{c} = \frac{5.1 \times 3.5}{2.1} = \frac{(3 \times 1.7) \times (5 \times 0.7)}{3 \times 0.7} = 1.7 \times 5 = 8.5 \right) \ .$$

Example 2: Sum the following numbers, working to 3SF:

$0.841, 0.00342, 0.670, 0.000232, 0.000483, 0.00178.$

Summing the numbers in the order they are given

$$0.841 + 0.00342 \ = 0.844 \ ,$$
$$0.844 + 0.670 \ \ \ \ = 1.51 \ ,$$
$$1.51 \ \ \ + 0.000232 = 1.51 \ ,$$
$$1.51 \ \ \ + 0.000483 = 1.51 \ ,$$
$$1.51 \ \ \ + 0.00178 \ = 1.51 \ .$$

Summing the numbers in increasing order of magnitude

$$0.000232 + 0.000483 = 0.000715 \ ,$$
$$0.000715 + 0.00178 \ \ = 0.00250 \ ,$$
$$0.00250 \ \ + 0.00342 \ = 0.00592 \ ,$$
$$0.00592 \ \ + 0.670 \ \ \ \ = 0.676 \ ,$$
$$0.676 \ \ \ \ + 0.841 \ \ \ \ = 1.52 \ .$$

The value 1.52 is the correct value of the sum to 3SF.

In general round-off errors will be decreased if numbers of the same order of magnitude are added together. This is not the case with subtraction, as the next example shows.

Example 3: Find the value of $\sqrt{2.02} - \sqrt{2.01}$, given that to 5 SF, $\sqrt{2.02} = 1.4213$ and $\sqrt{2.01} = 1.4177$.

By direct calculation

$$\sqrt{2.02} - \sqrt{2.01} = 1.4213 - 1.4177 = 0.0036 \ ,$$

a result containing 2SF.

But
$$\sqrt{2.02} - \sqrt{2.01} = (\sqrt{2.02} - \sqrt{2.01}) \left(\frac{\sqrt{2.02} + \sqrt{2.01}}{\sqrt{2.02} + \sqrt{2.01}} \right)$$

$$= \frac{2.02 - 2.01}{\sqrt{2.02} + \sqrt{2.01}}$$

$$= \frac{0.01}{2.8390}$$

$$= 0.0035224 \ ,$$

which is accurate to 5SF.

17.3 ROOTS OF EQUATIONS

Numerical procedures can be applied to the problem of finding the roots of an equation given by $f(x) = 0$, say. Here the task is to find values of x for which $f(x)$ is zero. If the function $f(x)$ is differentiable, Newton's method, as described in Section 8.4, can be employed. In this method a sequence $x_1, x_2, x_3, \ldots$ is obtained from the recurrence equation

$$x_{n+1} = x_n - f(x_n)/f'(x_n) \qquad n = 1, 2, 3, \ldots$$

which, normally, converges to a root of the equation $f(x) = 0$. This procedure is, in general, very good at locating a root once x_n gets near to that root. Therefore the process will be most efficient if the initial choice x_1 is near the root.

As indicated in Section 8.4, the value to take for x_1 may be obtained graphically. Alternatively, the whereabouts of a zero of a continuous function can be determined numerically by evaluating the function at a number of points and noting where it changes sign. Suppose $f(a) < 0$ and $f(b) > 0$ then a root of the equation $f(x) = 0$ lies between $x = a$ and $x = b$.

It is apparent, from Fig. 17.1, that the root lies close to the point at which the chord PQ, joining the points with coordinates $(a, f(a))$ and $(b, f(b))$, crosses the x-axis. The equation of the chord PQ, using (3.5), is given by

$$\frac{y - f(a)}{f(b) - f(a)} = \frac{x - a}{b - a} \ ,$$

so when $y = 0, x = a - \left(\frac{b - a}{f(b) - f(a)} \right) f(a) \ .$

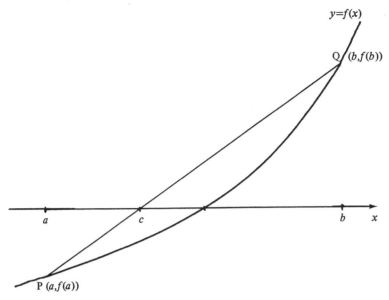

Fig. 17.1

This value of x can be taken as an approximation to the root and there-fore can be used as the starting value x_1 in Newton's method if the function $f(x)$ is differentiable. If $f(x)$ is not differentiable, or $f'(x)$ is difficult to obtain, this formula for x can itself be used as the basis of a root finding method known as **Regula falsi** or the **secant method**.

Let $$c = a - \left(\frac{b - a}{f(b) - f(a)}\right) f(a) \qquad (17.1)$$

and evaluate $f(c)$. The value of a or b which results in the function $f(x)$ taking the same sign as $f(c)$ is replaced by c, and the process is repeated to find a new value for c. The process is terminated when two successive values of c agree to the required accuracy and the corresponding values $f(c)$ are both close to zero.

Example: Find the positive root(s) of
$$x^2 - 1 = \sin x$$
correct to 4D.

Let $f(x) = x^2 - 1 - \sin x$. Taking integer values for x gives
$$f(0) = -1 \ ,$$
$$f(1) = -0.8415 \ ,$$
$$f(2) = \quad 2.0907 \ .$$

The function remains positive for x greater than 2, so the only positive root lies between $x = 1$ and $x = 2$. See Fig. 17.2 for a graph of $f(x)$.

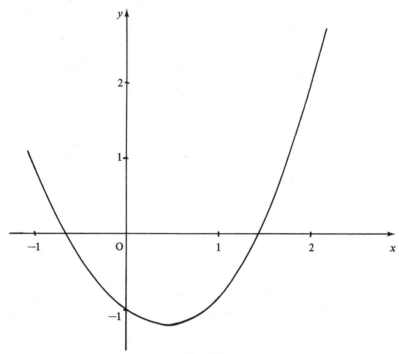

Fig. 17.2

To apply the secant method let $a = 1, b = 2$ then, by equation (17.1),

$$c_1 = 1 - \frac{(2-1)(-0.8415)}{2.0907 - (-0.8415)} = 1.287 .$$

Now $f(c_1) = -0.3036$, so choose $a = c_1 = 1.287, b = 2$ and reapply (17.1) to obtain

$$c_2 = 1.337 \text{ and } f(c_2) = -0.08515 < 0 .$$

The values obtained by continuing with the secant method are given in Table 17.1. The results show that the convergence to the root at 1.4096 is quite slow and that the original value of b has persisted throughout the process.

To speed up convergence to the root the secant method may be modified when a value of a or b persists for several iterations and $f(c)$ is closer to zero than $f(a)$ and $f(b)$. The modification consists of replacing the persistent value by c for the next iteration. Note that it is important to check that the new value of c generated by (17.1) is between the former value of c and the persistent value. If not, the persistent value must be retained for a further iteration.

Table 17.1

Secant Method				Modified Secant Method				Newton's Method	
a	b	c	$f(c)$	a	b	c	$f(c)$	x_n	$f(x_n)$
1	2	1.287	−0.3036	1	2	1.287	−0.3036	1.287	−0.3036
1.287	2	1.377	−0.08515	1.287	2	1.377	−0.08515	1.4193	−0.025866
1.377	2	1.4014	−0.021765	1.377	2	1.4014	−0.021765	1.4097	−0.000202
1.4014	2	1.40757	−0.005455	1.377	1.4014	1.40977	−0.000388	1.409624	−0.00000001
1.40757	2	1.40911	−0.001366	1.4014	1.40977	1.40962	−0.00001065	1.409624	−0.00000001
1.40911	2	1.40950	−0.000330	1.40962	1.40977	1.409624	−0.00000001		
1.40950	2	1.40959	−0.000090						
1.40959	2	1.40962	−0.00001065						

In the example to find the positive root of the equation $f(x) = x^2 - 1 - \sin x = 0$, the values, after the third iteration of the secant method has been performed, are

$$a = 1.377 \qquad f(a) = -0.08515 \ ,$$
$$b = 2 \qquad f(b) = 2.0907 \ ,$$
$$c = 1.4014 \qquad f(c) = -0.021765 \ .$$

The value for b has persisted throughout the three iterations and the value of $f(c)$ is the closest one to zero, so the modification to the secant method is applied. Let $a = 1.377$ and $b = 1.4014$, then the new value of c, given by (17.1), is found to be $c = 1.40977$, and this lies between 1.4014 and 2. Continuing with the standard secant method the values of c converge to the root 1.4096 faster than before, see Table 17.1.

As the function in this example is differentiable, Newton's method may also be used. The values obtained by Newton's method starting with the approximation given by the first application of the secant method are also included in Table 17.1. From this table it can be seen that Newton's method is superior to the other two. This observation holds true whenever Newton's method converges. Note that Newton's method requires the function to be differentiable whereas the secant methods do not.

17.4 THE EVALUATION OF DEFINITE INTEGRALS

There are situations in which an integral may require evaluating, but no function made up of a finite combination of the standard functions (listed in Appendix B) can be found having as its derivative the integrand. For example, the integral of e^{-x^2} cannot be found in terms of a finite number of standard functions. It is possible, however, to express the integral as the sum of an infinite number of functions by using the Maclaurin expansion

$$e^{-x^2} = 1 - x^2 + \frac{x^4}{2!} - \frac{x^6}{3!} + \ldots$$

and then to integrate each term in the series. Hence

$$\int e^{-x^2} \, dx = \int \left[1 - x^2 + \frac{x^4}{2!} - \frac{x^6}{3!} + \ldots \right] dx$$

$$= \left[x - \frac{x^3}{3} + \frac{x^5}{5.2!} - \frac{x^7}{7.3!} + \ldots \right] + C \ .$$

The value of the integral is thus given by an infinite sum which can only be evaluated approximately for given values of x.

An alternative approach is to approximate a definite integral by a finite sum directly. The value of a definite integral was shown in Section 11.6 to be given by

$$\int_a^b f(x)\,dx = \lim_{n \to \infty}\left(\sum_{k=1}^n Y_k\,\Delta x\right) \tag{17.2}$$

when the limit exists. In this expression $\Delta x = (b-a)/n$ and Y_k is the value taken by $f(x)$ using a value of x between $a + (k-1)\Delta x$ and $a + k\Delta x$. It is usual in numerical integration to let $h = \Delta x = (b-a)/n$ and $x_k = a + hk$ for $k = 0, 1, 2, \ldots, n$. Note that the limits of integration a and b must be finite and the integrand must remain finite for values of x between a and b if the sum in (17.2) is to exist. It can be shown that for finite limits of integration and a continuous integrand the definite integral does exist and can be obtained from equation (17.2). If the integrand is not continuous, but still finite, the integration must be carried out in sections the end points of which correspond to the points of discontinuity. Under these conditions an approximation to the integral can be obtained from (17.2) by using the expression

$$\sum_{k=1}^n Y_k\,\Delta x \ .$$

Notice that the approximation will usually be improved by taking larger values of n, that is more terms in the sum. If the integral is evaluated in sections there is no need for the value of Δx to be the same in each section.

Since a definite integral may be considered as giving the area under a curve when the integrand is positive, see Section 11.2, $Y_k\,\Delta x$ represents approximately, the area A_k of the kth strip, see Fig. 17.3. The total area under the curve

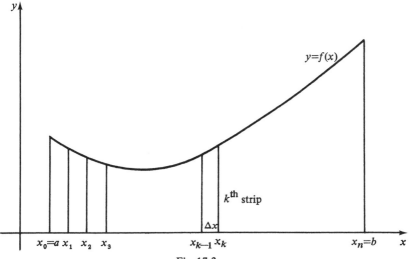

Fig. 17.3

consists of n strips each of width Δx of which the kth is typical. The various ways in which the area of the kth strip can be approximated give rise to different formulae for the approximate value of the definite integral. The value assigned to Y_k will, for a continuous integrand $f(x)$, often be formed from a linear combination of the quantities $y_i = f(x_i)$, $i = 0, 1, 2, \ldots, n$. The formulae which are obtained in this section are also valid when the integrand takes on negative values, although the geometrical interpretation of the integral as an area no longer holds.

The easiest way of approximating the area of the kth strip is by taking $Y_k = y_k$ (or $Y_k = y_{k-1}$) for $k = 1, 2, \ldots, n$. Thus the area under the curve is approximated by rectangles as shown in Fig. 17.4a (or 17.4b). In general this approach is not very accurate and so is rarely used.

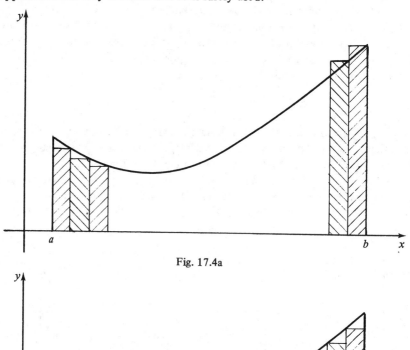

Fig. 17.4a

Fig. 17.4b

17.4.1 Trapezoidal Rule

A better approximation to the area of the kth strip under the curve, than that given by the rectangle shown in Fig. 17.4a (17.4b), can be obtained by joining the points P_{k-1} and P_k, with coordinates (x_{k-1}, y_{k-1}) and (x_k, y_k) respectively, by a straight line to produce a trapezium, see Fig. 17.5. Taking the area of the trapezium $X_{k-1}\ P_{k-1}\ P_k\ X_k$, which is $\frac{1}{2}(y_{k-1} + y_k)h$, as an approxi-

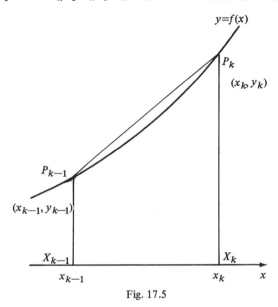

Fig. 17.5

mation to the area of the kth strip under the curve, Y_k is given by $Y_k = \frac{1}{2}(y_{k-1} + y_k)$. Hence the total area under the curve is given by

$$\int_a^b f(x)\,dx \cong \sum_{k=1}^{n} Y_k\, h = \frac{h}{2} \sum_{k=1}^{n} (y_{k-1} + y_k)$$

$$= \frac{h}{2}\,[(y_0 + y_1) + (y_1 + y_2) + \ldots + (y_{n-2} + y_{n-1}) + (y_{n-1} + y_n)]$$

$$= h\,[\tfrac{1}{2}(y_0 + y_n) + y_1 + y_2 + \ldots + y_{n-1}]\ . \tag{17.3}$$

This formula is the **trapezoidal rule** for evaluating a definite integral.

Example 1: Evaluate $\displaystyle\int_0^1 \frac{dx}{1 + x^2}$ by the trapezoidal rule, using 10 strips.

Here $a = 0$, $b = 1$, $y = f(x) = \dfrac{1}{1 + x^2}$ and $n = 10$.

So $h = (b - a)/n = 0.1$, and there are eleven ordinates $y_0, y_1, \ldots, y_{10}$ given by $y_k = \dfrac{1}{1 + x_k^2}$ where $x_k = 0.1\,k$ for $k = 0, 1, 2, \ldots, 10$.

The working may be set out in a table, thus:

k	x_k	$1 + x_k^2$	$y_k = 1/(1 + x_k^2)$ first and last ordinates	$y_k = 1/(1 + x_k^2)$ other ordinates
0	0.0	1.00	1.0000	
1	0.1	1.01		0.9901
2	0.2	1.04		0.9615
3	0.3	1.09		0.9174
4	0.4	1.16		0.8621
5	0.5	1.25		0.8000
6	0.6	1.36		0.7353
7	0.7	1.49		0.6711
8	0.8	1.64		0.6098
9	0.9	1.81		0.5525
10	1.0	2.00	0.5000	
	$h = 0.1$	Sums	1.5000	7.0998

Therefore, by the trapezoidal rule,

$$\int_0^1 \frac{dx}{1 + x^2} \cong 0.1\,[\tfrac{1}{2}(1.5000) + 7.0998] = 0.78498 \ .$$

This particular integral can be evaluated exactly, whence

$$\int_0^1 \frac{dx}{1 + x^2} = \left[\tan^{-1} x\right]_0^1 = \tan^{-1} 1 = \frac{\pi}{4} \cong 0.785398.$$

The numerical calculation is thus accurate to 3 significant figures.

17.4.2 Simpson's Rule

In the trapezoidal rule each section of the curve $y = f(x)$ was replaced by a straight line, viz. the chord. A more accurate integration procedure, known as Simpson's rule, can be obtained by approximating each section of the curve by a parabolic arc.

A parabolic arc is chosen to approximate the curve by requiring the parabola $y = \alpha x^2 + \beta x + \gamma$ to pass through the three points P_{k-1}, P_k and P_{k+1} on the curve described by $y = f(x)$, see Fig. 17.6. The points P_{k-1}, P_k and P_{k+1}

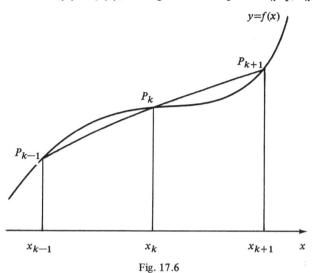

Fig. 17.6

have coordinates (x_{k-1}, y_{k-1}), (x_k, y_k) and (x_{k+1}, y_{k+1}) respectively, so that the values of the constants α, β and γ can be determined from the equations

$$\alpha(x_k - h)^2 + \beta(x_k - h) + \gamma = y_{k-1} \,, \tag{17.4}$$

$$\alpha x_k^2 \qquad\quad + \beta x_k \qquad\quad + \gamma = y_k \quad, \tag{17.5}$$

$$\alpha(x_k + h)^2 + \beta(x_k + h) + \gamma = y_{k+1} \,. \tag{17.6}$$

The area under the curve between the points P_{k-1} and $P_{k+1}, A_k + A_{k+1}$, can now be approximated by the area under the parabola $y = \alpha x^2 + \beta x + \gamma$ between $x = x_{k-1}$ and $x = x_{k+1}$. Thus

$$A_k + A_{k+1} \cong \int_{x_{k-1}}^{x_{k+1}} (\alpha x^2 + \beta x + \gamma)\,dx = \left[\alpha\frac{x^3}{3} + \beta\frac{x^2}{2} + \gamma x\right]_{x_{k-1}}^{x_{k+1}}$$

$$= \frac{\alpha}{3}\left[(x_k + h)^3 - (x_k - h)^3\right] + \frac{\beta}{2}\left[(x_k + h)^2 - (x_k - h)^2\right]$$

$$+ \gamma\left[(x_k + h) - (x_k - h)\right]$$

$$\text{since } x_{k-1} = x_k - h, \; x_{k+1} = x_k + h$$

$$= h\left[2\alpha x_k^2 + \frac{2}{3}\alpha h^2 + 2\beta x_k + 2\gamma\right]$$

$$= h\left[2y_k + \frac{2}{3}\alpha h^2\right] \quad \text{since } \alpha x_k^2 + \beta x_k + \gamma = y_k \;.$$

The value of α is found by eliminating β and γ from the equations (17.4), (17.5) and (17.6).

Subtracting (17.5) from (17.4) gives $-2\alpha h x_k + \alpha h^2 - \beta h = y_{k-1} - y_k$.

Subtracting (17.5) from (17.6) gives $2\alpha h x_k + \alpha h^2 + \beta h = y_{k+1} - y_k$.

Adding, to eliminate β, then gives $\quad 2\alpha h^2 = y_{k-1} - 2y_k + y_{k+1}$.

Hence $\quad A_k + A_{k+1} \cong h\left[2y_k + \dfrac{1}{3}(y_{k-1} - 2y_k + y_{k+1})\right]$

$$= \frac{h}{3}\left[y_{k-1} + 4y_k + y_{k+1}\right] \ .$$

Since the strips are considered in pairs (k^{th} with the $k+1^{\text{th}}$) the total area under the curve must be divided into an even number of strips in order that the integral may be evaluated by this procedure. Hence

$$\int_a^b f(x)\,dx = (A_1 + A_2) + (A_3 + A_4) + \ldots + (A_{n-1} + A_n)$$

$$\cong \frac{h}{3}\left[(y_0 + 4y_1 + y_2) + (y_2 + 4y_3 + y_4) + \ldots\right.$$

$$\left. + (y_{n-2} + 4y_{n-1} + y_n)\right]$$

$$= \frac{h}{3}\left[y_0 + y_n + 4(y_1 + y_3 + \ldots + y_{n-1})\right.$$

$$\left. + 2(y_2 + y_4 + \ldots + y_{n-2})\right] \ .$$

$$(17.7)$$

This formula is **Simpson's rule** for evaluating a definite integral, and it requires an odd number of y values.

Example 2: Evaluate $\displaystyle\int_0^1 \frac{dx}{1 + x^2}$ using Simpson's rule with 10 strips.

Here $a = 0, b = 1, f(x) = \dfrac{1}{1 + x^2}, n = 10$ so $h = 0.1$ and $y_k = \dfrac{1}{1 + x_k^2}$ where

$x_k = 0.1\,k, k = 0, 1, 2, \ldots, 10.$

Setting out the working in a table:

k	x_k	$1 + x_k^2$	$y_k = 1/(1 + x_k^2)$ first and last ordinates	$y_k = 1/(1 + x_k^2)$ odd ordinates	$y_k = 1/(1 + x_k^2)$ remaining even ordinates
0	0.0	1.00	1.0000		
1	0.1	1.01		0.9901	
2	0.2	1.04			0.9615
3	0.3	1.09		0.9174	
4	0.4	1.16			0.8621
5	0.5	1.25		0.8000	
6	0.6	1.36			0.7353
7	0.7	1.49		0.6711	
8	0.8	1.64			0.6098
9	0.9	1.81		0.5525	
10	1.0	2.00	0.5000		
	$h = 0.1$ Sums		1.5000	3.9311	3.1687

Therefore Simpson's rule gives

$$\int_0^1 \frac{dx}{1 + x^2} \cong \frac{0.1}{3} \left[1.5000 + 4(3.9311) + 2(3.1687) \right] \cong 0.78539 .$$

This compares favourably with the exact value of 0.785398.

Notice that the value given by Simpson's rule is more accurate than that obtained by using the trapezoidal rule, see example 1. This observation holds true for any integral with an integrand that is differentiable. It can be shown that the error in the value of the integral obtained by using the trapezoidal rule is of order h^2 whilst with Simpson's rule it is of order h^4. The more strips used (the larger the value of n) the more accurate both rules become, although errors due to round-off eventually begin to influence the accuracy.

Numerical integration is invaluable when the integral cannot be evaluated analytically. This is illustrated by the next example.

Example 3: Evaluate $\int_0^2 e^{-x^2/2} \, dx$.

Using Simpson's rule with 10 strips to evaluate the integral, the table of working is:

k	x_k	x_k^2	first and last ordinates	$\exp(-x_k^2/2)$ odd ordinates	remaining even ordinates
0	0.0	0.00	1.00000		
1	0.2	0.04		0.98020	
2	0.4	0.16			0.92312
3	0.6	0.36		0.83527	
4	0.8	0.64			0.72615
5	1.0	1.00		0.60653	
6	1.2	1.44			0.48675
7	1.4	1.96		0.37531	
8	1.6	2.56			0.27804
9	1.8	3.24		0.19790	
10	2.0	4.00	0.13534		
$h = 0.2$		Sums	1.13534	2.99521	2.41406

Thus

$$\int_0^2 e^{-\frac{x^2}{2}} \, dx \cong \frac{0.2}{3} [1.13534 + 4(2.99521) + 2(2.41406)] \cong 1.19628.$$

In some situations the integrand is known only as a series of numerical values — for example, experimental readings or observations — instead of as an analytic function of some variable. If the values are available at equally spaced intervals then an integral involving these values may be evaluated numerically by using Simpson's rule if there is an odd number of values; otherwise the trapezoidal rule for one strip may be applied to each interval.

Example 4: Use Simpson's rule to estimate the volume of revolution generated when the area under a curve, between $x = 0$ and $x = 6$, with values given by

x	0	1	2	3	4	5	6
$f(x)$	0.0	2.0	2.5	2.3	2.0	1.7	1.5

is rotated about the x-axis.

The volume of revolution is given by

$$V = \pi \int_0^6 [f(x)]^2 \, dx \quad .$$

Apply Simpson's rule to evaluate $I = \int_0^6 y \, dx$, where $y = f^2(x)$.

k	x_k	$f(x_k)$	first and last ordinates	$y_k = f^2(x_k)$ odd ordinates	remaining even ordinates
0	0	0.0	0.00		
1	1	2.0		4.00	
2	2	2.5			6.25
3	3	2.3		5.29	
4	4	2.0			4.00
5	5	1.7		2.89	
6	6	1.5	2.25		
$h = 1$		Sums	2.25	12.18	10.25

So $\qquad I \cong \dfrac{1}{3}[2.25 + 4(12.18) + 2(10.25)] = \dfrac{71.47}{3}$.

Therefore $\quad V = \pi I \cong \dfrac{71.47\pi}{3} \cong 74.85.$

17.4.3 Integrals with infinite limits of integration

When the limits of integration are not finite it is still possible, sometimes, to evaluate the integral numerically, but extra care must be taken. A simple test which ensures the existence of the integral $\int_a^\infty f(x)\,dx$ is that $\lim\limits_{x\to\infty} x\,f(x) = 0$, or if the integral is $\int_{-\infty}^b f(x)\,dx,\ \lim\limits_{x\to-\infty} x\,f(x) = 0.$

To obtain the value of $\int_a^\infty f(x)\,dx = \lim\limits_{B\to\infty}\int_a^B f(x)\,dx$, integrals of the form $\int_a^B f(x)\,dx$ must be evaluated for a range of values of B, where B is such that $|f(x)|$ is small for $x > B$. When a change in B makes no significant difference to the value obtained for the integral this is taken as an approximation to $\int_a^\infty f(x)\,dx$. Similarly for $\int_{-\infty}^b f(x)\,dx = \lim\limits_{A\to-\infty}\int_A^b f(x)\,dx$. Note that $\int_{-\infty}^\infty f(x)\,dx = \int_{-\infty}^0 f(x)\,dx + \int_0^\infty f(x)\,dx$. The procedure is illustrated in the next example with an integral which is of importance in statistics, see Section 22.6.

Example 5: Evaluate $\int_{-\infty}^{b} e^{-\frac{x^2}{2}} dx$ for (i) $b = 0$, (ii) $b = 2$, using Simpson's rule.

A sketch of the integrand, Fig. 17.7, shows that the curve $y = e^{-\frac{x^2}{2}}$ is symmetrical about the line $x = 0$ and that the contribution to the area under the curve due to the part to the left of -4 is likely to be very small.

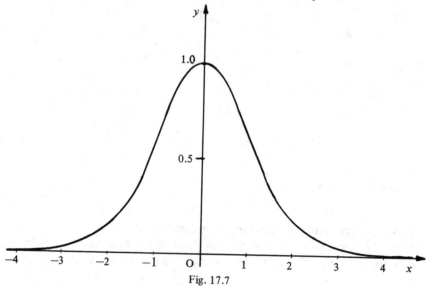

Fig. 17.7

Furthermore, $\lim_{x \to -\infty} xe^{-\frac{x^2}{2}} = \lim_{t \to \infty} (-\sqrt{2}.t^{\frac{1}{2}} e^{-t}) = 0$ on setting $x = -(2t)^{\frac{1}{2}}$ and using (12.9). Thus the integral does have a definite value.

By using the properties of the integral some of the working can be reduced. The evaluation of $\int_{-\infty}^{2} e^{-\frac{x^2}{2}} dx$ can be done in two stages since

$$\int_{-\infty}^{2} e^{-\frac{x^2}{2}} dx = \int_{-\infty}^{0} e^{-\frac{x^2}{2}} dx + \int_{0}^{2} e^{-\frac{x^2}{2}} dx \ .$$

The first integral on the right is that required by part (i) and the second integral was evaluated in example 3 and found to have the value 1.19628. Furthermore, symmetry about the line $x = 0$ means that

$$\int_{-2}^{0} e^{-\frac{x^2}{2}} dx \text{ equals } \int_{0}^{2} e^{-\frac{x^2}{2}} dx \text{, which is known.}$$

As $$\int_{-\infty}^{0} e^{-\frac{x^2}{2}} dx = \int_{-\infty}^{-2} e^{-\frac{x^2}{2}} dx + \int_{-2}^{0} e^{-\frac{x^2}{2}} dx$$

the only outstanding value required is that of

$$\int_{-\infty}^{-2} e^{-\frac{x^2}{2}} \, dx \ .$$

To evaluate this consider the integrals

$$I_A = \int_{-A}^{-2} e^{-\frac{x^2}{2}} \, dx$$

for values of A greater than or equal to 4.

When $A = 4$, Simpson's rule with $h = 0.2$ leads to a value for I_4.

k	x_k	$e^{-\frac{x_k^2}{2}}$ first and last ordinates	odd ordinates	remaining even ordinates
0	−4.0	0.00034		
1	−3.8		0.00073	
2	−3.6			0.00153
3	−3.4		0.00309	
4	−3.2			0.00598
5	−3.0		0.01111	
6	−2.8			0.01984
7	−2.6		0.03405	
8	−2.4			0.05613
9	−2.2		0.08892	
10	−2.0	0.13534		
	Sums	0.13568	0.13790	0.08348

Therefore $I_4 \cong \dfrac{0.2}{3} [0.13568 + 4(0.13790) + 2(0.08348)] \cong 0.056949$.

Evaluating $\displaystyle\int_{-4.4}^{-4} e^{-\frac{x^2}{2}} \, dx$ by Simpson's rule with $h = 0.2$:

k	x_k	$e^{-\frac{x_k^2}{2}}$ first and last ordinates	odd ordinates
0	−4.4	0.00006	
1	−4.2		0.00015
2	−4.0	0.00034	
	Sums	0.00040	0.00015

(Note that there are no remaining even terms to be included).

$$\int_{-4.4}^{-4} e^{-\frac{x^2}{2}} \, dx \cong \frac{0.2}{3} [0.00040 + 4(0.00015) + 2(0)] \cong 0.000067 \ .$$

So $\qquad I_{4.4} = I_4 + \int_{-4.4}^{-4} e^{-\frac{x^2}{2}} \, dx \cong 0.056949 + 0.000067 = 0.057016 \ .$

Evaluating $\int_{-4.8}^{-4.4} e^{-\frac{x^2}{2}} \, dx$ by Simpson's rule with $h = 0.2$:

k	x_k	$e^{-\frac{x_k^2}{2}}$ first and last ordinates	odd ordinates
0	−4.8	0.00001	
1	−4.6		0.00003
2	−4.4	0.00006	
	Sums	0.00007	0.00003

$$\int_{-4.8}^{-4.4} e^{-\frac{x^2}{2}} \, dx \cong \frac{0.2}{3} [0.00007 + 4(0.00003) + 2(0)] \cong 0.000013 \ .$$

Therefore

$$I_{4.8} = I_{4.4} + \int_{-4.8}^{-4.4} e^{-\frac{x^2}{2}} \, dx \cong 0.057016 + 0.000013 = 0.057029 \ .$$

As there is little difference in the values of the integrals $I_{4.4}$ and $I_{4.8}$ the latter is taken to be the approximate value of $\int_{-\infty}^{-2} e^{-\frac{x^2}{2}} \, dx$.

So $\qquad \int_{-\infty}^{-2} e^{-\frac{x^2}{2}} \, dx \cong I_{4.8} \cong 0.05703 \ .$

Then $\quad$ (i) $\int_{-\infty}^{0} e^{-\frac{x^2}{2}} \, dx = \int_{-\infty}^{-2} e^{-\frac{x^2}{2}} \, dx + \int_{-2}^{0} e^{-\frac{x^2}{2}} \, dx$

$$\cong 0.05703 + 1.19628 \cong 1.2533 \ ,$$

and $\quad$ (ii) $\int_{-\infty}^{2} e^{-\frac{x^2}{2}} \, dx = \int_{-\infty}^{0} e^{-\frac{x^2}{2}} \, dx + \int_{0}^{2} e^{-\frac{x^2}{2}} \, dx$

$$\cong 1.2533 + 1.1963 = 2.4496 \ .$$

The integral in (i) is known to have the exact value $\sqrt{(\pi/2)}$ (which is 1.2533 to 4D) and often arises in probability theory multiplied by $1/\sqrt{(2\pi)}$ (see Section 22.6)

More advanced methods for evaluating an integral can be found in the book by Williams (1972).

17.5 THE NUMERICAL SOLUTION OF FIRST ORDER ORDINARY DIFFERENTIAL EQUATIONS

The methods for solving first order differential equations described in Chapter 15 may not lead to a solution in terms of standard functions. For example, the linear equation $y' + y = 1/x$ with initial condition $y = 0$ when $x = 1$ cannot be solved completely by the method given in Section 15.2.4. The integrating factor e^x enables the equation to be written in the form

$$\frac{d}{dx}(e^x y) = \frac{e^x}{x} \ .$$

On integrating it is found that

$$e^x y = \int \frac{e^x}{x}\, dx + C \ .$$

Changing the variable of integration gives

$$e^x y = \int \frac{e^t}{t}\, dt \Bigg|_{t=x} + C = \int_a^x \frac{e^t}{t}\, dt$$

where the arbitrary limit of integration a in the definite integral allows for the arbitrary constant C. Using the initial condition $y(1) = 0$, a is found to be one. Hence the solution is given by

$$y = e^{-x} \int_1^x \frac{e^t}{t}\, dt \ .$$

This is the best that can be achieved by this approach. The integral cannot be evaluated in terms of known standard functions and therefore needs to be found numerically for each value of x.

Even when an analytical solution can be found to a differential equation it is sometimes difficult to find the actual numerical value of y corresponding to a given value of x. For example, the differential equation

$$\frac{dy}{dx} = \frac{x + y}{x - y}$$

has the general solution $y = x \tan\{\ln[C\sqrt{(x^2 + y^2)}]\}$ (as may be verified by substitution) but the value taken by y can only be found by solving numerically, using root finding methods, the implicit equation with fixed values of x and C.

In this section several methods for solving numerically first order differential equations of the form

$$y' = f(x, y) \qquad (17.8)$$

with $y(x_0) = y_0$ are given. The numerical solution of (17.8) consists of a list of the values taken by y at specific values of x starting from (x_0, y_0). Denote by y_1 the value of y obtained as the numerical solution to the equation when $x = x_1 = x_0 + h$ where h is the step length. For a fixed step length h let $x_k = x_0 + hk$, $k = 0, 1, 2, \ldots$, and let the corresponding value of y, as determined numerically by some formula, be denoted by y_k. In general at the point $x = x_k$ the numerical solution y_k will not be equal to the true solution $y(x_k)$ of the differential equation.

In solving a first order differential equation analytically an arbitrary constant appears in the general solution. By making the solution satisfy the given initial condition $y(x_0) = y_0$ the value of the arbitrary constant is determined. Thus once the general analytical solution is found only the value assigned to the arbitrary constant needs to be changed to obtain a particular solution having a specified value at $x = x_0$. In contrast, a numerical solution will have to be completely re-evaluated for each change in the initial condition.

The first method for solving the differential equation (17.8) is obtained from a consideration of the Taylor series expansion of y about $x = x_k$, see Section 7.2. Thus

$$y(x_{k+1}) = y(x_k + h) = y(x_k) + hy'(x_k) + 0(h^2) \, ,$$

therefore y_{k+1}, an approximation to $y(x_{k+1})$, can be obtained from the formula

$$y_{k+1} = y_k + hf(x_k, y_k) \qquad (17.9)$$

for $k = 0, 1, 2, \ldots$, using (17.8) to replace $y'(x_k)$ by $f(x_k, y_k)$. The scheme given by (17.9) to determine y_{k+1} for $k = 0, 1, 2, \ldots$, is called **Euler's method**. The first order recurrence equation (17.9) is often nonlinear so no general solution is usually available and it must be solved by successive evaluation starting from the known value y_0 at $x = x_0$ ($k = 0$).

Geometrically the true solution $y = y(x)$, between $x = x_0$ and $x = x_1$, is approximated by points on the tangent T_0 to the curve $y = y_0(x)$ at (x_0, y_0) by this method, see Fig. 17.8. Hence, in general, the computed value y_1 will not be the same as the true solution $y_0(x_1)$. As a result the next value found, y_2, will be in error on at least two counts, one due to the approximation of the curve by a straight line between x_1 and x_2, the other due to taking the starting point for the second step to be (x_1, y_1) and not the point $(x_1, y_0(x_1))$ on the solution curve. This results in the tangent T_1 to the curve $y = y_1(x)$ which satisfies the differential equation and passes through (x_1, y_1) being used and not the tangent T to the solution curve $y = y_0(x)$ at $(x_1, y_0(x_1))$. The first source of error is the truncation error resulting from the use of (17.9) and is of order h^2 because this is the next term in the Taylor series expansion. The second error is an example of a propagated error as it results from the use in the formula of values, here y_1, containing errors inherent in the previous step.

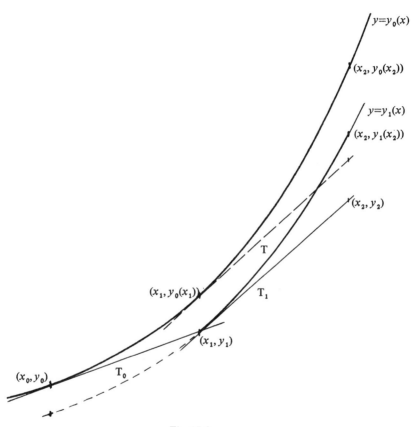

Fig. 17.8

Example 1: Given that $y = 1$ when $x = 0$ solve $y' = xy$ numerically using Euler's method with a step length of 0.1. Compare the numerical solution with the exact solution $y = \exp(x^2/2)$.

Here $h = 0.1, y_0 = 1, x_k = 0.1\,k$

and $\qquad y_{k+1} = y_k + hx_ky_k = (1 + hx_k)y_k, \quad k = 0, 1, 2, \ldots$

k	True $y(x_k)$	Calculated y_k	x_k	$1 + hx_k$	$y_{k+1} = (1 + hx_k)y_k$
0	1.0000	1.0000	0.0	1.00	1.0000
1	1.0050	1.0000	0.1	1.01	1.0100
2	1.0202	1.0100	0.2	1.02	1.0302
3	1.0460	1.0302	0.3	1.03	1.0611
4	1.0833	1.0611	0.4	1.04	1.1035
5	1.1331	1.1035	0.5	1.05	1.1587

The error is building up rapidly. Taking a smaller step size helps to reduce the error; with $h = 0.01$, $y_{50} = 1.1298$ while $y(x_{50}) = 1.1331$. Ultimately, decreasing the step size results in increased round-off error.

Another numerical procedure for solving $y' = f(x, y)$ can be obtained by integrating the differential equation with respect to x between x_k and x_{k+1}. Thus

$$\int_{x_k}^{x_{k+1}} y' \, dx = \int_{x_k}^{x_{k+1}} f(x, y) \, dx \ ,$$

so that integrating the integral on the left gives

$$\left[y(x) \right]_{x_k}^{x_{k+1}} = y(x_{k+1}) - y(x_k) = \int_{x_k}^{x_{k+1}} f(x, y) \, dx$$

$$\cong \frac{h}{2} [f(x_k, y(x_k)) + f(x_{k+1}, y(x_{k+1}))]$$

on approximating the remaining integral with the trapezoidal rule for one strip. Knowing an approximate value of $y(x_k)$, viz. y_k, this expression can be written as

$$y(x_{k+1}) \cong y_k + \frac{h}{2} [f(x_k, y_k) + f(x_{k+1}, y(x_{k+1}))] \ .$$

The value of $y(x_{k+1})$ can then be approximated by y_{k+1}, where y_{k+1} is found from the implicit relationship

$$y_{k+1} = y_k + \frac{h}{2} [f(x_k, y_k) + f(x_{k+1}, y_{k+1})] \ .$$

To obtain an explicit formula giving y_{k+1} directly, $f(x_{k+1}, y_{k+1})$ is replaced by $f(x_{k+1}, y_{k+1}^*)$. Here y_{k+1}^* is an approximation to y_{k+1} which is usually obtained from the formula

$$y_{k+1}^* = y_k + hf(x_k, y_k) \ .$$

Numerical methods for solving differential equations in which a simple approximation to y_{k+1} is used to obtain a better approximation are called **predictor-corrector methods**. The predictor used in this approach is

$$y_{k+1}^* = y_k + hf(x_k, y_k) \ , \tag{17.10}$$

whilst the corrector is

$$y_{k+1} = y_k + \frac{h}{2} [f(x_k, y_k) + f(x_{k+1}, y_{k+1}^*)] \ . \tag{17.11}$$

The error in each step of this method can be shown to be of order h^3. The closeness of the quantities y_{k+1} and y_{k+1}^* can be used to monitor errors. If the difference in the values is large a reduction in the step size is called for.

Example 2: Use the predictor-corrector method to solve the differential equation $y' = xy$ with $y(1) = 0$. Take a step length of 0.1.

The values needed are set out in Table 17.2 on page 332.

The values obtained agree with the true solution, differing only in the last decimal place when $k = 5$. The solution can also be compared with that obtained using Euler's method given in example 1.

One of the most widely used procedures for integrating a first order differential equation is the **fourth order Runge-Kutta method**. In this method y_{k+1} is calculated from the formula

$$y_{k+1} = y_k + \tfrac{1}{6} [a_k + 2b_k + 2c_k + d_k]$$

where
$$a_k = hf(x_k, y_k) \ ,$$
$$b_k = hf(x_k + \tfrac{1}{2}h, y_k + \tfrac{1}{2}a_k) \ ,$$
$$c_k = hf(x_k + \tfrac{1}{2}h, y_k + \tfrac{1}{2}b_k) \ ,$$
$$d_k = hf(x_k + h, y_k + c_k) \ .$$

The error in each step is of order h^5, so this Runge-Kutta method is more accurate than the predictor-corrector method described previously.

Example 3: Using the Runge-Kutta method with $h = 0.5$ find the numerical solution of $dy/dx = xy$ having $y = 1$ when $x = 0$.

Here $x_0 = 0, y_0 = 1, f(x, y) = xy$, so

$$a_0 = h x_0 y_0 = 0 \ ,$$
$$b_0 = h \left(x_0 + \frac{h}{2} \right) \left(y_0 + \frac{a_0}{2} \right) = 0.125 \ ,$$
$$c_0 = h \left(x_0 + \frac{h}{2} \right) \left(y_0 + \frac{b_0}{2} \right) = 0.1328 \ ,$$
$$d_0 = h(x_0 + h)(y_0 + c_0) = 0.2832 \ .$$

Table 17.2

k	y_k	x_k	$f_k = x_k y_k$	$y_{k+1}^* = y_k + h f_k$	x_{k+1}	$f_{k+1}^* = x_{k+1} y_{k+1}^*$	$\dfrac{h}{2}[f_k + f_{k+1}^*]$	$y_{k+1} = y_k + \dfrac{h}{2}[f_k + f_{k+1}^*]$
0	1.0000	0.0	0.0000	1.0000	0.1	0.1000	0.0050	1.0050
1	1.0050	0.1	0.1005	1.0150	0.2	0.2030	0.0152	1.0202
2	1.0202	0.2	0.2040	1.0406	0.3	0.3122	0.0258	1.0460
3	1.0460	0.3	0.3138	1.0774	0.4	0.4310	0.0372	1.0832
4	1.0832	0.4	0.4333	1.1265	0.5	0.5632	0.0498	1.1330
5	1.1330	0.5						

Therefore $y_1 = y_0 + \frac{1}{6} [a_0 + 2b_0 + 2c_0 + d_0] = 1.1331$, which agrees with the true value of y at $x = 0.5$.

Another advantage of the Runge-Kutta method is that it can also be applied to systems of first order equations, such as the Lotka-Volterra equations

$$\dot{x} = x(a_{11}x + a_{12}y + b_1) \ ,$$
$$\dot{y} = y(a_{21}x + a_{22}y + b_2) \ ,$$

given in Section 15.4. Furthermore, the second order equation $y'' = f(x, y, y')$ with initial conditions $y(x_0) = y_0$, $y'(x_0) = z_0$ can be rewritten as a pair of first order equations by setting $y' = z$. The equivalent first order system is then

$$y' = z, \qquad\qquad y(x_0) = y_0 \ ,$$
and $\qquad z' = f(x, y, z), \qquad z(x_0) = z_0 \ .$

The numerical solution of a second order differential equation having boundary conditions is more difficult than the corresponding initial value problem. For further details of numerical methods for solving differential equations see the book by Conte and Deboor (1972).

PROBLEMS

1. Obtain the two roots of the equation

    $$x^2 - 30x + 1 = 0$$

    correct to 4 SF, given that $\sqrt{224} = 14.967$.

2. Find the order of the truncation error involved in approximating

    $$e^x \ln(1 + x)$$

    near $x = 0$ by the polynomial

    $$x + \frac{x^2}{2} + \frac{x^3}{3} \ .$$

3. Show that the following equations each have a root lying in the given interval. Use the secant method to obtain the root correct to three decimal places.

    (a) $2x + \ln x = 1$ $(\frac{1}{2} < x < 1)$,

    (b) $x + \ln x = 3$ $(2 < x < 3)$,

    (c) $2x + e^x = 3$ $(0 < x < 1)$,

    (d) $2x - x^2 + e^{-x} = 0$ $(2 < x < 3)$.

4. Use Newton's method to determine the root of each equation given in question 3. Take as the initial value the lower value of x given.

5. Use the first two terms in the Maclaurin series expansion of $\ln(1 + \sin x)$ to find a first approximation to the root of the equation

$$\ln(1 + \sin x) + 4x = 1$$

which lies between 0 and $\pi/2$. Improve this approximation, using Newton's method, to obtain the root correct to three decimal places.

(Strathclyde, 1978)

6. Use the trapezoidal rule with four strips to evaluate, approximately, the following integrals. In each case check the accuracy of your answer by direct integration. (Work to four decimal places).

(a) $\displaystyle\int_0^1 \frac{x}{3}\, dx$,

(b) $\displaystyle\int_0^1 \frac{dx}{1 + \sqrt{x}}$ (Use the substitution $x = t^2$ to evaluate the integral),

(c) $\displaystyle\int_1^3 \frac{dx}{x + x\ln x}$ (Use the substitution $t = \ln x$ to evaluate the integral).

7. Sketch the area represented by the integral

$$\int_0^1 \frac{dx}{1 + x^3} .$$

Use the trapezoidal rule with five strips to find the approximate value of
(i) this area, and
(ii) the volume generated when this area is rotated, through 360°, about the x-axis.
(Work to four decimal places).

8. By definition $\ln x = \displaystyle\int_1^x \frac{1}{t}\, dt$ for $x > 1$. Show that $\ln 2$ is given approximately by

(i) 0.6970, using the trapezoidal rule with four strips,
(ii) 0.6933, using Simpson's rule with four strips.
(The value in mathematical tables for $\ln 2$ is 0.6931).

9. Evaluate $\displaystyle\int_0^1 x^2\, dx$ by

(i) direct integration,

(ii) using the trapezoidal rule with ten strips,

(iii) using Simpson's rule with four strips.

10. Use Simpson's rule with six strips to evaluate the following integrals approximately. In each case check the accuracy of your answer by direct integration.

(a) $\displaystyle\int_1^4 (3x - x^2)dx$,

(b) $\displaystyle\int_0^3 (4x^3 - 100)dx$,

(c) $\displaystyle\int_0^3 \frac{dx}{\sqrt{(1 + x)}}$,

(d) $\displaystyle\int_1^{2.5} e^{-x}\, dx$,

(e) $\displaystyle\int_0^\pi \frac{dx}{1 + \sin x}$ (Use the substitution $\sin x = \dfrac{2t}{1 + t^2}$, in which $t = \tan\dfrac{x}{2}$, to evaluate the integral),

(f) $\displaystyle\int_0^\pi x \sin x \, dx$.

11. Use Simpson's rule with six strips to find, approximately, the area enclosed by the curve $xy = 12$, the x-axis, and the lines $x = 1, x = 4$. By direct integration show that the exact value of the area differs from the value obtained by Simpson's rule in the second decimal place. (Work to three decimal places).

12. Obtain an expression for the length of the arc of the curve represented by the equation $y = \sin x$ between $x = 0$ and $x = \pi$ in the form of a definite integral. Show that Simpson's rule with six divisions gives the approximate value of the integral as 3.82.

13. By expanding $f(x)$ as a Taylor series about $x = a$ show that

(i) $\dfrac{1}{3} h[f(a + h) + 4 f(a) + f(a - h)] = 2h f(a) + \dfrac{1}{3} h^3 f''(a)$

$$+ \frac{1}{36} h^5 f^{iv}(a) + \ldots$$

and

(ii) $\displaystyle\int_{a-h}^{a+h} f(x)\, dx = 2h f(a) + \frac{1}{3} h^3 f''(a) + \frac{1}{60} h^5 f^{iv}(a) + \ldots$

Hence deduce that the error in evaluating $\displaystyle\int_{a-h}^{a+h} f(x)\, dx$ by using

Simpson's rule with three ordinates is, approximately, $\dfrac{h^5}{90} f^{iv}(a)$.

14. The reaction, r, in suitable units, of a patient to a given dose of a drug t hours after administration is shown in the following table

t	0	1	2	3	4	5	6
r	0.0	1.4	1.6	1.2	0.9	0.7	0.6 .

Find the average reaction to the drug during these first six hours as given

by $\frac{1}{6} \displaystyle\int_{0}^{6} r(t)\,dt$.

15. The girth, in metres, of a tree trunk measured at half metre intervals from one end was as follows:

distance, d	0	0.5	1.0	1.5	2.0	2.5	3.0
girth, g	1.92	2.12	1.85	1.71	1.46	1.18	0.93 .

By assuming that the tree trunk has a circular cross-section determine, approximately, the volume of wood that this tree trunk will produce if 15% of the total volume is wasted.

16. In a study of photosynthesis the area of a plant's leaf is required. Measurements of the width of the leaf, in cm, at intervals of 1 cm along the leaf's axis were taken and are set out in the following table:

axial distance	0	1	2	3	4	5	6	7	8	9	10
width	0	4.8	6.0	6.4	6.0	5.4	4.2	2.8	1.4	0.6	0

Use Simpson's rule to estimate the area of the leaf.

17. Calculate the mean blood pressure in an artery when the values obtained over a one second cycle are as follows:

time, t	0.0	0.1	0.2	0.3	0.4	0.5	0.6	0.7	0.8	0.9	1.0
pressure, t	70	78	88	100	114	120	100	107	92	80	70 .

18. The points (x_k, y_k), $k = 0, 1, 2, \ldots, n$, are the results of an experiment in which the values x_k have been ordered (that is, $x_0 < x_1 < x_2 < \ldots < x_n$). Show, by applying the trapezoidal rule to each of the subintervals $[x_k, x_{k+1}]$, $k = 0, 1, \ldots, n-1$, that

$$\int_{x_0}^{x_n} y \, dx \cong \tfrac{1}{2} \sum_{k=0}^{n-1} (x_{k+1} - x_k)(y_{k+1} + y_k) .$$

Hence estimate the work done by a gas in expanding from 0.1 m^3 to 1.0 m^3, using the following table:

volume, V (m^3)	0.1	0.2	0.3	0.6	0.7	0.9	1.0
pressure, p $(N m^{-2})$	100	57	38	18	15	12	11 .

(The work done is given by $\int p \, dV$.)

19. Evaluate $\displaystyle\int_1^{\infty} \frac{e^{-x}}{x} \, dx$, correct to three decimal places, by using Simpson's rule with a strip width of 0.25.

20. If $f(x) \leqslant g(x)$ for $a \leqslant x \leqslant b$ then $\displaystyle\int_a^b f(x) dx \leqslant \int_a^b g(x) dx$. Use this result and the inequality $e^{-x^2} \leqslant e^{-Mx}$ for $x \geqslant M$ to deduce that

$$\int_M^{\infty} e^{-x^2} \, dx \leqslant \frac{e^{-M^2}}{M} .$$

Find the least integer value of M such that the right-hand expression is less than 10^{-4}. Hence estimate the value of $\displaystyle\int_0^{\infty} e^{-x^2} \, dx$ correct to three decimal places by using Simpson's rule with a strip width of 0.25.

21. Use Euler's method with a step length of 0.1 to find the numerical solution to the differential equation

$$dy/dx = 2x + y ,$$

which satisfies $y(1) = 1$, for $1.0 \leqslant x \leqslant 1.5$. Compare your answer with the true solution. Work to three decimal places.

22. Find the numerical solution to the initial value problem

$$\frac{dy}{dx} = \frac{x}{1 + y}, \quad y(1) = 1$$

for $1.0 \leqslant x \leqslant 2.0$ by using

(i) Euler's method with a step length of 0.1,

(ii) the predictor-corrector method, using a step length of 0.2.
Compare your answers with the true solution. Work to three decimal places.

23. Solve numerically the differential equation

$$dy/dx = \sqrt{(1 + xy)}, \quad y(0) = 1$$

for $0.0 \leqslant x \leqslant 1.0$ by using the predictor-corrector method with a step length of (i) 0.5, (ii) 0.2, (iii) 0.1. Work to three decimal places.

24. Use the fourth order Runge-Kutta method to solve numerically

$$dy/dx = 4xy^{\frac{1}{2}}, \quad y(1) = 4$$

for $1.0 \leqslant x \leqslant 1.4$ with a step length of (i) 0.4, (ii) 0.2. Compare your answers with the true solution at $x = 1.4$. Work to three decimal places.

25. Solve numerically the initial value problem

$$dy/dx = \sqrt{(x + y^2)}, \quad y(1) = 1$$

for $1.0 \leqslant x \leqslant 1.6$ by using

(i) the predictor-corrector method with a step length of 0.1,

(ii) the fourth order Runge-Kutta method with a step length of 0.3. Work to three decimal places.

CHAPTER 18

Matrices

18.1 INTRODUCTION

Processes occurring in the biological sciences often require a large number of variables to describe them. For example, to study an ecological system in a river may need over one hundred variables related to the fish, insects, other aquatic animals, plants and the water's own characteristics, such as purity and temperature. In order to analyse such a complex process the mathematical model used to describe it must be kept as simple as possible.

The study of any problem mathematically involves the collection of data and the subsequent setting down of relationships between the variables. Often these relationships are of a linear form, and the coefficients of the variables introduce further numerical values. Handling all the variables and the associated data for the problem in a systematic manner is readily accomplished by using the mathematical entity known as a matrix. In particular, the rule for multiplying matrices enables the linear relationships governing the process to be written compactly.

In this chapter various types of matrices are introduced, algebraic operations with matrices are defined, and their properties listed. The solution of simultaneous linear equations, determinants and matrix eigenvalue problems are also covered. Practical applications drawn from a variety of situations are used as illustrations.

18.2 MATRICES

A **matrix** is a rectangular array of quantities (usually numbers) enclosed by brackets [ ] or ( ) and denoted by a bold capital letter. For example

$$\mathbf{M} = \begin{bmatrix} 1.0 & 4.1 & 0.0 \\ 2.3 & -3.8 & 7.4 \end{bmatrix}$$

is a matrix. The **size (order)** of a matrix is given by the dimensions of the rectangular array. A matrix having an array comprising m rows and n columns is called an $m \times n$ matrix. In the example **M** is a 2×3 matrix. If **A** is an $m \times n$ matrix then the **elements** forming the array are symbolised by a_{ij} for $1 \leqslant i \leqslant m$ and $1 \leqslant j \leqslant n$; i denotes the row in which the element appears and j the column. Thus

$$\mathbf{A} = \begin{bmatrix} a_{11} & a_{12} & a_{13} & \cdots & a_{1n} \\ a_{21} & a_{22} & a_{23} & \cdots & a_{2n} \\ \cdot & \cdot & \cdot & \cdots & \cdot \\ \cdot & \cdot & \cdot & \cdots & \cdot \\ a_{m1} & a_{m2} & a_{m3} & \cdots & a_{mn} \end{bmatrix} = [a_{ij}] \ 1 \leqslant i \leqslant m, 1 \leqslant j \leqslant n \ .$$

In the matrix **M**, $m_{11} = 1.0$, $m_{13} = 0.0$, $m_{22} = -3.8$, etc. Sometimes the elements of the matrix **A** are also denoted $[\mathbf{A}]_{ij}$, so $a_{ij} = [\mathbf{A}]_{ij}$.

Matrices that consist of just one row ($1 \times n$ matrix), or one column ($m \times 1$ matrix), of numbers are particularly important and are called **row**, or **column**, **vectors** respectively. Vectors are denoted by a bold lower case letter, for example

$$\mathbf{a} = [1 \quad 2 \quad 3], \ \mathbf{b} = \begin{bmatrix} 1 \\ 2 \\ 3 \end{bmatrix}, \ \mathbf{c} = \begin{bmatrix} c_1 \\ c_2 \\ \vdots \\ c_m \end{bmatrix} \ , \ \mathbf{r} = [r_1, r_2, \ldots, r_n] \ .$$

Notice that the notation makes no distinction between row and column vectors. In practice it will be clear from the context which is required. Sometimes commas are used to separate the elements in a row vector.

A matrix with an equal number of rows and columns ($m = n$) is called a **square matrix**. A square $n \times n$ matrix **A**, with all its elements zero except possibly the elements $[\mathbf{A}]_{11}$, $[\mathbf{A}]_{22}$, ..., $[\mathbf{A}]_{nn}$ which form the leading diagonal of the matrix, is called a **diagonal matrix**. The diagonal $n \times n$ matrix with each element on the diagonal one is called the $n \times n$ **identity, (or unit) matrix** and is written $\mathbf{I}_n$, or more often just **I**. The 1×1 matrix $[a_{11}]$ is called a **scalar** and the brackets are usually omitted.

18.3 THE ALGEBRA OF MATRICES

Equality

Two matrices **A**, **B** of the same size, say $m \times n$, are equal if all their corresponding elements are equal, that is, for all i,j

$$[\mathbf{A}]_{ij} = [\mathbf{B}]_{ij} \quad 1 \leqslant i \leqslant m, \ 1 \leqslant j \leqslant n .$$

Example 1: If the matrices $\begin{bmatrix} x^2 + x & y \\ y + z & x^2 + z \end{bmatrix}$ and $\begin{bmatrix} 2 & 3 \\ 0 & -2 \end{bmatrix}$ are equal determine x, y and z.

For equality $x^2 + x = 2$ (a) $y = 3$ (b)

$y + z = 0$ (c) $x^2 + z = -2$.   (d)

From (a) $x = 1$ or $x = -2$; from (b) $y = 3$.

Using (b) and (c), $z = -3$.   (e)

From (d), $x^2 = -z - 2 = 1$ using (e), so $x = \pm 1$.

For consistency with (a), $x = 1$. The solution is $x = 1, y = 3, z = -3$.

Addition

Two matrices of the same size may be added together. This is accomplished by adding together their corresponding elements. So, if A and B are $m \times n$ matrices then C = A + B is also of order $m \times n$ with $[C]_{ij} = [A + B]_{ij} = [A]_{ij} + [B]_{ij}$ $1 \leqslant i \leqslant m, 1 \leqslant j \leqslant n$.

For matrices of the same size the operation of addition satisfies the following rules:

(i) A + B = B + A ,

(ii) A + (B + C) = (A + B) + C .

Multiplication by a scalar

Any matrix may be multiplied by a scalar quantity (usually a number). Thus, if λ is a scalar and A is an $m \times n$ matrix then λA is an $m \times n$ matrix with elements

$$[\lambda A]_{ij} = \lambda [A]_{ij} \qquad 1 \leqslant i \leqslant m, 1 \leqslant j \leqslant n .$$

Two special cases arise.

(i) When $\lambda = 0$ a matrix with all its entries zero is obtained. Such a matrix, denoted **0**, is called a **zero matrix**. Notice that there is one zero matrix for each size of matrix. Also A + **0** = A where **0** is taken to be the same size as A.

(ii) Taking $\lambda = -1$ the matrix $-$A, having the property A + ($-$A) = **0**, is obtained. More often this is written as A $-$ A = **0**. This enables subtraction of matrices to be defined.

Example 2: Determine A + 3B $-$ I where

$$A = \begin{bmatrix} 2 & -2 & 3 \\ 1 & 0 & -1 \\ 3 & 2 & 4 \end{bmatrix} \text{ and } B = \begin{bmatrix} 1 & -1 & 0 \\ 0 & 2 & 1 \\ -1 & -2 & 3 \end{bmatrix} .$$

In order that the matrices can be combined take $I = I_3$. Then

$$A + 3B - I_3 = \begin{bmatrix} 2 & -2 & 3 \\ 1 & 0 & -1 \\ 3 & 2 & 4 \end{bmatrix} + 3 \begin{bmatrix} 1 & -1 & 0 \\ 0 & 2 & 1 \\ -1 & -2 & 3 \end{bmatrix} - \begin{bmatrix} 1 & 0 & 0 \\ 0 & 1 & 0 \\ 0 & 0 & 1 \end{bmatrix}$$

$$= \begin{bmatrix} 2+3-1 & -2-3-0 & 3+0-0 \\ 1+0-0 & 0+6-1 & -1+3--0 \\ 3-3-0 & 2-6-0 & 4+9-1 \end{bmatrix}$$

$$= \begin{bmatrix} 4 & -5 & 3 \\ 1 & 5 & 2 \\ 0 & -4 & 12 \end{bmatrix}.$$

Example 3: Evaluate $A - 3B$ where

$$A = \begin{bmatrix} 0 & 6 & 3 \\ -3 & -6 & 9 \end{bmatrix}, \quad B = \begin{bmatrix} 0 & 2 & 1 \\ -1 & -2 & 3 \end{bmatrix}.$$

$$A - 3B = \begin{bmatrix} 0 & 6 & 3 \\ -3 & -6 & 9 \end{bmatrix} - 3 \begin{bmatrix} 0 & 2 & 1 \\ -1 & -2 & 3 \end{bmatrix}$$

$$= \begin{bmatrix} 0-0 & 6-6 & 3-3 \\ -3+3 & -6+6 & 9-9 \end{bmatrix} = \begin{bmatrix} 0 & 0 & 0 \\ 0 & 0 & 0 \end{bmatrix} = 0 .$$

Multiplication

The multiplication of two matrices appears complicated but as later applications will show it is particularly suited to handling practical problems. Consider first the product of two of the simplest matrices, namely a row and column vector.

The **inner product** of the row vector $r = [r_1 \, r_2 \, \ldots r_n]$ with the column vector $c = \begin{bmatrix} c_1 \\ c_2 \\ \vdots \\ c_n \end{bmatrix}$ is defined as

$$r c = [r_1 r_2 \ldots r_n] \begin{bmatrix} c_1 \\ c_2 \\ \vdots \\ c_n \end{bmatrix} = r_1 c_1 + r_2 c_2 + \ldots + r_n c_n = \sum_{i=1}^{n} r_i c_i.$$

Notice the following points about this product:

(i) The two vectors must have the same number of elements.

(ii) The row vector must be to the left of the column vector.

(iii) The final value is a scalar quantity.

Because of (iii) this product is sometimes called the **scalar product** of two vectors.

Example 4: Find the inner products **a b** and **c d** where

$$\mathbf{a} = [1 \quad 2 \quad 3], \ \mathbf{b} = \begin{bmatrix} -2 \\ 3 \\ 1 \end{bmatrix}, \mathbf{c} = [-1 \quad 2 \quad 1 \quad 3], \mathbf{d} = \begin{bmatrix} 1 \\ -2 \\ -4 \\ 3 \end{bmatrix}.$$

$$\mathbf{a\,b} = [1 \quad 2 \quad 3] \begin{bmatrix} -2 \\ 3 \\ 1 \end{bmatrix} = 1(-2) + 2(3) + 3(1) = 7 \ .$$

$$\mathbf{c\,d} = [-1 \quad 2 \quad 1 \quad 3] \begin{bmatrix} 1 \\ -2 \\ -4 \\ 3 \end{bmatrix} = -1(1) + 2(-2) + 1(-4) + 3(3) = 0 \ .$$

The general rule for the product of two matrices can now be stated. The elements of the product matrix **AB**, $[\mathbf{AB}]_{ij}$, are scalars constructed by taking the ith row in the matrix **A** and forming the inner product with the jth column in the matrix **B**. For the inner product to be defined it is essential that the number of columns in **A** equals the number of rows in **B**. If this is so the matrices are said to be **conformable as a product AB**. The resultant product matrix **AB** has the same number of rows as **A** and columns as **B**. Thus if **A** is $m \times n$ and **B** is $n \times p$ the product **AB** is possible and is of the order $m \times p$. (It is only possible to form the product **BA** when $m = p$). From this it is clear that **BA** is quite different from **AB** as one may exist but not the other. When both **AB** and **BA** exist they are not, in general, of the same size. Even when the two products are the same size, as is the case when **A** and **B** are both $n \times n$, they may not be equal. Thus the order in which matrices are multiplied is important. In the event of **AB** equalling **BA** the matrices **A** and **B** are said to **commute**. In the product **AB**, **A** is said to premultiply **B** while **B** is said to postmultiply **A**.

Example 5: Evaluate **AB** where

$$\mathbf{A} = \begin{bmatrix} 2 & -2 & 3 \\ 3 & 2 & 4 \end{bmatrix}, \ \mathbf{B} = \begin{bmatrix} 1 & -1 & 0 \\ 0 & 2 & 1 \\ -1 & -2 & 3 \end{bmatrix}.$$

$$\mathbf{AB} = \begin{bmatrix} 2 & -2 & 3 \\ 3 & 2 & 4 \end{bmatrix} \begin{bmatrix} 1 & -1 & 0 \\ 0 & 2 & 1 \\ -1 & -2 & 3 \end{bmatrix}$$

$$= \begin{bmatrix} 2.1 + (-2)0 + 3(-1) & 2(-1) + (-2)2 + 3(-2) & 2.0 + (-2)1 + 3.3 \\ 3.1 + 2.0 + 4(-1) & 3(-1) + 2.2 + 4(-2) & 3.0 + 2.1 + 4.3 \end{bmatrix}$$

$$= \begin{bmatrix} 2 + 0 - 3 & -2 - 4 - 6 & 0 - 2 + 9 \\ 3 + 0 - 4 & -3 + 4 - 8 & 0 + 2 + 12 \end{bmatrix} = \begin{bmatrix} -1 & -12 & 7 \\ -1 & -7 & 14 \end{bmatrix}.$$

A is 2×3, **B** is 3×3, the product **AB** is 2×3.

Example 6: The chemical content, as percentages, of four soil types, denoted A, B, C, and D, is shown in the following table

		Nitrogen	Phosphate	Potash
	A	7	1	2
Soil	B	3	2	2
type	C	0	3	4
	D	2	0	3

If these soils are combined in the ratio 2 : 3 : 1 : 4 to form a loam, what will be its chemical composition?

The 4×3 matrix $\begin{bmatrix} 7 & 1 & 2 \\ 3 & 2 & 2 \\ 0 & 3 & 4 \\ 2 & 0 & 3 \end{bmatrix}$ contains the information on the chemicals in the four soil types; the row vector $[2, 3, 1, 4]$ gives the quantities of each soil type used to make ten units of the loam. The chemical content of one unit of the loam is given by

$$\frac{1}{10} [2 \quad 3 \quad 1 \quad 4] \begin{bmatrix} 7 & 1 & 2 \\ 3 & 2 & 2 \\ 0 & 3 & 4 \\ 2 & 0 & 3 \end{bmatrix} = [3.1 \quad 1.2 \quad 2.5] \ .$$

So the loam contains 3.1% nitrogen, 1.2% phosphate and 2.5% potash.

Rules satisfied by matrix multiplication are:

(i) $(A + B)C = AC + BC$,

(ii) $A(C + D) = AC + AD$,

(iii) $A(BC) = (AB)C$,

where A, B, C and D are matrices of a size for which the operations are defined;

(iv) post or premultiplication by any suitable zero matrix results in a zero matrix, that is, $A0 = 0$ and $0A = 0$,

(v) post or premultiplication by a suitable identity matrix preserves the original matrix, that is, if A is of order $m \times n$ then $AI_n = A = I_m A$.

Because of (iii) the index notation for positive integer powers of a square matrix can be used. Thus $A^3 = AAA = AA^2 = A^2 A$ where $A^2 = AA$, with the obvious extension to higher powers.

In addition to matrix multiplication being, in general, non-commutative it also possesses other unusual properties. For example, if $\mathbf{AB} = \mathbf{0}$ then one cannot automatically deduce that $\mathbf{A}$ or $\mathbf{B}$ is a zero matrix.

Example 7: Show that the matrices $\mathbf{A} = \begin{bmatrix} 1 & 2 & 1 \\ 3 & 0 & -1 \\ -1 & 1 & 1 \end{bmatrix}$ and $\mathbf{B} = \begin{bmatrix} 1 & -1 & 2 \\ -2 & 2 & -4 \\ 3 & -3 & 6 \end{bmatrix}$ do not commute.

$$\mathbf{AB} = \begin{bmatrix} 1 & 2 & 1 \\ 3 & 0 & -1 \\ -1 & 1 & 1 \end{bmatrix} \begin{bmatrix} 1 & -1 & 2 \\ -2 & 2 & -4 \\ 3 & -3 & 6 \end{bmatrix} = \begin{bmatrix} 0 & 0 & 0 \\ 0 & 0 & 0 \\ 0 & 0 & 0 \end{bmatrix}.$$

Here neither $\mathbf{A}$ nor $\mathbf{B}$ is a zero matrix.

$$\mathbf{BA} = \begin{bmatrix} 1 & -1 & 2 \\ -2 & 2 & -4 \\ 3 & -3 & 6 \end{bmatrix} \begin{bmatrix} 1 & 2 & 1 \\ 3 & 0 & -1 \\ -1 & 1 & 1 \end{bmatrix} = \begin{bmatrix} -4 & 4 & 4 \\ 8 & -8 & 8 \\ -12 & 12 & 12 \end{bmatrix}.$$

So that $\mathbf{AB} \neq \mathbf{BA}$.

An important application of matrices is in the representation and subsequent solution of simultaneous linear algebraic equations. A set of m linear equations in the n unknowns $x_1, x_2, \ldots, x_n$ may be written

$$a_{11} x_1 + a_{12} x_2 + \ldots + a_{1n} x_n = b_1$$
$$a_{21} x_1 + a_{22} x_2 + \ldots + a_{2n} x_n = b_2$$
$$\cdot \quad \cdot \quad \cdot \quad \cdot \cdot \cdot \cdot \quad \cdot \quad \cdot \quad \cdot$$
$$a_{m1} x_1 + a_{m2} x_2 + \ldots + a_{mn} x_n = b_m$$

where a_{ij} is the coefficient of x_j in the ith equation $1 \leqslant i \leqslant m, 1 \leqslant j \leqslant n$. The definition chosen for matrix multiplication enables such a set of equations to be written as the matrix equation $\mathbf{Ax} = \mathbf{b}$ where

$$\mathbf{A} = \begin{bmatrix} a_{11} & a_{12} & \cdots & a_{1n} \\ a_{21} & a_{22} & \cdots & a_{2n} \\ \cdot & \cdot & \cdots & \cdot \\ a_{m1} & a_{m2} & \cdots & a_{mn} \end{bmatrix}, \quad \mathbf{x} = \begin{bmatrix} x_1 \\ x_2 \\ \vdots \\ x_n \end{bmatrix}, \quad \mathbf{b} = \begin{bmatrix} b_1 \\ b_2 \\ \vdots \\ b_m \end{bmatrix}.$$

The most important case occurs when there are as many equations as unknowns, say n, then $\mathbf{A}$ will be an $n \times n$ matrix.

Example 8: Express the set of simultaneous equations

$$x + y + 2z = 3$$
$$x + 2y + 3z = 4$$
$$2x + 6y + 9z = 9$$

as a matrix equation.

Here the matrix of coefficients $\mathbf{A} = \begin{bmatrix} 1 & 1 & 2 \\ 1 & 2 & 3 \\ 2 & 6 & 9 \end{bmatrix}$, the column vector of

unknowns $\mathbf{x} = \begin{bmatrix} x \\ y \\ z \end{bmatrix}$, and the column vector containing the right-hand values of

the equations $\mathbf{b} = \begin{bmatrix} 3 \\ 4 \\ 9 \end{bmatrix}$. The simultaneous equation can then be written in the

form $\mathbf{Ax} = \mathbf{b}$, that is,

$$\begin{bmatrix} 1 & 1 & 2 \\ 1 & 2 & 3 \\ 2 & 6 & 9 \end{bmatrix} \begin{bmatrix} x \\ y \\ z \end{bmatrix} = \begin{bmatrix} 3 \\ 4 \\ 9 \end{bmatrix} .$$

18.3.1 Population matrices

The number of individuals in each age group within a population changes with the passage of time. The changes that occur in the age structure of a population can be studied conveniently by using a matrix representation.

Suppose a population is divided into n age groups with $p_i(t), i = 1, 2, \ldots, n$, members in each at a time t. After a fixed unit interval of time has elapsed t will have increased to $t + 1$ and the number of members in each age group will have changed. The number present in the ith age group at time $t + 1$ will be determined by the number of newcomers from an earlier age group and those that still remain in that age range. Other members who were formerly in the group will have moved up into the next age group or will have died. In the case of the youngest age group the newcomers will be due to the birth of offspring. To quantify these statements let m_i be the fraction of the ith age group moving into the $i + 1^{th}$ group; r_i be the proportion remaining in the ith group; and b_i be the number of births expected from each member of the ith group during the time interval t to $t + 1$. Note that b_i, m_i and r_i are all positive numbers and furthermore $m_i + r_i$, the fraction of survivors in the ith age group, is less than or equal to one. Thus $1 - (m_i + r_i)$ is the fraction of the ith group that die. At the end of the time interval the first or youngest age group is made up of those that remain in that age range and all the new births. Thus

$$p_1(t + 1) = r_1 p_1(t) + b_1 p_1(t) + b_2 p_2(t) + \ldots + b_n p_n(t) .$$

The numbers in the other age groups are given by

$$p_i(t + 1) = m_{i-1} p_{i-1}(t) + r_i p_i(t) \quad i = 2, 3, \ldots, n ;$$

made up of those that arrive from an earlier group and any that remain.

In matrix form the n equations can be written

$$
\begin{bmatrix}
p_1(t+1) \\
p_2(t+1) \\
p_3(t+1) \\
\vdots \\
p_n(t+1)
\end{bmatrix}
=
\begin{bmatrix}
b_1+r_1 & b_2 & b_3 & \cdots & b_{n-1} & b_n \\
m_1 & r_2 & 0 & \cdots & 0 & 0 \\
0 & m_2 & r_3 & \cdots & 0 & 0 \\
\vdots & \vdots & \vdots & \vdots\!\vdots\!\vdots & \vdots & \vdots \\
0 & 0 & 0 & \cdots & m_{n-1} & r_n
\end{bmatrix}
\begin{bmatrix}
p_1(t) \\
p_2(t) \\
p_3(t) \\
\vdots \\
p_n(t)
\end{bmatrix},
$$

or $p(t+1) = Tp(t),$

where the $n \times n$ matrix

$$
T =
\begin{bmatrix}
b_1+r_1 & b_2 & b_3 & \cdots & b_{n-1} & b_n \\
m_1 & r_2 & 0 & \cdots & 0 & 0 \\
0 & m_2 & r_3 & \cdots & 0 & 0 \\
\cdot & \cdot & \cdot & \cdots & \cdot & \cdot \\
\cdot & \cdot & \cdot & \cdots & \cdot & \cdot \\
0 & 0 & 0 & \cdots & m_{n-1} & r_n
\end{bmatrix}
$$

is an example of a **transition matrix**. On premultiplying the column vector $p(t)$ a transition matrix gives the state of the vector p after a unit of time has elapsed, that is, $p(t + 1)$.

In some problems the quantities b_i, m_i, r_i may depend on the time t so that the resulting transition matrix T has elements which will be functions of t.

Example 9: As a numerical example consider the matrix

$$
\begin{bmatrix}
0 & 1/6 & 1/2 & 1/6 & 0 \\
5/6 & 0 & 0 & 0 & 0 \\
0 & 5/6 & 1/3 & 0 & 0 \\
0 & 0 & 1/2 & 1/2 & 0 \\
0 & 0 & 0 & 1/3 & 1/3
\end{bmatrix}
$$

for a population divided into the five groups: children, teenagers, young adults, middle aged, old aged, and having a time interval of ten years. If the initial numbers in the groups, in millions, are $9, 9, 27, 18,$ and 9, respectively, find the numbers in the groups and the total population size after (i) 10 and (ii) 20 years.

The initial population numbers 72 million.

After 10 years the size of the age groups within the population is given by

$$\begin{bmatrix} 0 & 1/6 & 1/2 & 1/6 & 0 \\ 5/6 & 0 & 0 & 0 & 0 \\ 0 & 5/6 & 1/3 & 0 & 0 \\ 0 & 0 & 1/2 & 1/2 & 0 \\ 0 & 0 & 0 & 1/3 & 1/3 \end{bmatrix} \begin{bmatrix} 9 \\ 9 \\ 27 \\ 18 \\ 9 \end{bmatrix} = \begin{bmatrix} 18.0 \\ 7.5 \\ 16.5 \\ 22.5 \\ 9.0 \end{bmatrix}$$

resulting in a total population of 73.5 million after 10 years.

The number present in the age groups after a further 10 years is given by

$$\begin{bmatrix} 0 & 1/6 & 1/2 & 1/6 & 0 \\ 5/6 & 0 & 0 & 0 & 0 \\ 0 & 5/6 & 1/3 & 0 & 0 \\ 0 & 0 & 1/2 & 1/2 & 0 \\ 0 & 0 & 0 & 1/3 & 1/3 \end{bmatrix} \begin{bmatrix} 18.0 \\ 7.5 \\ 16.5 \\ 22.5 \\ 9.0 \end{bmatrix} = \begin{bmatrix} 13.25 \\ 15.0 \\ 11.75 \\ 19.5 \\ 10.5 \end{bmatrix} .$$

Hence the total population after 20 years is 70 million.

18.4 THE TRANSPOSED MATRIX

If all rows of an $m \times n$ matrix A are written, in the same order, as columns (or the columns as rows) the resulting $n \times m$ matrix is called the **transpose** of A, denoted A^T or A', that is,

$$[A^T]_{ij} = [A]_{ji} \quad 1 \leqslant i \leqslant n, \ 1 \leqslant j \leqslant m \ .$$

Thus row and column vectors are transposes of each other.

The following properties can be shown to hold:

     (i) $(A + B)^T = A^T + B^T$,

     (ii) $(AB)^T = B^T A^T$,

     (iii) $(A^T)^T = A$.

A square matrix A is said to be **symmetric** if it is equal to its transposed matrix, A^T, that is, $A = A^T$. If $A = -A^T$ then A is said to be **skew-symmetric**. It can be shown that any square matrix A can be written in the form $A = \frac{1}{2}(A + A^T) + \frac{1}{2}(A - A^T)$; the sum of a symmetric and a skew-symmetric matrix. Also, from (ii) and (iii), a symmetric matrix can be formed by multiplying the original matrix by its transpose. Thus $A^T A$ and AA^T are both symmetric. A square matrix A is said to be **orthogonal** if $AA^T = I$.

Example: If $A = \begin{bmatrix} 2 & 1 & 3 \\ -2 & 0 & 2 \end{bmatrix}$, $B = \begin{bmatrix} 1 & -2 \\ 3 & 2 \end{bmatrix}$, $C = \frac{1}{3} \begin{bmatrix} 1 & 2 & 2 \\ 2 & 1 & -2 \\ -2 & 2 & -1 \end{bmatrix}$ form the matrices $A^T, B^T, A^T A, AA^T, B + B^T, B - B^T, CC^T$ and state their type.

$$\mathbf{A}^T = \begin{bmatrix} 2 & -2 \\ 1 & 0 \\ 3 & 2 \end{bmatrix}, \quad \mathbf{B}^T = \begin{bmatrix} 1 & 3 \\ -2 & 2 \end{bmatrix}$$

$$\mathbf{A}^T\mathbf{A} = \begin{bmatrix} 2 & -2 \\ 1 & 0 \\ 3 & 2 \end{bmatrix}\begin{bmatrix} 2 & 1 & 3 \\ -2 & 0 & 2 \end{bmatrix} = \begin{bmatrix} 8 & 2 & 2 \\ 2 & 1 & 3 \\ 2 & 3 & 13 \end{bmatrix} \text{ symmetric,}$$

$$\mathbf{A}\mathbf{A}^T = \begin{bmatrix} 2 & 1 & 3 \\ -2 & 0 & 2 \end{bmatrix}\begin{bmatrix} 2 & -2 \\ 1 & 0 \\ 3 & 2 \end{bmatrix} = \begin{bmatrix} 14 & 2 \\ 2 & 8 \end{bmatrix} \text{ symmetric,}$$

$$\mathbf{B} + \mathbf{B}^T = \begin{bmatrix} 1 & -2 \\ 3 & 2 \end{bmatrix} + \begin{bmatrix} 1 & 3 \\ -2 & 2 \end{bmatrix} = \begin{bmatrix} 2 & 1 \\ 1 & 4 \end{bmatrix} \text{ symmetric,}$$

$$\mathbf{B} - \mathbf{B}^T = \begin{bmatrix} 1 & -2 \\ 3 & 2 \end{bmatrix} - \begin{bmatrix} 1 & 3 \\ -2 & 2 \end{bmatrix} = \begin{bmatrix} 0 & -5 \\ 5 & 0 \end{bmatrix} \text{ skew-symmetric.}$$

Notice that

$$\tfrac{1}{2}(\mathbf{B} + \mathbf{B}^T) + \tfrac{1}{2}(\mathbf{B} - \mathbf{B}^T) = \begin{bmatrix} 1 & 1/2 \\ 1/2 & 2 \end{bmatrix} + \begin{bmatrix} 0 & -5/2 \\ 5/2 & 0 \end{bmatrix}$$

$$= \begin{bmatrix} 1 & -2 \\ 3 & 2 \end{bmatrix} = \mathbf{B} .$$

$$\mathbf{C}\mathbf{C}^T = \frac{1}{9}\begin{bmatrix} 1 & 2 & 2 \\ 2 & 1 & -2 \\ -2 & 2 & -1 \end{bmatrix}\begin{bmatrix} 1 & 2 & -2 \\ 2 & 1 & 2 \\ 2 & -2 & -1 \end{bmatrix} = \begin{bmatrix} 1 & 0 & 0 \\ 0 & 1 & 0 \\ 0 & 0 & 1 \end{bmatrix} = \mathbf{I}$$

so **C** is orthogonal.

18.5 THE INVERSE MATRIX

Division in matrix algebra is not defined, but it is sometimes possible to find matrices which, through multiplication, have a similar effect. The most important of these matrices is the inverse matrix of a square matrix. For an $n \times n$ matrix **A** it may be possible to find an $n \times n$ matrix **B** satisfying $\mathbf{AB} = \mathbf{I}_n$ and $\mathbf{BA} = \mathbf{I}_n$. The matrix **B**, if there is one, is unique and is called the **inverse matrix** of **A**, denoted $\mathbf{A}^{-1}$, and **A** is said to be **non-singular**, or **invertible**. Thus

$$\mathbf{AA}^{-1} = \mathbf{I}_n = \mathbf{A}^{-1}\mathbf{A} .$$

If **A** does not have any inverse it is said to be **singular**. The inverse has the property that $(A^{-1})^{-1} = A$. Furthermore, if A and B are non-singular $n \times n$ matrices then

$$(AB)^{-1} = B^{-1}A^{-1} .$$

Example 1: Verify that the inverse of $A = \begin{bmatrix} 1 & 1 & 1 \\ 1 & -1 & 2 \\ 2 & 1 & 3 \end{bmatrix}$ is $B = \begin{bmatrix} 5 & 2 & -3 \\ -1 & -1 & 1 \\ -3 & -1 & 2 \end{bmatrix}$.

$$AB = \begin{bmatrix} 1 & 1 & 1 \\ 1 & -1 & 2 \\ 2 & 1 & 3 \end{bmatrix} \begin{bmatrix} 5 & 2 & -3 \\ -1 & -1 & 1 \\ -3 & -1 & 2 \end{bmatrix} = \begin{bmatrix} 1 & 0 & 0 \\ 0 & 1 & 0 \\ 0 & 0 & 1 \end{bmatrix}$$

$$BA = \begin{bmatrix} 5 & 2 & -3 \\ -1 & -1 & 1 \\ -3 & -1 & 2 \end{bmatrix} \begin{bmatrix} 1 & 1 & 1 \\ 1 & -1 & 2 \\ 2 & 1 & 3 \end{bmatrix} = \begin{bmatrix} 1 & 0 & 0 \\ 0 & 1 & 0 \\ 0 & 0 & 1 \end{bmatrix} .$$

Hence $B = A^{-1}$.

One way of finding the inverse of a square matrix A is to assume a general form for the inverse matrix and subsequently calculate the value of its elements from the requirement that $AA^{-1} = I$. This procedure is illustrated in the following examples.

Example 2: Find the inverse of $\begin{bmatrix} 2 & -3 \\ -1 & 2 \end{bmatrix}$.

Suppose $\begin{bmatrix} 2 & -3 \\ -1 & 2 \end{bmatrix}^{-1} = \begin{bmatrix} a & b \\ c & d \end{bmatrix}$ then

$$\begin{bmatrix} 2 & -3 \\ -1 & 2 \end{bmatrix} \begin{bmatrix} a & b \\ c & d \end{bmatrix} = \begin{bmatrix} 2a - 3c & 2b + 3d \\ -a + 2c & -b + 2d \end{bmatrix} = \begin{bmatrix} 1 & 0 \\ 0 & 1 \end{bmatrix} .$$

Equating elements $\left. \begin{array}{l} 2a - 3c = 1 \\ -a + 2c = 0 \end{array} \right\}$ and $\left. \begin{array}{l} 2b - 3d = 0 \\ -b + 2d = 1 \end{array} \right\}$.

These equations have the solution $a = 2, b = 3, c = 1, d = 2$.

Thus $\begin{bmatrix} 2 & 3 \\ 1 & 2 \end{bmatrix}$ is the required inverse. This may be checked by verifying

that $\begin{bmatrix} 2 & 3 \\ 1 & 2 \end{bmatrix} \begin{bmatrix} 2 & -3 \\ -1 & 2 \end{bmatrix} = \begin{bmatrix} 1 & 0 \\ 0 & 1 \end{bmatrix}$.

Example 3: Attempt to find the inverse of $\begin{bmatrix} 2 & 1 \\ 6 & 3 \end{bmatrix}$.

Suppose $\begin{bmatrix} 2 & 1 \\ 6 & 3 \end{bmatrix}^{-1} = \begin{bmatrix} a & b \\ c & d \end{bmatrix}$ then

$$\begin{bmatrix} 2 & 1 \\ 6 & 3 \end{bmatrix}\begin{bmatrix} a & b \\ c & d \end{bmatrix} = \begin{bmatrix} 2a + c & 2b + d \\ 6a + 3c & 6b + 3d \end{bmatrix} = \begin{bmatrix} 1 & 0 \\ 0 & 1 \end{bmatrix},$$

so
$$\left.\begin{matrix} 2a + c = 1 \\ 6a + 3c = 0 \end{matrix}\right\} \text{ and } \left.\begin{matrix} 2b + d = 0 \\ 6b + 3d = 1 \end{matrix}\right\}.$$

No solution of these simultaneous equations exists so the matrix $\begin{bmatrix} 2 & 1 \\ 6 & 3 \end{bmatrix}$ must be singular.

The procedure outlined above is not practicable for larger matrices; instead a method based on that used to solve a system of linear algebraic equations is employed, see Section 18.7.

18.6 SOLVING SIMULTANEOUS LINEAR EQUATIONS BY GAUSSIAN ELIMINATION

For the system

$$
\begin{aligned}
a_{11}\, x_1 + a_{12}\, x_2 + \ldots + a_{1n}\, x_n &= b_1 \\
a_{21}\, x_1 + a_{22}\, x_2 + \ldots + a_{2n}\, x_n &= b_2 \\
\cdot \quad \cdot \quad \cdot \quad \cdot \quad \cdot \cdot \cdot \quad \cdot \quad \cdot \quad \cdot \\
a_{n1}\, x_1 + a_{n2}\, x_2 + \ldots + a_{nn}\, x_n &= b_n
\end{aligned}
\qquad (18.1)
$$

of n simultaneous linear equations in n unknowns three possibilities may arise. The equations may have:

    (i) no solution ,

    (ii) one solution ,

or   (iii) more than one solution.

Case (ii) corresponds to the coefficient matrix, $\mathbf{A}$ say, being non-singular. Thus if the equations are written as $\mathbf{Ax} = \mathbf{b}$ premultiplying by $\mathbf{A}^{-1}$ would give

$$\mathbf{A}^{-1}\mathbf{Ax} = \mathbf{A}^{-1}\mathbf{b}, \text{ that is, } \mathbf{Ix} = \mathbf{x} = \mathbf{A}^{-1}\mathbf{b} \ .$$

In cases (i) and (iii) $\mathbf{A}$ is singular.

One way of solving a system of linear equations is to attempt to eliminate some of the unknowns by using one or more of the following operations:

    (a) adding a multiple of one equation to another ,

    (b) multiplying an equation by a non-zero constant ,

    (c) rearranging the order of the equations.

A similar procedure can be adopted for the matrix version of the problem; the operations become:

(a) adding a multiple of one row to another ,

(b) multiplying a row by a non-zero constant ,

(c) interchanging two rows.

As the solution of the system (18.1) depends only on the matrix A and the column vector b, the $n \times (n + 1)$ matrix

$$\begin{bmatrix} a_{11} & a_{12} & \cdots & a_{1n} & b_1 \\ a_{21} & a_{22} & \cdots & a_{2n} & b_2 \\ \cdot & \cdot & \cdots & \cdot & \cdot \\ a_{n1} & a_{n2} & \cdots & a_{nn} & b_n \end{bmatrix}$$

is introduced. This matrix is called the **augmented matrix** of the system (18.1) and is denoted $A|b$. The method of **Gaussian elimination** uses **row operations** to reduce the augmented matrix to a particularly simple form. First the elements below the diagonal of A are made zero, one column at a time, thereby making A **upper triangular**. The values of the unknowns $x_1, x_2, \ldots, x_n$ can now be found from the equations associated with the triangular form of the augmented matrix starting with the final equation and substituting back into preceding equations. In the following example Gaussian elimination and the standard method are shown in parallel. (Note that when manipulating augmented matrices it is usual to omit the matrix brackets and to separate the coefficient elements from the right-hand elements by a line).

Example 1: Solve the equations

$$x + y + 2z = 3$$
$$x + 2y + 3z = 4$$
$$2x + 6y + 9z = 9$$

for the unknowns x, y and z.

Equations   Augmented matrix

$$x + y + 2z = 3 \quad (R_1) \qquad \begin{matrix} 1 & 1 & 2 & | & 3 \end{matrix}$$
$$x + 2y + 3z = 4 \quad (R_2) \qquad \begin{matrix} 1 & 2 & 3 & | & 4 \end{matrix}$$
$$2x + 6y + 9z = 9 \quad (R_3) \qquad \begin{matrix} 2 & 6 & 9 & | & 9 \end{matrix}.$$

Reduction to triangular form

$$x + y + 2z = 3 \quad (R_1 := R_1) \qquad \begin{matrix} 1 & 1 & 2 & | & 3 \end{matrix}$$
$$y + z = 1 \quad (R_2 := R_2 - R_1) \qquad \begin{matrix} 0 & 1 & 1 & | & 1 \end{matrix}$$
$$4y + 5z = 3 \quad (R_3 := R_3 - 2R_1) \qquad \begin{matrix} 0 & 4 & 5 & | & 3 \end{matrix}.$$

The notation $R_2 := R_2 - R_1$ means the new row R_2 is formed by subtracting row R_1 from row R_2.

$$
\begin{array}{rl}
x + y + 2z = 3 & (R_1 := R_1) \\
y + z = 1 & (R_2 := R_2) \\
z = -1 & (R_3 := R_3 - 4R_2)
\end{array}
\qquad
\left[\begin{array}{ccc|c}
1 & 1 & 2 & 3 \\
0 & 1 & 1 & 1 \\
0 & 0 & 1 & -1
\end{array}\right].
$$

The equations/matrix are now in a triangular form.

Back substitution gives, from

$(R_3)\ z = -1$,

$(R_2)\ y = 1 - z = 1 + 1 = 2$,

$(R_1)\ x = 3 - y - 2z = 3 - 2 + 2 = 3$.

The solution to the equations is thus $x = 3, y = 2, z = -1$.

Example 2: Laboratory mice are to be given a diet containing 40 units of carbohydrate, 25 units of fat, and 25 units of protein. Three foods are available containing the following amounts of carbohydrate, fat and protein.

	Food		
	A	B	C
Carbohydrate	5	2	7
Fat	3	3	2
Protein	1	5	3

How should the foods be mixed in order to meet the requirements of the diet?

Let the amount of food A in the diet be x, that of food B be y, and that of food C be z. Then the following equations must be satisfied:

$$
\begin{array}{ll}
5x + 2y + 7z = 40 & \text{units of carbohydrate,} \\
3x + 3y + 2z = 25 & \text{units of fat,} \\
x + 5y + 3z = 25 & \text{units of protein.}
\end{array}
$$

The augmented matrix for this system is

$$
\left[\begin{array}{ccc|c}
5 & 2 & 7 & 40 \\
3 & 3 & 2 & 25 \\
1 & 5 & 3 & 25
\end{array}\right]
\begin{array}{l} R_1 \\ R_2 \\ R_3 \end{array}
$$

$$
\begin{array}{l} R_1 := R_3 \\ \\ R_3 := R_1 \end{array}
\left[\begin{array}{ccc|c}
1 & 5 & 3 & 25 \\
3 & 3 & 2 & 25 \\
5 & 2 & 7 & 40
\end{array}\right]
\qquad
\begin{array}{l} \\ R_2 := R_2 - 3R_1 \\ R_3 := R_3 - 5R_1 \end{array}
\left[\begin{array}{ccc|c}
1 & 5 & 3 & 25 \\
0 & -12 & -7 & -50 \\
0 & -23 & -8 & -85
\end{array}\right]
$$

$$
\begin{array}{l} \\ R_2 := R_3 - 2R_2 \\ \\ \end{array}
\left[\begin{array}{ccc|c}
1 & 5 & 3 & 25 \\
0 & 1 & 6 & 15 \\
0 & -23 & -8 & -85
\end{array}\right]
\qquad
\begin{array}{l} \\ \\ R_3 := R_3 + 23R_2 \end{array}
\left[\begin{array}{ccc|c}
1 & 5 & 3 & 25 \\
0 & 1 & 6 & 15 \\
0 & 0 & 130 & 260
\end{array}\right].
$$

The equations arising from the triangular form are

$$x + 5y + 3z = 25$$
$$y + 6z = 15$$
$$z = 2 .$$

By back substitution

$$y = 15 - 6z = 3 ,$$
$$x = 25 - 5y - 3z = 4 .$$

To supply the correct diet to the mice the food mixture must contain 4 units of food A, 3 units of food B, and 2 units of food C.

18.7 FINDING THE INVERSE OF A MATRIX

A similar procedure to Gaussian elimination can be used to find the inverse of an $n \times n$ matrix A. This time the $n \times 2n$ augmented matrix $A|I_n$ is considered. Using row operations this is changed into $I_n|B$ where B equals the inverse matrix A^{-1}. Carrying out row operations on A|I, as before, A is first made upper triangular, then the elements above the diagonal of A are made zero; finally the diagonal elements are made equal to one. If A is singular this method results in a zero row being produced in the left-hand part of the augmented matrix. When an inverse has been found it is good practice to check that AA^{-1} does result in the identity matrix.

Example 1: Find the inverse of $A = \begin{bmatrix} 1 & 1 & 2 \\ 1 & 2 & 3 \\ 2 & 6 & 9 \end{bmatrix}$, if it exists.

Form the augmented matrix A|I and carry out the row operations as indicated.

$$\begin{array}{ccc|ccc} 1 & 1 & 2 & 1 & 0 & 0 \\ 1 & 2 & 3 & 0 & 1 & 0 \\ 2 & 6 & 9 & 0 & 0 & 1 \end{array}$$

$$R_2 := R_2 - R_1, \ R_3 := R_3 - 2R_1$$

$$\begin{array}{ccc|ccc} 1 & 1 & 2 & 1 & 0 & 0 \\ 0 & 1 & 1 & -1 & 1 & 0 \\ 0 & 4 & 5 & -2 & 0 & 1 \end{array}$$

$$R_3 := R_3 - 4R_2$$

$$
\begin{array}{ccc|ccc}
1 & 1 & 2 & 1 & 0 & 0 \\
0 & 1 & 1 & -1 & 1 & 0 \\
0 & 0 & 1 & 2 & -4 & 1
\end{array}
$$

$R_1 := R_1 - 2R_3, \; R_2 := R_2 - R_3$

$$
\begin{array}{ccc|ccc}
1 & 1 & 0 & -3 & 8 & -2 \\
0 & 1 & 0 & -3 & 5 & -1 \\
0 & 0 & 1 & 2 & -4 & 1
\end{array}
$$

$R_1 := R_1 - R_2$

$$
\begin{array}{ccc|ccc}
1 & 0 & 0 & 0 & 3 & -1 \\
0 & 1 & 0 & -3 & 5 & -1 \\
0 & 0 & 1 & 2 & -4 & 1
\end{array} \; .
$$

This is now in the form $\mathbf{I}|\mathbf{B}$.

Therefore $\mathbf{B} = \begin{bmatrix} 0 & 3 & -1 \\ -3 & 5 & -1 \\ 2 & -4 & 1 \end{bmatrix}$ should be the inverse of $\mathbf{A} = \begin{bmatrix} 1 & 1 & 2 \\ 1 & 2 & 3 \\ 2 & 6 & 9 \end{bmatrix}$.

Check:
$$
\begin{bmatrix} 1 & 1 & 2 \\ 1 & 2 & 3 \\ 2 & 6 & 9 \end{bmatrix}
\begin{bmatrix} 0 & 3 & -1 \\ -3 & 5 & -1 \\ 2 & -4 & 1 \end{bmatrix}
=
\begin{bmatrix} 1 & 0 & 0 \\ 0 & 1 & 0 \\ 0 & 0 & 1 \end{bmatrix} \; .
$$

The matrix $\mathbf{A}$ in this example is the coefficient matrix for the system given in example 8 of Section 18.3. Using the inverse matrix $\mathbf{A}^{-1} = \mathbf{B}$ the solution to the system of equations is obtained from $\mathbf{x} = \mathbf{A}^{-1}\mathbf{b}$, that is,

$$
\begin{bmatrix} x \\ y \\ z \end{bmatrix}
=
\begin{bmatrix} 0 & 3 & -1 \\ -3 & 5 & -1 \\ 2 & -4 & 1 \end{bmatrix}
\begin{bmatrix} 3 \\ 4 \\ 9 \end{bmatrix}
=
\begin{bmatrix} 3 \\ 2 \\ -1 \end{bmatrix} \; .
$$

Example 2: Find the inverse of $\begin{bmatrix} 1 & 1 & 1 \\ -1 & 1 & 2 \\ 1 & 3 & 4 \end{bmatrix}$, if it exists.

Consider
$$
\begin{array}{ccc|ccc}
1 & 1 & 1 & 1 & 0 & 0 \\
-1 & 1 & 2 & 0 & 1 & 0 \\
1 & 3 & 4 & 0 & 0 & 1
\end{array}
$$

$R_2 := R_2 + R_1; \; R_3 := R_3 - R_1$

$$\begin{array}{ccc|ccc} 1 & 1 & 1 & 1 & 0 & 0 \\ 0 & 2 & 3 & 1 & 1 & 0 \\ 0 & 2 & 3 & -1 & 0 & 1 \end{array}$$

$R_3 := R_3 - R_2$

$$\begin{array}{ccc|ccc} 1 & 1 & 1 & 1 & 0 & 0 \\ 0 & 2 & 3 & 1 & 1 & 0 \\ 0 & 0 & 0 & -2 & -1 & 1 \end{array}.$$

As R_3 now contains zeros in the left-hand part the original matrix must be singular.

18.8 DETERMINANTS

Associated with every square matrix A, having numbers as elements, is a number called the **determinant** of A, written $|A|$ or det A. The determinant of a 2 × 2 matrix is defined by

$$\begin{vmatrix} a_{11} & a_{12} \\ a_{21} & a_{22} \end{vmatrix} = a_{11}a_{22} - a_{21}a_{12} .$$

For a 3 × 3 matrix the determinant is given by

$$\begin{vmatrix} a_{11} & a_{12} & a_{13} \\ a_{21} & a_{22} & a_{23} \\ a_{31} & a_{32} & a_{33} \end{vmatrix} = a_{11} \begin{vmatrix} a_{22} & a_{23} \\ a_{32} & a_{33} \end{vmatrix} - a_{12} \begin{vmatrix} a_{21} & a_{23} \\ a_{31} & a_{33} \end{vmatrix} + a_{13} \begin{vmatrix} a_{21} & a_{22} \\ a_{31} & a_{32} \end{vmatrix}$$

The determinant of an $n \times n$ matrix can be expressed similarly in terms of n, $(n - 1) \times (n - 1)$ determinants. The expansion consists of the sum of n terms each comprising the triple product of an element a_{1j} from the first row, the determinant formed by the elements remaining after the first row and the jth column containing the multiplying element have been deleted, and $(-1)^{j+1}$.

Example 1: Evaluate $\begin{vmatrix} 6 & 1 & 1 \\ 2 & 4 & -2 \\ 0 & 2 & 1 \end{vmatrix}$

$$\begin{vmatrix} 6 & 1 & 1 \\ 2 & 4 & -2 \\ 0 & 2 & 1 \end{vmatrix} = 6\begin{vmatrix} 4 & -2 \\ 2 & 1 \end{vmatrix} - 1\begin{vmatrix} 2 & -2 \\ 0 & 1 \end{vmatrix} + 1\begin{vmatrix} 2 & 4 \\ 0 & 2 \end{vmatrix}$$

$$= 6[4(1) - 2(-2)] - [2(1) - 0(-2)] + [2(2) - 0(4)]$$

$$= 48 - 2 + 4$$

$$= 50$$

Determinants have the following properties (A is $n \times n$):

(i) $|A^T| = |A|$.

(ii) $|A| = 0$ if and only if A is singular.

(iii) $|AB| = |A|\,|B|$ (B is $n \times n$).

(iv) If two rows (or two columns) are interchanged then the value of the determinant changes sign.

(v) If each element in a row (or column) is multiplied by a scalar then the value of the determinant is also multiplied by that scalar. Furthermore $|kA| = k^n|A|$.

(vi) If a multiple of one row is added to another the value of the determinant remains the same. (A similar result holds for columns).

(vii) The value of the determinant of an upper triangular matrix is given by the product of the elements on the leading diagonal.

Because of properties (iv) to (vii) an alternative, and frequently more efficient, way of calculating the value of a determinant is by reducing it to triangular form.

Example 2: Evaluate $\begin{vmatrix} 6 & 1 & 1 \\ 2 & 4 & -2 \\ 0 & 2 & 1 \end{vmatrix}$ by first reducing it to triangular form.

$$\begin{vmatrix} 6 & 1 & 1 \\ 2 & 4 & -2 \\ 0 & 2 & 1 \end{vmatrix} = \frac{1}{3}\begin{vmatrix} 6 & 1 & 1 \\ 6 & 12 & -1 \\ 0 & 2 & 1 \end{vmatrix} \qquad (R_2 := 3R_2)$$

$$= \frac{1}{3}\begin{vmatrix} 6 & 1 & 1 \\ 0 & 11 & -7 \\ 0 & 2 & 1 \end{vmatrix} \qquad (R_2 := R_2 - R_1)$$

$$= \frac{1}{3}\cdot\frac{1}{11}\begin{vmatrix} 6 & 1 & 1 \\ 0 & 11 & -7 \\ 0 & 22 & 11 \end{vmatrix} \qquad (R_3 := 11R_3)$$

$$= \frac{1}{33}\begin{vmatrix} 6 & 1 & 1 \\ 0 & 11 & -7 \\ 0 & 0 & 25 \end{vmatrix} \qquad \begin{array}{l}(R_3 := R_3 - 2R_2) \\ \text{(triangular form)}\end{array}$$

$$= \frac{1}{33}\cdot 6 \cdot 11 \cdot 25 = 50 \cdot$$

18.9 EIGENVALUES AND EIGENVECTORS

The product of an $n \times n$ matrix $\mathbf{A}$ with a column n vector $\mathbf{x}$ produces another column n vector, $\mathbf{y}$ say, that is $\mathbf{Ax} = \mathbf{y}$. For some vectors $\mathbf{x}$, $\mathbf{y}$ takes the form of a constant multiple of $\mathbf{x}$, that is, $\mathbf{Ax} = \lambda\mathbf{x}$. If such a vector $\mathbf{x}$ can be found which is not the zero vector then $\mathbf{x}$ is called an **eigenvector** of $\mathbf{A}$, and λ is the associated **eigenvalue**; λ is a number which may be real or complex. For an $n \times n$ matrix there are n eigenvalues.

To find λ and $\mathbf{x}$ the matrix eigenvalue equation

$$(\mathbf{A} - \lambda\mathbf{I}_n)\mathbf{x} = \mathbf{0} \tag{18.2}$$

must be solved. Since the solution $\mathbf{x}$ must be non-zero, the matrix $\mathbf{A} - \lambda\mathbf{I}$ must be singular. This requires, from property (ii) of determinants, that

$$|\mathbf{A} - \lambda\mathbf{I}| = 0 .$$

The expansion of this determinant will be a polynomial of degree n in λ called the **characteristic equation** of $\mathbf{A}$. The eigenvalues $\lambda_i, i = 1, 2, \ldots, n$, of $\mathbf{A}$ are the n roots of this equation. For each eigenvalue λ_i found, the matrix equation (18.2) is solved for $\mathbf{x}$, the corresponding eigenvector. Notice that because the matrix $\mathbf{A} - \lambda\mathbf{I}$ is singular the eigenvector associated with a particular eigenvalue will not be unique.

Example 1: Find the eigenvalues and eigenvectors associated with the matrix

$$\mathbf{A} = \begin{bmatrix} 4 & 1 & -5 \\ -5 & 2 & 5 \\ 1 & 1 & -2 \end{bmatrix}.$$

First the eigenvalues, λ, are found by solving $|\mathbf{A} - \lambda\mathbf{I}| = 0$. That is

$$\begin{vmatrix} 4-\lambda & 1 & -5 \\ -5 & 2-\lambda & 5 \\ 1 & 1 & -2-\lambda \end{vmatrix} = 0 .$$

Expanding the determinant gives

$$(4-\lambda)\left[(2-\lambda)(-2-\lambda)-5\right] - \left[-5(-2-\lambda)-5\right] - 5\left[-5-(2-\lambda)\right] = 0 ,$$

or

$$-\lambda^3 + 4\lambda^2 - \lambda - 6 = 0 ,$$

$$(\lambda + 1)(\lambda - 2)(3 - \lambda) = 0 .$$

The eigenvalues are thus $\lambda_1 = -1$, $\lambda_2 = 2$ and $\lambda_3 = 3$.

To find the eigenvector associated with each eigenvalue λ the equation $(\mathbf{A} - \lambda\mathbf{I})\mathbf{x} = \mathbf{0}$ must be solved for $\mathbf{x}$.

For $\lambda_1 = -1$ the solution of the equation $(\mathbf{A} + \mathbf{I})\mathbf{x} = \mathbf{0}$ is required. Applying Gaussian elimination to the augmented matrix

$$(A + I)|0 = \begin{bmatrix} 5 & 1 & -5 & | & 0 \\ -5 & 3 & 5 & | & 0 \\ 1 & 1 & -1 & | & 0 \end{bmatrix}$$

results in the matrix

$$\begin{bmatrix} 5 & 1 & -5 & | & 0 \\ 0 & 1 & 0 & | & 0 \\ 0 & 0 & 0 & | & 0 \end{bmatrix}.$$

Thus the elements of the eigenvector must satisfy the equations

$$5x_1 + x_2 - 5x_3 = 0 \,,$$

$$x_2 \qquad\quad = 0 \,.$$

So $x_2 = 0$ and $x_1 = x_3 = \alpha$ say. The eigenvector associated with $\lambda_1 = -1$ is therefore $x_1 = \begin{bmatrix} \alpha \\ 0 \\ \alpha \end{bmatrix}.$

When $\lambda_2 = 2$, $(A - 2I)|0 = \begin{bmatrix} 2 & 1 & -5 & | & 0 \\ -5 & 0 & 5 & | & 0 \\ 1 & 1 & -4 & | & 0 \end{bmatrix}.$

Using row operations this can be reduced to

$$\begin{bmatrix} 1 & 0 & -1 & | & 0 \\ 0 & 1 & -3 & | & 0 \\ 0 & 0 & 0 & | & 0 \end{bmatrix}.$$

From the associated equations

$$x_1 = x_3 = \beta \text{ say, and } x_2 = 3x_3 = 3\beta.$$

Therefore the eigenvector is $x_2 = \begin{bmatrix} \beta \\ 3\beta \\ \beta \end{bmatrix}.$

When $\lambda_3 = 3$, $(A - 3I)|0 = \begin{bmatrix} 1 & 1 & -5 & | & 0 \\ -5 & -1 & 5 & | & 0 \\ 1 & 1 & -5 & | & 0 \end{bmatrix}.$

Row operations give $\begin{bmatrix} 0 & 1 & -5 & | & 0 \\ 1 & 0 & 0 & | & 0 \\ 0 & 0 & 0 & | & 0 \end{bmatrix}.$

So $x_1 = 0$ and $x_2 = 5x_3$. Letting $x_3 = \gamma$ the eigenvector corresponding to $\lambda_3 = 3$ is $x_3 = \begin{bmatrix} 0 \\ 5\gamma \\ \gamma \end{bmatrix}$.

Properties of eigenvalues and eigenvectors:

(i) For a real symmetric matrix the eigenvalues and elements in the eigenvectors are real numbers.

(ii) The **trace** of the matrix **A**, denoted Tr(**A**), is defined to be the sum of the elements on the leading diagonal. This quantity also equals the sum of the n eigenvalues of **A**. Thus

$$\text{Tr}(\mathbf{A}) = \sum_{i=1}^{n} [\mathbf{A}]_{ii} = \sum_{i=1}^{n} \lambda_i = -c_{n-1} .$$

c_{n-1} is the coefficient of λ^{n-1} in the characteristic equation $\lambda^n + c_{n-1}\lambda^{n-1} + \ldots + c_1\lambda + c_0 = 0$. For relationships between roots and coefficients of a polynomial see Section 3.6. Notice that the coefficient of λ^n has been made equal to one.

(iii) The determinant of **A** equals the product of the n eigenvalues of **A**. Thus

$$|\mathbf{A}| = \prod_{i=1}^{n} \lambda_i = \lambda_1 \lambda_2 \ldots \lambda_n = (-1)^n c_0 .$$

(The symbol Π stands for the product of the terms indicated, cf Σ which is used to denote a sum of the terms.)

(iv) The eigenvector is not unique; any constant multiple of the vector will also be an eigenvector with the same eigenvalue.

$$\mathbf{A}(c\mathbf{x}) = c(\mathbf{A}\mathbf{x}) = c(\lambda\mathbf{x}) = \lambda(c\mathbf{x}) \quad (c \text{ a constant}) .$$

It is customary to arrange for the eigenvector **x** to satisfy $\mathbf{x}^T\mathbf{x} = 1$, by choosing a suitable constant multiplier. This is called **normalisation**.

(v) The eigenvalue with the greatest modulus is called the **dominant eigenvalue** and its associated eigenvector the **dominant eigenvector**.

Example 2: The transition matrix

$$\mathbf{L} = \begin{bmatrix} 0 & 7/3 & 4/3 \\ 1/3 & 0 & 0 \\ 0 & 1/2 & 0 \end{bmatrix}$$

characterises the yearly population growth of a species of beetle during each of three successive years. (As the diagonal elements in this transition matrix are all zero it is an example of a Leslie matrix, see Leslie (1945). Initially a population consists of 36 one-year old beetles. Estimate the numbers present in the population during the next sixteen years.

After one year the population distribution is given by

$$\begin{bmatrix} 0 & 7/3 & 4/3 \\ 1/3 & 0 & 0 \\ 0 & 1/2 & 0 \end{bmatrix} \begin{bmatrix} 0 \\ 36 \\ 0 \end{bmatrix} = \begin{bmatrix} 84 \\ 0 \\ 18 \end{bmatrix} .$$

After two years the population vector is given by

$$\begin{bmatrix} 0 & 7/3 & 4/3 \\ 1/3 & 0 & 0 \\ 0 & 1/2 & 0 \end{bmatrix}^2 \begin{bmatrix} 0 \\ 36 \\ 0 \end{bmatrix} = \begin{bmatrix} 0 & 7/3 & 4/3 \\ 1/3 & 0 & 0 \\ 0 & 1/2 & 0 \end{bmatrix} \begin{bmatrix} 84 \\ 0 \\ 18 \end{bmatrix} = \begin{bmatrix} 24 \\ 28 \\ 0 \end{bmatrix} .$$

In the succeeding years the population vectors are, respectively,

$$\begin{bmatrix} 65.3 \\ 8.0 \\ 14.0 \end{bmatrix}, \begin{bmatrix} 37.3 \\ 21.8 \\ 4.0 \end{bmatrix}, \begin{bmatrix} 56.2 \\ 12.4 \\ 10.9 \end{bmatrix}, \begin{bmatrix} 43.5 \\ 18.7 \\ 6.2 \end{bmatrix}, \begin{bmatrix} 51.9 \\ 14.5 \\ 9.4 \end{bmatrix}, \begin{bmatrix} 46.4 \\ 17.3 \\ 7.2 \end{bmatrix}, \begin{bmatrix} 50.0 \\ 15.5 \\ 8.6 \end{bmatrix},$$

$$\begin{bmatrix} 47.6 \\ 16.7 \\ 7.8 \end{bmatrix}, \begin{bmatrix} 49.4 \\ 15.9 \\ 8.4 \end{bmatrix}, \begin{bmatrix} 48.3 \\ 16.5 \\ 8.0 \end{bmatrix}, \begin{bmatrix} 49.2 \\ 16.1 \\ 8.2 \end{bmatrix}, \begin{bmatrix} 48.5 \\ 16.4 \\ 8.0 \end{bmatrix}, \begin{bmatrix} 48.9 \\ 16.2 \\ 8.2 \end{bmatrix}, \begin{bmatrix} 48.7 \\ 16.3 \\ 8.1 \end{bmatrix} .$$

From these vectors it can be seen that the population described in example 2 is tending towards a state in which the number present in each age group remains almost constant. A population distribution having such a property is said to be **stable**. If T is the transition matrix and p denotes a stable population vector $Tp = p$. This is similar to an eigenvalue equation with $\lambda = 1$. In fact the ultimate age distribution of a population can be shown to be governed by the dominant eigenvalue of the transition matrix T.

If $\lambda = \Lambda$ is the dominant eigenvalue of T then three cases can arise:

(i) $\Lambda = 1$ this is the case for a stable population to occur,

(ii) $\Lambda > 1$ the population grows without limit,

(iii) $\Lambda < 1$ the population ultimately becomes extinct.

In example 2 the eigenvalue equation is

$$(L - \lambda I)p = 0 .$$

The eigenvalues are given by

$$|\mathbf{L} - \lambda\mathbf{I}| = \begin{vmatrix} -\lambda & 7/3 & 4/3 \\ 1/3 & -\lambda & 0 \\ 0 & 1/2 & -\lambda \end{vmatrix} = 0 \ ,$$

or $9\lambda^3 - 7\lambda - 2 = (\lambda - 1)(3\lambda + 1)(3\lambda + 2) = 0$.

So $\lambda = 1, -1/3, -2/3$.

Notice that $\text{Tr}(\mathbf{L}) = 0 + 0 + 0 = 1 - \dfrac{1}{3} - \dfrac{2}{3} = 0$ (property (ii)).

The dominant eigenvalue is $\lambda = 1$, and consequently, as we have seen, the population stabilises.

The dominant eigenvector associated with $\lambda = 1$ is $\begin{bmatrix} 6\alpha \\ 2\alpha \\ \alpha \end{bmatrix}$ and the beetle

population is tending towards this age group distribution with $\alpha \cong 8.1$.

18.10 NUMERICAL METHODS

The efficient and accurate solution of problems involving matrices may require more advanced numerical procedures than those already described. This is particularly so when matrices of large order occur, as is the case when large systems of linear equations are to be solved; or when the elements in a matrix are not integers. Often the solving of such problems is performed by using a computer. The sizes of the matrices used in this chapter have been kept small for illustrative purposes, but, in principle, matrices of any order could be handled by the techniques given.

The numerical procedure known as Gaussian elimination has been employed to find the solution of a system of n simultaneous linear equations in n unknowns (Section 18.6), the inverse of a square matrix (Section 18.7), and to evaluate determinants (Section 18.8). A modification to Gaussian elimination is described in this section which avoids the build up of errors due to rounding off numbers. This is achieved by using row and column operations to rearrange the matrix elements in such a way that, of those elements not yet in triangular form, the one having the largest magnitude is positioned on the diagonal. A multiple of the row containing this element is then used to eliminate the other elements below it in that column. Note that when solving linear equations column interchanges result in the corresponding variables being interchanged. When column interchanges are applied in the matrix inversion procedure the corresponding rows in the final right-hand matrix must also be interchanged.

As a check against arithmetic errors in the calculation an extra column is sometimes incorporated. This column is formed by summing the terms in each

row of the matrix under consideration and is known as the **row-sum column.**
Row operations performed on the matrix are also applied to the row-sum
column. On comparing the values obtained in the row-sum column with the true
row sums a difference shows that an error is present in that row. A further use
for the row-sum column is in indicating the size of rounding errors.

In the following example these modifications to the basic Gaussian elimination procedure are illustrated.

Example: Evaluate, working to 2D, the determinant

$$\begin{vmatrix} 3.1 & 1.5 & 2.5 & -2.4 \\ -0.8 & -1.0 & 3.4 & -3.8 \\ 3.5 & 7.0 & -6.1 & 0.5 \\ 5.0 & 1.7 & 3.2 & -4.7 \end{vmatrix} .$$

As indicated in Section 18.8 the required value can be found by making the
determinant triangular, using row operations, and then finding the product of
the diagonal elements. Using the modified form of Gaussian elimination described above, the element 7.0, being the largest, is positioned in the first place
by interchanging rows 1 and 3 and then interchanging columns 1 and 2. Thus

$$\begin{vmatrix} 3.1 & 1.5 & 2.5 & -2.4 \\ -0.8 & -1.0 & 3.4 & -3.8 \\ 3.5 & 7.0 & -6.1 & 0.5 \\ 5.0 & 1.7 & 3.2 & -4.7 \end{vmatrix} = (-1) \begin{vmatrix} 3.5 & 7.0 & -6.1 & 0.5 \\ -0.8 & -1.0 & 3.4 & -3.8 \\ 3.1 & 1.5 & 2.5 & -2.4 \\ 5.0 & 1.7 & 3.2 & -4.7 \end{vmatrix}$$

$$= (-1)^2 \begin{vmatrix} 7.0 & 3.5 & -6.1 & 0.5 \\ -1.0 & -0.8 & 3.4 & -3.8 \\ 1.5 & 3.1 & 2.5 & -2.4 \\ 1.7 & 5.0 & 3.2 & -4.7 \end{vmatrix} .$$

The -1 appearing through the row and column interchanges.

Introducing the row-sum column gives

$$\begin{array}{|cccc|c} 7.0 & 3.5 & -6.1 & 0.5 & 4.9 \\ -1.0 & -0.8 & 3.4 & -3.8 & -2.2 \\ 1.5 & 3.1 & 2.5 & -2.4 & 4.7 \\ 1.7 & 5.0 & 3.2 & -4.7 & 5.2 \end{array} .$$

Eliminating the rest of the first column by the row operations

$$R_2 := R_2 + \frac{1.0}{7.0} R_1 = R_2 + 0.14 R_1,$$

$$R_3 := R_3 - \frac{1.5}{7.0} R_1 = R_3 - 0.21 R_1,$$

$$R_4 := R_4 - \frac{1.7}{7.0} R_1 = R_4 - 0.24 R_1$$

gives

$$\begin{vmatrix} 7.0 & 3.5 & -6.1 & 0.5 \\ 0 & -0.31 & 2.55 & -3.73 \\ 0 & 2.36 & 3.78 & -2.50 \\ 0 & 4.16 & 4.66 & \boxed{-4.82} \end{vmatrix} \begin{matrix} 4.9 \\ -1.49 \quad\vdots\quad -1.51 \\ 3.64 \quad\vdots\quad 3.67 \\ 4.00 \quad\vdots\quad 4.02 \end{matrix}$$

where the final column is the result of the row operation's action on the previous row-sum column. Examining the two row-sum columns we see that rounding errors have produced slight differences in the row-sum columns.

Before the next eliminations are carried out the element -4.82 must be positioned in the second place of the leading diagonal. This results in

$$\begin{vmatrix} 7.0 & 0.5 & -6.1 & 3.5 \\ 0 & -4.82 & 4.66 & 4.16 \\ 0 & -2.50 & 3.78 & 2.36 \\ 0 & -3.73 & 2.55 & -0.31 \end{vmatrix} \begin{matrix} 4.9 \\ 4.00 \\ 3.64 \\ -1.49 \end{matrix} \,.$$

Applying the row operations

$$R_3 := R_3 - \frac{2.50}{4.82} R_2 = R_3 - 0.52 R_2, \quad R_4 := R_4 - \frac{3.73}{4.82} R_2 = R_4 - 0.77 R_2, \text{ gives}$$

$$\begin{vmatrix} 7.0 & 0.5 & -6.1 & 3.5 \\ 0 & -4.82 & 4.66 & 4.16 \\ 0 & 0 & 1.36 & 0.20 \\ 0 & 0 & -1.04 & \boxed{-3.51} \end{vmatrix} \begin{matrix} 4.9 \\ 4.00 \\ 1.56 \quad\vdots\quad 1.56 \\ -4.55 \quad\vdots\quad -4.57 \end{matrix} \,.$$

Position -3.51 into the third place of the diagonal to obtain

$$\begin{vmatrix} 7.0 & 0.5 & 3.5 & -6.1 \\ 0 & -4.82 & 4.16 & 4.66 \\ 0 & 0 & -3.51 & -1.04 \\ 0 & 0 & 0.20 & 1.36 \end{vmatrix} \begin{matrix} 4.9 \\ 4.00 \\ -4.55 \\ 1.56 \end{matrix} \,.$$

The row operation

$$R_4 := R_4 + \frac{0.20}{3.51} R_3 = R_4 + 0.057\, R_3,\ \text{leads to the triangular form}$$

$$\begin{vmatrix} 7.0 & 0.5 & 3.5 & -6.1 \\ 0 & -4.82 & 4.16 & 4.66 \\ 0 & 0 & -3.51 & -1.04 \\ 0 & 0 & 0 & 1.30 \end{vmatrix} \begin{matrix} 4.9 \\ 4.00 \\ -4.55 \\ 1.30 \end{matrix} \quad 1.30\ .$$

Therefore the value of the determinant, as given by this method, is

$$(7.0)\,(-4.82)\,(-3.51)\,(1.30) = 153.96\ .$$

The exact value of the determinant is 155.52 and the value that is obtained without rearrangement, and working to 2D, is 149.53.

Numerical methods for determining eigenvalues and eigenvectors are also available and can be found in Williams (1972).

An important application of matrices involving probability theory, not covered in this book, is to **Markov processes**, see Jeffers (1978). Further information on matrix theory and the application of matrices to biological problems can be found in Searle (1966).

PROBLEMS

1. If $A = \begin{bmatrix} 4 & -4 & 1 \\ 2 & -2 & 2 \\ 3 & 2 & -4 \end{bmatrix}$ and $B = \begin{bmatrix} -4 & -8 & 2 \\ 4 & 1 & 4 \\ 6 & 4 & 0 \end{bmatrix}$

    find the matrices given by the following expressions:

    (a) $A + B$,   (c) $2A - B$,

    (b) $A - B$,   (d) $A - 2B + 4I$.

2. Find the values of x, y and z which satisfy

    $$\begin{bmatrix} x^2 + y & y^2 + z \\ z & y^2 - 2x^2 \end{bmatrix} = \begin{bmatrix} 10 & 25 \\ -11 & 4 \end{bmatrix}\ .$$

3. If $a = \begin{bmatrix} 1 & 3 & 2 \end{bmatrix}$, $b = \begin{bmatrix} 0 & -3 & 1 & 4 \end{bmatrix}$,

    $$c = \begin{bmatrix} 3 \\ -1 \\ 2 \end{bmatrix},\quad d = \begin{bmatrix} 4 \\ 4 \\ -6 \end{bmatrix},\quad e = \begin{bmatrix} 3 \\ 2 \\ 4 \\ 0 \end{bmatrix}$$

evaluate where possible, the following inner products:

(a) a b,   (c) a d,   (e) b c,

(b) a c,   (d) a e,   (f) b e.

4. If $A = \begin{bmatrix} 1 & 3 \\ 2 & 4 \end{bmatrix}$, $B = \begin{bmatrix} -2 & 3 \\ 2 & 0 \end{bmatrix}$, $C = \begin{bmatrix} 1 & -3 \\ -2 & -1 \end{bmatrix}$

evaluate the following matrix products:

(a) AB,   (e) BC,   (i) BCA,

(b) BA,   (f) CB,   (j) ACB,

(c) AC,   (g) ABC,   (k) CAB,

(d) CA,   (h) BAC,   (l) CBA.

5. If $A = \begin{bmatrix} 4 & 3 \end{bmatrix}$, $B = \begin{bmatrix} 1 & 2 \\ 4 & 3 \end{bmatrix}$, $C = \begin{bmatrix} 2 & -2 & 1 \\ -1 & 0 & 3 \end{bmatrix}$, $D = \begin{bmatrix} 0 & 5 \\ 1 & 3 \\ 2 & 2 \end{bmatrix}$

evaluate where possible, the following products:

(a) AB,   (e) AD,   (i) BD,

(b) BA,   (f) DA,   (j) DB,

(c) AC,   (g) BC,   (k) CD,

(d) CA,   (h) CB,   (l) DC.

6. If $A = \begin{bmatrix} 2 & 0 & 1 \\ 3 & -1 & 1 \\ -1 & 4 & 2 \end{bmatrix}$, $B = \begin{bmatrix} 1 & 1 & -1 \\ 3 & 2 & -3 \\ 4 & -2 & 3 \end{bmatrix}$, $C = \begin{bmatrix} -3 & 4 & 5 \\ 5 & -4 & 5 \\ 7 & 8 & 1 \end{bmatrix}$

evaluate the following matrix products:

(a) AB,   (c) AC,   (e) BC,

(b) BA,   (d) CA,   (f) CB.

7. From a study of the life-cycle of an insect the number of eggs, larvae, pupae and adults alive at the end of the spring, summer, autumn and winter, may be determined by post-multiplying the matrices S_1, S_2, S_3 and S_4, by a column vector containing the number of eggs, larvae, pupae and adults alive at the start of that season.

(Spring) $S_1 = \begin{bmatrix} 0 & 0 & 0 & 1000 \\ 0.2 & 0 & 0 & 0 \\ 0 & 0.2 & 0 & 0 \\ 0 & 0 & 0.2 & 0.4 \end{bmatrix}$

$$(\text{Summer}) \quad S_2 = \begin{bmatrix} 0 & 0 & 0 & 1600 \\ 0.3 & 0 & 0 & 0 \\ 0 & 0.2 & 0 & 0 \\ 0 & 0 & 0.2 & 0.4 \end{bmatrix}$$

$$(\text{Autumn}) \quad S_3 = \begin{bmatrix} 0.1 & 0 & 0 & 400 \\ 0.1 & 0.5 & 0 & 0 \\ 0 & 0 & 0.5 & 0 \\ 0 & 0 & 0 & 0.25 \end{bmatrix}$$

$$(\text{Winter}) \quad S_4 = \begin{bmatrix} 0.1 & 0 & 0 & 100 \\ 0 & 0.4 & 0 & 0 \\ 0 & 0 & 0.4 & 0 \\ 0 & 0 & 0 & 0.1 \end{bmatrix}.$$

Find the number in each age group at the end of the summer if there were 10^6, 2×10^5, 10^4 and 2.5×10^2, eggs, larvae, pupae and adults, respectively, present in a population after the winter. By the end of the next winter how many will there be in each age group?

8. If $A = \begin{bmatrix} -1 & 2 & 1 & a \\ 2 & 1 & b & 1 \\ 1 & c & 1 & -2 \\ d & 1 & -2 & -1 \end{bmatrix}$

    find the values of a, b, c and d such that A^2 is diagonal.

9. A non-zero square matrix is said to be **nilpotent** if powers higher than a certain integer value p give rise to a zero matrix, that is, A is nilpotent of order p if $A^k = 0$ for all $k \geqslant p$.

    Verify that (i) $\begin{bmatrix} -1 & -5 & -4 \\ 0 & -2 & -1 \\ 1 & 3 & 3 \end{bmatrix}$ and (ii) $\begin{bmatrix} 1 & -1 & -1 & -1 \\ 0 & 1 & 2 & 1 \\ -1 & 1 & 1 & 1 \\ 2 & -3 & -4 & -3 \end{bmatrix}$

    are nilpotent and find the order in each case.

10. A non-zero square matrix A is said to be **idempotent** if $A^2 = A$. Show that a matrix of the form $\begin{bmatrix} a & b & b \\ b & a & b \\ b & b & a \end{bmatrix}$

    is idempotent if (a, b) equals $(1, 0)$ $(\frac{1}{3}, \frac{1}{3})$ or $(\frac{2}{3}, -\frac{1}{3})$.

11.   The transition matrix

$$\begin{bmatrix} 0 & 0 & 16 \\ 1/4 & 0 & 0 \\ 0 & 1/2 & 0 \end{bmatrix}$$

characterises the annual movement between the three stages larvae, pupae and adults in the life cycle of an insect. Show that after three years the total population will have doubled and that the proportion in each stage of the life cycle is the same as it was initially.

12.   During an epidemic the transition between the stages of being healthy or sick the following day is described by the matrix

$$\begin{bmatrix} 5/8 & 1/8 \\ 3/8 & 7/8 \end{bmatrix}.$$

The elements of the first column denote the probabilities of a healthy person remaining healthy and becoming sick, respectively; the elements of the second column denote the probabilities of a person with the disease recovering or remaining sick, respectively. (A person that has had the disease can contract it again). (For details of probability see Chapter 21). On one day during the epidemic the population of a certain village consists of 1536 healthy and 512 sick individuals. How many will be sick one week later? What happens if initially the ratio of healthy to sick people is 1:3?

13.   If $A = \begin{bmatrix} 1 & a & -1 \\ 0 & 2 & b \end{bmatrix}$ and $A^T A = \begin{bmatrix} 1 & 3 & -c \\ 3 & 13 & c \\ -c & c & 5 \end{bmatrix}$

find the values of a, b and c.

14.   If $A = \begin{bmatrix} 2 & -2 \\ 1 & 3 \end{bmatrix}$, $B = \begin{bmatrix} -1 & 4 & 2 \\ 2 & 0 & -3 \end{bmatrix}$, $C = \begin{bmatrix} 1 & 2 \\ 3 & -1 \\ -2 & 0 \end{bmatrix}$, $d = [-1 \ \ 1 \ \ 2]$

evaluate the following expressions, when they exist:

(a) **AB**,	(g) **Bd**,	(l) **AC**^{T},
(b) **AC**,	(h) **Cd**,	(m) **Bd**^{T},
(c) **BA**,	(i) **dB**,	(n) **dB**^{T},
(d) **BC**,	(j) **dC**,	(o) **dCC**^{T}**d**^{T},
(e) **CA**,	(k) **B**^{T}**A**,	(p) 2**B**$-$**C**^{T}.
(f) **CB**,		

15. Show that

$$A_\theta = \begin{bmatrix} \cos\theta & \sin\theta \\ -\sin\theta & \cos\theta \end{bmatrix}$$

is orthogonal and has the property that $A_\theta A_\phi = A_{\theta+\phi}$.

16. Let $A = \begin{bmatrix} 1 & 0 & 3 \\ -1 & 2 & 4 \\ 2 & 3 & -2 \end{bmatrix}$ and $B = \begin{bmatrix} 1 & 2 \\ 3 & -4 \\ -2 & 1 \end{bmatrix}$.

(a) Verify that $(AB)^T = B^T A^T$.

(b) Find, for each matrix, the two symmetric matrices which can be formed from the given matrix and its transpose.

(c) Decompose A into the sum of a symmetric and a skew-symmetric matrix.

17. Write the following systems of linear equations in matrix form and then solve them by using Gaussian elimination:

(a) $x + 2y + 3z = 1$
     $x + y + 2z = -1$
     $x + 2y + z = 1,$

(b) $x + 3y - z = 12$
     $2x + z = 3$
     $x + 2y + 2z = 6,$

(c) $2x + 3y - 2z = 3$
     $7x + 3y - 3z = 7$
     $x - y - 3z = 3,$

(d) $w + x + y - z = 4$
     $w + 2x - y - 2z = 7$
     $2w - x + y + z = -1$
     $2w + 3x - 3y + 4z = 4.$

18. Three species of insect coexist in a laboratory environment supplied with three different foods each day. The average consumption of each of the three foods by an individual from each species is listed in the following table

		Insect species		
		X	Y	Z
	A	2	1	4
Food	B	4	2	2
	C	1	3	2 .

If 135, 120 and 105 units of the foods A, B and C, respectively, are supplied daily to the environment, and it is all consumed, how many of each species can coexist?

19. The trustees of a zoo have received a sum of money which will enable them to purchase, house and feed for the first year three new species. How many of each species should they acquire if:

(i) the total allocated for purchasing the animals is given by
$$5x + 4y + 2z = 20 ,$$

(ii) the total allocated to house the animals is given by
$$10x + 10y + 12z = 66 ,$$

(iii) the total allocated to feeding them is given by
$$3x + 2y + 2z = 14 ,$$

where x, y and z denote the number of each species?

20. Find, where possible, values of a, b, c and d in order that the matrix $\begin{bmatrix} a & b \\ c & d \end{bmatrix}$ becomes the inverse of the following matrices:

(a) $\begin{bmatrix} 2 & 1 \\ 5 & 3 \end{bmatrix}$

(c) $\begin{bmatrix} 2 & -1 \\ -4 & 2 \end{bmatrix}$

(b) $\begin{bmatrix} 2 & -5 \\ -2 & 6 \end{bmatrix}$

(d) $\begin{bmatrix} 4 & 1 \\ 5 & 2 \end{bmatrix}.$

21. Find the inverse matrix, when it exists, for each of the following matrices:

(a) $\begin{bmatrix} 2 & 3 \\ -1 & 1 \end{bmatrix}$

(e) $\begin{bmatrix} 2 & 4 & -3 \\ 1 & 1 & 2 \\ 2 & 3 & 0 \end{bmatrix}$

(b) $\begin{bmatrix} 1 & 2 & 3 \\ -1 & -1 & -2 \\ 2 & 6 & 9 \end{bmatrix}$

(f) $\begin{bmatrix} 1 & 4 & 8 \\ 2 & 10 & 13 \\ 1 & 6 & 5 \end{bmatrix}$

(c) $\begin{bmatrix} 5 & 2 & -3 \\ -1 & -1 & 1 \\ -3 & -1 & 2 \end{bmatrix}$

(g) $\begin{bmatrix} 1 & 1 & 1 \\ 1 & 2 & 3 \\ 3 & 5 & 6 \end{bmatrix}$

(d) $\begin{bmatrix} 4 & 0 & 5 \\ 0 & 1 & -6 \\ 3 & 0 & 4 \end{bmatrix}$

(h) $\begin{bmatrix} 1 & 1 & 1 & 1 \\ 1 & 2 & 2 & 2 \\ 1 & 2 & 3 & 3 \\ 1 & 2 & 3 & 4 \end{bmatrix}.$

22. By finding the inverse of the coefficient matrix solve the following systems of linear equations for each set of right-hand values, b_1, b_2, b_3 and b_4:

(a)
$$x + 2y + 3z = b_1$$
$$x + y + 2z = b_2$$
$$x + 2y + z = b_3$$

	(i)	(ii)	(iii)
	5	1	2
	4	−1	1
	1	1	0

(b)
$$x - y + z = b_1$$
$$2x - y - z = b_2$$
$$x - y + 2z = b_3$$

	(i)	(ii)	(iii)
	7	4	3
	9	3	2
	8	6	4

(c)
$$w + x - y + z = b_1$$
$$w - x + y + z = b_2$$
$$w + x + y - z = b_3$$
$$2w - 3x + y + z = b_4$$

	(i)	(ii)	(iii)
	4	10	2
	1	0	−2
	2	2	0
	1	6	4 .

23. Evaluate the following determinants:

(a) $\begin{vmatrix} 2 & 1 \\ 4 & 3 \end{vmatrix}$

(b) $\begin{vmatrix} 1 & 2 \\ 3 & 5 \end{vmatrix}$

(c) $\begin{vmatrix} 1 & 2 & 3 \\ 2 & 3 & 4 \\ 3 & 4 & 1 \end{vmatrix}$

(d) $\begin{vmatrix} 1 & 1 & 1 \\ 1 & 2 & 1 \\ 1 & 1 & -1 \end{vmatrix}$

(e) $\begin{vmatrix} 1 & 1 & 1 \\ 1 & 2 & 3 \\ 3 & 5 & 7 \end{vmatrix}$

(f) $\begin{vmatrix} 2 & 0 & 1 \\ 2 & 3 & -1 \\ -2 & 4 & -2 \end{vmatrix}$.

24. Use row (or column) operations to evaluate the following determinants:

(a) $\begin{vmatrix} 1 & 11 & 11 \\ 3 & 13 & 33 \\ 5 & 15 & 55 \end{vmatrix}$

(b) $\begin{vmatrix} 24 & 15 & 19 \\ 37 & 22 & 25 \\ 13 & 6 & 2 \end{vmatrix}$

(c) $\begin{vmatrix} 1 & 2 & 3 & 4 \\ 2 & 5 & 8 & 11 \\ 3 & 8 & 13 & 18 \\ 4 & 11 & 18 & 25 \end{vmatrix}$

(d) $\begin{vmatrix} 45 & 15 & -30 & 75 \\ 23 & 43 & 63 & 83 \\ 28 & -22 & 48 & 68 \\ 13 & 33 & 53 & 73 \end{vmatrix}$.

25. Find the eigenvalues λ, and the corresponding eigenvectors x, associated with the following matrices, A. Verify that $Ax = \lambda x$ in each case.

(a) $\begin{bmatrix} 3 & 1 \\ 2 & 4 \end{bmatrix}$

(e) $\begin{bmatrix} -2 & -1 & -5 \\ 1 & 2 & 1 \\ 3 & 1 & 6 \end{bmatrix}$

(b) $\begin{bmatrix} 7 & 5 \\ -4 & -2 \end{bmatrix}$

(f) $\begin{bmatrix} 2 & 1 & -3 \\ 0 & 1 & -1 \\ 0 & 0 & 0 \end{bmatrix}$

(c) $\begin{bmatrix} -2 & 2 & 3 \\ -3 & 3 & 3 \\ -2 & 0 & 5 \end{bmatrix}$

(g) $\begin{bmatrix} 2 & -2 & -1 \\ 1 & -1 & -1 \\ 1 & -2 & 0 \end{bmatrix}$

(d) $\begin{bmatrix} -1 & -1 & 5 \\ 3 & 1 & -1 \\ -2 & -1 & 6 \end{bmatrix}$

(h) $\begin{bmatrix} 2 & -2 & 3 \\ 1 & 1 & 1 \\ 1 & 3 & -1 \end{bmatrix}$.

26. For a particular species of insect the transition matrix

$$\begin{bmatrix} 1/3 & 1/3 & 1/3 \\ 2/3 & 1/6 & 1/2 \\ 0 & 1/2 & 1/6 \end{bmatrix}$$

describes the usual weekly pattern of movements for the insects between three sites. If initially all 972 insects are at the first site determine the number of insects present at each site (i) 3 weeks and (ii) 6 weeks later. (Assume no new insects arrive and that none die).

By determining the dominant eigenvector deduce that the insect population will ultimately settle down to a state in which 324, 405 and 243 insects are present at the three sites.

27. The matrix

$$\begin{bmatrix} 0 & 0 & 20 \\ 1/2 & 0 & 0 \\ 0 & 1/2 & 3/4 \end{bmatrix}$$

describes the annual transition between the age groups: fry, young, and mature fish in a fishery when no fish are being caught. Find the dominant eigenvalue and the associated eigenvector for this matrix. At what rate will a fish population with an age distribution proportional to the dominant eigenvector increase?

Verify that such a fish stock will be preserved at a constant level if half of the fish (from all three age groups) are caught each year.

28. Solve the system of linear equations

$$0.24\,x + 0.38\,y + 0.16\,z = 0.78$$
$$0.18\,x + 0.53\,y - 0.27\,z = 0.44$$
$$0.11\,x + 0.10\,y + 0.42\,z = 0.63$$

working to (i) two and (ii) three decimal places.
(The exact solution is $x = y = z = 1$).

29. Evaluate

$$\begin{vmatrix} 1.53 & 1.12 & -1.42 \\ 0.95 & -0.64 & 0.58 \\ 0.50 & 0.16 & -0.82 \end{vmatrix}.$$

Work to three decimal places. (The exact solution is 1.188).

30. The Leslie matrix

$$\begin{bmatrix} 0.05 & 0 & 0 & 1000 \\ 0.1 & 0 & 0 & 0 \\ 0 & 0.1 & 0 & 0 \\ 0 & 0 & 0.1 & 0 \end{bmatrix}$$

describes the transitions between the four stages: eggs, larvae, nymphs and adults, in the life cycle of a tick.

Show that the characteristic equation for the matrix is

$$\lambda^4 - 0.05\lambda^3 - 1 = 0.$$

Show that the equation has two real roots near $+1$ and -1. Use Newton's method to find these two roots correct to four decimal places.

Reduce the characteristic equation to the quadratic equation

$$\lambda^2 - 0.025\lambda + 1.000 = 0,$$

by dividing by the two real factors, and hence find the two complex eigenvalues.

Verify that the dominant eigenvalue is real and positive. (As this root is greater than one the population size will increase). Determine the dominant eigenvector and hence the ultimate proportion of eggs, larvae, nymphs and adults that will be present in the population.

CHAPTER 19

Curve fitting

19.1 INTRODUCTION

In the biosciences, mathematics is used to give quantitative descriptions of events or processes. The functions that occur show how one quantity or process depends upon other quantities or influences. Sometimes the form of the relationship between the quantities involved can be deduced from theoretical considerations. More often the function is obtained by studying the results obtained from experiments. This latter approach leads to an empirical formula for the relationship.

One way of finding out if a relationship between two variables exists is to use a graph. In general it is not possible, owing to experimental errors, to show that the observations (variables) fit a particular curve (function). However, if the relationship is linear, or approximately so, it is easy to spot this from the graph because the points will lie close to a straight line. In the latter case the equation of the line can be determined from the graph and hence the values of the constants in the relationship found. This process is known as fitting a straight line to the data.

In this chapter we illustrate how several types of functional relationship involving two variables may be verified; and also determine the value of the constants that appear in the function. A graphical approach is used initially, but the chapter concludes by introducing the method of least squares, which enables the equation of a straight line through the points to be determined numerically.

19.2 LINEAR RELATIONSHIPS

If the data being examined satisfy a linear relationship the points, when plotted on a graph, will lie close to a straight line. Hence the variables, x and y say, may be taken to be related by the equation

$$y = mx + c ,$$

where m and c are constants.

The values of m and c can be obtained easily from the graph (see Fig. 19.1): m is the slope of the straight line, c the intercept on the y-axis.

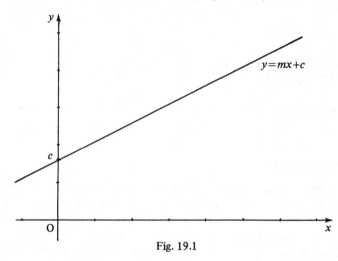

Fig. 19.1

To ensure the greatest accuracy there are several points which must be observed.

(i) The largest possible graph should be drawn with the points spread throughout. This may mean that the axes (the lines $y = 0$ and $x = 0$) are not included. Hence c may have to be calculated.

(ii) The straight line should be drawn so that the plotted points are distributed evenly on each side.

(iii) Two widely spaced points on the straight line should be used to find the slope m. If the points have coordinates (x_1, y_1) and (x_2, y_2) then

$$m = \frac{y_2 - y_1}{x_2 - x_1},$$

and the equation of the line is given, from equation (3.5), by

$$y - y_1 = \frac{y_2 - y_1}{x_2 - x_1} (x - x_1).$$

Example 1: The following table shows the sulphur content in the leaves of a cucumber plant. The leaves are counted from the bottom of the plant and the sulphur content is expressed as a percentage of the dry leaf weight.

Leaf	L	1	3	5	7	9	11
% dry weight of sulphur	S	1.52	1.41	1.29	1.21	1.09	0.98

Show that the results support the hypothesis of a linear sulphur gradient between the bottom and top of the plant and find an equation for the relationship.

If there is a linear relationship the points, when plotted on a graph of S against L, should lie close to a straight line. The ranges of the variables are $1 \leqslant L \leqslant 11$ and $0.9 < S < 1.6$. From the graph, Fig. 19.2, it is seen that the points do follow a straight line, hence the hypothesis is accepted.

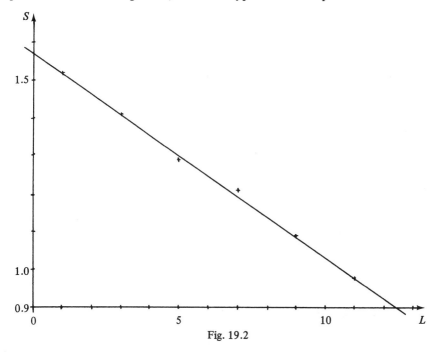

Fig. 19.2

Using the two points $(0, 1.57)$ and $(11, 0.98)$ the equation of the line is

$$S - 1.57 = \frac{0.98 - 1.57}{11 - 0} (L - 0) = -\frac{0.59}{11} L ,$$

or $\qquad S = -0.054 \, L + 1.57 .$

Sometimes a relationship can be derived from theoretical considerations. It is then necessary to validate the relationship, using experimental data, and subsequently find the value of any constants. For example, in the study of enzymes acting as catalysts in a chemical reaction the velocity v with which a single substrate, having concentration S, changes into a single product, for a fixed amount of enzyme, is given by the Michaelis-Menten equation

$$v = \frac{aS}{b+S} , \qquad\qquad (19.1)$$

where a and b are physical constants which depend upon the substances involved. This equation is non-linear, but can be rearranged, by inverting, to give

$$\frac{1}{v} = \frac{b+S}{aS} = \frac{b}{a}\frac{1}{S} + \frac{1}{a}. \tag{19.2}$$

Replacing $\frac{1}{v}$ and $\frac{1}{S}$ by y and x respectively, the equation is seen to be of the form $y = mx + c$ with $m = \frac{b}{a}$ and $c = \frac{1}{a}$. The equation in x and y represents a straight line which can be verified graphically. Once m and c are determined from the graph, a and b may be found.

In general a relationship between the variables s and t of the form

$$s^\alpha = mt^\beta + c ,$$

where α and β are known constants, and m and c are unknown constants, can be converted into the linear relationship

$$y = mx + c$$

by putting $y = s^\alpha$ and $x = t^\beta$.

Example 2: The following data were obtained in an experiment with an enzyme-catalysing reaction

S	5	7	10	15	20	30
v	5.5	7.0	8.5	10.0	11.0	12.5

where the concentration of the substrate S is measured in units of 10^{-5} moles/litre and the velocity of the reaction v, is given in units of 10^{-9} moles/litre/min.

Verify that the Michaelis-Menten equation (19.1) holds, and find the constants a and b.

If the Michaelis-Menten equation holds, the data should fit equation (19.2), that is, the straight line given by $\frac{1}{v} = \frac{b}{a}\frac{1}{S} + \frac{1}{a}$.

The quantities required for a graph are set down in the table:

S	5	7	10	15	20	30
$1/S$	0.200	0.143	0.100	0.067	0.050	0.033
$1/v$	0.182	0.143	0.118	0.100	0.091	0.080
v	5.5	7.0	8.5	10.0	11.0	12.5

The intervals $0.0 < 1/S \leqslant 0.2$ and $0.05 < 1/v < 0.20$ determine the portions of the axes to be used for the graph.

From the graph of $1/v$ against $1/S$, Fig. 19.3, it can be seen that a straight line is a good fit to the points. Hence the reaction proceeds according to the Michaelis-Menten equation.

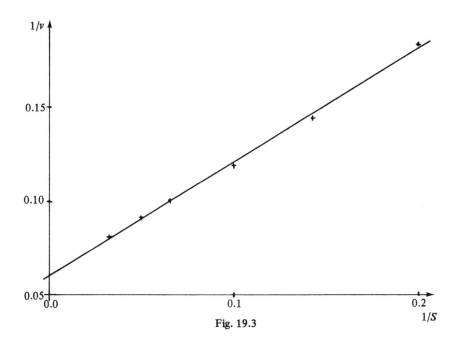

Fig. 19.3

The equation of the line, obtained by using the points $(0.0, 0.06)$ and $(0.2, 0.18)$ from the graph, is given by

$$\frac{1}{v} - 0.06 = \frac{0.18 - 0.06}{0.2 - 0.0}\left(\frac{1}{S} - 0\right) = 0.6\frac{1}{S}$$

so

$$\frac{1}{v} = 0.6\frac{1}{S} + 0.06 \ .$$

On comparing this with equation (19.2), $b/a = 0.6$ and $1/a = 0.06$, hence

$$a = 16.67 \quad \text{and} \quad b = 10 \ .$$

19.3 POWER LAWS

In another situation it may be expected that experimental results conform to a relationship between the variables x and y of the form

$$y = Ax^{\alpha}$$

where A and α are constants to be found. This kind of relationship would be difficult to determine by plotting y against x, since curves of the form shown in Fig. 19.4 would result.

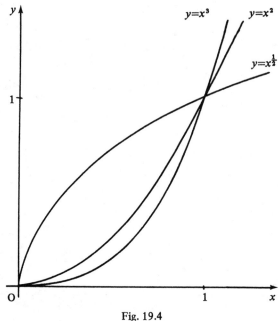

Fig. 19.4

It is much easier, in this case, to transform the relationship $y = Ax^\alpha$ into a linear form by taking logarithms, thus

$$\log y = \log A + \alpha \log x .$$

On writing $Y = \log y$, $C = \log A$ and $X = \log x$ this becomes

$$Y = \alpha X + C .$$

This is a linear relationship between X and Y, and so represents a straight line with slope α on a graph of Y against X. From a graph, α and C — and hence A — may be found. Note that although logarithms to any base could be used it is usual to use logarithms to base 10. To save finding the logarithms of the data special log-log graph paper could be used enabling the points to be plotted directly.

Example: The body weight w, in kilograms, and the sitting height h, in centimetres, of five people are given in the following table

w	50	65	80	100	120
h	84.7	91.2	97.2	104.7	111.0

Show that these readings satisfy a law of the form $h = Aw^{\alpha}$ and determine A and α from a graph.

As the law is thought to be of the form $h = Aw^{\alpha}$, take logarithms to base 10 to obtain

$$\log h = \log A + \alpha \log w .$$

Letting $H = \log h$, $C = \log A$ and $W = \log w$ this becomes

$$H = \alpha W + C .$$

So, by plotting a graph of H against W, a straight line should be a good fit to the points if the law is correct.

w	50	65	80	100	120
$W = \log_{10} w$	1.6990	1.8129	1.9031	2.0000	2.0792
$H = \log_{10} h$	1.9279	1.9600	1.9877	2.0170	2.0453
h	84.7	91.2	97.2	104.7	111.0

Now $1.6 < W < 2.1$ and $1.9 < H < 2.1$ so, taking these sections of the axes only, the points will be spread across the graph paper. Since the H-axis is not included on the graph C will have to be calculated.

From the graph, Fig. 19.5, it can be seen that a straight line is a good fit to the points. Hence, the data do fit a law of the form $h = Aw^{\alpha}$.

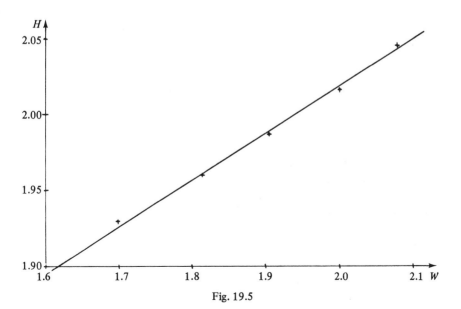

Fig. 19.5

Using the points $(1.615, 1.90)$ and $(2.100, 2.05)$ the slope of the line,

$$\alpha = \frac{2.05 - 1.90}{2.100 - 1.615} = \frac{0.15}{0.485} = 0.03093 .$$

At $W = 2.0, H = 2.02$ so

$$C = H - \alpha W = 2.02 - (0.03093)2$$
$$= 2.02 - 0.06186 = 1.95814 .$$

Therefore, $A = $ antilog $C = $ antilog $1.95814 = 90.81$. The law relating body weight w to sitting height h is thus given, approximately, by $h = 90.8 \, w^{0.03}$.

19.4 EXPONENTIAL LAWS

Exponential laws arise frequently in the biosciences. Here the variables x and y satisfy a relationship of the form

$$y = Ae^{bx} \tag{19.3}$$

where A and b are constants to be determined.

Note that a relationship of the form $y = Ac^x$ can also be rewritten in the form given in (19.3) by letting $c = e^b$.

Once again a linear relationship can be obtained by taking logarithms. This time, using logarithms to base e since this is suggested by equation (19.3), we get

$$\ln y = \ln A + bx .$$

Letting $Y = \ln y$ and $a = \ln A$ this becomes the straight line equation

$$Y = bx + a .$$

(Alternatively, logarithms to base 10, or indeed any base, could have been used. Equation (19.3) then becomes

$$\log y = \log A + bx \log e .$$

Letting $Y = \log y$, $C = \log A$ and $m = b \log e$　　we obtain

$$Y = mx + C.)$$

Once again special semi-log graph paper is available enabling the data points to be plotted directly.

Example: It is thought that the removal of a drug from the circulatory system follows a law of the form $C = Ae^{-kt}$ where C is the concentration of the drug in the blood and t is the time after administration. Verify this for the given data and determine A and k graphically. Find the predicted concentration of the drug in the blood after 8 hours.

t hrs	1	2	3	4	5
$C \, \mu$g/ml	70	52	39	29	21

If $C = Ae^{-kt}$ then $\ln C = \ln A - kt$. Letting $c = \ln C$ and $a = \ln A$ this becomes $c = -kt + a$.

t	1	2	3	4	5
$c = \ln C$	4.248	3.951	3.664	3.367	3.045
C	70	52	39	29	21

So $1 \leqslant t \leqslant 5$ and $3.0 < c < 4.3$.

From the graph, Fig. 19.6, a straight line is seen to be a good fit to the points, so the given data do fit a law of the form $C = Ae^{-kt}$.

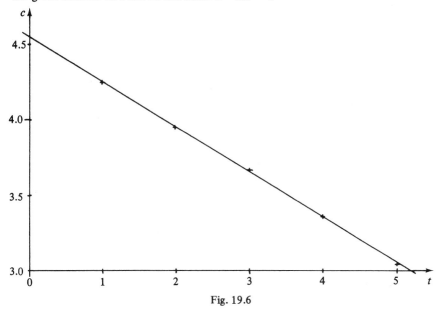

Fig. 19.6

The equation of the line, obtained by using the values at $t = 0$ and $t = 5$, is

$$c - 4.55 = \frac{3.05 - 4.55}{5 - 0} (t - 0)$$

$$= -\frac{1.5}{5} t = -0.3 t .$$

Therefore $c = -0.3t + 4.55$.   (19.4)

Now $k = 0.3$ and $a = \ln A = 4.55$ so $A = 94.63$.

The concentration of the drug at a time t hours after administration is thus given by

$$C = 94.6 \exp(-0.3t) \mu g/ml .$$

When $t = 8$, equation (19.4) gives,

$$c = -(0.3)8 + 4.55 = 2.15 \ .$$

Thus $\ln C = c = 2.15$, hence $C = 8.58 \ \mu g/ml$.

19.5 THE METHOD OF LEAST SQUARES

The graphical approach for determining the equation relating the variables relies heavily on the individual's choice of the best straight line through the points. However, the equation of a straight line through the n points (x_1, y_1), $(x_2, y_2), \ldots, (x_n, y_n)$ can be determined without, necessarily, having to plot the points on a graph. The equation of the best straight line through the points is found by the **method of least squares**.

    Let the equation of the line of best fit be

$$y = mx + c \ .$$

The point P with data values (x_i, y_i) will not, in general, lie on the line $y = mx + c$. Let Q be the point on this line with coordinates $(x_i, mx_i + c)$. If d_i measures the difference between the y value of the point P and the straight line at Q, see Fig. 19.7, then $d_i = y_i - (mx_i + c)$. Data points above the line give a positive d_i, those below a negative d_i.

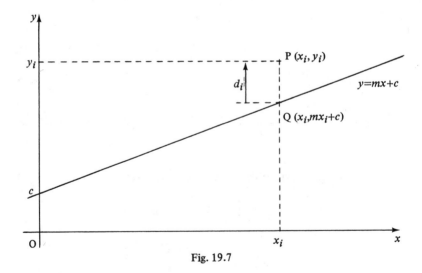

Fig. 19.7

    Considering all the points in turn n values of d will be obtained. The line taken as being the best is that for which the sum of the d's squared is a minimum.

Let $z = \sum\limits_{i=1}^{n} (y_i - mx_i - c)^2 = \sum\limits_{i=1}^{n} d_i^2 \ .$

The values of m and c which minimise z are, from Section 9.5, given by

$$\frac{\partial z}{\partial m} = -2 \sum_{i=1}^{n} x_i(y_i - mx_i - c) = 0 \; ,$$

and

$$\frac{\partial z}{\partial c} = -2 \sum_{i=1}^{n} (y_i - mx_i - c) = 0 \; .$$

Solving these equations for m and c we find, on suppressing the subscript i, that

$$m = \frac{n\Sigma xy - \Sigma x \Sigma y}{n\Sigma x^2 - (\Sigma x)^2} \; , \tag{19.5}$$

and

$$c = (\Sigma y - m\Sigma x)/n \; , \tag{19.6}$$

or, alternatively,

$$c = \frac{\Sigma y \Sigma x^2 - \Sigma x \Sigma xy}{n\Sigma x^2 - (\Sigma x)^2} \; .$$

Example 1: In example 2 of Section 19.2 the equation of a straight line relating $1/v$ to $1/S$ had to be found in order that the physical constants a and b could be determined. Letting $x = 1/S$ and $y = 1/v$ the data points

x	0.200	0.143	0.100	0.067	0.050	0.033
y	0.182	0.143	0.118	0.100	0.091	0.080

are to be fitted to the straight line $y = mx + c$.

The terms required by the method of least squares may be set out in tabular form, thus,

x^2	x	y	xy
0.0400	0.200	0.182	0.0364
0.0204	0.143	0.143	0.0204
0.0100	0.100	0.118	0.0118
0.0045	0.067	0.100	0.0067
0.0025	0.050	0.091	0.0046
0.0011	0.033	0.080	0.0026
Σ 0.0785	0.593	0.714	0.0825

$$n = 6$$

From (19.5)

$$m = \frac{6(0.0825) - (0.593)(0.714)}{6(0.0785) - (0.593)^2} = 0.59989 \ ,$$

and from (19.6),

$$c = [(0.714) - (0.59989)(0.593)]/6 = 0.059629 \ .$$

The graphical approach used in example 2 of Section 19.2 resulted in the values 0.6 and 0.06, respectively.

The method of least squares will fit a straight line through *any* set of points even when the relationship between the variables is not linear. What is now required is a means of telling whether the data points actually do lie close to a straight line. This could be decided by drawing a graph, but an alternative numerical way is to examine the **correlation coefficient**. The correlation coefficient, r, is calculated using

$$r = \frac{n\Sigma xy - \Sigma x \Sigma y}{\sqrt{([n\Sigma x^2 - (\Sigma x)^2] [n\Sigma y^2 - (\Sigma y)^2])}} \ , \qquad (19.7)$$

or $\qquad r = m \sqrt{\left(\dfrac{n\Sigma x^2 - (\Sigma x)^2}{n\Sigma y^2 - (\Sigma y)^2}\right)} \ ,$

where m is given by (19.5). The value of r will lie between -1 and $+1$ and have the same sign as m, the slope of the straight line given by the method of least squares. The closer r is to -1 or $+1$ the more likely it is that the data can be represented by a straight line. In practice fitting a straight line through n points should only be considered when the correlation coefficient satisfies

$$\frac{2}{\sqrt{(n-1)}} \leqslant |r| \leqslant 1 \ .$$

In example 1, $\Sigma y^2 = 0.0921$ and the value of the correlation coefficient r is 1.00. (A value close to one was to be expected after the comparison made with the graphical solution.)

Example 2: Examine the data

x	0.5	1.0	1.5	2.0	2.5	3.0	3.5	4.0
y	200	100	66	50	40	33	28	25

to see if a linear relationship exists between, x, the time after midday and, y, the pollen count.

The sums required to calculate the correlation coefficient r, given by (19.7), are obtained from the table

x^2	x	xy	y	y^2
0.25	0.5	100	200	40000
1.00	1.0	100	100	10000
2.25	1.5	99	66	4356
4.00	2.0	100	50	2500
6.25	2.5	100	40	1600
9.00	3.0	99	33	1089
12.25	3.5	98	28	784
16.00	4.0	100	25	625
$n = 8$ Σ 51.00	18.0	796	542	60954

Then from (19.7)

$$r = \frac{n\Sigma xy - \Sigma x \Sigma y}{\sqrt{([n\Sigma x^2 - (\Sigma x)^2] [n\Sigma y^2 - (\Sigma y)^2])}}$$

$$= \frac{8.796 - 18.542}{\sqrt{([8.51 - 18^2] [8.60945 - 542^2])}}$$

$$= -0.84 \ .$$

This is not all that close to -1 so a straight line is unlikely to be a good fit; however, as $2/\sqrt{(n-1)} = 0.76$ a linear relationship cannot be ruled out. The points are shown in Fig. 19.8, and, from the table's xy-column, the relationship between the variables is probably given by $xy = 100$.

By making use of the distinctive character of a straight line it has been shown how it is possible to verify that experimental data conform to one of several types of relationship. The equation of the best straight line fitting the data can be found from a graph or, more accurately, by the method of least squares, enabling physical constants in the relationship to be determined.

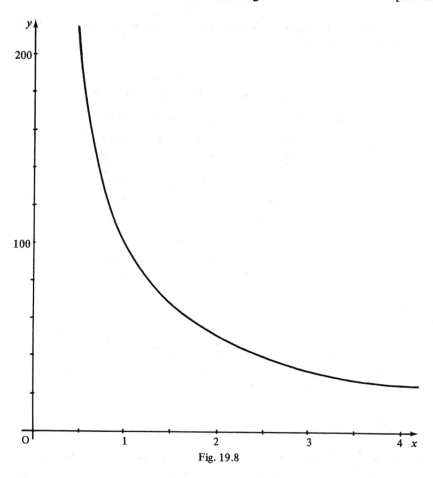

Fig. 19.8

PROBLEMS

1. The following table gives levels of iodine-131 found in the milk of cows on five dates approximately two months after a nuclear test.

x Date in December	4	6	8	10	12
y Content of iodine	280	230	170	135	80

Plot the points on a graph and verify that a linear relationship between x and y probably exists. Find the equation giving y in terms of x.

(Strathclyde 1972, amended)

2. The amount of food required by an animal each day depends linearly on its live weight. Verify this statement for deer, for which the following data are available.

Live weight W, kg.	36	41	45	52	56
Food intake F, kg. (dry weight)	2.1	2.8	3.5	4.0	4.6

Determine the equation giving the food intake F as a function of the deer's live weight W.

3. (i) Part of a table of values of y for certain values of x is:

x	0	1	4
y	2.5	2.0	0.5 .

Show that y could be a linear function of x. Hence estimate the value of y when x is 2.

(ii) A more complete extract of the table is:

x	0	1	4	6
y	2.5	2.0	0.5	0.25 .

Show that y cannot now be a linear function of x; but, by writing $u = x^2$ and $v = 1/y$, show that there could be a relation between x and y of the form $y = 1/(a + bx^2)$. What value of y is predicted by this relationship when x is 2? (Edinburgh 1975, amended)

4. Show that the discrete form of logistic growth, given by $$N_{t+1} = \frac{K\alpha N_t}{K + (\alpha - 1)N_t},$$ can be written as the linear equation $y = ax + b$ where $y = \dfrac{1}{N_{t+1}}$, $x = \dfrac{1}{N_t}$, $a = \dfrac{1}{\alpha}$ and $b = \dfrac{\alpha - 1}{K\alpha}$. Hence verify that the following data, giving the area covered by a colony of bacteria at daily intervals, exhibit logistic growth

t, days	0	1	2	3	4	5	6	7	8	9
A_t, area cm^2.	1.0	2.4	5.8	12.6	24.0	37.4	48.3	54.7	57.8	59.1 .

Estimate the final area covered by the colony.

5. The yield from a crop of tomatoes grown in 10 m^2 plots depends on the number of plants in the plot. The seedsman claims that the maximum yield for this particular variety is 20 kg/m^2.

Verify, by means of a suitable straight line graph, that the data given below support a relationship between the yield, y, of a plot and the density of planting, d, given by

$$y = 200 - A(B - d)^4$$

where A and B are constants.

d	8	9	10	11	12	13
y	94	184	199	172	195	144

Find appropriate values of A and B from the graph. What number of plants should be grown in each plot in order that the maximum yield be attained? Comment on the observed and predicted yield when $d = 11$.

6. The live weight w, in kilograms, of six sheep and the surface area s, in square metres, of their bodies are given in the following table:

w	41.0	44.5	48.5	52.0	56.0	60.5
s	1.20	1.24	1.30	1.36	1.42	1.50 .

It is believed that s and w are related by a formula of the form $s = kw^\alpha$, where k and α are constants. Verify this graphically and find the appropriate values of k and α.

7. The following recordings of the number of bacteria, N, present in a culture at time, t hours, were made:

t	0	2	4	6	8
N	1000	1400	2000	2800	3600 .

Show that these readings satisfy a law of the form $N = Ae^{bt}$ and determine A and b graphically. Predict the number of bacteria present when $t = 16$ hours.

8. The spread of a tree fungus was observed over a period of three hundred days. The percentage area of the tree covered at different times was as follows:

Time (days)	20	40	60	90	120	180	240	300
Percentage area covered by fungus	1.0	1.3	1.6	2.2	4.7	11.4	33.6	95.7

By drawing graphs of

(i) the percentage area against time,

(ii) the logarithm of the percentage area against time,

determine which of these variables is more likely to be linearly related to time and find the associated linear relationship. Use this relationship to estimate the percentage area of the tree covered by the fungus after one hundred days.   (Strathclyde, 1979)

9. Each day newly spawned fertile eggs are placed in an incubation tank at a fish hatchery. The number of fish N which hatch t days later are recorded below:

t	8	10	12	14	16	18	20	22
N	1	20	410	2040	2050	400	15	0 .

Theoretically the number of fish hatching should be given by a relationship of the form

$$N = b \exp[-c(t - a)^2]$$

where a, b and c are positive constants.

Plot the value of N against t on a graph and hence estimate a, the day on which most fish hatch. Show that if the relationship between N and t is given by the form above, then a plot of $\ln N$ against $(t - a)^2$ will lie on a straight line. Verify this for the given data and determine appropriate values for b and c.

10. The fraction f of a virus surviving a dose d of radiation is given by the **sigmoid survival function**

$$f = 1 - [1 - e^{-kd}]^\alpha$$

where k and α are positive constants. This function incorporates the feature that doses below a certain level have little effect on the virus, unlike that predicted by exponential decay.

Expand $[1 - e^{-kd}]^\alpha$ in terms of e^{-kd}, using the binomial theorem, and deduce that, for large doses,

$$f \cong \alpha e^{-kd} .$$

The values of the constants k and α can thus be found by examining the linear form

$$\ln f = \ln \alpha - kd$$

for large values of d.

For a particular virus the fraction surviving a dose d of radiation is given in the following table:

d	0	50	100	150	200	250	300	350	400
f	1.00	0.96	0.78	0.57	0.38	0.25	0.16	0.096	0.059 .

Verify that a graph of $\ln f$ against d is linear, for large values of d, and hence determine the appropriate values of k and α.

11. The following table shows the flowering dates, over a period of seven years, of Cox's Orange Pippins measured D days after the 31st March, and the corresponding monthly mean temperature, T, for the December preceding flowering.

D days	48	57	56	61	60	65	67
$T\,°C$	0.0	1.7	2.1	2.3	3.6	3.7	3.8

Calculate the correlation coefficient and comment on the relationship it suggests.

Find, using the method of least squares, the best straight line of the form $D = aT + b$ through the points. Use the resulting linear relationship to estimate the flowering date corresponding to a December which has a mean temperature of $3°C$.

Briefly suggest other factors that might influence the flowering date.

(Strathclyde, 1977)

12. Five predator mites were transferred to new habitats. During the first day each mite was kept at a different temperature and its area of discovery recorded.

Area of discovery, A cm^2	31	49	42	53	71
Temperature, $T\,°C$	10	12	14	16	18

Calculate the correlation coefficient and comment on the suitability of a linear relationship between A and T. Determine, by the method of least squares, a linear relationship between T and A, and hence find the predicted area of discovery at $11°C$.

(Strathclyde, 1976)

13. The growth of a bacterial colony was observed over a period of seven hours. The number of counts per cubic centimetre in successive samples during its growth was recorded.

Incubation time (mins.)	20	40	60	90	120	180	240	300	360	420
Bacterial count	47	62	73	103	220	537	1580	4500	9200	12800

Plot

(i) time of incubation against bacterial count,

(ii) time of incubation against the logarithm of the bacterial count, on separate graphs. Which of these two do you consider to be closest to a linear relationship? Calculate the best straight line through the points you have chosen (either count or log(count)) by the method of least squares.

(Strathclyde, 1977)

14. In a study of the theory of biological clocks fruit fly pupae were exposed to a single pulse of blue light. The rate of emergence of the adult flies was then recorded for various temperatures:

E, emergence rate/hr	6	7	10	16	23	32
T, temperature °C	10	15	20	25	30	35

Use the method of least squares to fit a relationship of the form $E = ae^{bT}$ to these data and estimate the constants a and b. Evaluate the correlation coefficient and comment on the suitability of the relationship.

(Strathclyde, 1977)

15. In an experiment designed to investigate and compare the clotting times of blood plasma from normal subjects and plasma from subjects suffering from a blood-clotting disease, five concentrations of each of the two types of plasma were made up. The clotting time, in seconds, of each concentration of plasma was measured, and the results for the two types of plasma are shown below.

(A concentration of 100% corresponds to undiluted plasma.)

Plasma from Normal Patients		Plasma from Diseased Patients	
Concentration	Clotting time	Concentration	Clotting time
%	secs.	%	secs.
0.01	75	0.01	95
0.1	62	0.1	82
1.0	50	1.0	71
10.0	41	10.0	58
100.0	30	100.0	52

For each plasma type, plot the clotting time against the logarithm of the concentration, confirming that the logarithmic transformation is a suitable one, and then calculate, using the method of least squares, the best straight lines through these points.

Comment briefly on the lines obtained and estimate the concentration of the diseased plasma which would produce a clotting time equal to that of 0.5% normal plasma.   (Strathclyde, 1978)

CHAPTER 20

Permutations and combinations

20.1 INTRODUCTION

The theory of permutations and combinations is used extensively in probability and statistics. **Permutations** of a set of different objects are the (number of) ways of arranging the objects where the order is important. More often subsets of a required size are selected from the given objects in order to make different permutations. If we have three objects a, b and c then the permutations of the objects taken

(i) two at a time will be : ab, ba, ac, ca, bc, cb,

(ii) three at a time will be : abc, bca, cab, cba, bac, acb.

On the other hand, **combinations** of different objects are the (number of) collections or groups that can be made from the objects irrespective of order. Thus for the three objects a, b and c the combinations taken

(i) two at a time will be the groups : ab, ac, bc,

(ii) three at a time will be the group : abc.

Experimental situations may require the number of combinations or permutations of objects. For example, if a large number of fish are kept under observation in a tank and, during a period of observation, the experimenter is interested in which groups of three fish arrive at an algae source, then the possible outcomes of the experiment would be the combinations of the fish taken three at a time. If the experimenter was interested in the order in which the fish arrive to make groups of three fish then the possible outcomes would be the permutations of the fish taken three at a time.

20.2 PERFORMING SUCCESSIVE OPERATIONS

Suppose we wish to perform a set of r operations consecutively, and the first operation can be performed in one of n_1 ways, the second in one of n_2 ways and the rth in one of n_r ways. Then the total number of different ways to choose from in order to perform the operation will be

$$n_1 \times n_2 \times n_3 \times \ldots \ldots \times n_r \; . \tag{20.1}$$

This can be demonstrated by an example. In a study of development rates, the development times of the crustacean *Acanthocylops* during larval and adult stages were recorded. The larvae were kept at one of four different temperatures L_1, L_2, L_3 and L_4 until the adult stage was reached, then the adults were kept at one of three different temperatures A_1, A_2, and A_3 until death. The total number of ways of performing the experiment are shown schematically:

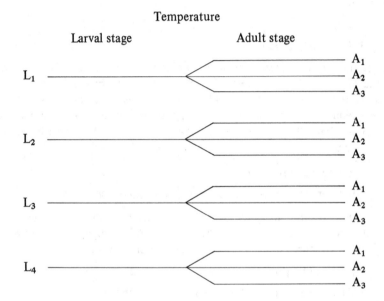

Temperature

Larval stage   Adult stage

Instead of listing all possible arrangements, application of formula (20.1) with $n_1 = 4$ and $n_2 = 3$ gives 12 as the total number of possible arrangements.

20.3 PERMUTATIONS AND COMBINATIONS

Using the formula for successive operations it is possible to derive an expression for the number of permutations obtained when r objects are to be chosen from n. The first object may be selected from n, the second from the $(n-1)$ remaining objects, and so on until the rth object will be selected from $(n-r+1)$ remaining objects. Therefore the total number of ways of arranging the r objects in order is

$$n \times (n-1) \times (n-2) \times \ldots \ldots \times (n-r+1) = \frac{n!}{(n-r)!} \; .$$

This final expression is denoted by the shorthand notation ^nP_r, thus

$$^nP_r = \frac{n!}{(n-r)!} \ . \tag{20.2}$$

The number of combinations of n different objects taken r at a time, denoted by the shorthand notation ^nC_r, is given by

$$^nC_r = \frac{n!}{r!(n-r)!} \ . \tag{20.3}$$

The quantity ^nC_r is identical to $\binom{n}{r}$, the coefficient of $x^r y^{n-r}$ in the binomial expansion of $(x+y)^n$. This is to be expected since when $(x+y)^n$ is multiplied out the number of times x^r will combine with y^{n-r} will be the number of ways of choosing r objects from n.

Notice that ^nC_r differs from ^nP_r by division by a factor $r!$ This can be explained as follows. Suppose we have the particular arrangement

$$a_1 a_2 a_3 \ \ldots \ldots \ a_r$$

then these r objects may be arranged in $r!$ ways, each way containing the same objects in a different order. For instance when $r = 3$ then $a_1 a_2 a_3$ contains the same objects as the arrangements $a_1 a_3 a_2$, $a_2 a_1 a_3$, $a_2 a_3 a_1$, $a_3 a_1 a_2$ and $a_3 a_2 a_1$. Thus if ^nC_r is the number of groups of r objects, multiplying this by $r!$ must give ^nP_r. Hence

$$^nP_r = r! \ ^nC_r \ ,$$

and rearranging gives

$$^nC_r = \frac{^nP_r}{r!} = \frac{n!}{r!(n-r)!} \ .$$

Example 1: During its life cycle a parasite must use a snail as its first host, a fish as its intermediate host, and man as the final host. If three species of snail and six species of fish are available as hosts in how many ways can a parasite complete its life cycle?

Since there are three ways of completing the intra-snail stage and six ways of completing the intra-fish stage then $6 \times 3 = 18$ will be the total number of ways of completing the cycle.

Example 2: Mice who have had tumours removed are to be given a treatment consisting of two courses of radiotherapy at different intensities. If there are ten possible intensities available and the order in which intensities are administered is important, how many mice will be needed if each treatment is to be carried out?

How many mice are needed when the order in which the intensities is given is not important?

In the first case the order of intensity is important, that is, an intensity i_1 followed by an intensity i_2 is not the same as intensity i_2 followed by intensity i_1. Therefore the total number of mice required is $^{10}P_2 = \dfrac{10!}{8!} = 90$, the number of ways of arranging two intensities out of a possible ten intensities.

When the order is unimportant, that is, intensity i_1 followed by intensity i_2 is the same as i_2 followed by i_1, then the number of mice required is $^{10}C_2 = \dfrac{10!}{2!8!} = 45$, the number of ways of combining two intensities out of ten.

20.4 LIKE AND UNLIKE OBJECTS

Suppose we have n objects composed of p objects of one type and q objects of another. It is required to find the number of different ways of arranging the objects in a row. The total number of arrangements if the objects were all different would be

$$n \times (n-1) \times (n-2) \times \ldots 3 \times 2 \times 1 = n! \ .$$

A particular arrangement might be: $a_1 a_2 \ldots a_p b_1 b_2 \ldots b_q$.

If the type b objects are fixed then the p objects of type a may be interchanged in $p!$ ways. If the type a objects are fixed then the q objects of type b may be interchanged in $q!$ ways. Thus if the a's are all the same and the b's are all the same then these $(p+q)$ objects can be interchanged in $p!q!$ ways to give the same arrangement of a's and b's. For example, the arrangement $a_1 a_2 b_1 b_2$ may be interchanged to give $a_2 a_1 b_1 b_2$, $a_1 a_2 b_2 b_1$ and $a_2 a_1 b_2 b_1$ all of which give the same arrangement when the a's are equal and the b's are equal. Since this is true for every one of the $n!$ arrangements when the n objects are different, then $n!/(p!q!)$ will be the number of different ways of arranging the objects when the a's are the same and the b's are the same.

The same approach can be extended to the case of n objects which are composed of p objects of one kind, q of another, r of another and so on to give

$$\frac{n!}{p!q!r! \ \ldots\ldots}$$

as the number of different arrangements.

Example 1: Out of twenty tulip bulbs to be planted in a row along a border six are yellow flowering, eight are red flowering, and six orange flowering. How many colour patterns could be created?

The number of colour patterns that can be obtained by varying the planting order will be

$$\frac{20!}{6!8!6!} = 116\ 396\ 280 \ .$$

Example 2: In a clinic there is a specialist for ear patients and a specialist for nose patients. If during a day four patients are to arrive, three for ear treatment and one for nose treatment, in how many ways can the duty rota for the specialists be arranged? List the arrangements.

$$\text{Number of ways of arranging duty rota} = \frac{4!}{3!1!} = 4.$$

Let E and N denote an ear and a nose patient respectively, then possible arrangements are

NEEE, ENEE, EENE, EEEN .

PROBLEMS

1. Determine the numerical value of the following expressions:

    (i) $^7P_4$,   (v) $^7P_0$,

    (ii) $^7P_3$,   (vi) $^7P_7$,

    (iii) $^7C_4$,   (vii) $^7C_0$,

    (iv) $^7C_3$,   (viii) $^7C_7$.

2. A foraging ant when placed in a maze must pass through passages labelled A, B, C in turn in order to reach the exit. The maze is designed in such a way that there are seven ways to reach A, ten ways to reach B from A, four ways to reach C from B, and six ways to reach the exit from C. In how many different ways is it possible for the ant to reach the exit?

3. At a regeneration point a cell divides to produce four new cells. Starting with one cell in a population, how many will be present after five regeneration points?

4. In a field trial on crop yield, each of the chemicals phosphorus, nitrogen and potassium can be applied to plots at one of three different concentrations. How many plots will be needed if the crop yields for every combination of phosphorus, nitrogen and potassium concentrations are to be examined?

5. Of twenty-one patients undergoing treatment nine are males and twelve are females. Five of the patients are to be selected for further treatment in one of two ways, either

    (i) one male followed by four females,

    or   (ii) two males followed by three females.

    In each case find how many ways this can be achieved.

6. Prior to examination, the packed cell volume (*PCV*) values in six Zebu and nine N'Dama cattle are measured. In how many ways can the examination take place if

    (i) all the cattle are to be examined in order of increasing *PCV* value,

    (ii) the Zebu cattle only are to be examined in order of increasing *PCV* value,

    (iii) the N'Dama cattle are to be examined in order of increasing *PCV* value followed by the Zebu cattle in order of increasing *PCV* value?

7. In a coding system for poisonous mushrooms details of each species are to be stored in a computer under a code number made up from the digits 0, 1, 2, . . ., 9. If no number is allowed to be repeated and only 3-figure numbers are to be used for each species, find the maximum number of species' details that can be stored in the computer.

8. It is believed that some of the species of birds on a land mass subsequently colonised a neighbouring island. If the order in which the species arrived on the island is important and there were twenty-seven species of birds present on the land mass, in how many different ways could ten of these species have colonised the island?

9. From a group of forty patients given chemotherapy treatment a sample of ten people are to be chosen for investigation into possible side effects. In how many different ways can such a sample be chosen?

10. Over a period of one year the growth of seven insect populations present in a grassland was monitored. If the growth rate of each population is to be compared with the growth rate of every other population, how many comparisons will be made?

11. A neuron can either respond or not respond to a signal. If six signals are transmitted in how many ways is it possible for the neuron to respond to two and only two of the signals? If *n* signals are transmitted in how many ways is it possible for the neuron to make *r* and only *r* responses?

CHAPTER 21

Probability

21.1 INTRODUCTION

The progress of bioscience has been closely linked to successful experimentation. The early scientists confined their attention mainly to experiments of the **deterministic** type. These are experiments where the observable consistently gives the same value each time the experiment takes place under the same conditions. For example, during photosynthesis the volume of carbon dioxide given off by a square centimetre of tomato plant leaf will always be the same provided the light intensity, temperature and atmospheric carbon dioxide concentration are kept constant.

The other type of experiment is the **random** or **non-deterministic** one, where each time the experiment is performed a different outcome may be obtained although the conditions are the same. The interpretation of such experiments remained undeveloped until the last century, when the techniques of probability and statistics emerged. Examples of random experiments are:

(i) the number of eggs laid by a bird each time it reproduces,

(ii) the recovery times of patients given the same treatment.

To describe a random experiment the following terminology is used:

The **random variable**, usually denoted by a capital letter, is the name given to the observable in the experiment, and it may take the value of one of several **outcomes** in a particular experiment. For example, the recovery times of patients measured in days represents a random variable which might be denoted by T, and a recovery time of nine days for a particular patient would be written as $T = 9$.

A **continuous random variable** is one which may take any value in a given range, whereas a **discrete random variable** can only take certain point values in a given range. (This may not appear to be an important distinction but will become so later, when the properties of the observable are examined.)

In our example T, the random variable denoting the number of days to recovery is discrete since it can only take integer values.

The **sample space** is the name given to the set of all possible outcomes for the random variable in the experiment. In many cases the entire sample space of the random variable cannot be listed, as the random variable may, in theory, take values up to infinity.

For the random variable T, the sample space will be the set of integers $\{1, 2, 3, \ldots\}$.

An **event** may be one, or a combination, of several of the outcomes. For instance, the event of a recovery time less than four days might be of interest. This would be a combination of the four outcomes: a recovery time of (i) one day, (ii) two days, (iii) three days and (iv) four days.

21.1.1 Historical probability

Although it is impossible to predict the exact behaviour of the random variable in experiments to be performed, its probable behaviour can be determined by observing its past. If the experiment is performed a large number of times, N, and the frequency, n, of the occurrence of an event A counted, then n/N will be a reasonable estimate of the proportion of times the event A is likely to occur in future experiments. For large values of N it is this proportion that is interpreted as the probability of the event A:

$$\text{Probability of the event A} = P(\text{event A}) = \lim_{N \to \infty} \frac{n}{N}.$$

21.1.2 Probability when the outcomes are equally likely

The historical definition of probability requires a knowledge of the past history of the observable to obtain the probability of events. However, in those experiments in which each outcome is equally likely the probability of an event A can be arrived at by taking the ratio of the number of outcomes in favour of A to the total number of possible outcomes:

$$P(\text{event A}) = \frac{\text{number of outcomes favourable to A}}{\text{total number of possible outcomes}}.$$

In these definitions the probability will be a number between 0 and 1, an event which is certain not to occur having probability zero, and an event which is certain to occur having probability one.

In an experiment with equally likely outcomes it is possible to evaluate the probabilities by using the historical probability definition. This presents no problem as the probabilities will be the same irrespective of how they are evaluated.

The following are some examples of random experiments.

Example 1: Serum of a standard concentration is repeatedly diluted by adding equal volumes of saline to give solutions with successive dilution factors of 1, 2, 4, 8, 16, ... At each level of dilution the solution is tested for an 'end-point'.

The random variable will be the dilution factor at which an end-point is recorded. It will be discrete since it can only take prescribed point values, and the sample space will be $\{1, 2, 4, 8, 16, \ldots\}$.

Example 2: A snail population is introduced to a new habitat, and daily changes in the biomass of the population are recorded as a percentage of the biomass of the original population.

The random variable will be the daily percentage change. Since the population biomass may increase indefinitely or decrease to size zero, then the random variable will be continuous with sample space the interval $(-100, \infty)$. (The idea of a percentage change as high as infinity is only for completeness since in practice such a result is unlikely!) The negative values denote a decrease in the biomass of the population.

Example 3: In a pancreatitis study 300 out of 1000 patients admitted to hospital had blood clotting times (B.C.T.) in excess of 2 minutes.

In this case the B.C.T. of a patient is the random variable. Denoting the random variable by X, the sample space of X will be the set of values in the interval $(0, \infty)$, and since X can take any value in this interval it will be a continuous random variable. From the information given and the definition of historical probability we can write

P (patient has a B.C.T. in excess of 2 minutes)

$$= P(X > 2)$$

$$= \frac{300}{1000}$$

$$= 0.3 \ .$$

Example 4: A receptor cell has six possible sites, labelled $1, 2, 3, 4, 5$ and 6, such that a free antibody is equally likely to attach itself to any site.

During the process of attachment, the label of the site that the antibody becomes attached to will be the random variable. Denoting the random variable by S, it will be discrete, having sample space $\{1, 2, 3, 4, 5, 6\}$. An attachment to site 4 will have one favourable outcome out of a possible six, therefore from the definition of probability for events that are equally likely,

P(attachment to site 4)

$$= P(S = 4) = \frac{1}{6} \ .$$

The event of the antibody attaching itself to an even-numbered site will have three favourable outcomes out of a possible six, therefore

P (attachment to an even-numbered site)

$$= P(S = 2 \text{ or } 4 \text{ or } 6) = \frac{3}{6} = \frac{1}{2} \ .$$

Example 5: Two mice are to be selected from a group of fifteen female and seven male mice. If each mouse is equally likely to be chosen what is the probability that the two mice selected will be

(i) both female?

(ii) both male?

In this case the random variable is the sex of the pair of mice selected. If F and M denote female and male mice respectively then the sample space will be the possible combinations of the sexes of the mice {FF, MF, MM}.

(i) P(both mice are female)

$$= \frac{\text{number of events in favour of both being female}}{\text{total number of possible events}} = \frac{^{15}C_2}{^{22}C_2} = \frac{5}{11}$$

since, as explained in Section 20.3, $^{15}C_2$ and $^{22}C_2$ are the number of ways of selecting 2 female mice from 15 female mice and 2 mice from 22 mice respectively.

(ii) P(both mice are male)

$$= \frac{\text{number of events in favour of both being male}}{\text{total number of possible events}} = \frac{^7C_2}{^{22}C_2} = \frac{1}{11}.$$

21.2 LAWS OF PROBABILITY

The following laws of probability are the building blocks of even the most complicated probability models. Their proofs are explained on the basis of the definition of probability when the outcomes are equally likely, but the laws also hold for historical probability.

1. In an experiment, if $P(A)$ is the probability of an event A, then the probability of the event "not A" is

$$P(\bar{A}) = 1 - P(A) ,\tag{21.1}$$

where $\bar{A}$ is used to denote the event "not A".

If the event A is favoured n_1 times out of a possible n outcomes then

$$P(A) = \frac{n_1}{n} .$$

However, the event "not A" must be favoured on the other $n - n_1$ outcomes, so

$$P(\bar{A}) = \frac{n - n_1}{n} = 1 - \frac{n_1}{n} = 1 - P(A) .$$

This is sometimes referred to as the **law of complementary probability**.

2. In an experiment, if A and B are two mutually exclusive events then the probability of either event A or event B occurring is

$$P(A \text{ or } B) = P(A) + P(B) \ . \tag{21.2}$$

By **mutually exclusive events** it is meant that if one event A occurs during the experiment then this excludes event B from occurring. For instance when a baby is born there are two possible outcomes, male or female, but if a male baby is the outcome then it is impossible for a female baby to be the outcome.

Let event A be favoured on n_1 outcomes and event B on n_2 outcomes out of a possible n outcomes. Then

$$P(A) = \frac{n_1}{n} \text{ and } P(B) = \frac{n_2}{n} \ .$$

The event A or B must be favoured on $n_1 + n_2$ outcomes therefore

$$P(A \text{ or } B) = \frac{n_1 + n_2}{n} = \frac{n_1}{n} + \frac{n_2}{n} = P(A) + P(B) \ .$$

If A and B are not mutually exclusive events then

$$P(A \text{ or } B) = P(A) + P(B) - P(A \text{ and } B) \ . \tag{21.3}$$

As an example of two events which are not mutually exclusive consider sowing a pea plant seed in which the new plant may be tall or dwarfed and its peas may be smooth or wrinkled. If A is the event of the plant being tall and B is the event of the plant having smooth peas, then it is possible to have a plant which is tall and has smooth peas, so that event A occurring does not exclude event B. The events A and B are therefore not exclusive.

To arrive at equation (21.3) consider three possible events: A occurring, B occurring, and the event A and B occurring simultaneously, with respective numbers of outcomes in favour n_1, n_2 and n_3.

Now the event A or B can be arrived at by one of three mutually exclusive events. These are: A occurring without B, B occurring without A, and A and B occurring simultaneously. Therefore adding these probabilities gives

$$P(A \text{ or } B) = \frac{n_1 - n_3}{n} + \frac{n_2 - n_3}{n} + \frac{n_3}{n}$$

$$= \frac{n_1}{n} + \frac{n_2}{n} - \frac{n_3}{n}$$

$$= P(A) + P(B) - P(A \text{ and } B) \ .$$

The above laws, in equations (21.2) and (21.3), are sometimes referred to as the **addition laws of probability**.

3. Let A and B be events in an experiment such that the occurrence of B influences the occurrence of A. Then the probability of A occurring when it is known that B has occurred is

$$P(A|B) = \frac{P(A \text{ and } B)}{P(B)} . \tag{21.4}$$

Note the introduction of the vertical bar to denote A given B (or A conditional on B).

Let the three events: A occurring, B occurring, and A and B occurring simultaneously, have respectively n_1, n_2 and n_3 outcomes in their favour, so that

$$P(A) = \frac{n_1}{n}, \quad P(B) = \frac{n_2}{n}, \quad P(A \text{ and } B) = \frac{n_3}{n} .$$

Now if B has occurred then for A to occur it can only do so in n_3 ways out of the n_2 possible ways in which B occurred, therefore

$$P(A|B) = \frac{n_3}{n_2} = \frac{n_3/n}{n_2/n} = \frac{P(A \text{ and } B)}{P(B)} .$$

This is the **law of conditional probability** and it is sometimes rearranged to give

$$P(A \text{ and } B) = P(A|B) P(B) . \tag{21.5}$$

Note that if A and B are mutually exclusive, then A cannot happen when B has occurred, thus $P(A|B) = 0$, and in this case we get

$$P(A \text{ and } B) = 0 .$$

4. The previous law of conditional probability is also valid when the situation is extended to two experiments and the probability of A occurring in one experiment and B occurring in the other is required. For the case that the event B does not influence the occurrence of event A then $P(A|B) = P(A)$ and the events are called **independent**. Thus, from equation (21.5) the probability of A and B occurring becomes

$$P(A \text{ and } B) = P(A) P(B) .$$

At first it may be found difficult to formulate the probability of an event when given the probabilities of the outcomes. This can be overcome if care is taken to write out a probability statement, stating clearly the combination of outcomes leading to the event. To such a statement the laws of probability can be applied, depending on the wording. In general the word *OR* between combinations of outcomes will lead to the addition of the separate probabilities of outcome, and the word *AND* to their multiplication. This approach is adopted in the following applications.

21.3 APPLICATIONS OF THE PROBABILITY LAWS

Example 1: Ten aquatic snails of species X and fifteen of species Y are released into a pond containing fifty snails of species Z. The snails are recaptured one at a time for inspection. If each snail is equally likely to be captured and after capture it is returned to the pond, calculate the following probabilities:

   (i) capturing a snail of species X,

   (ii) capturing a snail of species Y,

   (iii) capturing a snail of species Z,

   (iv) capturing a snail of species X or Y.

On two successive captures what is the probability of

   (v) each snail being of species X,

   (vi) one snail being of species X and the other of species Z?

If a snail is no longer returned to the pond after recapture what is the probability on two successive captures of

(vii) each snail being of species X,

(viii) one snail being of species X and the other of species Z?

(i) $P(\text{capturing a snail of species X}) = \dfrac{\text{number of snails of species X}}{\text{total number of snails}}$

$$= \frac{10}{10 + 15 + 50} = \frac{10}{75} = \frac{2}{15} .$$

(ii) $P(\text{capturing a snail of species Y}) = \dfrac{15}{75} = \dfrac{1}{5} .$

(iii) $P(\text{capturing a snail of species Z}) = \dfrac{50}{75} = \dfrac{2}{3} .$

(iv) In this case the addition law of probability for mutually exclusive events is used.

$P(\text{capturing a snail of species X OR Y})$

$\qquad = P(\text{capturing a snail of species X}) +$

$\qquad P(\text{capturing a snail of species Y})$

$$= \frac{10}{75} + \frac{15}{75} = \frac{1}{3} .$$

(v) P(each snail being of species X)

$= P$(first snail captured is of species X AND
second snail captured is of species X)

$= P$(first snail captured is of species X) $\times$

P(second snail captured is of species X)

$= \dfrac{10}{75} \cdot \dfrac{10}{75} = \dfrac{4}{225}$.

Note the AND formulation and the independence of the two events which leads to the use of the multiplication law of probability.

(vi) P(one snail being of species X AND the other of species Z)

$= P$(first snail captured is of species X AND
second snail captured is of species Z OR
first snail captured is of species Z AND
second snail captured is of species X)

$= P$(first snail captured is of species X AND
second snail captured is of species Z) $+$

P(first snail captured is of species Z AND
second snail captured is of species X)

$= P$(first snail captured is of species X) $\times$

P(second snail captured is of species Z) $+$

P(first snail captured is of species Z) $\times$

P(second snail captured is of species X)

$= \dfrac{10}{75} \cdot \dfrac{50}{75} + \dfrac{50}{75} \cdot \dfrac{10}{75} = \dfrac{8}{45}$.

(vii) This case is similar to that in part (v), only it must be taken into consideration that the snail is not returned after capture. Conditional probability is used when the probabilities are multiplied:

P(each snail being of species X)

$= P$(first snail captured is of species X) $\times$

P(second snail captured is of species X | first snail captured
is of species X)

$= \dfrac{10}{75} \cdot \dfrac{9}{74} = \dfrac{3}{185}$.

Note that in the evaluation of conditional probability there are only a total of 74 possible snails for capture since one snail was removed from the pond after the first capture.

(viii) Once again using the law of conditional probability, when the probabilities are multiplied this gives

P(one snail being of species X AND the other of species Z)

$= P$(first snail captured is of species X) $\times$

P(second snail captured is of species Z | first snail captured is of species X) $+$

P(first snail captured is of species Z) $\times$

P(second snail captured is of species X | first snail captured is of species Z)

$$= \frac{10}{75} \cdot \frac{50}{74} + \frac{50}{75} \cdot \frac{10}{74} = \frac{20}{111}.$$

Example 2: The probability of detecting *Heterodera* cysts within a soil sample using a flotation technique is 0.7. What is the probability of not detecting the cysts if

(i) one soil sample is taken?

(ii) two soil samples are taken?

(i) P(not detecting cysts in one soil sample)

$= 1 - P$(detecting cysts in one soil sample)

$= 1 - 0.7 = 0.3.$

(ii) P(not detecting cysts in two soil samples)

$= P$(not detecting cysts in first soil sample AND not detecting cysts in second soil sample)

$= P$(not detecting cysts in first soil sample) $\times$

P(not detecting cysts in second soil sample)

$= 0.3 \times 0.3 = 0.09.$

Example 3: In the previous problem how good must the detection process be in a soil sample to ensure that the probability of not detecting cysts in two soil samples is less than 0.16?

Let p be the probability of detection in a soil sample then

P(not detecting cysts in two soil samples) $< 0.16.$

Therefore $(1 - p)(1 - p) < 0.16.$

Choosing the positive square root of the R.H.S. we get

$(1 - p) < 0.4,$

and after rearrangement

$p > 0.6.$

Example 4: The probability of a late spring frost causing blossom damage to an apple tree is 0.6. In a season two frosts are experienced; what is the probability of an apple tree being injured in this period?

This can be solved by one of two methods. Either the addition law of probability for events which are not mutually exclusive can be used:

P(tree is injured)

  $= P$(tree is injured by the first frost OR
    tree is injured by the second frost)

  $= P$(tree is injured by first frost) $+$

   P(tree is injured by second frost) $-$

   P(tree is injured by first and second frost)

  $= 0.6 + 0.6 - 0.6 \times 0.6 = 0.84.$

Alternatively, using the law of complementary probability:

P(tree is injured)

  $= 1 - P$(tree is not injured)

  $= 1 - P$(tree is not injured in first frost AND
     not injured in second frost)

  $= 1 - P$(tree is not injured in first frost) $\times$

   P(tree is not injured in second frost)

  $= 1 - 0.4 \times 0.4 = 0.84.$

Example 5: From past experience it is known that the independent observables B_1, B_2 and B_3 from a patient's blood sample successfully predict recurring cancer with probability 0.8, 0.7 and 0.4 respectively. What is the detection rate of recurrence for a given patient if

(i) only B_1 and B_2,

(ii) all three observables

are recorded?

From the data given we can deduce

  $P(B_1$ does not predict recurrence)

   $= 1 - P(B_1$ does predict recurrence)

   $= 1 - 0.8 = 0.2.$

Similarly $P(B_2$ does not predict recurrence) $= 0.3$

and $P(B_3$ does not predict recurrence) $= 0.6.$

  (i) P(detection of recurrence using only B_1 and B_2)

   $= 1 - P$(no detection of recurrence)

$$= 1 - P(\text{B}_1 \text{ does not predict recurrence AND}$$
$$\text{B}_2 \text{ does not predict recurrence})$$

$$= 1 - P(\text{B}_1 \text{ does not predict recurrence}) \times$$
$$P(\text{B}_2 \text{ does not predict recurrence})$$

$$= 1 - 0.2 \times 0.3 = 0.94.$$

(ii) $P(\text{detection of recurrence using all three observables})$

$$= 1 - P(\text{no detection of recurrence})$$

$$= 1 - P(\text{B}_1 \text{ does not predict recurrence AND}$$
$$\text{B}_2 \text{ does not predict recurrence AND}$$
$$\text{B}_3 \text{ does not predict recurrence})$$

$$= 1 - 0.2 \times 0.3 \times 0.6$$

$$= 1 - 0.036 = 0.964.$$

Example 6: A carrier comes into contact with three susceptibles. If the probability of transmitting the infection is 0.9, what is the probability of all three susceptibles being infected?

$P(\text{all three susceptibles are infected})$

$$= P(\text{first susceptible is infected AND}$$
$$\text{second susceptible is infected AND}$$
$$\text{third susceptible is infected})$$

$$= P(\text{first susceptible is infected}) \times$$

$$P(\text{second susceptible is infected}) \times$$

$$P(\text{third susceptible is infected})$$

$$= 0.9^3 = 0.729.$$

Example 7: A spermatozoon which comes into contact with an egg has probability p of fertilising it. What is the probability that the egg is not fertilised until the fourth spermatazoon comes into contact with it?

$P(\text{egg is fertilised by fourth spermatazoon})$

$$= P(\text{egg is not fertilised by the first AND}$$
$$\text{not by the second AND not by the third AND}$$
$$\text{egg is fertilised by the fourth spermatazoon})$$

$$= P(\text{egg is not fertilised by the first}) \times$$

$$P(\text{egg is not fertilised by the second}) \times$$

$$P(\text{egg is not fertilised by the third}) \times$$

$$P(\text{egg is fertilised by the fourth})$$

$$= (1 - p)(1 - p)(1 - p)p$$

$$= (1 - p)^3 p.$$

PROBLEMS

1. In a study of the mating habits of pigeons, the total number of eggs, T, laid by six female pigeons is to be recorded. State whether T is discrete or continuous. If each pigeon can lay a maximum of three eggs what will be the sample space of T?

2. Give the sample space and type of random variable in each of the following experiments:

   (i) recording the diameters of tumours in mice seven days after the mice have been injected with tumour cells,

   (ii) counting the number of unacceptable pineapples in boxes each containing twenty pineapples,

   (iii) measuring the weight change in calves fed on a milk substitute diet,

   (iv) recording the ratio of female ticks to the total number of ticks found on sheep.

3. On average 80% of the tomato seeds in a packet will germinate. Find the probability that

   (i) one seed taken from a packet will germinate,

   (ii) none of five seeds taken from a packet will germinate,

   (iii) at least one of five seeds taken from a packet will germinate.

4. At cell division it is known that one and only one of the six genes: A, B, C, D, E and F on a chromosome will have a mutation defect. If four such chromosomes undergo cell division what is the probability that

   (i) the gene defects occur with gene A on each chromosome,

   (ii) one only of the gene defects occurs with gene E?

5. The probability of an ultrasonic scan detecting twins during the early stages of pregnancy is 0.9. What is the probability that at least one of two such scans will correctly detect twins?

6. The probabilities of three components in a dialyser being replaced within a year are $\frac{1}{9}$, $\frac{1}{12}$ and $\frac{1}{15}$. If the components act independently of each other, what is the probability of replacing

   (i) at least one component,

   (ii) all three components,

   during a year?

7. The probability of a weather ballon being recovered is $\frac{1}{9}$. If three balloons are released what is the probability of recovering

    (i) one balloon only,

    (ii) all three balloons,

    (iii) at least one balloon?

8. A phagocytic white blood cell will attack three different types of bacteria with probability $\frac{1}{4}$, $\frac{1}{6}$ and $\frac{1}{8}$. If the cell is exposed to these three types of bacteria and it acts independently, calculate the probability that it attacks only two of the three bacteria.

9. A nerve cell when stimulated responds by sending a signal of type A or type B to the brain. If a type A signal is sent with probability 0.3 and a type B signal with probability 0.7, calculate the probability that the first type B signal to be received by the brain occurs at

    (i) the second stimulus,

    (ii) the fourth stimulus,

    (iii) the nth stimulus.

10. There are m molecules of substance A and n molecules of substance B in a unit volume. Owing to collisions, molecules are ejected at random and are unable to re-enter the unit volume. Find the probability that

    (i) the first ejected molecule is of substance A,

    (ii) the first two ejected molecules are of substance A,

    (iii) the second ejected molecule is of substance B, given that the first was a substance A molecule,

    (iv) the third ejected molecule is of substance A.

11. (i) A rat can be one of three genotypes A, B and C. Twin rats can therefore be of six types (A,A), (A,B), (A,C), (B,B), (B,C) or (C,C), and the probabilities associated with these pairings are respectively

    $$\frac{9}{64}, \frac{3}{16}, \frac{1}{32}, \frac{5}{16}, \frac{3}{16} \text{ and } \frac{9}{64} .$$

    Calculate the probabilities that of two twin rats

    (a) at least one of the two rats is of type A,

    (b) at least one of the two rats is either of type A or of type B,

    (c) the two rats are of the same type.

(ii) The probability that a female of a certain species shows a positive reaction to a drug is $\frac{1}{2}$ and the corresponding probability for a male is $\frac{1}{3}$. If each member of a group of three females and one male is administered the drug, what is the probability that exactly two members of the group show a positive reaction?

(iii) Articles are produced by independent operations of a machine, and at any operation the probability is 0.1 that the article is defective.

What is the probability that the first defective article is produced

(a) at the fifth operation,

(b) before the fifth operation,

(c) subsequent to the fifth operation?

(Strathclyde, 1971)

12. On average 4 out of 5 eggs of species A and 7 out of 10 eggs of species B will hatch within one day when kept in an incubator.

(i) If one egg of each species is incubated for a day, what are the probabilities that neither, just one, at least one and both will hatch?

(ii) If four eggs (two of each species) are placed in the incubator what are the probabilities that just one of each species, the same number of each species, at least one of each species hatch in one day?

(iii) If one egg of each species is placed in the incubator, and just one of them hatches in a day, what is the probability that it is of species A?

(Strathclyde, 1972)

CHAPTER 22

Populations and probability distributions

22.1 INTRODUCTION

In this chapter further use is made of probability as a means of describing results from experiments. Before commencing it is important to be aware of the distinction between a population and a sample. Suppose an experimenter wishes to discover the reaction of mice to transplanted tumour cells. The observable to be measured to detect a reaction is to be the white blood cell (W.B.C.) count in a fixed volume of plasma taken from a treated mouse. Ideally the experimenter would like to take every mouse that exists and record the W.B.C. count. This collection of results would then represent a **population** of W.B.C. counts. However, to use every mouse that exists would be impossible, and the best the experimenter can hope for is to obtain W.B.C. counts from some mice. The collection of results then represents a **sample** of W.B.C. counts.

It is important to remember that in each such experimental situation, it is the behaviour of the population that is uppermost in the mind of the experimenter. It is for this reason that the behaviour of the population is examined in this chapter under the assumption that the population values are known. In the next chapter the importance of the sample in relation to the population is discussed.

22.2 DESCRIBING A POPULATION

The description of a population depends on whether the values are those of a discrete or a continuous random variable. In this section the discrete case is considered first and then the continuous case.

Let the observable in the experiment be a discrete random variable X which can take values $\{x_1, x_2, \ldots, x_n\}$. For example, in the above experiment on mice X would be the W.B.C. count having sample space $\{0, 1, 2, 3, \ldots\}$.

Since the values of X in the population are known, the frequency with which any particular value of X, say x_i, has occurred is known, and it becomes possible to evaluate the probability of x_i:

$$P(X = x_i) = p(x_i) = \frac{\text{number of times } x_i \text{ has occurred in population}}{\text{total number in population}} .$$

(22.1)

Here $p(x_i)$ is used as an abbreviation for $P(X = x_i)$ and will always be greater than or equal to zero.

If the probabilities are plotted for each value of x_i then the **probability distribution** of X will be of the form illustrated in Fig. 22.1. The length of the ordinate line corresponds to the probability.

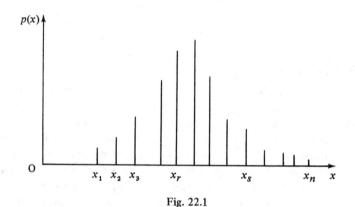

Fig. 22.1

It also follows from equation (22.1) that the sum of all the probabilities must equal 1, that is

$$p(x_1) + p(x_2) + \ldots + p(x_n) = 1 .$$

This can be expressed more briefly by using the summation notation

$$\sum_{i=1}^{n} p(x_i) = 1 ,$$

(22.2)

and holds for all probability distributions of discrete random variables.

If the probability that X takes a value between x_r and x_s inclusive is required then using the addition law of probability this will be

$$P(x_r \leqslant X \leqslant x_s) = P(x_r \text{ OR } x_{r+1} \text{ OR } x_{r+2} \ldots \text{OR } x_s)$$
$$= p(x_r) + p(x_{r+1}) + p(x_{r+2}) + \ldots + p(x_s)$$
$$= \sum_{i=r}^{s} p(x_i) .$$

(22.3)

A special function which is found useful is the **cumulative distribution** function $F(x_r)$. This denotes the probability that X takes a value less than or equal to some value x_r,

$$F(x_r) = P(X \leqslant x_r)$$
$$= p(x_1) + p(x_2) + \ldots + p(x_r)$$
$$= \sum_{i=1}^{r} p(x_i) . \tag{22.4}$$

When the random variable observed in the experiment is not discrete but continuous with sample space the interval a to b, then the probabilities must be evaluated for very small intervals of X. This leads to the distribution curve not having the structure shown in Fig. 22.1 but becoming smooth with the area under the curve equal to one. For example, if instead of recording the W.B.C. count in the experiment on mice, the volume occupied by the white blood cells is recorded then this new random variable X would be continuous having sample space the interval 0 to ∞. (Once again it is emphasized that infinity is used in the theory for completeness, since in practice a volume of size infinity is not possible!) The distribution curve of X would then take the form shown in Fig. 22.2.

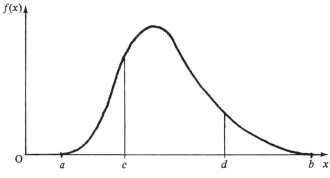

Fig. 22.2

Here $f(x)$ is used to indicate that the probability distribution of a continuous random variable is being dealt with and it will always be a quantity greater than or equal to zero.

Since the area under a curve can be obtained by integration then

$$\int_{a}^{b} f(x)\mathrm{d}x = 1 . \tag{22.5}$$

The probability of X taking a value between c and d is the area under the distribution curve between c and d, so

$$P(c \leqslant X \leqslant d) = \int_{c}^{d} f(x)\mathrm{d}x . \tag{22.6}$$

The cumulative distribution function, $F(c)$, is the probability that X takes a value less than or equal to c and therefore is the area under the curve to the left of c. Hence

$$P(c \leqslant X \leqslant d) = \text{area under curve between } c \text{ and } d$$

$$= \text{area under curve to left of } d$$

$$- \text{area under curve to left of } c$$

$$= \int_a^d f(x)dx - \int_a^c f(x)dx$$

$$= F(d) - F(c) . \tag{22.7}$$

This demonstrates the convenience of the cumulative distribution function as a means of evaluating probabilities over required ranges.

Note that a clear distinction has emerged in the way in which a discrete and a continuous random variable is handled. In the former case the summation operation, Σ, is used, whereas in the latter case the integration operation, $\int dx$, is used.

Frequently, for theoretical reasons or by inspection of the population values, the probability distribution of the random variable can be described by a simple mathematical expression. When this is possible it can simplify the task of finding probabilities, as is demonstrated in the following examples.

Example 1: A cell when it multiplies can give birth to a maximum of four daughter cells. The probability of x daughter cells being formed by a cell which has just multiplied is given by the following probability distribution,

x	1	2	3	4
$P(X = x) = p(x)$	$\dfrac{1}{4}$	$\dfrac{3}{8}$	$\dfrac{1}{8}$	$\dfrac{1}{4}$

Verify that $p(x)$ is a probability distribution. Using the cumulative distribution function find the probability of at least two cells being formed at birth.

In this example X, the number of cells formed at birth, is the random variable and it is discrete with sample space $\{1, 2, 3, 4\}$. It is easily seen that $p(x)$ is a probability distribution since

$$\sum_{x=1}^{4} p(x) = p(1) + p(2) + p(3) + p(4)$$

$$= \frac{1}{4} + \frac{3}{8} + \frac{1}{8} + \frac{1}{4} = 1 .$$

The cumulative distribution function for each value of x is obtained using equation (22.4):

$$F(1) = P(X \leqslant 1) = \frac{1}{4} \, ,$$

$$F(2) = P(X \leqslant 2) = \frac{1}{4} + \frac{3}{8} = \frac{5}{8} \, ,$$

$$F(3) = P(X \leqslant 3) = \frac{1}{4} + \frac{3}{8} + \frac{1}{8} = \frac{6}{8} \, ,$$

$$F(4) = P(X \leqslant 4) = \frac{1}{4} + \frac{3}{8} + \frac{1}{8} + \frac{1}{4} = 1 \, .$$

The probability of at least two cells being formed at birth is given by

$$
\begin{aligned}
P(2 \leqslant X \leqslant 4) &= P(X \leqslant 4) - P(X < 2) \\
&= P(X \leqslant 4) - P(X \leqslant 1) \\
&= F(4) - F(1) \\
&= 1 - \frac{1}{4} = \frac{3}{4} \, .
\end{aligned}
$$

Example 2: In an experiment carried out in a glass-house, the probability of a whitefly, which is x days old, being destroyed by a parasitic wasp is given by the probability distribution

$$p(x) = q^{x-1}p$$

where p and q are constants. What proportion of the whiteflies will be destroyed before they are four days old if $p = 0.2$ and $q = 0.8$?

The random variable, X, in this example is the age in days at which a whitefly is destroyed by a wasp. Therefore X will be discrete with sample space $\{1, 2, 3, \ldots\}$. The probability of a whitefly being destroyed before it is four days old will be

$$
\begin{aligned}
P(X \leqslant 3) &= P(X = 1) + P(X = 2) + P(X = 3) \\
&= p(1) + p(2) + p(3) \\
&= p + qp + q^2p \, .
\end{aligned}
$$

Substituting for p and q we get

$$P(X \leqslant 3) = 0.2 + 0.16 + 0.128 = 0.488 \, .$$

Thus almost half of the whitefly population in the glass-house will be dead by the fourth day.

Example 3: An examination of circadian rhythms of scorpion spiders suggests that the probability distribution for X, the fraction of a day spent feeding, is given by

$$f(x) = 2 - 2x$$

where X can take any value in the interval 0 to 1. Show that $f(x)$ is a probability distribution and calculate the probability of the fraction, X, taking a value between $\frac{1}{4}$ and $\frac{1}{2}$.

In this case X is a continuous random variable and since

$$\int_0^1 f(x)\,dx = \int_0^1 (2 - 2x)\,dx$$

$$= \left[2x - \frac{2x^2}{2}\right]_0^1 = 1 \ ,$$

it is a probability distribution.

The probability of the fraction, X, being between $\frac{1}{4}$ and $\frac{1}{2}$ is

$$P\left(\frac{1}{4} \leqslant X \leqslant \frac{1}{2}\right) = P\left(X \leqslant \frac{1}{2}\right) - P\left(X \leqslant \frac{1}{4}\right)$$

$$= F\left(\frac{1}{2}\right) - F\left(\frac{1}{4}\right)$$

$$= \int_0^{\frac{1}{2}} (2 - 2x)\,dx - \int_0^{\frac{1}{4}} (2 - 2x)\,dx$$

$$= \frac{3}{4} - \frac{7}{16} = \frac{5}{16} \ .$$

22.3 INTERPRETING PROBABILITY DISTRIBUTIONS

Although the probability distribution tells us all we need to know about the random variable X, we often need to summarise its properties. This is possible by using suitable measures which reflect the essential characteristics of the distribution. The two most commonly used measures are the **mean**, μ, and the **variance**, σ^2. Sometimes the **standard deviation**, σ, which is simply the positive square root of the variance, is used. The mean gives a measure of the central location of the probability distribution curve, and the variance (and standard deviation) gives a measure of the spread about the mean. This is illustrated in Fig. 22.3 and Fig. 22.4 for symmetrical distributions. In Fig. 22.3 the distribution curves of three continuous random variables with different means but the same standard deviations are shown. In Fig. 22.4 the distribution curves of two continuous random variables with the same mean but different standard deviations are shown. In Fig. 22.4 note that as the spread of X increases so does the standard deviation (and variance), a large spread occurring when the probabilities of X are high for a wide range of values of X.

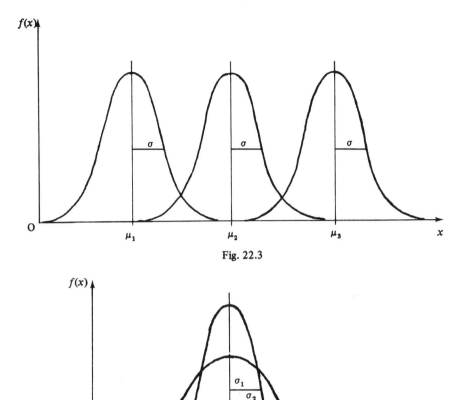

Fig. 22.3

Fig. 22.4

Finally, before giving the formulae used to calculate the mean and variance of a probability distribution, it should be clearly understood that μ and σ^2 are not the only measures used to reflect the properties of the distribution of a random variable. Other possible measures of central location are the **median** and **mode**. The median, m_o, is the value of X which divides the probability distribution into two equal areas and so the probability of an X value less than m_o is $\frac{1}{2}$ and also is equal to the probability of an X value greater than m_o. For a symmetrical distribution the mean and median are identical. The mode is the value of X which occurs most frequently. Another measure of spread sometimes used is the **range** which is the difference between the largest and the smallest value of X. However, the range is not a very sensitive measure of spread.

22.3.1 Calculating the mean and variance

As the name suggests, the mean gives the average value of the random variable X. If X is discrete and takes values $x_1, x_2, \ldots, x_n$ then the mean is calculated using

$$\mu = x_1 p(x_1) + x_2 p(x_2) + \ldots + x_n p(x_n)$$

$$= \sum_{i=1}^{n} x_i p(x_i) . \tag{22.8}$$

The explanation of why μ is calculated this way is best demonstrated by a particular example. Suppose the number of insects with 3, 5, or 7 black markings is n_1, n_2, and n_3 respectively in a population of size $n = n_1 + n_2 + n_3$. The average number of black markings on an insect is therefore

$$\mu = \frac{(3n_1 + 5n_2 + 7n_3)}{n}$$

$$= \frac{3n_1}{n} + \frac{5n_2}{n} + \frac{7n_3}{n} . \tag{22.9}$$

Alternatively, if X denotes the number of black markings on the insect then X will have sample space $\{x_1 = 3, x_2 = 5, x_3 = 7\}$ and

$$P(X = 3) = n_1/n \ ,$$
$$P(X = 5) = n_2/n \ ,$$
$$P(X = 7) = n_3/n \ .$$

Substituting for n_1/n, n_2/n and n_3/n into equation (22.9) then gives

$$\mu = 3P(X = 3) + 5P(X = 5) + 7P(X = 7)$$

$$= 3p(3) + 5p(5) + 7p(7)$$

$$= \sum_{i=1}^{3} x_i p(x_i) \ ,$$

which is consistent with the formula given for the mean in equation (22.8).

If X is a continuous random variable which can take values in the interval a to b then the mean is calculated using

$$\mu = \int_a^b x f(x) \mathrm{d}x . \tag{22.10}$$

The variance is the average of the squares of the difference between the mean and each value of X. For a discrete random variable it is given by

$$\sigma^2 = (x_1 - \mu)^2 p(x_1) + \ldots + (x_n - \mu)^2 p(x_n)$$

$$= \sum_{i=1}^{n} (x_i - \mu)^2 p(x_i) . \tag{22.11}$$

This is more commonly rearranged to give

$$\sigma^2 = \sum_{i=1}^{n} (x_i^2 - 2\mu x_i + \mu^2)p(x_i)$$

$$= \sum x_i^2 p(x_i) - 2\mu\sum x_i p(x_i) + \mu^2 \sum p(x_i)$$

$$= \sum x_i^2 p(x_i) - \mu^2 ,$$ (22.12)

since $\sum_{i=1}^{n} x_i p(x_i) = \mu$ and $\sum_{i=1}^{n} p(x_i) = 1$. The form of equation (22.12) is simpler than equation (22.11) for calculation. The variance of a continuous random variable is similarly given by

$$\sigma^2 = \int_a^b (x - \mu)^2 f(x) dx$$ (22.13)

which simplifies to

$$\sigma^2 = \int_a^b x^2 f(x) dx - \mu^2 .$$ (22.14)

Example 1: Using the probability distribution for X, the number of cells produced at birth, given in example 1 of Section 22.2, calculate the mean, variance and standard deviation of X.

For the mean

$$\mu = \sum_{x=1}^{4} xp(x) = 1.\frac{1}{4} + 2.\frac{3}{8} + 3.\frac{1}{8} + 4.\frac{1}{4} = 2.375 .$$

For the variance

$$\sigma^2 = \sum_{x=1}^{4} x^2 p(x) - \mu^2$$

$$= 1^2.\frac{1}{4} + 2^2.\frac{3}{8} + 3^2.\frac{1}{8} + 4^2.\frac{1}{4} - (2.375)^2$$

$$= 1.234$$

and hence the standard deviation is

$$\sigma = \sqrt{1.234}$$

$$= 1.111 .$$

Example 2: The blood clotting times, T, of patients with a particular disease are thought to have the probability distribution

$$f(t) = \frac{1}{4}e^{-t/4} \qquad 0 \leqslant t < \infty$$

where t is measured in minutes. Show that the mean and variance of the clotting times are 4 minutes and 16 minutes^2 respectively.

For the mean

$$\mu = \frac{1}{4} \int_0^\infty t\,e^{-t/4}\,dt \ .$$

This integration is similar to that in example 5(i) of Section 12.4 using $a = \frac{1}{4}$ and gives $\mu = 4$. For the variance

$$\sigma^2 = \frac{1}{4} \int_0^\infty t^2\,e^{-t/4}\,dt - (4)^2$$

and using the integration technique demonstrated in example 5(ii) of Section 12.4 gives $\sigma^2 = 16$.

Example 3: Calculate the mean and standard deviation of the fraction of the day spent feeding by the scorpion spiders described in example 3 of Section 22.2. The probability distribution of the fraction of the day spent feeding of a second species of scorpion spiders is given by

$$f(x) = 2x \qquad 0 \leqslant x \leqslant 1 \ .$$

Compare the means and variances of the two species of scorpion spiders.
    For the scorpion spiders described in example 3 of Section 22.2:

$$\mu = \int_0^1 xf(x)dx = \int_0^1 x(2 - 2x)dx$$

$$= \left[x^2 - \frac{2x^3}{3} \right]_0^1$$

$$= 1 - \frac{2}{3} = \frac{1}{3} \ ,$$

$$\sigma^2 = \int_0^1 x^2 f(x)dx - \mu^2$$

$$= \int_0^1 x^2(2 - 2x)dx - \left(\frac{1}{3}\right)^2$$

$$= \left[\frac{2x^3}{3} - \frac{2x^4}{4} \right]_0^1 - \frac{1}{9}$$

$$= \frac{2}{3} - \frac{1}{2} - \frac{1}{9} = \frac{1}{18} \ .$$

For the second species of scorpion spider:

$$\mu = \int_0^1 xf(x)\mathrm{d}x = \int_0^1 2x^2\mathrm{d}x$$

$$= \left[\frac{2x^3}{3}\right]_0^1 = \frac{2}{3} \, ,$$

$$\sigma^2 = \int_0^1 x^2 f(x)\mathrm{d}x - \mu^2$$

$$= \int_0^1 2x^3\mathrm{d}x - \left(\frac{2}{3}\right)^2$$

$$= \left[\frac{x^4}{2}\right]_0^1 - \frac{4}{9}$$

$$= \frac{1}{2} - \frac{4}{9} = \frac{1}{18} \, .$$

Therefore the mean fraction of the day spent feeding is higher for the second species although both species have the same variance.

In this section the properties of some simple probability distributions that are used to describe the behaviour of a population of observables have been examined. In the following section three probability distributions that are widely applicable will be introduced. Their mathematical forms can be arrived at on a theoretical basis, but their importance lies in their value in describing a wide range of experimental situations. It is important not only to be familiar with the mean, variance, and the evaluation of probabilities for each distribution, but to recognise the experimental process to which they can be applied. The fixed constants of the probability distributions, often called the **parameters**, are assumed to be known. The method of estimating these parameters, when they are unknown, is dealt with in Chapter 23.

22.4 BINOMIAL DISTRIBUTION

This distribution arises in an experiment where n independent trials take place, each trial having one of two outcomes, say success and failure. In each of the n trials the probability of success is a constant p and the probability of failure $q = 1 - p$. Each time the experiment takes place the number of successes observed, R, will be an integral number between 0 and n. Hence R is a discrete random variable with sample space $\{0, 1, 2, \ldots, n\}$. If the experiment is repeated continually then it will be found that the probability distribution of R will be

$$P(R = r) = p(r) = {}^{n}C_r\, p^r\, q^{n-r}. \tag{22.15}$$

Since n is specified at the beginning of the experiment, and p (and hence q) is known, then each probability can be evaluated. Sometimes a binomial distribution is described by the abbreviated notation **B(n, p)**.

Example 1: The probability of a bacterium being infected with a phage is 0.4. If four bacteria are examined under a microscope what is the probability of

(i) no bacteria being infected,

(ii) three bacteria being infected,

(iii) at least one bacterium being infected?

In this example observing each bacterium represents a trial, and therefore $n = 4$. Let R denote the number of infected bacteria. In each trial the probability of discovering an infected bacterium is 0.4 and hence $p = 0.4$ and $q = 0.6$.

(i) P(no bacteria are infected)

$$= P(R = 0)$$
$$= {}^{n}C_0\, p^0\, q^{n-0}$$
$$= q^4 = (0.6)^4 = 0.1296\ .$$

(ii) P(three bacteria are infected)

$$= P(R = 3)$$
$$= {}^{n}C_3\, p^3\, q^{n-3}$$
$$= {}^{4}C_3\, p^3\, q^1$$
$$= 4(0.4)^3(0.6) = 0.1536\ .$$

(iii) P(at least one bacterium is infected)

$$= P(R \geqslant 1)$$
$$= P(R = 1 \text{ OR } R = 2 \text{ OR } R = 3 \text{ OR } R = 4)$$
$$= P(R = 1) + P(R = 2) + P(R = 3) + P(R = 4)$$
$$\text{(using the addition law of probability)}$$
$$= {}^{4}C_1\, pq^3 + {}^{4}C_2\, p^2 q^2 + {}^{4}C_3\, p^3 q + {}^{4}C_4\, p^4$$
$$= 4(0.4)(0.6)^3 + 6(0.4)^2(0.6)^2 + 4(0.4)^3(0.6) + (0.4)^4$$
$$= 0.8704\ .$$

Alternatively the solution can be arrived at using

$$P(R \geqslant 1) = 1 - P(R = 0)$$
$$= 1 - (0.6)^4 = 0.8704\ .$$

Example 2: In a drug trial, the probability of a patient having a preference for drug A instead of drug B is 0.9. If fifty patients are entered into a trial what is the probability of forty-eight of them preferring drug A?

Let R denote the number of patients preferring drug A. For $n = 50$, $p = 0.9$ and $q = 0.1$ then

$P(48$ patients prefer drug A$)$

$$= P(R = 48)$$

$$= {}^{50}C_{48} \, p^{48} \, q^{50-48}$$

$$= 1225(0.9)^{48}(0.1)^2$$

$$= 0.0779 \ .$$

22.4.1 Mathematical basis of the binomial distribution

The derivation of the binomial distribution is not difficult using the multiplication and addition laws of probability. Let S denote a success and F a failure in each of the n trials, then if the trials are performed consecutively some of the events leading to exactly r successes will be

$$E_1 \equiv \underbrace{SS \ \ldots\ldots\ldots\ S}_{r \text{ trials}} \ \underbrace{FF \ \ldots\ldots\ldots\ F}_{(n-r) \text{ trials}} \ ,$$

$$E_2 \equiv \underbrace{SS \ \ldots\ldots\ S \ FS}_{(r-1) \text{ trials}} \ \underbrace{FF \ \ldots\ldots\ldots\ F}_{(n-r-1) \text{ trials}} \ ,$$

$$E_3 \equiv \underbrace{SS \ \ldots\ldots\ S \ FFSS}_{(r-2) \text{ trials}} \ \underbrace{FF \ \ldots\ldots\ldots\ F}_{(n-r-2) \text{ trials}} \ .$$

Using the multiplication law of probability the probability of each of these events is the same since

$$P(E_1) = p^r \, q^{n-r},$$

$$P(E_2) = p^{r-1} \, q \, p \, q^{n-r-1} = p^r \, q^{n-r},$$

$$P(E_3) = p^{r-2} \, q^2 \, p^2 \, q^{n-r-2} = p^r \, q^{n-r},$$

and hence the probability of r successes,

$$P(R = r) = P(E_1 \text{ OR } E_2 \text{ OR } E_3 \text{ OR } \ldots\ldots)$$

$$= P(E_1) + P(E_2) + P(E_3) + \ldots\ldots$$

$$\text{(using the addition law of probability)}$$

$$= p^r \, q^{n-r} + p^r \, q^{n-r} + p^r \, q^{n-r} + \ldots\ldots$$

All that remains is to determine the number of events leading to exactly r successes and hence the number of terms of the type $p^r q^{n-r}$ that have to be added. This is simply ^nC_r, the number of ways of selecting r items out of n (see Section 20.2), and thus

$$P(R = r) = {}^nC_r\, p^r\, q^{n-r}\ .$$

The binomial distribution is so called because the probabilities are the terms in the binomial expansion of $(p + q)^n$, see equation 2.18,

$$(p + q)^n = {}^nC_0\, p^0\, q^n + {}^nC_1\, p^1\, q^{n-1} + \ldots + {}^nC_r\, p^r\, q^{n-r}$$
$$+ \ldots + {}^nC_n\, p^n\, q^{n-0}\ .$$

Using the above equation it is easy to show that the probabilities of the binomial distribution sum to 1,

$$\sum_{r=0}^{n} p(r) = \sum_{r=0}^{n} {}^nC_r\, p^r\, q^{n-r}$$
$$= {}^nC_0\, p^0\, q^n + {}^nC_1\, p^1\, q^{n-1} + \ldots + {}^nC_n\, p^n\, q^0$$
$$= (p + q)^n$$
$$= 1 \text{ (since } p + q = 1)\ .$$

22.4.2 Mean and variance of the binomial distribution

The mean and variance are derived using the general formulae given in equations (22.8) and (22.12). The derivation is not very important but the results are important.

For the mean

$$\mu = \sum_{r=0}^{n} rp(r)$$
$$= \Sigma r\, {}^nC_r\, p^r\, q^{n-r}$$
$$= \Sigma r\, \frac{n!}{r!(n - r)!}\, p^r\, q^{n-r}$$
$$= np \sum_{r=1}^{n} \frac{(n-1)!}{(r-1)!\,(n-r)!}\, p^{r-1}\, q^{n-r}$$
$$= np\, \Sigma\, {}^{n-1}C_{r-1}\, p^{r-1}\, q^{n-r}$$
$$= np\, (p + q)^{n-1}$$
$$= np\ . \tag{22.16}$$

Hence the average value of R to be expected if the experiment is repeated many times is np.

For the variance

$$\sigma^2 = \sum_{r=0}^{n} r^2 p(r) - \mu^2$$

$$= \Sigma(r^2 - r + r)p(r) - \mu^2$$

$$= \Sigma r(r - 1)p(r) + \Sigma rp(r) - \mu^2$$

$$= \Sigma r(r - 1)\frac{n!}{r!(n - r)!}p^r q^{n-r} + \mu - \mu^2$$

$$= n(n - 1)p^2 \sum_{r=2}^{n} {}^{n-2}C_{r-2}\, p^{r-2}\, q^{n-r} + np - (np)^2$$

$$= n(n - 1)p^2 (p + q)^{n-2} + np - (np)^2$$

$$= n(n - 1)p^2 + np - (np)^2$$

$$= npq \; . \tag{22.17}$$

(The device used in the second line of the derivation, of adding in r and subtracting out r, is often useful in evaluating variances of discrete random variables).

A graph of the binomial probability distribution for the case $n = 8$ and $p = 0.5$ is shown in Fig. 22.5.

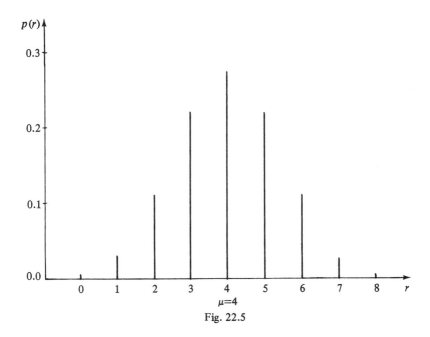

Fig. 22.5

Example 3: Each offspring from a pair of mice inherits a blood defect with probability 0.06. If several pairs of mice each give birth to twenty offspring, calculate the mean and standard deviation of the number of offspring in each family with the defect.

Let R denote the number of mice with the defect. Since there are 20 births and the probability of a blood defect in each mouse is 0.06 then $n = 20$, $p = 0.06$ and $q = 0.94$. Hence

$$\mu = np = 20 \times 0.06 = 1.2$$

and $\qquad \sigma = \sqrt{(npq)} = \sqrt{(20 \times 0.06 \times 0.94)} = 1.06$.

22.5 POISSON DISTRIBUTION

This distribution is widely used in ecology to describe the way in which shrubs, trees, insects, etc. are spread over areas. If objects are scattered over a large area it is almost certain that they will not be found at equally spaced intervals but are likely to be clumped in some places and sparse in others. Suppose the large area is divided into many small areas of equal size, then R, the number of objects in each of these small areas, will be a discrete random variable with sample space $\{0, 1, 2, \ldots\}$. If R has a Poisson distribution then the probability distribution of R will be

$$P(R = r) = p(r) = \lambda^r \frac{e^{-\lambda}}{r!} \tag{22.18}$$

where λ is a parameter (fixed constant) which depends on the size of the small area chosen. Such a distribution is sometimes referred to by the abbreviated notation **Po(λ)**.

In section 22.4 it was shown that if the probability of success in a trial is p then the probability of r successes in n independent trials is given by the binomial distribution. Suppose that there are n small areas each of equal size and that an object is found independently in such an area with probability p. Using the binomial distribution, the expression for the probability of r objects being present in any small area may be rearranged to give

$$P(R = r) = {}^nC_r\, p^r q^{n-r}$$

$$= \frac{n!}{r!(n-r)!}\, p^r(1-p)^{n-r}$$

$$= n(n-1)(n-2) \ldots (n-r+1)\, \frac{p^r}{r!}\,(1-p)^{n-r} .$$

Expanding $(1-p)^{n-r}$ by the binomial expansion gives

$$P(R = r) = n(n-1)(n-2) \ldots (n-r+1)\, \frac{p^r}{r!}\, \Big\{ 1 + (n-r)\,(-p)$$

$$+ \frac{(n-r)\,(n-r-1)\,(-p)^2}{2!} + \frac{(n-r)\,(n-r-1)\,(n-r-2)\,(-p)^3}{3!} + \ldots \Big\}.$$

Removing a factor n from each term then gives

$$P(R = r) = 1(1 - \frac{1}{n})(1 - \frac{2}{n})\ldots(1 - \frac{r-1}{n})\frac{n^r p^r}{r!}\{1$$

$$+ (1 - \frac{r}{n})(-np) + (1 - \frac{r}{n})(1 - \frac{r+1}{n})\frac{(-np)^2}{2!}$$

$$+ (1 - \frac{r}{n})(1 - \frac{r+1}{n})(1 - \frac{r+2}{n})\frac{(-np)^3}{3!} + \ldots\}.$$

As n is large and p is small, so that np is not too large, for small values of r the terms $(1 - \frac{1}{n}), (1 - \frac{2}{n}), \ldots, (1 - \frac{r-1}{n}), \ldots$, which occur in the expression, can be approximated by 1. Thus

$$P(R = r) \simeq \frac{(np)^r}{r!}[1 + (-np) + \frac{(-np)^2}{2!} + \frac{(-np)^3}{3!} + \ldots]$$

and using the expression for e^x given in equation 12.8, with $x = -np$, this becomes

$$P(R = r) \simeq \frac{(np)^r}{r!} e^{-np}.$$

This is the Poisson distribution with λ assigned the value np. The distribution is not confined to areas but can be used for any quantity which can be subdivided into a large number of smaller units of equal size. The Poisson distribution and its ecological applications are further discussed by Pielou (1969).

Example 1: If insect egg clusters on the leaves of a tree have a Poisson distribution with parameter $\lambda = 0.5$ calculate the probability of a leaf having

(i) no egg clusters,

(ii) at least one egg cluster.

Let R denote the number of egg clusters on a leaf.

(i) $P(\text{no egg clusters on leaf}) = P(R = 0)$

$$= \frac{\lambda^0 e^{-\lambda}}{0!}$$

$$= e^{-\lambda}$$

$$= e^{-0.5} = 0.6066 \ .$$

(ii) P(at least one egg cluster on leaf) $= P(R \geqslant 0)$

$$= 1 - P(R = 0)$$

$$= 1 - 0.6066 = 0.3934 \ .$$

Example 2: If the distribution of attached female ticks on each square centimetre of the skin of a sheep is distributed with Poisson frequency, calculate the parameter value such that the probability of a square centimetre containing at least one tick is greater than 0.5.

Let R denote the number of ticks on a square centimetre. The parameter λ must be found such that

P(at least one tick on a square centimetre) $\geqslant 0.5$.

This leads to

$$P(R \geqslant 1) = 1 - P(R = 0) \geqslant 0.5$$

and so

$$1 - e^{-\lambda} \geqslant 0.5 \ .$$

Rearranging gives

$$e^{-\lambda} \leqslant 0.5$$

and taking logarithms of both sides leads to

$$\lambda \geqslant 0.693 \ .$$

22.5.1 Mean and variance of the Poisson distribution

Since R is discrete, the mean

$$\mu = \sum_{r=0}^{\infty} rp(r)$$

$$= \Sigma r \frac{\lambda^r e^{-\lambda}}{r!}$$

$$= \lambda e^{-\lambda} \sum_{r=1}^{\infty} \frac{\lambda^{r-1}}{(r-1)!}$$

$$= \lambda e^{-\lambda} e^{\lambda} \quad \left(\text{since } \sum_{r=1}^{\infty} \frac{\lambda^{r-1}}{(r-1)!} = e^{\lambda}\right)$$

$$= \lambda \ . \tag{22.19}$$

Thus the parameter λ is also the mean of the distribution. For the variance

$$\sigma^2 = \sum_{r=0}^{\infty} r^2 p(r) - \mu^2$$

and, using a method similar to that described in Section 22.4.2, this gives

$$\sigma^2 = \lambda \ . \tag{22.20}$$

Thus the parameter λ is also the variance, and so the mean and variance are always equal for a Poisson distribution. A graph of a Poisson distribution with parameter λ equal to 3 is shown in Fig. 22.6.

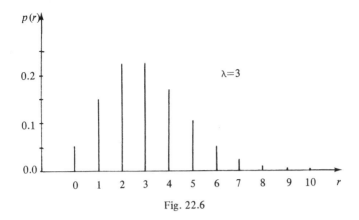

Fig. 22.6

Earlier it was mentioned that λ depended on the size of the area considered. It can be shown that a convenient property of a Poisson distribution with a parameter λ equal to a is that if the area is increased or decreased by a factor k then the parameter associated with the new area is ka.

Example 3: Colonies of bacteria are distributed over the surface of a food according to a Poisson process. A microscope reveals a mean of 5 colonies on areas of size one tenth of the total surface. What is the mean and variance of the number of colonies over areas of size one thirtieth of the total surface?

Let R denote the number of colonies on areas of size one thirtieth of the total surface. Since $\lambda = 5$ for areas of size one tenth of the total surface, the new value for λ for areas of size one thirtieth of the total surface will be 5/3, since the area has been changed by a factor of 1/3. Hence the mean and variance of R is 5/3.

So far the use of the Poisson distribution has been discussed in connection with spatial patterns. However, the distribution is found equally useful in describing how events occur in time. The probability of r events occurring in a time t is given by

$$P(R = r) = p(r) = \frac{(\lambda t)^r}{r!} e^{-\lambda t} \tag{22.21}$$

where the parameter λ is now the mean number of events occurring in unit time.

Example 4: The mean number of impulses received by a nerve cell during a unit of time is 2.5. Assuming a Poisson distribution, calculate the probability of fewer than 3 impulses arriving in five units of time.

Let R denote the number of impulses to arrive in five units of time,

$$P(\text{fewer than 3 impulses}) = P(R \leqslant 2)$$

$$= P(R = 0) + P(R = 1) + P(R = 2)$$

$$= e^{-\lambda t} + (\lambda t)e^{-\lambda t} + \frac{(\lambda t)^2}{2!}e^{-\lambda t} .$$

Since $\lambda = 2.5$ and $t = 5$ then

$$P(R \leqslant 2) = 0.0000037 + 0.0000466 + 0.0002911$$

$$= 0.0003414 .$$

Example 5: A predator catches on average one prey a day. If the prey are caught according to a Poisson process what is the probability of the predator catching exactly ten prey within one week?

Let R denote the number of prey caught by the predator in one week. The parameter λt associated with R will be 7.

$$P(\text{catching 10 of the prey population}) = P(R = 10)$$

$$= \frac{(\lambda t)^{10} e^{-\lambda t}}{10!}$$

$$= \frac{7^{10} e^{-7}}{10!} = 0.071 .$$

22.6 NORMAL DISTRIBUTION

This is a very important distribution that describes many continuous random variables. If the observable X in an experiment has a normal distribution then the probability distribution is given by

$$f(x) = \frac{1}{\sqrt{(2\pi)}\sigma} e^{-\frac{(x-\mu)^2}{2\sigma^2}} \tag{22.22}$$

where μ and σ are two parameters. The sample space of X is the interval $-\infty$ to ∞, therefore allowing the random variable to take both negative and positive values. In applications where the random variable does not take negative values such as the heights of plants, or the reaction times of antibodies to antigen, the normal distribution is still widely used, as the probability assigned to negative values is so small that it can be neglected. Although it is not proved here, it can be shown, using formulae (22.10) and (22.14), that the parameters μ and σ correspond to the mean and standard deviation of X, and for convenience the distribution is sometimes referred to by the notation $N(\mu,\sigma^2)$. Figure 22.7 shows the symmetrical bell shape which is the characteristic shape of the normal probability distribution.

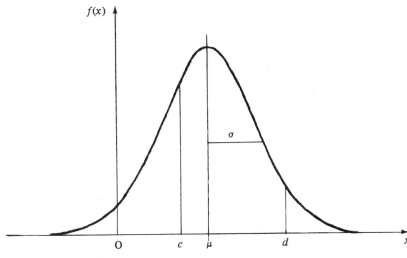

Fig. 22.7

The probability that X takes a value between c and d is the area under the curve between c and d, and so

$$P(c \leqslant X \leqslant d) = \int_c^d \frac{1}{\sqrt{(2\pi)}\sigma} e^{-\frac{(x-\mu)^2}{2\sigma^2}} dx$$

$$= \int_{-\infty}^d \frac{1}{\sqrt{(2\pi)}\sigma} e^{-\frac{(x-\mu)^2}{2\sigma^2}} dx - \int_{-\infty}^c \frac{1}{\sqrt{(2\pi)}\sigma} e^{-\frac{(x-\mu)^2}{2\sigma^2}} dx$$

$$= F(d) - F(c)$$

where F denotes the cumulative distribution. To evaluate $F(d)$ and $F(c)$ the substitution $z = \frac{x-\mu}{\sigma}$ is used in each of the integrals to give

$$P(c \leqslant X \leqslant d) = \int_{-\infty}^{\frac{d-\mu}{\sigma}} \frac{1}{\sqrt{(2\pi)}} e^{-z^2/2} dz - \int_{-\infty}^{\frac{c-\mu}{\sigma}} \frac{1}{\sqrt{(2\pi)}} e^{-z^2/2} dz$$

$$= \int_{-\infty}^{\frac{d-\mu}{\sigma}} f(z) dz - \int_{-\infty}^{\frac{c-\mu}{\sigma}} f(z) dz \qquad (22.23)$$

where $\quad f(z) = \frac{1}{\sqrt{(2\pi)}} e^{-z^2/2}$.

Inspection of $f(z)$ shows that it is the equation of a normal probability distribution which has zero mean ($\mu = 0$) and unit variance ($\sigma^2 = 1$), that is,

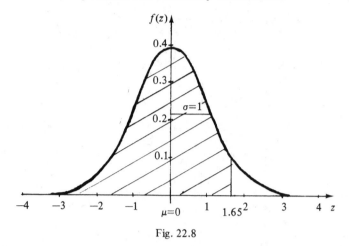

Fig. 22.8

$N(0, 1)$. Thus for any variable with a normal distribution, subtracting the mean and dividing by the standard deviation leads to the variable Z which has a normal distribution with mean 0 and variance 1. For this reason Z is called the **standard normal variate** and $f(z)$ the **standard normal distribution**. A variate which has a standard normal distribution has approximately 95% of its values lying within two standard deviations of the mean, and so the area under $f(z)$ between z equal to -2 and 2 is approximately 0.95. Furthermore over 99% of the values lie within three standard deviations of the mean. A graph of $f(z)$ is shown in Fig. 22.8. The cumulative distribution function of the standard normal distribution is denoted by the special symbol $\Phi(a)$ and is the area under the distribution curve between $-\infty$ and a. The area shaded in Fig. 22.8 denotes the value of $\Phi(1.65)$.

The integrals of $f(z)$ from $-\infty$ to $(d - \mu)/\sigma$ and from $-\infty$ to $(c - \mu)/\sigma$ required in equation (22.23) are $\Phi\left(\dfrac{d-\mu}{\sigma}\right)$ and $\Phi\left(\dfrac{c-\mu}{\sigma}\right)$ respectively. Hence

$$P(c \leqslant X \leqslant d) = \Phi\left(\frac{d-\mu}{\sigma}\right) - \Phi\left(\frac{c-\mu}{\sigma}\right). \qquad (22.24)$$

Unfortunately integration of $f(z)$ to get Φ is difficult and must be performed numerically. In example 4 of Section 17.4.3 the related integral $\displaystyle\int_{-\infty}^{b} e^{-x^2/2}\,dx$ was evaluated numerically for $b = 0$ and $b = 2$. To avoid having to use numerical integration each time a value is required, the table of the standard normal integral given in Appendix E provides the values of $\Phi(z)$ obtained by numerical integration for a range of z. For example, if $\Phi(1.65)$ is required then from the row labelled 1.6 and column headed 0.05 of the table this value is 0.9505. In the interests of compactness the table gives only the value of $\Phi(z)$ for positive z. When the argument is negative the value of $\Phi(-z)$ is obtained by subtracting from 1 the value obtained for $\Phi(z)$, since the symmetrical shape of the standard normal curve indicates that

$$\Phi(-z) = 1 - \Phi(z) .$$

For example, $\Phi(-0.88) = 1 - \Phi(0.88)$

$$= 1 - 0.8106 = 0.1894 .$$

Thus to find the probability that a normally distributed random variable with mean μ and variance σ^2 lies within an interval, the problem is changed by substitution to finding the area under the standard normal curve. The value of this area is then available from the table of the standard normal integral.

Example 1: A population of marine gastropods have shell lengths which are normally distributed with mean 7 millimetres and variance 2 millimetres^2. What proportion of the population will have a shell length between 5 and 9 millimetres?

Let X denote the shell length.

$$P(\text{shell length is between 5 and 9}) = P(5 \leqslant X \leqslant 9)$$
$$= F(9) - F(5)$$
$$= \Phi\left(\frac{9 - 7}{\sqrt{2}}\right) - \Phi\left(\frac{5 - 7}{\sqrt{2}}\right)$$
$$= \Phi(1.41) - \Phi(-1.41)$$
$$= 0.9207 - 0.0793 = 0.8414 .$$

Hence approximately 84% of the population will have a shell length in the range 5 to 9 millimetres.

Example 2: If the time taken for an eyelid to close after a foreign particle strikes the surface of the eyeball is normally distributed with mean 5 microseconds, calculate the standard deviation in order that the closing time is less than 4 microseconds with probability 0.3.

Let T be the closing time. A value of σ must be found such that

$$P(T \leqslant 4) = 0.3 .$$

This implies

$$\Phi(z) = 0.3$$

where $z = (4 - 5)/\sigma .$

From the table of the standard normal integral the value of z which makes $\Phi(z)$ equal to 0.3 is -0.53 and hence

$$-0.53 = -\frac{1}{\sigma}$$

and so $\sigma = 1.88$ microseconds.

22.7 APPROXIMATION OF THE BINOMIAL DISTRIBUTION BY THE NORMAL DISTRIBUTION

When using the binomial distribution for large n the calculation of $P(X \leqslant a)$ requires the sum of all probabilities from 0 to a. This can be very tedious. In such cases the binomial distribution can be replaced by a normal curve as shown in Fig. 22.9 and the probability evaluated as if X were normally distributed.

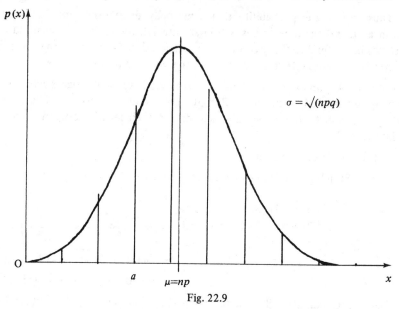

Fig. 22.9

The values of the mean, np, and variance, npq, of the binomial distribution are used in the normal distribution for μ and σ^2 respectively,

$$P(X \leqslant a) = F(a)$$

$$\simeq \Phi\left(\frac{a - np}{\sqrt{(npq)}}\right). \tag{22.25}$$

A correction factor of $\frac{1}{2}$ is sometimes used to improve the approximation, when $a \neq np$, resulting in

$$P(X \leqslant a) = F(a)$$

$$\simeq \begin{cases} \Phi\left(\dfrac{a + \frac{1}{2} - np}{\sqrt{(npq)}}\right) & \text{when } a \neq np \\[4mm] \Phi(0) & \text{when } a = np \end{cases} \tag{22.26}$$

The following examples demonstrate the convenience of the approximation.

Example 1: Each spore released from a spore capsule of a fern has a probability

0.2 of becoming a plant. If a capsule contains 100 spores what is the probability of at least 25 of them becoming plants?

If X is the number of spores which become plants then X will have a binomial distribution with $n = 100$ and $p = 0.2$.

$$P(\text{at least 25 spores become plants}) = P(X \geqslant 25)$$
$$= 1 - P(X \leqslant 24)$$
$$= 1 - (P(X = 0) + P(X = 1) + \ldots + P(X = 24)) .$$

Using the normal approximation the calculation of the above individual probabilities can be avoided:

$$P(X \geqslant 25) = 1 - P(X \leqslant 24)$$
$$= 1 - F(24) .$$

Since np equals 20, which is not equal to 24, a correction factor of $\frac{1}{2}$ must be added:

$$P(X \geqslant 25) \simeq 1 - \Phi\left(\frac{24 + \frac{1}{2} - np}{\sqrt{(npq)}}\right)$$
$$= 1 - \Phi\left(\frac{24 + \frac{1}{2} - 20}{4}\right)$$
$$= 1 - \Phi(1.12) = 0.1314 .$$

Example 2: The probability of a tree being infested with larvae of a species of beetle is 0.3. How likely is it that six or more trees will be infested if forty trees are examined?

X, the number of trees infested, will be distributed according to the binomial distribution with mean $np = 40 \times 0.3 = 12$ and variance $npq = 40 \times 0.3 \times 0.7 = 8.4$. Using the normal approximation for X with $\mu = 12$ and $\sigma^2 = 8.4$ gives

$$P(\text{6 or more trees infested}) = P(X \geqslant 6)$$
$$= 1 - P(X \leqslant 5)$$
$$= 1 - \Phi\left(\frac{5 + \frac{1}{2} - 12}{\sqrt{8.4}}\right)$$
$$= 1 - \Phi(-2.24)$$
$$= 1 - 0.0125 = 0.9875 .$$

A useful book for further reading on probability distributions and their uses is Sokal and Rohlf (1971).

PROBLEMS

1.   The number of shoots on a branch is a random variable X which takes values x with probability $p(x) = kx$, $x = 1, 2, \ldots, 10$. Find the value of k in order that $p(x)$ is a probability distribution. Show that the probability

that a branch has two shoots is 2/55 and that the mean number of shoots on a branch will be 7.

2. In an appetite suppressant experiment, the number of meals, N, a rat eats per day is observed. The maximum number of meals available each day is 2. If N has probability distribution

$$P(N = n) = \tfrac{4}{7} (\tfrac{1}{2})^n, \quad n = 0, 1, 2$$

find the mean and variance of the number of meals eaten each day.

3. The efficiency, X, of a digestive enzyme can be described by the probability density function

$$f(x) = \tfrac{1}{4} (5 - 3x^2), \quad 0 \leqslant x \leqslant 1 .$$

Find the probability of the enzyme having an efficiency greater than 50%. What is the mean and variance of the efficiency?

4. A beam of X-rays when fired at a protein crystal is reflected by the atoms in the molecules of the protein. The intensity of reflection, R, is recorded on a continuous scale from 0 to 3 and its associated probability distribution for a particular crystal is given by

$$f(r) = \frac{2}{3} - \frac{2}{9}r, \quad 0 \leqslant r \leqslant 3 .$$

Find the mean and variance of the intensity of reflection.

5. The inter-arrival times of bees returning to a hive is a random variable T which takes values t with probability

$$f(t) = ke^{-\frac{1}{3}t}, \quad 0 \leqslant t \leqslant \infty .$$

Prove that k must equal 1/3 for $f(t)$ to be a probability distribution. Show that T has mean 3 and variance 9.

   What is the probability of an inter-arrival time taking a value between 1 and 2?

6. The genetic features of two adult mice are such that the probability of an offspring being an albino is 0.2. If the mice give birth to six offspring calculate the probability of

   (i) no albinos,

   (ii) one albino only,

   (iii) two or more albinos.

If several of these mice give birth to a total of fifty offspring use the normal distribution to give an estimate of the probability of fifteen or more offspring being albinos.

7. The probability of a cow in a dairy herd showing symptoms of having mastitis is 0.2. How likely is it that five or more animals in the herd will show symptoms if

    (i) a sample of six animals are chosen at random,

    (ii) a sample of sixty animals are chosen at random?

8. A forecasting system for plant disease is such that a wet day is defined as a day during which 4 mm or more of rain occurs. The rainfall on any day is independent of rainfall on other days, and the following table gives the probability distribution for x mm of rain on any day,

	$0 \leqslant x < 1$	$1 \leqslant x < 2$	$2 \leqslant x < 3$	$3 \leqslant x < 4$	$4 \leqslant x < 5$	$5 \leqslant x < 6$	$x \geqslant 6$
$p(x)$	0.05	0.20	0.20	0.20	0.15	0.15	0.05

    Calculate the probability of

    (i) a wet day,

    (ii) three successive wet days.

    Using the binomial distribution find the probability of at least two wet days in any week.

9. The mean number of mud snails of a rare species found in a square metre of pasture is 0.035. If the snails are distributed on the pasture with a Poisson frequency, calculate the probabilities that in a search over 100 square metres

    (i) no snails are found,

    (ii) more than three snails are found.

10. The distribution of parasites in one cubic centimetre of an animal's liver behaves according to a Poisson model with the mean number of parasites per cc equal to 0.02.

    What is the probability of finding no parasites in a liver of volume 50 cc?

    If the mean number of parasites per cc varies from animal to animal, what value must this mean take if the probability of finding at least one parasite in a liver of 50 cc is not to exceed 0.05?

11. The development time of green mould in oranges (caused by *Penicillium digitatum*) maintained at a temperature of 20°C is normally distributed with a mean of 7 days and a variance of 4 days^2. At the same temperature the development time of sour rot (caused by *Geotrichum candidum*) is normally distributed with a mean of 10 days and a variance of 2 days^2. If an orange, contaminated with spores of both fungi, is kept at 20°C for 9 days what is the probability that

    (i) a green mould will develop,

    (ii) a sour rot will develop,

    (iii) a green mould and a sour rot will develop,

assuming that there is no interaction between the two fungi?

12. It has been found, from extensive research, that for patients with disease A, the urinary secretion rate (mg/24 hr) is described by a $N(5, 1)$ variable whereas for B the description is given by a $N(7, 4)$ variable.

    It is proposed to measure the secretion rate of a patient who has either disease A or B, and if it is less than 5.5 mg/24 hr to classify the patient as having disease A. Otherwise the patient would be classified as having disease B.

    Obtain the probabilities of misclassifying

    (i) a patient with disease A,

    (ii) a patient with disease B.

An alternative proposal is to classify a patient as having disease A if the secretion rate is less than 6 mg/24 hr, and as disease B otherwise. Find the misclassification probabilities and compare this decision rule with the original rule.

    If it is necessary that the probability of misclassifying a patient with disease A should be no greater than 0.2, find the level at which the decision dividing line should be drawn.

13. The time taken after planting for winter wheat crops grown on the Great Plains to reach maturity is normally distributed with a mean 183 days and variance 100 days^2. If the crops are planted on 1st October what percentage will mature

    (i) before April,

    (ii) during April,

    (iii) after May?

(Assume the month of February contains twenty-eight days.)

CHAPTER 23

Samples and sampling distributions

23.1 INTRODUCTION

At the start of the previous chapter it was assumed that the population values were known but experimental limitations usually prevent values for the entire population being known. The best that can be hoped for is that a collection of observations taken from the population will enable good estimates of the unknown population mean, variance, median etc. to be determined. When the observations are collected in such a way that each population value has an equal chance of being included, the sample is said to be **random** and the estimates deduced from the sample are called the **sample estimates**. In the following text it is assumed that the samples are always random.

Example 1: The temperature at which *Corynebacterium hofmanni* commences growth on agar is to be determined.

Since it would be impossible to obtain the temperature for all colonies of *C. hofmanni* which exist, the experimenter must settle for studying a sample of colonies. Using the temperatures obtained in the sample, the sample mean and sample variance can be calculated as estimates for the true population mean and true population variance.

Example 2: A pharmaceutical company measures the shelf-life of drugs by the number of days they remain above 90% potency when stored in containers under fixed conditions of temperature and humidity. The median and range of the shelf-life of a drug stored in a newly designed jar are to be determined.

Once again the experimenter has to rely on a sample of newly designed jars containing the drug. From the sample, the sample median and sample range can be calculated as estimates for the median and range of the shelf-life of the drug.

The following section shows how the sample estimates are calculated.

23.2 INTERPRETATION OF THE SAMPLE

Let X be a random variable from which a sample of n observations $x_1, x_2, \ldots, x_n$ have been collected. The **sample mean** is denoted by $\bar{x}$ and calculated using

$$\bar{x} = \frac{x_1 + x_2 + \ldots + x_n}{n}$$

$$= \frac{\sum\limits_{i=1}^{n} x_i}{n}. \tag{23.1}$$

The **sample variance** is denoted by s^2 and calculated using

$$s^2 = \frac{(x_1 - \bar{x})^2 + (x_2 - \bar{x})^2 + \ldots + (x_n - \bar{x})^2}{n - 1}$$

$$= \frac{\sum\limits_{i=1}^{n} (x_i - \bar{x})^2}{n - 1}. \tag{23.2}$$

The divisor in the expression for the sample variance is not n as might be expected but $n - 1$, as this can be shown to give a better estimate of the population variance. Expanding and rearranging (23.2) leads to

$$s^2 = \frac{\Sigma(x_i^2 - 2\bar{x}x_i + \bar{x}^2)}{n - 1}$$

$$= \frac{\Sigma x_i^2 - 2\bar{x}\Sigma x_i + n\bar{x}^2}{n - 1}$$

$$= \frac{\sum\limits_{i=1}^{n} x_i^2 - n\bar{x}^2}{n - 1}, \tag{23.3}$$

which is more convenient for calculation. By taking the positive square root of s^2 the **sample standard deviation**, s, is obtained,

$$s = \sqrt{\left(\frac{\Sigma x_i^2 - n\bar{x}^2|}{n - 1}\right)}. \tag{23.4}$$

If the observations are arranged in ascending order, then the **sample median** is the middle observation when n is an odd number, and the average of the two middle observations when n is an even number. The **sample range** is the difference between the largest and smallest observations in the sample.

The **sample mode** is that value which occurs in the sample most often. When each observation occurs only once in the sample then the mode does not exist.

Example 1: The number of heartbeats per minute of a patient recorded on ten successive days was as follows

$$73, \quad 72, \quad 73, \quad 74, \quad 76, \quad 70, \quad 71, \quad 72, \quad 72, \quad 74 \ .$$

Calculate the sample estimates.

$$\text{Sample mean}, \ \bar{x} \ = \ \frac{\sum\limits_{i=1}^{10} x_i}{10} \ = \ \frac{73 + 72 + \ldots + 74}{10} \ = \ 72.7 \ .$$

$$\text{Sample variance}, s^2 \ = \ \frac{\sum\limits_{i=1}^{10} x_i^2 - n\bar{x}^2}{n-1}$$

$$= \ \frac{(73^2 + 72^2 + \ldots + 74^2) - 10 \times (72.7)^2}{9}$$

$$= \ 2.9 \ .$$

Sample standard deviation, $s = \sqrt{2.9} = 1.7$.

Arranging the observations in ascending order

$$70, \quad 71, \quad 72, \quad 72, \quad 72, \quad 73, \quad 73, \quad 74, \quad 74, \quad 76,$$

the sample median will be $\dfrac{72 + 73}{2} = 72.5$, and the sample range will be $76 - 70 = 6$.

Since the value 72 occurs most often it is the sample mode.

Example 2: The following observations are the body temperatures measured in °C of twelve squirrels during hibernation,

$$6.0, \ 5.5, \ 6.1, \ 6.2, \ 7.5, \ 5.9, \ 6.4, \ 6.9, \ 6.7, \ 5.7, \ 6.3, \ 5.1 \ .$$

Calculate estimates for the population mean, variance, median and mode.

Estimates for the population mean μ and variance σ^2 are the sample mean

$$\bar{x} = 6.19 \ ,$$

and the sample variance

$$s^2 = 0.415 \ .$$

The population median will be estimated by the sample median which is 6.15.

An estimate for the population mode is not possible since, no value occurs more frequently than the others.

In each of the previous examples the number of observations in the sample was small. When the number of observations becomes large the sample estimates are easier to calculate if the observations are grouped into **class intervals**. If the n

observations are grouped into c class intervals and X_j denotes the middle of the jth class interval and f_j the number of observations which fall into the jth class interval, then the data can be concisely recorded in a **frequency table**. In practice the class intervals can be of any size but are preferably chosen to give at least six equal class intervals over the range of observations. A graphical illustration of the frequency table, with frequency plotted along the y-axis and the class intervals along the x-axis, gives a frequency distribution **histogram**.

Example 3: Fifty plants were examined on the tenth day after planting and their daily increment in height recorded in mm as follows:

3.15,   3.54,   3.69,   3.80,   3.57,   3.66,   4.27,   3.92,   3.97,   3.52,

3.55,   3.68,   3.88,   4.08,   4.00,   4.03,   3.86,   3.31,   3.35,   3.51,

3.51,   3.47,   3.40,   3.59,   3.64,   3.79,   3.89,   3.73,   3.88,   3.79,

4.06,   4.03,   3.87,   3.74,   3.67,   3.42,   3.60,   3.70,   3.89,   4.13,

3.60,   3.73,   3.74,   3.97,   3.96,   3.75,   3.47,   3.41,   3.61,   3.75.

Use a frequency table with suitably chosen class intervals to illustrate the data and plot the histogram associated with the frequency distribution.

The observations range from 3.15 to 4.27, which is a range of 1.12, and so it is convenient to choose six class intervals each of size 0.20. The frequency table becomes:

Mid-point of class interval X_j	3.20	3.40	3.60	3.80	4.00	4.20
Frequency f_j	1	7	15	16	9	2

Thus, for example, the second class interval ranges from 3.30 to 3.50 and will contain all values from 3.30 to 3.49 inclusive. The mid-point X_2 is equal to 3.40 and since this class interval contains the height increments 3.31, 3.35, 3.47, 3.40, 3.47, 3.41 and 3.42 the frequency f_2 is equal to 7. The histogram for the data is shown in Fig. 23.1.

For observations tabulated in a frequency table the sample estimates can be calculated using the mid-points of the class intervals and their associated frequency. The sample mean

$$\bar{x} = \frac{f_1 X_1 + f_2 X_2 + \ldots + f_c X_c}{n}$$

$$= \frac{\sum_{j=1}^{c} f_j X_j}{n}. \tag{23.5}$$

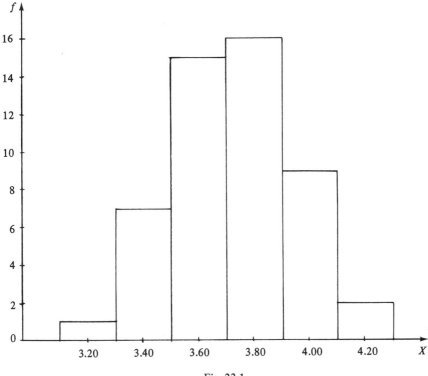

Fig. 23.1

The sample variance

$$s^2 = \frac{f_1(X_1 - \bar{x})^2 + f_2(X_2 - \bar{x})^2 + \ldots + f_c(X_c - \bar{x})^2}{n - 1}$$

$$= \frac{\sum\limits_{j=1}^{c} f_j(X_j - \bar{x})^2}{n - 1} , \qquad (23.6)$$

or in a more convenient form for calculation

$$s^2 = \frac{\sum\limits_{j=1}^{c} f_j X_j^2 - n\bar{x}^2}{n - 1} . \qquad (23.7)$$

The sample standard deviation is the square root of the sample variance s^2. The sample median, range and mode are as previously defined, but as shown in the following example these estimates can be obtained easily by inspection of the frequency table.

Example 4: For the treatment of superficial skin infections, solutions which should consist of 1000 units of sodium penicillin per ml in sterile water are bottled. The penicillin content was analysed in a sample of 200 bottles and the following frequency table obtained:

Penicillin units X_j	880	910	940	970	1000	1030	1060	1090
Frequency f_j	10	16	45	52	31	25	15	6

Calculate the sample mean and variance. Contrast these estimates with the sample median and range. Calculate the sample mode. What conclusion do the results suggest?

$$\text{Sample mean, } \bar{x} = \frac{\sum\limits_{j=1}^{8} f_j X_j}{200} = \frac{10 \times 880 + 16 \times 910 + \ldots + 6 \times 1090}{200}$$

$$= 976.45 \ .$$

Sample variance,

$$s^2 = \frac{\sum\limits_{j=1}^{8} f_j X_j^2 - 200 \bar{x}^2}{199}$$

$$= \frac{10 \times 880^2 + 16 \times 910^2 + \ldots + 6 \times 1090^2 - 200 \times (976.45)^2}{199}$$

$$= 2495.37 \ .$$

The sample median will be the average of the 100th and 101st observations when the observations are ranked in ascending order. Accumulating the frequencies from left to right in the frequency table, the 100th and 101st values will both lie in the class interval with mid-point 970, and hence 970 is the sample median. The sample range is simply $1090 - 880 = 210$.

The sample mean and median are close since the observations are almost symmetrically distributed around the mean. The large values for the variance and range reflect the wide spread about the mean.

The sample mode is 970 since most observations fall into the class interval with mid-point 970.

The results suggest that the mean penicillin content is not up to the expected level of 1000 units, and furthermore the wide spread suggests that many bottles will not contain the required concentration.

23.3 SAMPLING DISTRIBUTIONS

It has already been emphasised that if X is a random variable with mean μ and variance σ^2, then the sample mean $\bar{x}$ and variance s^2, calculated from a sample of size n taken from the population of X, will be estimates for μ and σ^2 respectively. This is illustrated in Fig. 23.2 where the shaded area represents the sample.

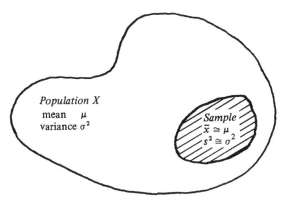

Fig. 23.2

However, the value of $\bar{x}$ will vary depending on which sample of n observations is chosen. If, for example, five samples of size n are taken, each one is likely to give a different value $\bar{x}_1, \bar{x}_2, \ldots, \bar{x}_5$, all of which are estimates for μ. This is illustrated in Fig. 23.3.

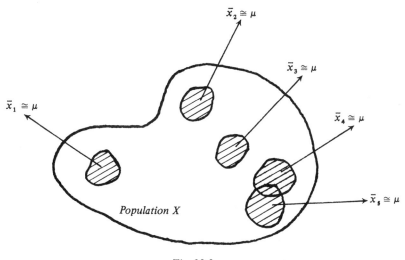

Fig. 23.3

Thus, just as X is a random variable with a probability distribution, mean μ and variance σ^2, so $\bar{X}$, denoting the collection of sample means $\{\bar{x}_1, \bar{x}_2, \ldots\}$ will be a random variable with some distribution, mean $\mu_{\bar{x}}$ and variance $\sigma^2_{\bar{x}}$. The positive square root of $\sigma^2_{\bar{x}}$ is $\sigma_{\bar{x}}$, the standard deviation of $\bar{X}$, which is often given the special name the **standard error**. The distribution of $\bar{X}$ is important since if the standard error is large this suggests that each time a single sample is taken and $\bar{x}$ calculated, then the chances are high that $\bar{x}$ will not be close to μ and therefore not a good estimate for μ.

Similarly the sample variance and all other sample estimates will be different for each sample of size n, and each will have a probability distribution. These probability distributions are called **sampling distributions**. For most elementary applications of statistics to experimental data a knowledge of the sampling distribution of $\bar{X}$ is sufficient.

23.4 SAMPLING DISTRIBUTION OF $\bar{X}$

The question that arises is: if $\bar{X}$ is the random variable denoting the sample mean calculated from samples of size n drawn from a population X with mean μ and variance σ^2, what is its probability distribution, mean and variance? Fortunately statisticians considered this question many years ago and the answer is as follows:

(a) If X has any non-normal probability distribution with mean μ and variance σ^2, then the distribution of $\bar{X}$ approaches the normal distribution with mean $\mu_{\bar{x}} = \mu$ and variance $\sigma^2_{\bar{x}} = \sigma^2/n$ as the sample size n increases.

(b) If X has a normal probability distribution with mean μ and variance σ^2, then the distribution of $\bar{X}$ has a normal distribution with mean $\mu_{\bar{x}} = \mu$ and variance $\sigma^2_{\bar{x}} = \sigma^2/n$.

The first part follows from a famous theorem known as the **Central Limit Theorem** and is particularly fascinating since X can have any probability distribution and yet the sampling distribution of $\bar{X}$ will always be close to a normal distribution for large n. Using the properties of the normal distribution it is then possible to calculate the probability of a sample mean lying in any interval. In practice by large n we usually mean $n \geqslant 30$. In the next chapter it will be seen that this is why it is most important in experimental situations to use as large a sample size as possible in order that the distribution of $\bar{X}$ will be close to the normal distribution.

As a demonstration of the Central Limit Theorem consider a mutant population of female flies which is kept in captivity. Suppose the ovarian period is recorded for each member of the population and found to have mean μ equal to 48 hours, variance σ^2 equal to 25 hours^2, and probability distribution as shown in Fig. 23.4(a). The Central Limit Theorem states that if samples of size, say 30, are taken and the sample mean calculated for each sample, then the probability distribution of the sample mean will be almost normal, with mean $\mu_{\bar{x}}$ equal to 48 hours and variance $\sigma^2_{\bar{x}}$ equal to 25/30 hours^2, as shown in Fig. 23.4(b).

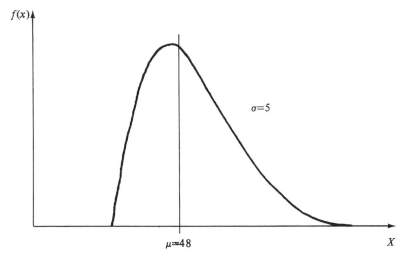

Fig. 23.4(a)

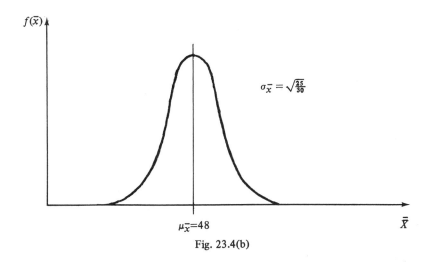

Fig. 23.4(b)

Furthermore, using the distribution of $\bar{X}$, it is seen that if a sample of size 30 is taken, the sample mean will lie between 46 and 50 hours with probability 0.97 since

$$P(46 \leqslant \bar{X} \leqslant 50) = P(\bar{X} \leqslant 50) - P(\bar{X} \leqslant 46) \ ,$$

and taking the distribution of $\bar{X}$ to be normal,

$$P(\bar{X} \leqslant 50) - P(\bar{X} \leqslant 46) = \Phi\left(\frac{50 - \mu_{\bar{x}}}{\sigma_{\bar{x}}}\right) - \Phi\left(\frac{46 - \mu_{\bar{x}}}{\sigma_{\bar{x}}}\right)$$

$$= \Phi\left(\frac{50 - 48}{\sqrt{(\frac{25}{30})}}\right) - \Phi\left(\frac{46 - 48}{\sqrt{(\frac{25}{30})}}\right)$$

$$= \Phi(2.19) - \Phi(-2.19)$$

$$= 0.97 \ .$$

Example 1: The diameter, D, of a population of *staphylocci* bacteria have mean 1 micro mm and variance 0.25 micro mm². What is the probability of a bacterium having a diameter between 0.75 and 1.25? If a sample of 36 bacteria is measured what is the probability of the sample mean lying between 0.75 and 1.25?

Since D has mean $\mu = 1$ and variance $\sigma^2 = 0.25$ the sample mean $\bar{D}$ will have mean $\mu_{\bar{D}} = 1$ and variance $\sigma^2_{\bar{D}} = 0.25/36$.

Assuming that the diameters are normally distributed,

$P(\text{bacterium has diameter between 0.75 and 1.25}) = P(0.75 \leqslant D \leqslant 1.25)$

$$= P(D \leqslant 1.25) - P(D \leqslant 0.75)$$

$$= \Phi\left(\frac{1.25 - \mu}{\sigma}\right) - \Phi\left(\frac{0.75 - \mu}{\sigma}\right)$$

$$= \Phi\left(\frac{1.25 - 1}{0.5}\right) - \Phi\left(\frac{0.75 - 1}{0.5}\right)$$

$$= \Phi(0.5) - \Phi(-0.5)$$

$$= 0.3830 \ .$$

$P(\text{sample mean has diameter between 0.75 and 1.25}) = P(0.75 \leqslant \bar{D} \leqslant 1.25)$

$$= P(\bar{D} \leqslant 1.25) - P(\bar{D} \leqslant 0.75)$$

$$= \Phi\left(\frac{1.25 - \mu_{\bar{D}}}{\sigma_{\bar{D}}}\right) - \Phi\left(\frac{0.75 - \mu_{\bar{D}}}{\sigma_{\bar{D}}}\right)$$

$$= \Phi(3) - \Phi(-3)$$

$$= 0.9972 \ .$$

Hence on approximately 99.7% of the occasions that a sample of size 36 is drawn the sample mean will have a value in the interval 0.75 to 1.25, although only 38.3% of the bacteria have a diameter in this range.

Example 2: In an experiment a sample is to be drawn from a population of 10-day old chicks whose weights, W, are normally distributed with mean 175 gm and standard deviation 15 gm. What size should the sample be in order that the sample mean will lie within

(i) 160 and 190 gm with probability 0.95,

(ii) 5% of the mean with probability 0.95?

Since W is normally distributed with mean 175 gm and standard deviation 15 gm, the sample mean $\bar{W}$ will also be normally distributed with mean 175 gm and standard deviation $15/\sqrt{n}$ gm.

(i) The value of n is to be found that satisfies the equation

$$P(160 \leqslant \bar{W} \leqslant 190) = 0.95 \text{ and so}$$

$$\Phi\left(\frac{190 - 175}{15/\sqrt{n}}\right) - \Phi\left(\frac{160 - 175}{15/\sqrt{n}}\right) = 0.95$$

$$\Phi(\sqrt{n}) - \Phi(-\sqrt{n}) = 0.95$$

$$\Phi(\sqrt{n}) - (1 - \Phi(\sqrt{n})) = 0.95$$

$$\Phi(\sqrt{n}) = 0.975 \ .$$

Using the standard normal table the above equation is true when $\sqrt{n} = 1.96$, and since n must be an integer a sample of size 4 must be used.

(ii) The value of n is to be found that satisfies the equation

$$P(\mu - 0.05\,\mu \leqslant \bar{W} \leqslant \mu + 0.05\,\mu) = 0.95 \ .$$

Since $\mu = 175$ then n must be found so that

$$P(166.25 \leqslant \bar{W} \leqslant 183.75) = 0.95 \ .$$

Therefore

$$\Phi\left(\frac{183.75 - 175}{15/\sqrt{n}}\right) - \Phi\left(\frac{166.25 - 175}{15/\sqrt{n}}\right) = 0.95$$

and so

$$\Phi(0.58\,\sqrt{n}) - \Phi(-0.58\,\sqrt{n}) = 0.95 \ .$$

Simplifying gives

$$\Phi(0.58\,\sqrt{n}) = 0.975 \ ,$$

which is true when $n = 12$.

23.5 CONFIDENCE INTERVALS

In the Central Limit Theorem given in Section 23.4 it was stated that if samples of size n are collected from a population which has mean μ and variance σ^2, the distribution of the sample means will be approximately normal with mean μ and variance σ^2/n. From this it follows that 95% of the sample means obtained will lie in the interval $\mu - 1.96\,\sigma/\sqrt{n}$ to $\mu + 1.96\,\sigma/\sqrt{n}$ since

$$P(\mu - 1.96\,\sigma/\sqrt{n} \leqslant \bar{X} \leqslant \mu + 1.96\,\sigma/\sqrt{n})$$

$$= \Phi\left(\frac{\mu + 1.96\,\sigma/\sqrt{n} - \mu}{\sigma/\sqrt{n}}\right) - \Phi\left(\frac{\mu - 1.96\,\sigma/\sqrt{n} - \mu}{\sigma/\sqrt{n}}\right)$$

$$= \Phi(1.96) - \Phi(-1.96)$$

$$= 0.95 \ .$$

It would be useful to know the above interval, as the smaller it is, the more likely it is a sample will provide a sample mean close to the population mean μ. However, since μ and σ^2 are unknown, it can be estimated for large n using $\bar{x} \pm 1.96\,s/\sqrt{n}$, where $\bar{x}$ and s^2 are estimates for μ and σ^2 respectively, obtained from experimental data. This estimate is known as the **95% confidence interval** and will include the true mean with probability 0.95. Other confidence intervals can similarly be derived and in particular the **99% confidence interval** is found to be $\bar{x} \pm 2.57\,s/\sqrt{n}$.

When the sample size is small the estimate for σ^2 is not so reliable and the constants 1.96 and 2.57 must be replaced by numbers obtained from a t-distribution. This distribution is discussed further in the following chapter. The 95% confidence interval becomes $\bar{x} \pm t_\nu\,(0.025)\,s/\sqrt{n}$ where the value for $t_\nu\,(0.025)$ is found by reading across row $\nu = n - 1$ and under column 0.025 of the table of t-values in Appendix F. The 99% confidence interval is similarly determined using $\bar{x} \pm t_\nu\,(0.005)\,s/\sqrt{n}$.

23.6 CHOICE OF SAMPLE SIZE IN A SINGLE EXPERIMENT

One of the most important questions in experimental design is how large a sample should be taken so that $\bar{x}$ will be a good estimate for the population mean μ. This question cannot be answered unless the experimenter specifies

(i) how accurate the estimate for μ is required to be, and

(ii) how often he is prepared to accept that the estimate will not have the required accuracy if he were to repeat the experiment many times.

Once these questions are answered the optimal sample size can be determined. The discussion on the behaviour of the sample mean in Section 23.4 stated that when many samples of size n are taken from a population with mean μ and variance σ^2 the sample means, $\bar{X}$, have mean μ and variance σ^2/n. Therefore as n becomes large the values of $\bar{X}$ have less spread, as illustrated in Fig. 23.5.

In practice the experimenter only takes one sample from which $\bar{x}$ is calculated as an estimate for μ. It follows from the probability curves of $\bar{X}$ shown in Fig. 23.5 that as n becomes large it is more likely that $\bar{x}$ will be close to μ. In

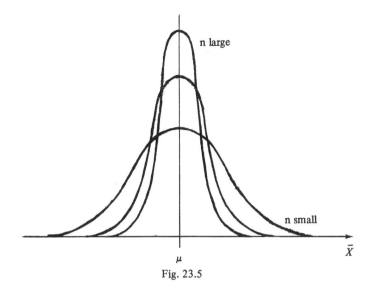

Fig. 23.5

order that 95% of the $\bar{X}$ values lie within a units of the population mean μ, that is, lie between $\mu - a$ and $\mu + a$, the value of n is chosen using the formula

$$n = 1.96^2 \frac{\sigma^2}{a^2} .$$

In practice σ^2 is not likely to be known, so a pilot experiment must be carried out from which s^2 can be obtained as an estimate for σ^2, and n is found using

$$n = 1.96^2 \frac{s^2}{a^2} .$$

Alternatively, if it is important to lie within a units of μ on 99% of the occasions the experiment is repeated, then n is given by

$$n = 2.576^2 \frac{s^2}{a^2} .$$

These values for n are only guidelines, and in practice, if possible, the choice of n should be made larger than that calculated.

Example: In a pilot experiment the standard deviation, s, of the number of *N. brazilenses* worms in infected rats was found to be 287. If in 95% of the future experiments $\bar{x}$ is to lie within 250 worms of the true mean worm count, how many rats should be used?

The number of rats, n, is given by

$$n = 1.96^2 \frac{s^2}{a^2} = 1.96^2 \frac{287^2}{250^2} = 5.06 .$$

The number of rats to be chosen should be six. Thus if the experiment is repeated a large number of times using six rats then in 95% of these experiments the sample mean calculated will lie within 250 of the true mean.

If this is not accurate enough, and it is required to be within 100 of the true mean, then the sample size must be increased considerably:

$$n = 1.96^2 \frac{287^2}{100^2} = 31.64$$

and thirty two rats are required.

23.7 CHOICE OF SAMPLE SIZES FOR EXPERIMENTS WITH SEVERAL TREATMENTS

In many experimental situations the experimenter has a fixed number of animals, or people, to allocate to the control and treated groups. Suppose there are two groups – control X and treated Y. A pilot experiment gives s_x^2 and s_y^2 as estimates for σ_x^2 and σ_y^2 respectively. Assuming s_y^2 is greater than s_x^2, this suggests that when samples of size n are taken from each group then $\bar{x}$ is more likely to be closer to μ_x than $\bar{y}$ will be to μ_y, since $s_x/\sqrt{n}$ will be smaller than $s_y/\sqrt{n}$. This is illustrated in Fig. 23.6.

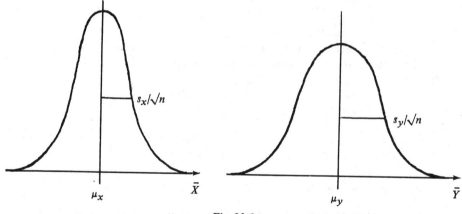

Fig. 23.6

If the accuracy of the estimates for μ_x and μ_y is to be the same, then, taking a sample of size n_x from the control group and one of size n_y from the treated group, the distributions of $\bar{X}$ and $\bar{Y}$ must have the same spread:

$$\frac{s_x}{\sqrt{n_x}} = \frac{s_y}{\sqrt{n_y}} \ .$$

Rearranging gives $\dfrac{n_x}{n_y} = \dfrac{s_x^2}{s_y^2}$ and so $n_x : n_y = s_x^2 : s_y^2$.

Thus n_x and n_y should be chosen in the same ratio as the sample variances. (In most cases it is found that the sample variance of the treated group is larger than the control group. This leads to n_y being larger than n_x, and so more animals are assigned to the treated group.)

The same procedure can be extended to more than two groups. For three groups X, Y and Z the sample sizes are determined using

$$\frac{s_x}{\sqrt{n_x}} = \frac{s_y}{\sqrt{n_y}} = \frac{s_z}{\sqrt{n_z}}$$

and so $n_x : n_y : n_z = s_x^2 : s_y^2 : s_z^2$.

Example 1: One hundred animals are to be allocated to two experiments A and B in which pre-patency periods of a disease are to be compared. A pilot experiment gives the sample variances of pre-patency periods for treatments A and B to be 9.4 days^2 and 27.3 days^2 respectively. How many animals should be allocated to each treatment?

Here $n_A : n_B = s_A^2 : s_B^2$

$$= 9.4 : 27.3 .$$

The ratio can be approximated by

$$n_A : n_B = 1 : 3 ;$$

therefore $n_A = \dfrac{1 \times 100}{(1 + 3)} = 25$,

and $n_B = \dfrac{3 \times 100}{(1 + 3)} = 75$.

Example 2: If a third treatment C is introduced in example 1, in which the sample variance of pre-patency period is found to be 40 days^2, how many animals should be allocated to each treatment?

Here $n_A : n_B : n_C = s_A^2 : s_B^2 : s_C^2$.

Approximating the ratios gives

$$n_A : n_B : n_C = 1 : 3 : 4\tfrac{1}{2}$$

$$= 2 : 6 : 9 .$$

Therefore $n_A = \dfrac{2 \times 100}{(2+6+9)} = \dfrac{2 \times 100}{17} \cong 12$,

$n_B = \dfrac{6 \times 100}{(2+6+9)} = \dfrac{6 \times 100}{17} \cong 35$,

$n_C = \dfrac{9 \times 100}{(2+6+9)} = \dfrac{9 \times 100}{17} \cong 53$.

So 12, 35 and 53 animals should be allocated to the three groups A, B and C respectively.

PROBLEMS

1. The following periods for biological rhythms were observed in fifteen people living in underground bunkers and isolated from all knowledge of solar time:

   Period (hrs)

   26.0   25.5   26.5   24.3   24.2   26.5   27.4   26.6   25.3

   26.1   25.9   25.4   26.2   25.1   27.1.

   Calculate the sample mean and variance.

2. The following results show the height in metres at which birds from species A and birds from species B were observed to forage.

   Foraging height

   Species A   17 10 13 12 13 11 13 16 17 19

   Species B   20 22 22 23 20 21 22 29 25 21 29 20

   Calculate the mean and standard deviation of the foraging heights for each species. Comment on the results.

   Determine the median and range for each species. Which of the two statistics, the standard deviation or the range, is a better reflection of the variability in the foraging height?

3. The length of stay in a hospital, measured in days, of a sample of patients being treated for femoral fractures was as follows:

   17 23 67 60 24 27 47 55 42 30 32 37 42 37 .

   Calculate the mean and variance of the length of stay in hospital. Do you think that the age of the patients might influence the length of stay and if so what precautions should be taken when collecting the sample?

4. In a study of a tropical savannah ecosystem, the number of termite nests in each of forty one-hectare areas was counted and classified as follows:

Number of nests	0–	10–	20–	30–	40–	50–	60–	70–
Frequency	0	1	2	2	5	19	11	0

   Calculate the sample mean and variance. What is the mode of the distribution? Plot the corresponding histogram and state if you think it would be reasonable to regard the number of termite nests per hectare to be normally distributed.

5. During a cloud seeding project, an aircraft was used to seed the clouds with silver iodide on sixty days on which certain meteorological condi-

tions occurred. The intensity of precipitation that followed was then classified from the highest level, 1, to the lowest level, 6.

Level	1	2	3	4	5	6
Frequency (days)	26	14	10	5	5	0

What are the median, range and mode of the data?

6. The following table shows the number of sedge *Carex flacca* plants found in 1000 quadrats:

Number of plants per quadrat	0	1	2	3	4	5	6	7
Frequency	368	366	185	61	15	3	1	1

   (a) Calculate the sample mean and variance. Comment on these results.

   (b) Assuming the plants are distributed with Poisson frequency, what would be the mean and variance of the number of plants per quadrat if the size of the quadrat was quadrupled?

7. Ultra-violet radiation, of wavelength less than 2000 ÅU, produces a pigment in the lower living cells of human skin which then causes an inflammation which is helpful in the treatment of some skin diseases. For thirty patients with the same skin disease the wavelengths, measured in ÅU, found to be successful in treating the disease were recorded:

Wavelengths	1400–	1450–	1500–	1550–	1600–	1650–
Number of patients	1	3	19	4	2	1

Calculate the sample mean $\bar{x}$ and variance s^2 of the wavelengths. If it is assumed that the wavelengths are normally distributed, then using $\bar{x}$ and s^2 as estimates of the population mean μ and variance σ^2, calculate the probability of a skin disease being successfully treated using radiation of wavelength greater than 1550 ÅU. How does this result compare with the observed estimate of probability?

8. The residues of fungicide, measured in parts per million, in a random sample of fifty fresh apples harvested forty days after receiving the last of six sprays of the fungicide, were as follows:

1.63	1.40	1.64	1.30	1.49	1.58	1.03	1.06	1.33	1.20
1.52	1.87	1.83	1.97	1.62	1.21	1.01	1.14	1.58	1.76
1.43	1.41	1.51	1.15	1.61	1.10	1.03	1.84	1.61	1.54
1.71	1.32	1.29	1.82	1.99	1.43	1.53	1.56	1.48	1.58
1.82	1.81	1.21	1.73	1.59	1.99	1.34	1.23	1.65	1.99.

Arrange the results in a grouped frequency table and calculate the sample mean and variance from this table.

9. In a nervous system the speed in metres per second at which impulses travel in axons of the same diameter were as follows:

   62   64   61   69   60   62   71   78   73   64   69   76

   70   70   62   60   61   75   70   77   76   73   73   72

   69   68   73   78   62   71   75   73   59   58   66   69

   63   68   65   67   77   73   65   66   69   71   65   62 .

   Using suitable statistics contrast these results with the following speeds for impulses observed in axons with a larger diameter :

   68   68   73   67   70   74   68   70   66   69

   79   64   74   72   69   79   82   76   64   70

   78   73   69   65   69   68   71   71   70   63

   70   70   71   67   68   61   66   65   68   71

   74   69   66   68   74   70   69   73   72   68 .

10. The following data represent the age at onset of anorexia nervosa in a random sample of thirty-five patients:

    12.4   11.9   11.5   13.3   14.2   11.4   13.2   13.7   12.1   11.7

    13.2   12.7   14.5   14.3   14.4   13.7   11.9   12.1   14.7   11.7

    13.7   11.5   14.2   11.4   13.7   14.0   15.0   11.6   11.2   13.9

    11.7   14.1   11.9   12.9   14.8   13.6   12.6   13.2   12.8   13.7

    12.0   12.2   13.0   11.9   14.0   12.5   14.3   11.8   13.4   11.7.

    Choosing suitable class intervals, find the sample mean and the sample standard deviation. Draw a histogram of the data.

11. The angle between two adjacent toes was measured from radiographs of the affected feet of fifty young adults undergoing treatment for a foot abnormality:

    Angle between toes (degrees)

    42   32   33   33   29   31   33   29   40   31

    27   30   29   43   34   29   34   29   28   30

    36   46   30   41   45   31   30   33   29   29

    33   35   37   27   29   43   32   27   32   32

    39   41   44   32   35   29   31   28   28   29 .

Similar measurements were made on the feet of forty normal young adults:

Angle between toes (degrees)

12	18	13	15	16	12	15	18	15	15
17	15	16	17	17	16	18	13	12	15
14	15	12	14	14	18	17	18	12	14
13	12	12	14	17	16	12	16	15	13 .

Choosing a suitable class interval, arrange each set of results in a frequency table. Compare the means and standard deviations of the sets of data.

12. If pears of a certain cultivar are known to come from a population with a mean storage life of 6.25 weeks and variance 2.25 weeks^2, what is the probability in a sample of size thirty-six that the sample mean will lie between 6.5 and 6.75 weeks?

13. The amount of hay, X, consumed by a calf during a day can be taken to be normally distributed with mean 9 kg and standard deviation 1.5 kg. Find the probability that

   (i) X lies between 8.5 kg and 10.25 kg,

   (ii) $\bar{X}$, the mean of samples of size 9, lies between 8.5 kg and 10.25 kg.

14. Portions of prepared meat should have pH values with mean 5.6 and standard deviation 1.1. The portions will be rejected if the pH value exceeds 6.6, in order to minimise the risk of bacteria spoiling them. In a consignment of such meat, assuming the pH values are normally distributed, what percentage of the portions will be rejected if each portion is inspected? If a random sample of ten portions is taken from the consignment and the consignment rejected if the pH value of the sample mean exceeds 6.6, what is the probability of the consignment being rejected? What is the probability if the sample size is increased to twenty?   (Strathclyde, 1980)

15. The transit times, T, of impulses across a membrane are measured in μ sec. and distributed according to a $N(1860, 4624)$. Find

   (i) $P(T \geqslant 2000)$,

   (ii) $P(T \leqslant 1750)$.

   If samples of four impulses are observed, find

   (iii) $P(\bar{T} \geqslant 2000)$,

   (iv) $P(\bar{T} \leqslant 1750)$,

   where $\bar{T}$ is the mean transit time of the sample.

16. The rate of water flow at a certain location on a river is estimated to have a mean of 62 cubic metres per second and a variance of 49. What can you say about the probability distribution of the mean of samples of forty measurements taken from the location?

17. The levels of radiation from terrestrial background sources such as buildings, rocks etc. absorbed by ten individuals were as follows:

Radiation levels (units per year)

95   107   103   90   123   100   162   76   123   176 .

Calculate the sample mean and variance. If the sample mean is to be within 10 units of the true mean in 95% of such samples collected, how large a sample should be taken?

18. In an experiment, the viral activity in the bloodstream of inoculated mice is to be determined. A previous experiment gave a mean activity level of 248 units and a standard deviation of 27 units. In how many mice should the activity be measured if the sample mean is to be within

(i) 15 units,

(ii) 30 units,

of the true mean in 99% of such experiments?

19. If you wish to estimate the mean of a normal distribution with variance 9, how large a sample should you take so that the probability is 0.9 that your estimate will not depart from the mean by more than 0.5?

20. The time until the detection of parasitaemia in infected Zebu and N'Dama cattle is to be measured. A pilot study indicates that the standard deviations of these times in the Zebu and N'Dama cattle are 5.6 days and 10.2 days respectively. A total of forty cattle can be bought for the experiment. How many of each type of cattle should be purchased in order to assess the time until detection in each type of cattle with the same accuracy?

21. The quantity of radiant solar energy available to plants kept under one of two glass-houses, A and B, of different design, is to be investigated. A preliminary experiment estimated the variance of the energy levels to be 0.26 units^2 in glass-house A and 0.78 units^2 in glass-house B where a unit is 10^8 cal/m^2/year. If a total of 100 observations are to be taken, how many observations should be taken in each glass-house if the mean energy levels are to be determined with the same accuracy?

22. The severity of a patients' illness can be classified as mild, moderate
     or severe. In a trial, the percentage packed cell volume (P.C.V.) in their
     blood is to be measured for each patient on admission to hospital. The
     variances of the percentage P.C.V. values for mild, moderate and severe
     illnesses are estimated to be 2.2, 2.3 and 7.0 respectively. If one hundred
     and fifty patients are to be entered and the mean percentage P.C.V. values
     assessed with the same accuracy for each type of illness, estimate how
     many patients should be within each group.

CHAPTER 24

Parametric tests

24.1 INTRODUCTION

In statistics, **parametric tests** are those which involve inference about the population parameters. The tests contained in this chapter deal specifically with the population mean parameter. Since the sample mean gives an estimate of the population mean it is important to be familiar with the probability distribution of the sample mean as stated in the Central Limit Theorem in Section 23.4. Central to statistical testing is the formulation and testing of **hypotheses**. In parametric tests, a hypothesis is a statement about what is believed to be the value of the population parameter before experimental data is presented. After presentation and analysis of the data the hypothesis will be accepted as true or rejected. A discussion of hypothesis testing is given in detail for the one sample z- and t-tests. In subsequent tests the hypotheses are formulated and the main steps for carrying out the tests are listed. Before applying any of the tests the validity of the tests should be considered carefully.

24.2 ONE SAMPLE z-TEST

A sample of n observations is drawn from a population X with known variance σ^2. The population mean μ is unknown but is thought to be μ_0. It has to be decided which of the two alternatives $\mu = \mu_0$ or $\mu \neq \mu_0$ is to be accepted.

Rationale: The alternatives $\mu = \mu_0$ and $\mu \neq \mu_0$ in statistical terms are called the **Null hypothesis, H_0**, and the **Alternative hypothesis, H_1**, respectively, and are written

$$H_0 : \mu = \mu_0$$
$$H_1 : \mu \neq \mu_0 \ .$$

If the Null hypothesis is correct and the sample size is large (usually greater than 30) then, from the Central Limit Theorem, the sample mean $\bar{X}$ will have a normal distribution with mean μ_0 and variance σ^2/n. It follows from the discussion on the normal distribution in Section 22.6 that the transformed variable

$Z = \dfrac{\bar{X} - \mu_0}{\sigma/\sqrt{n}}$ will have a standard normal distribution, as shown in Fig. 24.1, with total area under the curve equal to 1.

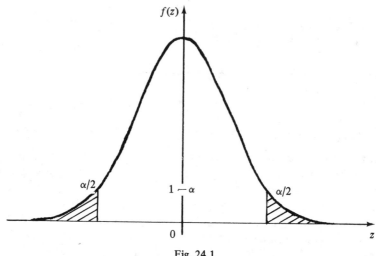

Fig. 24.1

From the probability curve it is seen that most values of Z will lie close to 0. This suggests that a simple criterion to decide which hypothesis is reasonable is to obtain a sample mean $\bar{x}$ and if $z = \dfrac{\bar{x} - \mu_0}{\sigma/\sqrt{n}}$ is close to 0 then accept H_0. As a measure of closeness, if z takes a value corresponding to the unshaded area then H_0 is accepted, and if z takes a value corresponding to the shaded area then H_1 is accepted on the basis that z is too far away from 0. For this reason z is called the **test statistic** and the shaded area is called the **critical region**. The area of the critical region is called the **significance level** of the test and is sometimes denoted by α.

The choice of significance level for a test is not easy, and by convention is often chosen before analysis of the data to be either 5% or 1%. An alternative approach that is sometimes used is to calculate the test statistic, and then determine the significance level in order that this value just falls in the shaded area. For simplicity, in the examples which follow, the former approach will be adopted.

For a 5% significance level the critical region has total area 0.05 with an area of 0.025 in each **tail** of the distribution. From the table of the standard normal integral in Appendix E, the values of z at which the critical region commences, moving from the origin, are -1.96 and 1.96. Thus if the z-statistic obtained from the data is less than -1.96 or greater than 1.96 the Null hypothesis is rejected and the Alternative hypothesis accepted. For a 1% significance level the critical region will have area 0.01 consisting of 0.005 in each tail. The corresponding values of z at which the critical region commences are found from the standard normal table to be -2.57 and 2.57.

It should be understood that the decision process described is not entirely satisfactory, since on some occasions a value of z falling in the critical region will lead to the Null hypothesis being rejected when it is in fact correct. The chance of this mistake happening will depend on the significance level. If the significance level is 5% then a mistake will be made on average 5 times in every 100 times the test is carried out. For a 1% significance level it is only likely to happen once in every 100 times. It is for this reason that it may appear desirable to choose a small significance level. However, the decision process can also lead to the Alternative hypothesis being rejected on some occasions when it is correct and the smaller the significance level of the test the more likely this type of mistake will occur. It can only be hoped that the decision arrived at in a single test is not one of those occasions when a mistake is made.

Sometimes the Alternative hypothesis is changed and the hypotheses become

$$H_0 : \mu = \mu_0 \qquad\qquad H_0 : \mu = \mu_0$$
$$\text{or}$$
$$H_1 : \mu > \mu_0 \qquad\qquad H_1 : \mu < \mu_0 .$$

In both cases if the Null hypothesis is correct then the distribution of $Z = \dfrac{\bar{X} - \mu_0}{\sigma/\sqrt{n}}$ will be a standard normal distribution as before. For acceptance of $H_1 : \mu > \mu_0$ the z-statistic must be much greater than 0 and the critical region will only consist of one tail. Using a significance level of 5%, the one tail of the critical region will have an area of 0.05 and from the table of the standard normal integral the z-statistic must be greater than 1.65 for the Alternative hypothesis to be accepted, as shown in Fig. 24.2(a). For acceptance of $H_1 : \mu < \mu_0$ the z-statistic must be much less than 0, and at 5% significance level the critical region commences at -1.65 as shown in Fig. 24.2(b). The critical regions at other significance levels can similarly be found by using the table of the standard normal integral.

When carrying out such a test a suggested procedure is:

1. State the Null and Alternative hypotheses.

2. Choose the significance level of the test.

3. Calculate the sample mean, $\bar{x}$, from the data.

4. Calculate the statistic $z = (\bar{x} - \mu_0)/(\sigma/\sqrt{n})$.

5. Decide if z falls into the critical region, using the table of the standard normal integral.

6. State hypothesis to be accepted.

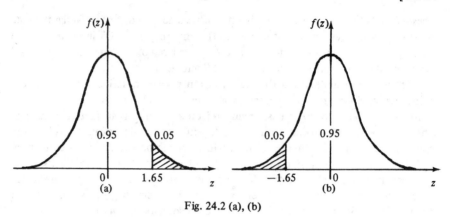

Fig. 24.2 (a), (b)

Validity: The z-test is valid for all large samples ($n > 30$) and for small samples when the probability distribution of X is normal.

Example 1: A forensic scientist knows that glass manufactured in 1935 has a hydrated surface of mean thickness 0.5 μm and variance 0.11 μm^2. Eight glass fragments recovered from the scene of a crime gave a mean thickness of 0.83 μm.

    (a) Is it likely that these fragments were manufactured during 1935?

    (b) If the thickness of the hydrated surface increases with age is it likely that the fragments were manufactured before 1935?

Let μ be the mean thickness of the glass surface and assume that the thickness of the hydrated surface is normally distributed.

    (a) The Null and Alternative hypotheses will be

$$H_0 : \mu = 0.5$$
$$H_1 : \mu \neq 0.5 \ .$$

Choosing a 1% significance level then, since the variance $\sigma^2 = 0.11$ and the sample mean $\bar{x} = 0.83$,

$$z = \frac{\bar{x} - \mu_0}{\sigma/\sqrt{n}}$$

$$= \frac{0.83 - 0.5}{\sqrt{(0.11/8)}} = 2.81 \ .$$

From the table of the standard normal integral the critical regions commence at -2.57 and 2.57, as shown in Fig. 24.3. Since $z = 2.81$ exceeds 2.57 the test statistic falls in the critical region and the Null hypothesis is rejected, and it is believed that the glass was not manufactured in 1935.

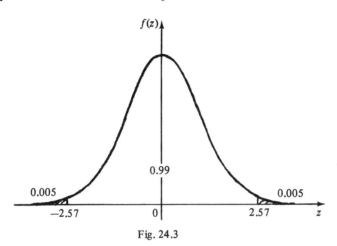

$f(z)$

0.99

0.005

0.005

-2.57

0

2.57

z

Fig. 24.3

(b) If the glass were manufactured before 1935 then the mean thickness would be expected to be greater than 0.5 μm. The hypotheses become

$$H_0 : \mu = 0.5$$
$$H_1 : \mu > 0.5 .$$

The critical region will now only be one-tailed and for a 1% significance level will commence at 2.326. The z-value will still be 2.81, and since 2.81 is greater than 2.326 the Null hypothesis is rejected and it is accepted that the fragments were likely to be manufactured before 1935.

Example 2: 'Cambridge Favourite' strawberry plantations give a mean marketable fruit yield of 17.5 tonne per hectare with variance 0.106 (tonne per hectare)^2 over a two year period. A sample of seven plantations in which the 'Cambridge Rival' variety was grown gave the following yields:

17.60   17.55   17.40   17.00   16.86   17.50   16.90

over a two year period. Assuming that yield is normally distributed, is the yield from the 'Cambridge Favourite' variety greater than that from the 'Cambridge Rival' variety?

Let μ be the mean yield from the 'Cambridge Rival' variety. The hypotheses will be

$$H_0 : \mu = 17.5$$
$$H_1 : \mu < 17.5 .$$

The critical region for a 5% significance level will be one-tailed and commence at -1.645 as shown in Fig. 24.4.

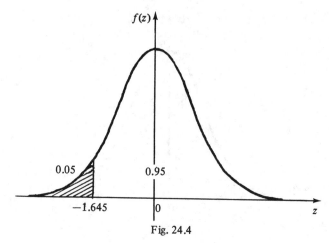

Fig. 24.4

The sample mean, $\bar{x} = \sum_{i=1}^{7} \dfrac{x_i}{7} = 17.26$,

and the test statistic takes the value

$$z = \frac{\bar{x} - \mu_0}{\sigma/\sqrt{n}}$$

$$= \frac{17.26 - 17.5}{0.326/\sqrt{7}} = -1.95 \ .$$

Since $z = -1.95$ does fall in the critical region the Null hypothesis is rejected and there is evidence of the 'Cambridge Favourite' variety giving a better yield.

In this example the significance level was chosen to be 5% and the Null hypothesis rejected. However, if the test were carried out using a 1% significance level the critical region commences at -2.33 and the Null hypothesis would be accepted. This demonstrates the sensitivity of the test to the choice of significance level. Often it is difficult to decide on a satisfactory significance level, and it may be advisable to collect more data when the z-value falls close to the start of the critical region in order to reach a firm conclusion.

24.3 ONE SAMPLE t-TEST

A sample of n observations is drawn from a population X with unknown variance σ^2 and unknown mean μ. The population mean is thought to be μ_0. It has to be decided which of the hypotheses

$$H_0 : \mu = \mu_0$$
$$H_1 : \mu \neq \mu_0$$

is to be accepted.

Rationale: In this case the population variance, σ^2, is unknown and it must be estimated by the sample variance, s^2, obtained from the data.

When the sample size is small and the Null hypothesis is correct the t-statistic,

$$t = \frac{\bar{x} - \mu_0}{s/\sqrt{n}}$$

has a probability distribution known as Student's t-distribution. This new distribution is symmetrical and closely resembles the standard normal distribution shown in Fig. 24.1. The precise shape of the curve depends on the parameter $\nu = n - 1$. The values at which the critical region commences can be read from the table of the t-distribution given in Appendix F. The column heading specifies the area in the upper tail of the distribution and the row heading specifies the parameter ν. For example if the sample size is 10 and the critical region has two tails each of area 0.025 then under the column 0.025 and across the row $\nu = 9$ the value of t at which the **upper tail** critical region commences is seen to be 2.262. By symmetry the **lower tail** critical region commences at -2.262. Thus if the t-statistic obtained from the data is less than -2.262 or greater than 2.262 then the Null hypothesis is rejected and the Alternative hypothesis accepted.

Critical regions can similarly be deduced for other significance levels and for different Alternative hypotheses.

When the sample size is large s^2 will be a good estimate for σ^2, and instead of using the t-statistic the z-test may be carried out, using the z-statistic,

$$z = \frac{\bar{x} - \mu_0}{\sigma/\sqrt{n}}$$

with σ replaced by s.

Suggested steps for carrying out the t-test are:

1. State the Null and Alternative hypotheses.

2. Choose the significance level of the test.

3. Calculate the sample mean, $\bar{x}$, and variance, s^2, from the data.

4. Calculate the statistic $t = (\bar{x} - \mu_0)/(s/\sqrt{n})$.

5. Decide if t falls into the critical region, using the table of the t-distribution with $\nu = n - 1$.

6. State hypothesis to be accepted.

Validity: The t-test is used instead of the z-test when the sample size is small and the population variance has to be estimated by the sample variance. The population X should have a normal distribution.

Example 1: The mean time taken for mice to die when injected with one thousand leukaemia cells is known to be 12.5 days. When the injection dose was doubled in a sample of ten mice, the survival times were

$$10.5, \quad 11.2, \quad 12.9, \quad 12.7, \quad 10.3, \quad 10.4, \quad 10.9, \quad 11.3, \quad 10.6, \quad 11.7.$$

If the survival times are normally distributed do the results suggest that the increased injection dose has decreased survivorship?

If the mean survival time of the mice for double the injection dose is μ, then the hypotheses will be

$$H_0 : \mu = 12.5$$
$$H_1 : \mu < 12.5 .$$

Choosing a 1% significance level the critical region will have one tail because of H_1.

The sample mean, $\bar{x} = \sum_{i=1}^{10} \frac{x_i}{10} = 11.25$.

The sample variance, $s^2 = \dfrac{\sum_{i=1}^{10} x_i^2 - 10\bar{x}^2}{9} = 0.86.$

The value of the test statistic will be $\quad t = \dfrac{11.25 - 12.5}{0.93/\sqrt{10}} = -4.25 .$

From the t-distribution table using $\nu = 9$ the critical region commences at -2.821 and is as shown in Fig. 24.5.

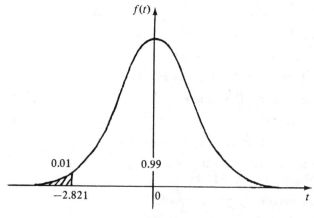

Fig. 24. 5

Clearly the value -4.25 falls in the critical region, and H_0 is rejected. The results suggest that a larger injection dose reduces the life expectancy of mice.

Example 2: The number of wing beats per second of sixteen male house flies were as follows

$$194.7, \quad 191.5, \quad 187.0, \quad 189.7, \quad 190.2, \quad 197.0, \quad 189.9, \quad 188.9,$$

$$197.2, \quad 191.4, \quad 193.1, \quad 186.9, \quad 189.3, \quad 185.2, \quad 193.1, \quad 196.6 \ .$$

If the mean number of wing beats per second of female flies is 190, do the wings of the males beat with a different frequency?

It is assumed that the number of beats per second is normally distributed. If μ is the mean number of wing beats per second of male flies the hypotheses will be

$$H_0 : \mu = 190$$

$$H_1 : \mu \neq 190 \ .$$

The critical region will be two-tailed as shown in Fig. 24.6.

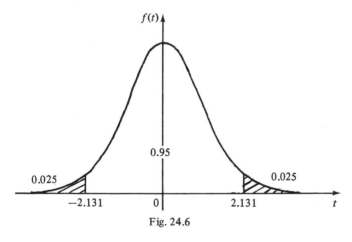

Fig. 24.6

A 5% significance level will be used.

The sample mean, $\bar{x} = \sum\limits_{i=1}^{16} \dfrac{x_i}{16} = 191.35$.

The sample variance, $s^2 = \dfrac{\sum\limits_{i=1}^{16} x_i^2 - 16\bar{x}^2}{15} = 13.60$.

The test statistic takes the value $t = \dfrac{191.35 - 190}{3.69/\sqrt{16}} = 1.46$.

From the *t*-distribution table with $\nu = 15$ the critical region commences at -2.131 and 2.131.

Since 1.46 is less than 2.131 and greater than -2.131 the Null hypothesis is accepted, and it is concluded that there is no difference in wing speeds between male and female flies.

24.4 TWO SAMPLE z-TEST

A sample of n_x observations is drawn from a population X with known variance σ_x^2 and a sample of n_y observations is drawn from a population Y with known variance σ_y^2. It has to be decided if the population means μ_x and μ_y are the same.

Rationale: The Null and Alternative hypotheses will be

$$H_0 : \mu_x = \mu_y$$

$$H_1 : \mu_x \neq \mu_y \ .$$

It can be shown that if the Null hypothesis is correct and the sample sizes are large (n_x and $n_y > 30$), the statistic

$$z = \frac{\bar{x} - \bar{y}}{\sqrt{\left(\dfrac{\sigma_x^2}{n_x} + \dfrac{\sigma_y^2}{n_y}\right)}}$$

has a standard normal probability distribution. The critical regions of the tests for the various Alternative hypotheses are obtained from the standard normal tables as described for the one sample z-test.

Suggested steps for carrying out the two sample z-test are:

1. State the Null and Alternative hypotheses.

2. Choose the significance level of the test.

3. Calculate the sample means, $\bar{x}$ and $\bar{y}$, from the data.

4. Calculate the statistic, $z = \dfrac{\bar{x} - \bar{y}}{\sqrt{\left(\dfrac{\sigma_x^2}{n_x} + \dfrac{\sigma_y^2}{n_y}\right)}}$.

5. Decide if z falls into the critical region, using the table of the standard normal integral.

6. State the hypothesis to be accepted.

Validity: The two sample z-test is valid for all sample sizes if the probability distributions of X and Y are normal. If X and Y have non-normal probability distributions then the sample sizes must be large.

Example: The bacterial count in the mouths of ten patients just admitted to hospital was as follows:

1570, 2275, 1194, 7006, 9993, 4034, 8608, 7976, 7280, 6337.

A second group of twelve patients who had spent six days in hospital gave the following counts

9709, 9847, 5292, 7751, 9038, 4030,

4011, 7325, 7054, 5877, 8074, 5247 .

If the bacterial counts are known to have a population standard deviation of 2500 for each group of patients, do the results indicate that the mean bacterial count is influenced by a stay in hospital? Assume the data to be normally distributed.

Let μ_x and μ_y be the mean bacterial count for those patients just admitted and those who have spent six days in hospital respectively.

Setting up the hypotheses gives

$$H_0 : \mu_x = \mu_y$$
$$H_1 : \mu_x \neq \mu_y .$$

At a 1% significance level the z-statistic must lie in the interval between -2.57 and 2.57 for H_0 to be accepted.

The sample means of each group are

$$\bar{x} = \sum_{i=1}^{10} \frac{x_i}{10} = 5627.3 ,$$

and

$$\bar{y} = \sum_{i=1}^{12} \frac{y_i}{12} = 6937.9 .$$

The z-statistic becomes

$$z = \frac{\bar{x} - \bar{y}}{\sqrt{\left(\dfrac{\sigma_x^2}{n_x} + \dfrac{\sigma_y^2}{n_y}\right)}}$$

$$= \frac{5627.3 - 6937.9}{2500\sqrt{\left(\dfrac{1}{10} + \dfrac{1}{12}\right)}} = -1.224 .$$

Hence the Null hypothesis is accepted and no evidence is found in support of a period in hospital influencing bacterial counts.

24.5 TWO SAMPLE _t_-TEST

A sample of n_x observations is drawn from a population X with unknown mean, μ_x, and unknown variance, σ_x^2. A sample of n_y observations is drawn from a population Y with unknown mean, μ_y, and unknown variance, σ_y^2. It has to be decided if the population means are the same.

Rationale: The two sample t-test replaces the two sample z-test when the sample sizes are small and the population variances σ_x^2 and σ_y^2 are estimated by the sample variances s_x^2 and s_y^2.

The Null and Alternative hypotheses will be

$$H_0 : \mu_x = \mu_y$$

$$H_1 : \mu_x \neq \mu_y \ .$$

If the Null hypothesis is correct, the statistic

$$t = \frac{\bar{x} - \bar{y}}{s\sqrt{\left(\dfrac{1}{n_x} + \dfrac{1}{n_y}\right)}}, \text{ where } s = \sqrt{\left(\frac{(n_x - 1)s_x^2 + (n_y - 1)s_y^2}{n_x + n_y - 2}\right)},$$

has a Student's t-distribution with $\nu = n_x + n_y - 2$, and the critical region is obtained from the t-table.

When the sample sizes are large the two sample z-test can be used with σ_x^2 being replaced by s_x^2 and σ_y^2 replaced by s_y^2 in the equation for the z-statistic.

Suggested steps for carrying out the two sample t-test are:

1. State the Null and Alternative hypotheses.

2. Choose the significance level of the test.

3. Calculate the sample means, $\bar{x}$ and $\bar{y}$, and the sample variances, s_x^2 and s_y^2, from the data.

4. Calculate the statistic, $t = \dfrac{\bar{x} - \bar{y}}{s\sqrt{\left(\dfrac{1}{n_x} + \dfrac{1}{n_y}\right)}}$,

$$\text{where } s = \sqrt{\left(\frac{(n_x - 1)s_x^2 + (n_y - 1)s_y^2}{n_x + n_y - 2}\right)} \ .$$

5. Decide if t falls into the critical region, using the table of the t-distribution with $\nu = n_x + n_y - 2$.

6. State hypothesis to be accepted.

Validity: The test should be used when the sample sizes are small and the probability distributions of X and Y are normal. Another important requirement is that the variances of the two populations, σ_x^2 and σ_y^2, should be equal. In practice, as evidence of this, the sample standard deviations, s_x and s_y, should be nearly the same.

Example 1: To compare Brussels sprout plants derived from stem callus cultures with those from seedlings, twenty plants from each origin were grown under field conditions and the following results obtained,

Origin	Mean Height (cm)	±	Standard Deviation (cm)
Callus	88.04	±	8.15
Seedling	78.77	±	5.70

If the heights are normally distributed, are they significantly different?

Let μ_x and μ_y denote the mean height of plants from callus and seedling origins respectively.

The hypotheses will be

$$H_0 : \mu_x = \mu_y$$
$$H_1 : \mu_x \neq \mu_y .$$

A 1% significance level is chosen.

Since $\bar{x} = 88.04$, $s_x^2 = 8.15^2$, $\bar{y} = 78.77$ and $s_y^2 = 5.70^2$ then

$$s = \sqrt{\left(\frac{19(8.15)^2 + 19(5.70)^2}{38}\right)} = 7.03 .$$

The *t*-statistic,

$$t = \frac{\bar{x} - \bar{y}}{s\sqrt{\left(\frac{1}{n_x} + \frac{1}{n_y}\right)}}$$

becomes

$$t = \frac{88.04 - 78.77}{7.03\sqrt{(\frac{2}{20})}} = 4.17 .$$

From the table of the *t*-distribution with $\nu = 38$, by linear interpolation, the *t*-statistic must be between -2.721 and 2.721 for H_0 to be accepted. Since $4.17 > 2.721$, H_0 is rejected and it is believed that there is a difference in height depending on origin.

Example 2: A flock of eighty ewes was divided into two equal groups, one group was given the usual type A flukicide and the other a new flukicide, type B. Both groups had grazed the same pasture for the same length of time. At a post-mortem the mean and variance of the fluke count per ewe for those ewes given type A flukicide were 130 and 60 respectively, and for those ewes given type B flukicide they were 124 and 45 respectively. Is type A flukicide less efficient than type B?

Let μ_x and μ_y be the mean fluke count per ewe for ewes given type A and type B flukicides respectively.

The hypotheses are

$$H_0 : \mu_x = \mu_y$$
$$H_1 : \mu_x > \mu_y .$$

Although the population variances are unknown, the sample sizes are large and the z-test can be used instead of the t-test with σ_x^2 and σ_y^2 replaced by s_x^2 and s_y^2 respectively. At a 5% significance level the z-statistic must be less than 1.645 for H_0 to be accepted. Calculating z gives

$$z = \frac{\bar{x} - \bar{y}}{\sqrt{\left(\dfrac{s_x^2}{n_x} + \dfrac{s_y^2}{n_y}\right)}}$$

$$= \frac{130 - 124}{\sqrt{(\tfrac{60}{40} + \tfrac{45}{40})}} = 3.70$$

which falls in the critical region, and the Alternative hypothesis is therefore accepted. Flukicide A is less efficient than flukicide B.

24.6 PAIRED COMPARISONS

When observations collected from two populations are influenced by a common factor, then the experiment can be designed so that for each observation drawn from population X an observation is drawn from population Y which has the same value of the common factor. The data consists of n pairs of observations $(x_1, y_1), (x_2, y_2), \ldots, (x_n, y_n)$, and it has to be decided if the population means are the same.

Rationale: If μ_x and μ_y are the same, the probability distribution of the differences, $d_i = x_i - y_i$, will have mean $\mu_d = 0$. The Null and Alternative hypotheses will be

$$H_0 : \mu_d = 0$$
$$H_1 : \mu_d \neq 0 \ ,$$

and if H_0 is correct and n is large, then the statistic

$$z = \frac{\bar{d}}{s_d/\sqrt{n}} \ ,$$

where $\bar{d}$ and s_d^2 are the sample mean and variance of the differences, will have a standard normal distribution.

When n is small the statistic

$$t = \frac{\bar{d}}{s_d/\sqrt{n}} \ ,$$

will have a Student's t-distribution with $\nu = n - 1$.

Suggested steps for carrying out the paired comparison test are:

1. State the Null and Alternative hypotheses.

2. Choose the significance level of the test.

3. Calculate the differences, $d_i = x_i - y_i$, for each pair.

4. Calculate the sample mean $\bar{d}$ and variance s_d^2.

5. Calculate the statistic $\bar{d}/(s_d/\sqrt{n})$.

6. If n is large use the standard normal table, and if n is small use the table of Student's t-distribution, to decide if the statistic falls into the critical region.

7. State the hypothesis to be accepted.

Validity: The distribution of X and Y must be normal when the sample size is small and the t-statistic used. The z-test is valid for all large samples.

Example: Oleoresins are extracted from conifers by tapping the resin ducts in the sapwood. A suggested new technique is to apply sulphuric acid to the undamaged surface of the sapwood to act as a flow stimulant. The following results show the oleoresin content as a percentage of the dry weight for different species, using the two methods. Assuming that the observations are normally distributed is the new technique an improvement?

Species	Flow stimulant	Tapping
Abies grandis	1.74	1.57
Abies nobilis	0.47	0.54
Larix decidva	1.52	0.93
Picae abies	0.63	0.57
Pinus elliotti	0.75	0.68
Pinus pinaster	5.01	4.90
Pinus radiata	1.40	1.72
Pinus sylvestris	0.79	0.75

Since each technique has been applied to each of the different species the data can be treated as eight pairs of observations. The differences, d_i, will be

$$+0.17, \ -0.07, \ +0.59, \ +0.06, \ +0.07, \ +0.11, \ -0.32, \ +0.04 \ .$$

The hypotheses will be

$$H_0 : \mu_d = 0$$
$$H_1 : \mu_d > 0 \ .$$

Since there are only eight pairs the t-statistic is used with $\nu = n - 1 = 7$. From the t-distribution table at 1% significance level the critical region occurs for t greater than 2.998. Calculating the sample mean and variance of the differences gives

$$\bar{d} = \sum_{i=1}^{8} \frac{d_i}{8} = 0.081$$

and
$$s_d^2 = \frac{\sum\limits_{i=1}^{8} d_i^2 - 8\bar{d}^2}{7} = 0.065 \ .$$

The t-statistic becomes

$$t = \frac{\bar{d}}{s_d/\sqrt{n}} = 0.9 \ .$$

This value of t clearly falls outside the critical region, and the Null hypothesis, that there is no difference between the treatments, is accepted.

24.7 TRANSFORMING DATA

In many of the previous tests, if the sample size is small, it is necessary that the observations used in the samples should come from normal populations. This can be a serious drawback to parametric testing but can sometimes be overcome if the observations are transformed so that they are approximately normally distributed. Unfortunately, to determine the correct transformation requires large sample sizes. However, in some cases, before data are analysed similar results from previous experiments may provide the correct transformation. Suppose several previous groups of observations are available and the sample mean, $\bar{x}$, and variance, s^2, are calculated for each group. When the sample mean is plotted against the sample variance a relationship may exist between $\bar{x}$ and s^2. If the mean and variance are related then the observations cannot come from a normal distribution. By fitting the relationship

$$s^2 = a(\bar{x})^b$$

to the curve the index b may be determined (see Chapter 19) and then substituting b into the integral

$$\int x^{-b/2} \, dx$$

the correct transformation can be obtained.

For example, if after fitting the curve,

  (i) $b = 1$ then the integral is $\int x^{-\frac{1}{2}} \, dx = \frac{1}{2} \sqrt{x}$ and a suitable transformation is $\sqrt{x}$ (the constant is unimportant).

  (ii) $b = 2$ then the integral is $\int x^{-1} \, dx$ which leads to the transformation $\ln x$.

  (iii) $b = 4$ then the integral is $\int x^{-2} \, dx$ which leads to the transformation $1/x$.

For further discussion on suitable transformations see Quenouille (1966).

Example: In previous experiments four groups of patients were given different initial doses of a drug. The time taken for 50% of the initial drug dose to be eliminated from the blood system was recorded in hours for each patient as shown below:

	Elimination times
Group A :	2.9, 2.6, 1.1, 1.5, 0.8, 1.8
Group B :	3.9, 2.7, 2.4, 1.4, 4.3, 2.1
Group C :	4.8, 3.5, 5.0, 6.3, 2.2, 2.5
Group D :	4.5, 7.9, 7.4, 2.6, 6.0, 1.7 .

What transformation is likely to give normally distributed data?

The sample mean and variance for each group are as follows:

	mean, $\bar{x}$	variance, s^2
Group A :	1.78	0.69
Group B :	2.80	1.22
Group C :	4.05	2.53
Group D :	5.02	6.41

From the graph shown in Fig. 24.7 a relationship appears to exist between $\bar{x}$ and s^2.

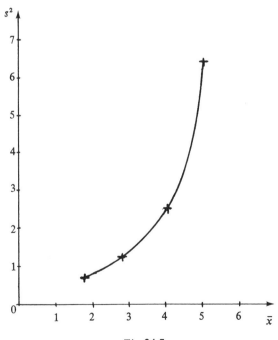

Fig. 24.7

A reasonable description of the relationship is found to be

$$s^2 = 0.18 \, \bar{x}^{2.05} \quad ,$$

using the straight line fitting technique in Section 19.5.

Using $b = 2$ the transformation is $\int x^{-1} \, dx = \ln x$. The transformed data and the mean and variance for each group will be

	Transformed elimination times	Mean	Variance
Group A :	1.065, 0.960, 0.095, 0.405, −0.223, 0.587	0.481	0.246
Group B :	1.361, 0.993, 0.875, 0.336, 1.459, 0.742	0.961	0.171
Group C :	1.568, 1.252, 1.609, 1.840, 0.788, 0.916	1.329	0.173
Group D :	1.504, 2.067, 2.000, 0.955, 1.791, 0.531	1.475	0.378

From the graph in Fig. 24.8 of the mean and variance of the transformed data no relationship between the mean and variance now exists.

The transformed data would now be in a suitable form for use in tests requiring normally distributed data.

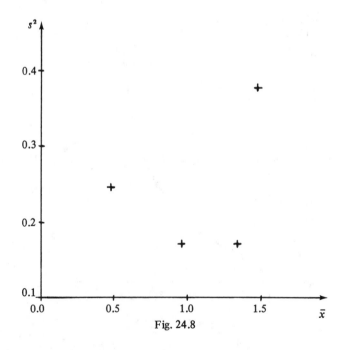

Fig. 24.8

PROBLEMS

1. In a population of non-cancer bearing patients the areas (measured in arbitrary planimetric units) observed in leucocyte migration tests have a mean of 145 and a standard deviation of 120. A random sample of eighty cancer bearers gave a mean migration area of 185.2. Do the results indicate that the migration areas of cancer bearers are different from non-cancer bearers?

2. The mean and variance of the number of contaminants in a bacterial vaccine preserved with phenol are 60 and 25 respectively. A sample of twenty bacterial vaccines in a different preservative gave the following contamination counts:

    67   62   57   55   54   61   51   59   54   57
    57   60   50   66   68   54   53   52   58   56 .

   Assuming that the counts are normally distributed, has the preservative significantly changed the contamination counts?

3. In a population the weight of babies at birth is normally distributed with mean 3200 g. A group of fifteen babies in which a low protein level was detected in mothers during pregnancy gave the following weights at birth:

    2305   2908   2577   2275   2941   3680   2997   2403
    3089   3176   3280   3337   2847   2927   2751 .

   Determine if mothers with a low protein level give birth to babies having a lower birth weight than those from the whole population.

4. The distance travelled (cm) by one constituent of a dye on a paper chromatograph is described approximately by a normal distribution with mean 6.0 cm. The distance travelled by a second constituent of the dye was measured on a sample of 12 paper chromatographs. The mean distance travelled was found to be 5.75 cm and the sample variance was 0.44. Is there any evidence to suggest that the mean distance travelled by the second constituent differs from the mean distance travelled by the first constituent?

5. In September of last year a sample of forty adult damsel flies were collected from a pond. The mean length of these damsel flies was calculated as 12 mm. During the following summer the pond was overpopulated and food availability was limited. In September of this year forty adult damsel flies gave a mean length of 9 mm. If the sample variance for both samples was 4 mm^2, is it likely that overpopulation has arrested the development of the adults?

6. The mean and standard deviation of blood glucose level (mg per 100 ml) in a sample of fifty fish kept for twenty days in water with a known cadmium concentration were 30.0 and 5.0 respectively. When the cadmium concentration was doubled another sample of sixty fish had a mean blood glucose level of 32.0 and a standard deviation of 7.0 after twenty days. Is blood glucose level in the fish influenced by the cadmium concentration?

7. Two samples of observations on the diameters of mycelial colonies produced the following results. Sample A, of 11 observations, gave a mean value of 6.65 and a variance of 3.89, sample B of 16 observations, gave a mean value of 4.28 and a variance of 2.53. Test whether the samples could have come from distributions with the same mean.

8. To discover if there is a difference in yield between two cultivars of tomatoes the following observations were randomly collected from plots

	Cultivar 1	Cultivar 2
Number of plots in sample	10	17
Mean yield per plot (kg)	5.3	7.2
Variance (kg^2)	3.4	2.8

Determine whether the difference between yields of cultivars is significant at a 5% level.

9. The activity of an enzyme (units/gram protein) in twelve liver tissues infected with hepatitis and eighteen normal liver tissues was as follows:

Hepatitis liver tissue	4.15	4.48	4.22	3.94	4.52	3.70
	4.77	4.03	3.90	4.86	3.16	3.33
Normal liver tissue	3.15	4.23	3.12	2.70	3.99	4.40
	3.86	3.86	3.16	4.27	4.34	3.79
	4.28	4.63	4.98	3.52	2.77	3.18

Is there a significant difference in enzyme activity?

10. In a trial the drug gentamicin was administered to six people, and in a later trial the same quantity of pentobarbitone was given to the same people. The results below show the concentration of each drug in the blood stream one hour after administration:

Person	1	2	3	4	5	6
Gentamicin (μg/ml)	2.8	5.8	4.3	3.8	3.5	4.0
Pentobarbitone (μg/ml)	3.1	6.7	4.2	3.6	2.9	4.4

Is the concentration of pentobarbitone significantly greater than that of gentamicin?

11. The following results show the total number of leaves produced by tomato seedlings forty days after planting in late January and grown in two environments, namely a normal atmosphere and a CO_2-enriched atmosphere:

	Recorded number of leaves on plants	
Planting date	Normal atmosphere	CO_2-enriched atmosphere
16th January	6	7
18th January	7	6
20th January	5	6
22nd January	8	11
24th January	8	9
26th January	12	12
28th January	6	11
30th January	9	10

Do the results support the hypothesis that a CO_2-enriched atmosphere affects the total number of leaves produced?

12. The quantity of calcium in a serum sample from each of nine adult patients with disease R is measured:

Patient number　　1　2　3　4　5　6　7　8　9

Calcium (mmol/l)　2.09 1.80 1.97 2.35 2.08 1.90 2.06 2.30 2.35.

In normal adults serum calcium is normally distributed with mean 2.4 mmol/1.

Show that there is a significant difference between the mean serum calcium of patients with disease R and normal adults.

Each of the nine patients was given the same quantity of dietary supplement for a period of six months. The serum calcium measurement was then repeated in all nine subjects and the results are given below in the same order as the original observations:

Calcium (mmol/l) 2.15 2.13 2.27 2.52 2.11 2.24 2.26 2.34 2.68.

If it is known that the dietary supplement cannot decrease the serum calcium, has the supplement had a beneficial effect on the patients?

(Strathclyde, 1979)

13. The hourly rate of production of demersal plankton was measured in cc per m^2 on six occasions at each of five sites on a reef substrate, with the following results:

Site 1	Site 2	Site 3	Site 4	Site 5
0.86	0.43	0.24	1.37	1.05
0.90	0.49	0.14	1.24	1.20
0.88	0.70	0.41	1.29	0.96
0.99	0.40	0.28	1.06	1.21
0.59	0.27	0.25	1.17	1.04
0.64	0.41	0.30	1.27	0.62

Calculate the mean and variance of the production rate at each site. Find a suitable transformation to give normally distributed data. What will be the values of the transformed data?

14. Monitoring the level of parasitaemia in infected mice shows that the parasitaemia level peaks every three days. The results show the parasitaemia levels in each of six mice at each of these peaks over an eighteen day period.

1st peak	2nd peak	3rd peak	4th peak	5th peak	6th peak
623	400	244	120	120	40
90	350	558	310	220	78
760	44	440	360	160	340
257	396	20	700	400	80
8	86	80	11	64	100
681	800	590	170	260	96

Calculate the mean, $\bar{x}$, and variance, s^2, at each peak. Fit a relationship of the form $s^2 = a\bar{x}^b$, and hence find a transformation which will make the data normally distributed. Using the transformation, determine if there is a significant difference between the parasitaemia levels at the first and second peaks.

CHAPTER 25

Non-parametric tests

25.1 INTRODUCTION

In order to apply the parametric tests discussed in the previous chapter, when the sample size is small, it must be assumed that the observations come from a population with a normal distribution. The tests to be discussed in this chapter are called **non-parametric** or **distribution free tests**, as parameters such as the mean and variance do not need to be estimated, and assumptions about the probability distribution of the observations are unnecessary. These tests offer a simpler approach and are often valid when parametric tests are not. For small sample sizes, the Mann-Whitney and Wilcoxon tests are alternatives to the two sample and paired comparison t-tests respectively. The Mann-Whitney and Wilcoxon tests can be extended to deal with large sample sizes, but this will not be dealt with in this text. The non-parametric approach is extended to give the Kruskal-Wallis and the Friedman tests as methods suitable for detecting differences between several groups of observations. Finally, contingency tables are presented as a method of handling categorised data.

25.2 ONE SAMPLE BINOMIAL OR SIGN TEST

This test is not widely used but serves to illustrate the simple approach of non-parametric statistics. A sample of n observations is drawn from a population X. It has to be decided if the observations come from a population with median m_0.

Rationale: The Null and Alternative hypotheses will be

H_0 : population has median m_0

H_1 : population does not have median m_0 .

If the Null hypothesis is correct the median, m_0, will divide the probability distribution of X into two equal areas, and an observation drawn at random from the population will have probability $\frac{1}{2}$ of not exceeding the median. The sample will then be expected to contain an equal number of observations below and above the median. As evidence to support or reject the Null hypothesis, the statistic used is T, the number of observations below m_0. The collection of each observation from the population represents an independent trial in which the probability of not exceeding m_0 is $\frac{1}{2}$. Since there are n observations, T will have a binomial probability distribution with parameter $p = q = \frac{1}{2}$:

$$P(T = r) = {}^nC_r \, p^r \, q^{n-r} \quad r = 0, 1, 2, \ldots, n$$
$$= {}^nC_r \, (\tfrac{1}{2})^n \; .$$

The values of T which lie in the critical region can easily be calculated using the binomial distribution. For a 5% significance level, the critical region will consist of the two tails of the binomial distribution, with area 0.025 in each. The critical region for low values of T will consist of all values less than t_L, where t_L is determined from the equation

$$P(T = 0) + P(T = 1) + \ldots + P(T = t_L) = 0.025 \; .$$

The sum of these binomial probabilities may not add up to 0.025 exactly, so in practice t_L is chosen to make the sum close to, but not exceeding 0.025. The critical region for high values of T will consist of all values greater than t_H, where t_H satisfies

$$P(T = t_H) + P(T = t_H + 1) + \ldots + P(T = n) = 0.025 \; .$$

However, this calculation is unnecessary as $t_H = n - t_L$. Thus, if the T value obtained from the data is less than or equal to t_L, or greater than or equal to t_H, the Null hypothesis is rejected.

For a 1% significance level the critical region will have area 0.01, consisting of 0.005 in each tail. The values t_L and t_H are calculated using

$$P(T = 0) + P(T = 1) + \ldots + P(T = t_L) = 0.005 \; ,$$

and $\quad\quad t_H = n - t_L \; .$

Critical regions for one-tailed tests where the Alternative hypothesis is either that the observations come from a population with median greater than m_0 or less than m_0 can similarly be deduced.

Suggested steps for carrying out the Sign test are:

1. State the Null and Alternative hypotheses.

2. Choose the significance level of the test.

3. Calculate T, the number of observations less than m_0.

4. Calculate t_L and t_H.

5. If T is less than t_L or greater than t_H reject the Null hypothesis.

Several points should be remembered when using the Sign test:

(i) If an observation is identical to the median m_0 it must be omitted from the sample and the sample size reduced by 1.

(ii) The hypotheses as stated concern the median but can be expressed in terms of the population mean when X has a symmetrical probability distribution, since the mean and median are then identical.

(iii) Instead of using the number of observations below m_0 as the test statistic, the number of observations above m_0 may be used, the critical region remaining the same.

Example: The time taken, in days, to diagnose leukaemia in ten animals infected with leukaemic cells was as follows:

$$239 \quad 119 \quad 265 \quad 278 \quad 257 \quad 227 \quad 286 \quad 279 \quad 228 \quad 145 \ .$$

Is the median time to diagnosis in a population of such animals likely to be 200 days?

$$H_0 : \text{population median } m_0 = 200 \text{ days}$$

$$H_1 : \text{population median } m_0 \neq 200 \text{ days} \ .$$

The number of observations less than 200, T, is equal to 2. For a 5% significance level,

$$P(T = 0) + P(T = 1) = {}^{10}C_0(\tfrac{1}{2})^{10} + {}^{10}C_1(\tfrac{1}{2})^{10}$$
$$= 0.01074 \ ,$$

which is less than 0.025, and

$$P(T = 0) + P(T = 1) + P(T = 2) = {}^{10}C_0(\tfrac{1}{2})^{10} + {}^{10}C_1(\tfrac{1}{2})^{10} + {}^{10}C_2(\tfrac{1}{2})^{10}$$
$$= 0.05468 \ ,$$

which is greater than 0.025, so that $t_L = 1$. By subtraction

$$t_H = n - t_L = 10 - 1 = 9 \ .$$

The critical region will be those T values less than or equal to 1 and greater than or equal to 9, and so the Null hypothesis is accepted.

25.3 TWO SAMPLE MANN-WHITNEY TEST

This test is widely used to compare observations from a control and a treated group. It is often suitable in cases where the two sample t-test cannot be used. Samples of size n_x and n_y are drawn from populations X and Y respectively. It has to be decided if the populations have the same probability distribution.

Rationale: The Null and Alternative hypotheses are

$$H_0 : X \text{ and } Y \text{ have the same distribution}$$

$$H_1 : X \text{ and } Y \text{ have different distributions.}$$

If the Null hypothesis is correct, arranging all the observations in ascending order should lead to the X observations being interspersed among the Y observations. It will be unlikely that all of the X observations will precede the Y

observations or all the Y observations will precede the X observations. Let the observations be expressed in terms of **ranks** by assigning the score 1 to the first ordered observation, the score 2 to the second ordered observation, and so on until the last ordered observation, which will be assigned the score $(n_x + n_y)$. Observations which are equal are called **ties** and the rank is obtained by assigning to each the average of the scores that would be assigned if the observations were not tied. As a measure of the level of interspersion of the X and Y values, the ranks of the X values are identified and summed to give S_x. When all the X observations precede all the Y observations this S_x will be small, whereas when all the Y observations precede all the X observations S_x will be large. For all other arrangements of the X and Y observations, S_x will lie between the two previous values. This is demonstrated for sample sizes $n_x = 3$ and $n_y = 5$. For the X observations preceding the Y's, the arrangement in ascending order and assignment of scores gives the ranks

$$\text{Arrangement} \quad x \quad x \quad x \quad y \quad y \quad y \quad y \quad y$$
$$\text{Ranks} \quad 1 \quad 2 \quad 3 \quad 4 \quad 5 \quad 6 \quad 7 \quad 8$$

and for the Y's preceding the X's,

$$\text{Arrangement} \quad y \quad y \quad y \quad y \quad y \quad x \quad x \quad x$$
$$\text{Ranks} \quad 1 \quad 2 \quad 3 \quad 4 \quad 5 \quad 6 \quad 7 \quad 8$$

where x and y denote observations from the X and Y populations respectively. The sum of the ranks of the X values in the first case is $S_x = 1 + 2 + 3 = 6$, and in the second case $S_x = 6 + 7 + 8 = 21$. For all other arrangements S_x must lie between 6 and 21. Although it would be possible to use S_x as the test statistic it turns out to be more convenient to use

$$U = S_x - \frac{n_x (n_x + 1)}{2} .$$

The last term of this equation is the sum of the integers from 1 to n_x. This gives a test statistic which takes values from 0 upwards. In the above example $n_x (n_x + 1)/2 = 6$ and so U will lie in the range 0 to 15. By considering the possible realisations, the probability distribution of U is not difficult to calculate,

$$P(U = 0) = P(S_x = 6)$$
$$= P(\text{arrangement } x\, x\, x\, y\, y\, y\, y\, y)$$
$$= \frac{\text{number of events in favour of arrangement}}{\text{total number of possible events}}$$

In choosing the first x in the arrangement there are three possible ways, for the second x there are two possible ways and so on. Using the approach given in Section 20.2 this gives

$$P(U = 0) = \frac{3.2.1.5.4.3.2.1}{8!} = \frac{3! \, 5!}{8!} = 0.017857 \ .$$

$$P(U = 1) = P(S_x = 7)$$

$$= P(\text{arrangement } x \, x \, y \, x \, y \, y \, y \, y)$$

$$= \frac{3.2.5.1.4.3.2.1}{8!} = \frac{3! \, 5!}{8!} = 0.017857 \ .$$

$$P(U = 2) = P(S_x = 8)$$

$$= P(\text{arrangement } x \, x \, y \, y \, x \, y \, y \, y \text{ OR } x \, y \, x \, x \, y \, y \, y \, y)$$

$$= \frac{3.2.5.4.1.3.2.1}{8!} + \frac{3.5.2.1.4.3.2.1}{8!}$$

$$= \frac{3! \, 5!}{8!} + \frac{3! \, 5!}{8!} = 0.035714 \ .$$

Similar calculations lead to the probability distribution of U, shown in Fig. 25.1.

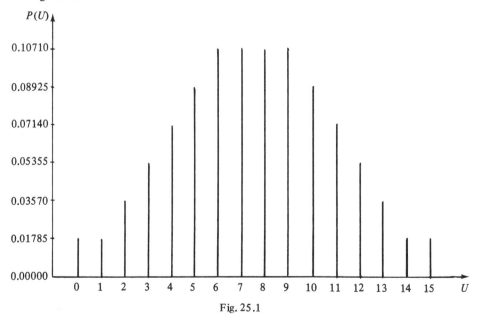

Fig. 25.1

Using a 5% significance level, and working from the middle of the probability distribution, the values of U denoting the start of the lower and upper tail critical regions will be 1 and 14 since

$$P(U < 1) \leqslant 0.025 \ ,$$

and $\qquad P(U > 14) \leqslant 0.025 \ .$

Fortunately, in practice, it is not necessary to calculate the probability distribution of U to find the critical values of U, as the Mann-Whitney table given in Appendix G provides these for various combinations of sample sizes. It will be seen from the table that no readings are given for n_x values less than n_y. This is to avoid repetition as the readings are the same, using the values of n_x and n_y interchanged. For $n_x = 3$ and $n_y = 5$ the table is read under column $n_x = 5$ and across row $n_y = 3$ to get

$$\begin{array}{|c|} \hline 0 \\ \hline 1 \\ \hline 2 \\ \hline \end{array}$$

The numbers 0, 1 and 2 denote the start of the lower tail region for a two-tailed test at significance levels 2%, 5% and 10% respectively. The start of the upper tail critical region is obtained by subtracting the start of the lower tail critical region from $n_x n_y$. Therefore, working with a 5% significance level, U must either be less than 1 or greater than $n_x n_y - 1 = 14$ if the Null hypothesis is to be rejected.

Suggested steps for carrying out the Mann-Whitney test are:

1. State the Null and Alternative hypotheses.
2. Choose the significance level of the test.
3. Arrange all the observations in ascending order.
4. Obtain the ranks of the ordered observations.
5. Sum the ranks of the X observations, S_x.
6. Calculate the test statistic

$$U = S_x - \frac{n_x (n_x + 1)}{2} .$$

7. Determine, with the use of the Mann-Whitney table, whether U falls in the critical region, and if so reject the Null hypothesis.

The use of S_x, the sum of the ranks of the X observations, to reflect the level of interspersion of the X and Y observations can be replaced by S_y, the sum of the ranks of the Y observations. The test statistic then becomes

$$U = S_y - \frac{n_y (n_y + 1)}{2}$$

and the critical region remains unchanged.

Example: The data show the length of time in days that control mice, X, and inoculated mice, Y, live.

X	20	23	28	30	31	32	44	33		
Y	20	29	23	48	41	32	36	43	42	46 .

Is there a difference in survivorship between control and inoculated mice?
Here $n_x = 8, n_y = 10$ and the hypotheses are

H_0 : X and Y have the same distribution of life times

H_1 : X and Y have different distributions of life times.

Arranging the data in ascending order gives

20 20 23 23 28 29 30 31 32 32 33 36 41 42 43 44 46 48

and by assigning scores the corresponding ranks for the X and Y values become

x	y	x	y	x	y	x	x	x	y	x	y	y	y	y	x	y	y
$1\frac{1}{2}$	$1\frac{1}{2}$	$3\frac{1}{2}$	$3\frac{1}{2}$	5	6	7	8	$9\frac{1}{2}$	$9\frac{1}{2}$	11	12	13	14	15	16	17	18 .

Summing the ranks of X,

$$S_x = 1\frac{1}{2} + 3\frac{1}{2} + 5 + 7 + 8 + 9\frac{1}{2} + 11 + 16 = 61\frac{1}{2} .$$

Calculating the test statistic,

$$U = S_x - \frac{n_x (n_x + 1)}{2} = 61\frac{1}{2} - \frac{8 \times 9}{2} = 25\frac{1}{2} .$$

From the Mann-Whitney table for $n_x = 8$ and $n_y = 10$ the start of the lower tail critical region at a 5% significance level is 18. The upper tail critical region will start at $n_x n_y - 18 = 62$. Since $U = 25\frac{1}{2}$ does not fall in the critical region the Null hypothesis is accepted, and it is believed that control and inoculated mice have the same survivorship.

25.4 WILCOXON PAIRED COMPARISON TEST

An experimental design in which each observation collected from population X is paired with an observation collected from population Y was introduced in Section 24.6 on *Paired comparisons*. From the n pairs of observations (x_1, y_1), ..., (x_n, y_n) it has to be decided if populations X and Y have the same distribution.

Rationale: The Null and Alternative hypotheses are

H_0 : X and Y have the same distribution

H_1 : X and Y have different distributions .

Subtracting the paired observations will lead to n differences $d_1, d_2, \ldots, d_n$, where $d_1 = x_1 - y_1, d_2 = x_2 - y_2, \ldots, d_n = x_n - y_n$. If H_0 is correct these differences are likely to consist of positive and negative numbers of similar magnitude. If H_1 is correct either the X observations are likely to exceed the Y observations, leading to mostly positive differences; or the Y observations are likely to exceed the X observations, giving mostly negative differences. A suitable statistic to decide which hypothesis is reasonable, is obtained by arranging the differences in ascending order IRRESPECTIVE of their signs, giving rank 1 to the first ordered difference, rank 2 to the second ordered difference etc. and using $S\_$, the sum of the ranks associated with the negative differences. A value of $S\_$ close to 0 or $n(n + 1)/2$, will be evidence supporting the Alternative hypothesis, as the differences will be mostly positive or negative. The $S\_$ value is $n(n + 1)/2$ when all the differences are negative, as it is the sum of the ranks from 1 to n, and $S\_ = 0$ when all the differences are positive.

The use of $S\_$ as the test statistic will now be illustrated. For a sample containing 6 pairs of observations from populations X and Y, subtracting will give the differences d_1, d_2, d_3, d_4, d_5 and d_6.

If the differences are all positive, arranging them in ascending order irrespective of signs and ranking will lead to $S\_ = 0$. If the differences are all negative it will lead to $S\_ = 1 + 2 + 3 + 4 + 5 + 6 = 21$. In all other cases the value of $S\_$ will lie between 0 and 21. The probability distribution of $S\_$ can easily be evaluated as shown.

$$P(S\_ = 0) = \frac{\text{number of events in favour of } S\_ = 0}{\text{total number of possible events}}$$

$$= \frac{\text{number of events in which ranked differences are all positive}}{\text{total number of possible events}}$$

$$= \frac{1}{2^6},$$

since there is only the one event:

$$+ + + + + +$$

in favour of $S\_ = 0$ and the total number of possible events is 2^6.

$$P(S\_ = 1) = \frac{\text{number of events in favour of } S\_ = 1}{\text{total number of possible events}} = \frac{1}{2^6},$$

since only the one event:

$$- + + + + +$$

will lead to $S\_ = 1$.

Similarly, the associated events and probabilities for other $S\_$ values will be:

$$P(S\_ = 2) = \frac{\text{event } (+ - + + + +)}{2^6} = \frac{1}{2^6},$$

$$P(S\_ = 3) = \frac{\text{event } (+ + - + + +) \text{ or event } (- - + + + +)}{2^6} = \frac{2}{2^6},$$

$$P(S\_ = 4) = \frac{\text{event } (+ + + - + +) \text{ or event } (- + - + + +)}{2^6} = \frac{2}{2^6},$$

and so on. The probability distribution of S_- will be symmetric, as shown in Fig. 25.2.

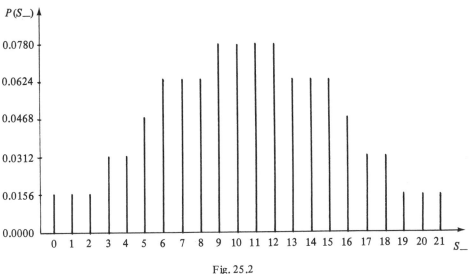

Fig. 25.2

The values of S_- denoting the start of the lower tail and upper tail critical regions can be found from the probability distribution; but to simplify this task the Wilcoxon table given in Appendix H gives the lower tail value for different significance levels across the row n equal to 6. The upper tail value is determined by subtracting the lower tail value from twice the value listed under the column 0.50. For example, for a 5% significance level the table gives 1 as the lower tail value, and the upper tail value is calculated as $2 \times 10.5 - 1 = 20$. Therefore S_- must be less than 1 or greater than 20 if the Null hypothesis is to be rejected.

Suggested steps for carrying out the Wilcoxon test are:

1. State the Null and Alternative hypotheses.

2. Choose the significance level of the test.

3. Subtract the paired observations to get the n differences.

4. Arrange the differences in ascending order IRRESPECTIVE of sign.

5. Obtain the ranks of the differences.

6. Calculate the test statistic S_-, the sum of the ranks of the negative differences.

7. From the Wilcoxon table determine if S_- falls in the critical region, and if so reject the Null hypothesis.

The following points should be remembered when using the test:

   (i) If a pair of observations gives a zero difference, the pair must be removed and the number of paired observations reduced by 1.

  (ii) If tied differences are encountered the corresponding rank for each of these differences is the average of the ranks that would be assigned if the differences were not tied.

 (iii) The test statistic S_+, the sum of the ranks of the positive differences, may be used instead of S_- and the critical region for S_+ is the same as that for S_-.

Example: Primary infections of *H. contortus* parasites in sheep were terminated by using an anthelmintic, and the animals were subsequently reinfected to see if there was evidence of a change in the immune status of the animals.

Animal	1	2	3	4	5	6	7
Worm burden with primary infection	1633	1364	1639	1721	1103	840	1400
Worm burden with secondary infection	2617	1986	984	1671	1684	1462	2398
Differences, d,	984	622	−655	−50	581	622	998

      H_0 : worm burdens are the same for primary and secondary infections

      H_1 : worm burdens are different for primary and secondary infections.

Arranging the differences in order IRRESPECTIVE of signs:

    −50   581   622   622   −655   984   998 .

Assigning scores gives the corresponding ranks

      1   2 $3\tfrac{1}{2}$ $3\tfrac{1}{2}$ 5   6   7 .

Taking the sum of ranks of the negative differences

      $S_- = 1 + 5 = 6$.

At a 5% significance level the value of S_- denoting the start of the lower tail critical region is found in the Wilcoxon table under the column 0.025 and across the row $n = 7$ to be 3. The upper tail value is calculated as $(2 \times 14) - 3 = 25$. Since the S_- data value does not fall in the critical region it is believed that there is no difference in worm burdens.

25.5 KRUSKAL-WALLIS TEST FOR SEVERAL TREATMENTS

This test, which is an extension of the method used in the two sample Mann-Whitney test, allows several treatments to be compared. Since it is concerned with examining treatment effects only, it is sometimes referred to as a **one-way analysis**. Suppose there are c treatments $T_1, \ldots, T_c$ and there are n_1 observations from treatment 1, n_2 from treatment 2 and so on. The total number of observations will be $N = n_1 + n_2 + \ldots + n_c$. It has to be decided if at least one of the treatments has a different effect from the others.

Rationale: The hypotheses will be

H_0 : treatment effects are the same

H_1 : treatment effects are different .

If treatments are the same, then pooling the observations and arranging them in ascending order will lead to the observations from different treatments being interspersed among each other. Assigning scores will lead to the ranks associated with the observations in each treatment being made up of small and large numbers. Tied observations are given ranks as in the Mann-Whitney and Wilcoxon tests. The totals of the ranks for each treatment $C_1, C_2, \ldots, C_c$, when divided by $n_1, n_2, \ldots, n_c$ respectively, should be close as evidence of no treatment effect. As a measure of closeness the chi-squared test statistic

$$\chi^2 = \frac{12}{N(N+1)} \left(\frac{C_1^2}{n_1} + \frac{C_2^2}{n_2} + \ldots + \frac{C_c^2}{n_c} \right) - 3(N+1)$$

can be used. The χ^2 value is always positive and its probability distribution is known to be of the form shown in Fig. 25.3.

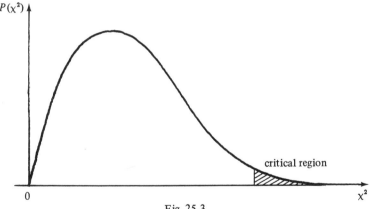

Fig. 25.3

When the rank treatment totals are very different the χ^2 value will be large, and this is evidence of a treatment effect being different, whereas when the rank treatment totals are close the χ^2 value will be small, supporting the Null hypothesis of no difference in treatment effects. The critical region will therefore consist only of high values of χ^2 as illustrated by the shaded area in Fig. 25.3. The χ^2 table in Appendix I gives the value of χ^2 at which the critical region commences, for various significance levels. Like the t-table used in parametric t-tests, the parameter v in the table must be specified before using the χ^2 table, and v is equal to $c - 1$ for the Kruskal-Wallis test.

Suggested steps for carrying out the Kruskal-Wallis test are:

1. State the Null and Alternative hypotheses.

2. Choose a significance level for the test.

3. Arrange all the observations in ascending order.

4. Obtain the ranks of the ordered observations.

5. For each treatment replace each observation by its rank and sum the ranks to get $C_1, C_2, \ldots, C_c$.

6. Calculate the test statistic:

$$\chi^2 = \frac{12}{N(N+1)} \left(\frac{C_1^2}{n_1} + \frac{C_2^2}{n_2} + \ldots + \frac{C_c^2}{n_c} \right) - 3(N+1)$$

$$\text{where } N = n_1 + n_2 + \ldots + n_c \;.$$

7. Using the χ^2 table, with $v = c - 1$, obtain the critical region.

8. Accept or reject the Null hypothesis.

Example: The cattle disease, East Coast Fever, is transmitted by the tick *D. marginatus*. Data, as shown below, were collected to see if the time spent feeding by female ticks was affected by photoperiodic conditions.

Simulated hours of daylight per day

	6	12	18	24
Time (days) spent feeding by individual ticks	7.1	8.6	12.1	9.1
	14.3	11.0	13.9	14.5
	13.4	9.0	14.1	11.5
	10.7	12.6	8.6	12.7
		14.8	13.2	11.7
				11.2

$n_1=4$ $n_2=5$ $n_3=5$ $n_4=6$

Here $N = n_1 + n_2 + n_3 + n_4 = 20$.

H_0 : no photoperiodic effect on time spent feeding

H_1 : photoperiodicity affects time spent feeding .

Arranging data in ascending order and ranking:

	7.1	8.6	8.6	9.0	9.1	10.7	11.0	11.2	11.5	11.7
Ranks	1	$2\frac{1}{2}$	$2\frac{1}{2}$	4	5	6	7	8	9	10

	12.1	12.6	12.7	13.2	13.4	13.9	14.1	14.3	14.5	14.8
Ranks	11	12	13	14	15	16	17	18	19	20 .

Replacing data by their ranks:

Simulated hours of daylight per day

6	12	18	24
1	$2\frac{1}{2}$	11	5
18	7	16	19
15	4	17	9
6	12	$2\frac{1}{2}$	13
	20	14	10
			8
$C_1=40$	$C_2=45\frac{1}{2}$	$C_3=60\frac{1}{2}$	$C_4=64$

Calculating the test statistic:

$$\chi^2 = \frac{12}{N(N+1)} \left(\frac{C_1^2}{n_1} + \frac{C_2^2}{n_2} + \frac{C_3^2}{n_3} + \frac{C_4^2}{n_4} \right) - 3(N+1)$$

$$= \frac{12}{20 \times 21} \left(\frac{40^2}{4} + \frac{(45\frac{1}{2})^2}{5} + \frac{(60\frac{1}{2})^2}{5} + \frac{64^2}{6} \right) - 3 \times 21$$

$$= 0.68 .$$

At 5% significance level, and with $v = c - 1 = 3$, the χ^2 table value is 7.81.

Since the χ^2 value of 0.68 from the data does not exceed 7.81 the Null hypothesis that photoperiodicity is not influenced by feeding times is accepted.

25.6 FRIEDMAN TEST

In the Kruskal-Wallis test it is assumed that there can only be differences in treatment effects. It is possible that the repeated observations taken within each treatment are influenced by another factor. An observation may then be influenced by a treatment effect, a factor effect, or by both effects. The Friedman two-way analysis test provides a method of testing for these effects. Suppose there are c treatments and r factors, and let the observations be set out in an array form as follows:

$$T_1 \qquad T_2 \ldots \ldots T_c$$

The total number of observations will be $r \times c$. Two questions may be posed:

(i) Taking into consideration factor effects, is there a difference in treatment effects?

(ii) Taking into consideration treatment effects, is there a difference in factor effects?

Rationale: Often only one of the above questions may be of interest, but the two sets of hypotheses, one concerning differences in treatment effects and one concerning differences in factor effects, will be considered here. The first Null and Alternative hypotheses will be

H_0 : treatment effects are the same

H_1 : treatment effects are different .

If there is no treatment effect then ranking the observations for each factor INDIVIDUALLY and replacing the observations by their ranks should give an equal distribution of high and low ranks for each treatment. The totals of these ranks for each treatment $C_1, C_2, \ldots, C_c$ should be close. If one treatment effect is different then it is likely to give consistently a high (or low) rank, and the total of the ranks for this treatment will be high (or low) compared with the rank totals for the other treatments. As a measure of the closeness of the rank totals the χ^2 test statistic, used in Section 25.5, only calculated differently, can be used:

$$\chi^2 = \frac{12}{rc(c+1)} (C_1^2 + C_2^2 + \ldots + C_c^2) - 3r(c+1) .$$

The greater the difference in the rank totals the larger the χ^2 value becomes and the more likely it is that the test statistic will fall in the critical region, as evidence in support of rejecting the Null hypothesis. The critical region will be those values of χ^2 greater than the value given in the table in Appendix I under the column specifying the significance level and across the row $\nu = c - 1$.

The hypotheses concerning the factors will be

H_0 : factor effects are the same

H_1 : factor effects are different .

An approach similar to that used for treatment effects can be followed. The observations are ranked INDIVIDUALLY within each treatment. Replacing the observations by their ranks, the totals of the ranks for each factor $R_1, R_2, \ldots, R_r$ are obtained. The χ^2 test statistic is calculated using

$$\chi^2 = \frac{12}{cr(r+1)}(R_1^2 + R_2^2 + \ldots + R_r^2) - 3c(r+1) .$$

If the χ^2 value exceeds the value given in the table in Appendix I under the required significance level and with $v = r - 1$, then the Null hypothesis is rejected.

Suggested steps for carrying out the Friedman test for evidence of treatment effects are:

1. State the Null and Alternative hypotheses concerning treatment effects.

2. Choose a significance level for the test.

3. Obtain the ranks of the observations within each factor.

4. Sum the ranks within each treatment to get $C_1, C_2, \ldots, C_c$.

5. Calculate the test statistic:

$$\chi^2 = \frac{12}{rc(c+1)}(C_1^2 + C_2^2 + \ldots + C_c^2) - 3r(c+1) .$$

6. Using the χ^2 table, with $v = c - 1$, decide which hypothesis is to be accepted.

Suggested steps for carrying out the Friedman test for evidence of factor effects are:

1. State the Null and Alternative hypotheses concerning factor effects.

2. Choose a significance level for the test.

3. Obtain the ranks of the observations within each treatment.

4. Sum the ranks within each factor to get $R_1, R_2, \ldots, R_r$.

5. Calculate the test statistic:

$$\chi^2 = \frac{12}{cr(r+1)}(R_1^2 + R_2^2 + \ldots + R_r^2) - 3c(r+1) .$$

6. Using the χ^2 table, with $v = r - 1$, decide which hypothesis is to be accepted.

Example: The following table shows the yield (kg/hectare) of soya bean cultivars with different maturing dates when sown at different times of the year. Decide whether the influence of the maturing date, or the date of sowing, is important.

Date of sowing	Maturing date of cultivars		
	early	medium	late
October	2154	2160	2744
November	2343	2139	2287
December	1934	1690	1514
January	1296	1065	1749

Here $r = 4$ and $c = 3$.

Analysing the data for maturing date effects on yield first:

H_0 : maturing date of cultivars does not affect yield

H_1 : maturing date of cultivars affects yield .

Ranking the observations for each month, and summing the ranks for each maturing date, gives:

	Early	Medium	Late
October	1	2	3
November	3	1	2
December	3	2	1
January	2	1	3
	$C_1=9$	$C_2=6$	$C_3=9$

Calculating the test statistic:

$$\chi^2 = \frac{12}{rc(c + 1)}(C_1^2 + C_2^2 + C_3^2) - 3r(c + 1)$$

$$= \frac{12}{4 \times 3 \times 4}(81 + 36 + 81) - 3 \times 4 \times 4 = 1.5 .$$

At a 5% significance level and with $\nu = c - 1 = 2$ the χ^2 table value is 5.99. Since the χ^2 value obtained from the data is less than 5.99 it is concluded that the maturing date of cultivars does not affect yield.

Next analysing the data for effects in sowing dates,

H_0 : sowing date does not affect yield

H_1 : sowing date does affect yield .

Ranking the observations for each maturing date, and summing the ranks for each sowing date, gives:

	Early	Medium	Late	
October	3	4	4	$R_1 = 11$
November	4	3	3	$R_2 = 10$
December	2	2	1	$R_3 = 5$
January	1	1	2	$R_4 = 4$

Calculating the test statistic:

$$\chi^2 = \frac{12}{cr(r+1)} (R_1^2 + R_2^2 + R_3^2 + R_4^2) - 3c(r+1)$$

$$= \frac{12}{3 \times 4 \times 5} (11^2 + 10^2 + 5^2 + 4^2) - 3 \times 3 \times 5 = 7.4 .$$

This is only just not significant at a 5% significance level as the table gives 7.8 with $\nu = r - 1 = 3$, and the date of sowing is thought not to influence yield.

Since we almost obtain significance it would be advisable to collect more data to reach a firm conclusion.

25.7 TESTING FOR INTER-RELATIONSHIPS, USING CONTINGENCY TABLES

Frequently, when a sample of size n is collected, the observations are classified in terms of two variables. The classification gives a **contingency table** with values of one variable as the column headings and values of the second variable as the row headings. The entries in the table are the number of observations from the sample which take the values given in the row and column headings. Suppose that the row variable, V, consists of r classes, and the column variable, W, of c classes. The contingency table will resemble a matrix in layout and take the form

	W_1	W_2	....	W_c	
V_1	O_{11}	O_{12}	....	O_{1c}	R_1
V_2	O_{21}	O_{22}	....	O_{2c}	R_2
	. . .	. . .	....	. . .	. . .
V_r	O_{r1}	O_{r2}	....	O_{rc}	R_r
	C_1	C_2	....	C_c	

where $O_{11}, O_{12}, \ldots, O_{rc}$ are the **observed frequencies** of observations. $R_1, R_2,$ $\ldots, R_r$ are the row totals of the entries, and $C_1, C_2, \ldots, C_c$ are the column totals of the entries. From the data it has to be decided if there is a relationship between the row and column variables.

Rationale: The Null and Alternative hypotheses are

H_0 : the row and column variables are not related

H_1 : the row and column variables are related.

An estimate of the probability of an observation being in class W_1 will be

$$\frac{\text{number of observations in sample in class } W_1}{\text{total number of observations in sample}} = \frac{C_1}{n} \ .$$

Similarly, the probability of an observation being in class V_1 will be estimated by,

$$\frac{\text{number of observations in sample in class } V_1}{\text{total number of observations in sample}} = \frac{R_1}{n} \ .$$

If the row and column variables are not related then, using the multiplication law of probability for independent events,

$$P(\text{A and B}) = P(\text{A})P(\text{B}) \ ,$$

the probability of an observation falling in classes W_1 and V_1 will be

$$\frac{C_1}{n} \times \frac{R_1}{n} \ .$$

The expected number of observations, E_{11}, in classes W_1 and V_1 in a sample of size n will be this probability multiplied by n:

$$E_{11} = \frac{R_1}{n} \times \frac{C_1}{n} \times n = \frac{R_1 C_1}{n} \ .$$

Similarly, other expected numbers can be evaluated by using the column and row totals. For example

$$E_{12} = \frac{R_1C_2}{n}, E_{13} = \frac{R_1C_3}{n}, \ldots, E_{23} = \frac{R_2C_3}{n}, \ldots, E_{rc} = \frac{R_rC_c}{n}.$$

This leads to a table of **expected frequencies** calculated on the basis of the Null hypothesis being correct:

	W_1	W_2	. . . .	W_c
V_1	E_{11}	E_{12}	. . . .	E_{1c}
V_2	E_{21}	E_{22}	. . . .	E_{2c}
$\vdots$	$\cdot$	$\cdot$	. . . .	$\cdot$
V_r	E_{r1}	E_{r2}	. . . .	E_{rc}

In practice, it is not necessary to calculate all of the $r \times c$ expected values as shown above. For each row, all but one of the expected values are calculated, and the sum of these values is subtracted from the row total, to get the remaining expected value. A similar approach can be used for the columns. For a contingency table with r rows and c columns only $(r-1) \times (c-1)$ values need be calculated explicitly. Comparing the observed and expected values, if these values are close the Null hypothesis is likely to be correct. As a measure of closeness the χ^2 statistic, calculated using

$$\chi^2 = \frac{(O_{11} - E_{11})^2}{E_{11}} + \frac{(O_{12} - E_{12})^2}{E_{12}} + \ldots + \frac{(O_{rc} - E_{rc})^2}{E_{rc}}$$

can be used. Large deviations between the observed and expected values will lead to a large χ^2 value. If the χ^2 value obtained exceeds the χ^2 table value, given in Appendix I using $v = (r-1) \times (c-1)$, the Null hypothesis is rejected and an inter-relationship between V and W is believed to exist.

The formula for calculating the χ^2 value from the data can be rearranged to give

$$\chi^2 = \frac{O_{11}^2}{E_{11}} + \frac{O_{12}^2}{E_{12}} + \ldots + \frac{O_{rc}^2}{E_{rc}} - n,$$

and this may be used as an alternative for determining the χ^2 value.

Example 1: Flies of the species *G. morsitans* which feed on the blood of animals infected with trypanosomes become carriers of the trypanosomes. Transmission of the disease occurs if these flies subsequently feed on uninfected animals. A sample of 350 flies were caught. Each fly was classified according to sex and the type of animal from which its last blood meal originated. The data are to be analysed to see if flies of one sex have a preference for the type of animal from which they feed.

Origin of blood meal

	Avian	Ovine	Bovine	
Male	O_{11} 49	O_{12} 64	O_{13} 84	197
Female	O_{21} 33	O_{22} 25	O_{23} 95	153
	82	89	179	

H_0 : origin of blood meal is not related to the sex of the flies

H_1 : origin of blood meal is related to the sex of the flies .

Since the contingency table has 2 rows and 3 columns it is only necessary to calculate 2 expected values as the others can be obtained by subtraction from column and row totals:

$$E_{11} = \frac{R_1 C_1}{n} = \frac{197 \times 82}{350} = 46.2 \ ,$$

$$E_{12} = \frac{R_1 C_2}{n} = \frac{197 \times 89}{350} = 50.1 \ .$$

By subtraction

$$E_{21} = 82 - E_{11} = 35.8 \ ,$$

$$E_{22} = 89 - E_{12} = 38.9 \ ,$$

$$E_{13} = 197 - E_{11} - E_{12} = 100.7 \ ,$$

$$E_{23} = 179 - E_{13} = 78.3 \ .$$

The table of expected values will be

E_{11} 46.2	E_{12} 50.1	E_{13} 100.7
E_{21} 35.8	E_{22} 38.9	E_{23} 78.3

Calculating the χ^2 statistic gives

$$\chi^2 = \frac{(O_{11} - E_{11})^2}{E_{11}} + \frac{(O_{12} - E_{12})^2}{E_{12}} + \frac{(O_{13} - E_{13})^2}{E_{13}}$$

$$+ \frac{(O_{21} - E_{21})^2}{E_{21}} + \frac{(O_{22} - E_{22})^2}{E_{22}} + \frac{(O_{23} - E_{23})^2}{E_{23}}$$

$$= \frac{2.8^2}{46.2} + \frac{13.9^2}{50.1} + \frac{16.7^2}{100.7} + \frac{2.8^2}{35.8} + \frac{13.9^2}{38.9} + \frac{16.7^2}{78.3}$$

$$= 0.17 + 3.86 + 2.77 + 0.22 + 4.97 + 3.56$$

$$= 15.55 \ .$$

From the χ^2 table, with $v = (r-1)(c-1) = 2$, the critical region for a 1% significance level starts at 9.21. The χ^2 value obtained from the data clearly exceeds the table value, so the hypothesis H_0 is rejected and it is believed that flies of a certain sex have a preference for feeding on certain types of animal. Examination of the contributions to the χ^2 data value can be helpful. The largest component to contribute to the χ^2 value comes from the deviation between the observed value of 25 and the expected value of 38.9. This suggests that female flies have least preference for ovine type animals.

Before using a contingency table, each expected value should be greater than or equal to 5. When this does not happen the table may be rearranged by combining columns or rows to meet this requirement. The next example demonstrates this situation.

Example 2: The three common types of trypanosomes are *T. vivax, T. congolense* and *T. brucei*. A sample of 100 flies were classified according to the types of trypanosome they carried and the origin of their last blood meal. It is required to determine whether or not the type of infection is related to the animal origin.

	Avian	Ovine	Bovine	
T. vivax	O_{11} 7	O_{12} 16	O_{13} 20	$R_1 = 43$
T. congolense	O_{21} 9	O_{22} 8	O_{23} 22	$R_2 = 39$
T. brucei	O_{31} 4	O_{32} 3	O_{33} 11	$R_3 = 18$
	$C_1 = 20$	$C_2 = 27$	$C_3 = 53$	

H_0 : type of infection is not related to animal origin

H_1 : type of infection is related to animal origin .

Calculating the expected values:

$$E_{11} = \frac{R_1 C_1}{n} = \frac{43 \times 20}{100} = 8.6 \ ,$$

$$E_{12} = \frac{R_1 C_2}{n} = \frac{43 \times 27}{100} = 11.6 \ ,$$

$$E_{21} = \frac{R_2 C_1}{n} = \frac{39 \times 20}{100} = 7.8 \ ,$$

$$E_{22} = \frac{R_2 C_2}{n} = \frac{39 \times 27}{100} = 10.5 \ .$$

By subtraction

$$E_{13} = 22.8, \ E_{23} = 20.7, \ E_{31} = 3.6, \ E_{32} = 4.9, \ E_{33} = 9.5 \ .$$

The expected values in tabular form are

	Avian	Ovine	Bovine
T. vivax	E_{11} 8.6	E_{12} 11.6	E_{13} 22.8
T. congolense	E_{21} 7.8	E_{22} 10.5	E_{23} 20.7
T. brucei	E_{31} 3.6	E_{32} 4.9	E_{33} 9.5

Unfortunately, some of the expected values are less than 5, so two of the row classes or two of the column classes must be combined. If the birds were to be found mostly in those areas grazed by sheep then it would be useful to combine the avian and ovine classes to give the reduced table of expected values:

	Avian and Ovine	Bovine
T. vivax	E_{11} 20.2	E_{12} 22.8
T. congolense	E_{21} 18.3	E_{22} 20.7
T. brucei	E_{31} 8.5	E_{32} 9.5

These expected values are now all greater than or equal to 5. The corresponding reduced table of observed values will be:

	Avian and Ovine	Bovine
T. vivax	O_{11} 23	O_{12} 20
T. congolense	O_{21} 17	O_{22} 22
T. brucei	O_{31} 7	O_{32} 11

The χ^2 statistic becomes

$$\chi^2 = \frac{O_{11}^2}{E_{11}} + \frac{O_{12}^2}{E_{12}} + \ldots + \frac{O_{32}^2}{E_{32}} - n$$

$$= \frac{23^2}{20.2} + \frac{20^2}{22.8} + \frac{17^2}{18.3} + \frac{22^2}{20.7} + \frac{7^2}{8.5} + \frac{11^2}{9.5} - 100$$

$$= 1.41 \ .$$

For $v = (r-1)(c-1) = 2$, and a 5% significance level, the χ^2 table gives the critical region as $\chi^2 > 5.99$. Hence the χ^2 statistic is not in the critical region, and so the hypothesis H_0 is accepted, namely that the type of infection is not influenced by the origin of the blood meal.

25.8 NON-PARAMETRIC AND PARAMETRIC TESTS

A common feature of the non-parametric tests presented in Sections 25.3 to 25.6 is the replacement of observations by ranks. This simplifies calculations and is one reason why, in non-parametric tests, the data need not be normally distributed. The method of ranking also enables analysis of results from experiments in which the observations can only be measured relative to each other on some arbitrary scale. It should be remembered that information will be wasted when rank methods are used, and this is undesirable in the analysis of any data. When deciding which type of test to use, a simple guideline is to use parametric tests when the sample sizes are large. When the sample sizes are small, unless there is evidence to support the validity of the use of a parametric test, the corresponding non-parametric tests should be used. Often the analysis of data by both methods will lead to the same conclusion.

Non-parametric and parametric tests are discussed by Sokal and Rohlf (1971). However, Siegel (1956) and Conover (1971) give excellent discussions on the use of non-parametric tests exclusively.

PROBLEMS

1. In the treatment of psoriasis, nine patients were irradiated with ultra-violet light for three minutes each day, and the healing times were recorded:

    Healing time (weeks) 5.0 5.7 6.0 5.9 6.1 5.6 5.9 6.3 5.7 .

    The median time from the start of treatment to being healed is claimed to be 5.5 weeks. Is this claim substantiated?

2. Air temperatures measured in degrees centigrade at the same time each day are known to be distributed about a median value of 14.5. A sample of ten observations of soil temperature measured at 10 cm depth was as follows:

    12.8 12.5 14.5 12.6 13.2 14.2 17.6 16.2 12.3 14.1 .

    Are the soil temperatures significantly different from the air temperatures?

3. In a comparison of Brussels sprout plants derived from stem callus cultures and from seedlings, the diameters of eight sprouts from each source were measured. Use a Mann-Whitney test to decide if the diameters are different.

Origin	Diameters (cm)							
Callus	2.1	2.3	2.4	2.1	2.6	1.9	2.5	1.8
Seedling	1.7	2.6	1.8	2.0	2.1	2.2	1.6	2.3 .

4. Fish, which were spawned on the same day, were randomly divided into one group of ten and another of eight. The first group was maintained at a temperature of 25°C and the second group at 30°C, and the percentage of days on which eggs were subsequently laid by each fish recorded:

Temperature	Percentage of days on which eggs were laid									
25°C	44	48	94	72	39	62	43	41	54	71
30°C	76	81	48	52	68	73	41	88 .			

    Has the temperature influenced the frequency of egg laying?

5. The generation times of six cultures of microorganisms kept stationary were compared with those of six cultures aerated by shaking.

    Generation times of individual cultures (hrs)

Stationary	8.20	5.64	7.91	5.57	7.10	6.45
Aerated	5.54	4.81	3.68	8.09	5.23	5.02

    Does aeration have a shortening effect on the generation time of the organisms?

6. Pre- and post-treatment observations on the lymphocyte counts in ten patients were as follows:

Patient	1	2	3	4	5	6	7	8	9	10
Pre-	1720	1661	1808	1627	1274	1801	1686	1417	1665	1111
Post-	1543	1608	1935	1781	1101	1262	1488	1378	1337	1067

    Use the Wilcoxon paired comparison test to determine if the lymphocyte counts in the patients have changed after treatment.

7. The intensity of infection in mussels was assessed on a scale from 1 to 10 by inspection. Each mussel was then immersed in a treatment tank for one day and the intensity of infection reassessed:

Mussel	1	2	3	4	5	6	7	8
Before immersion	9	10	1	3	2	9	7	6
After immersion	4	2	1	2	1	7	5	1

    Has the treatment reduced the intensity of infection?

8. The rate at which honey-eater birds consume nectar was measured at sunrise and sunset for each of eight birds:

    Consumption rate (cm^3/hr)

Sunrise	0.9	1.6	1.4	1.2	1.6	1.1	0.8	1.0
Sunset	0.9	1.1	1.2	1.3	1.1	1.0	0.7	0.8

    Do the results suggest that the consumption rates at sunrise and sunset are significantly different?

9. Experiments on the effect of looping on the outgrowth of laterals of willows gave the following measurements for the extension growth of apical shoots for the three treatments shown. Each treatment was used on eight different plants.

| | Treatments | |
Double loop (cm)	Single loop (cm)	Vertical (cm)
39.8	126.0	121.0
27.0	123.0	135.9
43.4	127.0	136.8
36.9	121.8	112.7
41.1	137.1	119.3
53.2	139.4	117.2
38.2	124.0	129.2
34.1	112.6	123.8

Use the Kruskal-Wallis test to test for differences between the treatments.

10. Four groups of avocado pears of the same cultivar were each kept under one of four different controlled-atmospheres and the storage life recorded for each pear. Analyse these data to see if the nature of the controlled-atmosphere affects storage life.

| | Groups | | | |
	1	2	3	4
Storage life	3.5	4.2	5.7	2.7
of individual	4.1	6.8	5.3	3.4
pears (weeks)	3.7	7.6	6.1	4.7
	3.8	5.2	5.1	2.4
	4.2	7.1	5.4	3.1
	4.1	7.6		
		6.2		
		4.8		

11. The daily food intake of pigs fed on diets containing different levels of cotton seed flour was as follows:

| | Level of cotton seed flour in diet | | |
	4%	8%	12%
	0.66	0.73	0.61
Food intake	0.69	0.82	0.91
of individual	0.63	0.84	0.64
pigs	0.65	0.78	0.71
	0.74	0.61	0.74
	0.81	0.52	0.84

Does the cotton seed content of the diet influence the food intake of the pigs?

12. The table below shows the antigen levels (ng/ml) in serum from patients before treatment, 14 days after a first treatment, and 14 days after a second treatment:

Antigen levels

Patient	Before treatment	14 days after 1st treatment	14 days after 2nd treatment
1	171.0	211.0	196.0
2	24.9	22.0	12.0
3	10.6	10.9	7.4
4	9.0	10.6	13.9
5	145.0	95.0	125.0
6	57.0	61.0	31.0
7	18.7	20.4	20.6

Use the Friedman test to analyse the data for a difference in treatment effect.

13. The yield obtained from a tomato crop is known to be influenced by the intensity of potato root eelworm in the soil. In an attempt to control the parasite and improve yield, plots containing three different intensities of viable eelworm eggs were exposed to different chemical sterilants. The results below show the yield in kg from 12 plots each half a square metre in area:

Sterilant

		Control	Metham- sodium	Chloropicrin	Carbon Disulphide
Intensity of eelworm eggs	high	3.4	10.1	11.4	3.6
	medium	5.9	10.2	12.6	7.2
	low	6.3	11.3	12.3	8.7

From analysis of the above results what conclusions would you draw about the effects of eelworm intensity and sterilants on yield?

14. Plant nematodes removed from soil samples exposed to three different ranges of sub-zero temperatures were examined for viability. From the

following contingency table determine whether or not viability is related to temperature conditions:

	$-15°C$ to $-10°C$	$-10°C$ to $-5°C$	$-5°C$ to $0°C$
Viable	53	78	46
Non-viable	32	16	65

15. In an experiment on peas, performed by Mendel, the following results relating surface and colour were obtained:

		Colour	
		Blue	Green
Surface	Round	316	109
	Wrinkled	102	33

Do the results suggest a significant relationship between colour and surface?

16. The following contingency table shows the effect of chilling on the number of shoots of blackcurrant that develop. Three treatments were used, chilling the plants for 8 hours/day, 12 hours/day, and 16 hours/day for a period of eight weeks. The shoots were then classified according to size, the sizes ranging from size 1, the longest, to size 4, the shortest.

		Size classification			
		1	2	3	4
Treatment	8 hours/day	45	33	23	15
	12 hours/day	43	29	17	25
	16 hours/day	32	30	21	37

Use a χ^2-test to test for a relationship between size classification and treatments. State clearly your conclusions.

17. Sacks of wheat containing insect pests were sprayed with three types of fumigant. One week later the insect mortality for each type of fumigant was assessed, as shown below, by counting the number of adult- and larval-stage insects still alive.

Larval-stage

	Type of fumigant		
	A	B	C
Number of insects alive	84	67	51
Number of insects dead	96	26	34

Adult-stage

	Type of fumigant		
	A	B	C
Number of insects alive	6	12	83
Number of insects dead	10	25	90

Is the type of fumigant important if:

(i) only the larval-stage insects are to be killed,

(ii) both larval- and adult-stage insects are to be killed?

Appendices A - I

APPENDIX A

TRIGONOMETRIC IDENTITIES

$\tan \theta = \sin \theta / \cos \theta,$	A 1
$\cot \theta = \cos \theta / \sin \theta,$	A 2
$\sec \theta = 1/\cos \theta,$	A 3
$\operatorname{cosec} \theta = 1/\sin \theta,$	A 4
$\sin (-\theta) = -\sin \theta,$	A 5
$\cos (-\theta) = \cos \theta,$	A 6
$\tan (-\theta) = -\tan \theta,$	A 7

$$\sin \left(\frac{\pi}{2} - \theta \right) = \cos \theta, \qquad \text{A 8}$$

$$\cos \left(\frac{\pi}{2} - \theta \right) = \sin \theta, \qquad \text{A 9}$$

$$\tan \left(\frac{\pi}{2} - \theta \right) = \cot \theta, \qquad \text{A 10}$$

$\sin (\pi - \theta) = \sin \theta,$	A 11
$\cos (\pi - \theta) = -\cos \theta,$	A 12
$\tan (\pi - \theta) = -\tan \theta,$	A 13
$\sin 2\theta = 2 \sin \theta \cos \theta,$	A 14
$\cos 2\theta = \cos^2\theta - \sin^2\theta = 2 \cos^2\theta - 1 = 1 - 2 \sin^2\theta,$	A 15
$\sin^2\theta = \frac{1}{2}(1 - \cos 2\theta),$	A 16
$\cos^2\theta = \frac{1}{2}(1 + \cos 2\theta),$	A 17
$\sin^2\theta + \cos^2\theta = 1,$	A 18
$1 + \tan^2\theta = \sec^2\theta,$	A 19
$1 + \cot^2\theta = \operatorname{cosec}^2\theta,$	A 20
$\sin (A \pm B) = \sin A \cos B \pm \cos A \sin B,$	A 21
$\cos (A \pm B) = \cos A \cos B \mp \sin A \sin B,$	A 22

$$\tan(A \pm B) = \frac{\tan A \pm \tan B}{1 \mp \tan A \tan B},$$ A 23

$$\sin A + \sin B = 2\sin\left(\frac{A+B}{2}\right)\cos\left(\frac{A-B}{2}\right),$$ A 24

$$\sin A - \sin B = 2\sin\left(\frac{A-B}{2}\right)\cos\left(\frac{A+B}{2}\right),$$ A 25

$$\cos A + \cos B = 2\cos\left(\frac{A+B}{2}\right)\cos\left(\frac{A-B}{2}\right),$$ A 26

$$\cos A - \cos B = 2\sin\left(\frac{A+B}{2}\right)\sin\left(\frac{B-A}{2}\right),$$ A 27

$$2\sin A \cos B = \sin(A+B) + \sin(A-B),$$ A 28

$$2\cos A \cos B = \cos(A+B) + \cos(A-B),$$ A 29

$$2\sin A \sin B = -\cos(A+B) + \cos(A-B).$$ A 30

APPENDIX B
STANDARD DERIVATIVES AND INTEGRALS

Differentiation	Integration
$\dfrac{d}{dx}x^\alpha = \alpha x^{\alpha-1}$	$\displaystyle\int x^\alpha dx = \dfrac{1}{\alpha+1}x^{\alpha+1} + C, \quad (\alpha \neq -1)$
	$\displaystyle\int \dfrac{1}{x}\,dx = \ln|x| + C$
$\dfrac{d}{dx}e^{ax} = a\,e^{ax}$	$\displaystyle\int e^{ax}dx = \dfrac{1}{a}e^{ax} + C$
$\dfrac{d}{dx}a^x = a^x \ln a, \qquad (a>0)$	$\displaystyle\int a^x\,dx = \dfrac{1}{\ln a}a^x + C, \quad (a>0)$
$\dfrac{d}{dx}\sin(ax) = a\cos(ax)$	$\displaystyle\int \sin(ax)\,dx = -\dfrac{1}{a}\cos(ax) + C$
$\dfrac{d}{dx}\cos(ax) = -a\sin(ax)$	$\displaystyle\int \cos(ax)dx = \dfrac{1}{a}\sin(ax) + C$
$\dfrac{d}{dx}\tan(ax) = a\sec^2(ax)$	$\displaystyle\int \tan(ax)dx = \dfrac{1}{a}\ln|\sec(ax)| + C$
	$\displaystyle\int \sec^2(ax)dx = \dfrac{1}{a}\tan(ax) + C$
$\dfrac{d}{dx}\sec x = \sec x \tan x$	$\int \sec x\,dx = \ln|\sec x + \tan x| + C$
	$\int \sec x \tan x\,dx = \sec x + C$
$\dfrac{d}{dx}\operatorname{cosec} x = -\operatorname{cosec} x \cot x$	$\int \operatorname{cosec} x\,dx = -\ln|\operatorname{cosec} x + \cot x| + C$
	$\qquad\qquad = \ln\left|\tan\dfrac{x}{2}\right| + C$
	$\int \operatorname{cosec} x \cot x\,dx = -\operatorname{cosec} x + C$
$\dfrac{d}{dx}\cot x = -\operatorname{cosec}^2 x$	$\int \cot x\,dx = \ln|\sin x| + C$
	$\int \operatorname{cosec}^2 x\,dx = -\cot x + C$
$\dfrac{d}{dx}\ln|x| = \dfrac{1}{x}$	$\int \ln|x|\,dx = x\ln|x| - x + C$
$\dfrac{d}{dx}\sin^{-1}\left(\dfrac{x}{a}\right) = \dfrac{1}{\sqrt{(a^2-x^2)}}, \left(\left|\dfrac{x}{a}\right|\leqslant 1\right)$	$\displaystyle\int \dfrac{1}{\sqrt{(a^2-x^2)}}\,dx = \sin^{-1}\left(\dfrac{x}{a}\right) + C, \left(\left|\dfrac{x}{a}\right|\leqslant 1\right)$
$\dfrac{d}{dx}\tan^{-1}\left(\dfrac{x}{a}\right) = \dfrac{a}{a^2+x^2}$	$\displaystyle\int \dfrac{1}{a^2+x^2}\,dx = \dfrac{1}{a}\tan^{-1}\left(\dfrac{x}{a}\right) + C$

The constant a is non-zero.

APPENDIX C

VALUES OF NATURAL LOGARITHMS

	0	1	2	3	4	5	6	7	8	9	10	Mean Proportional Parts
												1 2 3 4 5 6 7 8 9 10
1.0	0.0000	0099	0198	0296	0392	0488	0583	0677	0770	0862	0.0953	10 19 29 38 48 58 67 77 86 96
1.1	0.0953	1044	1133	1222	1310	1398	1484	1570	1655	1740	0.1823	9 17 26 35 44 52 61 70 78 87
1.2	0.1823	1906	1989	2070	2151	2231	2311	2390	2469	2546	0.2624	8 16 24 32 40 48 56 64 72 80
1.3	0.2624	2700	2776	2852	2927	3001	3075	3148	3221	3293	0.3365	7 15 22 30 37 44 52 59 67 74
1.4	0.3365	3436	3507	3577	3646	3716	3784	3853	3920	3988	0.4055	7 14 21 28 35 41 48 55 62 69
1.5	0.4055	4121	4187	4253	4318	4383	4447	4511	4574	4637	0.4700	6 13 19 26 32 39 45 52 58 65
1.6	0.4700	4762	4824	4886	4947	5008	5068	5128	5188	5247	0.5306	6 12 18 24 30 36 42 48 55 61
1.7	0.5306	5365	5423	5481	5539	5596	5653	5710	5766	5822	0.5878	6 11 17 23 29 34 40 46 51 57
1.8	0.5878	5933	5988	6043	6098	6152	6206	6259	6313	6366	0.6419	5 11 16 22 27 32 38 43 49 54
1.9	0.6419	6471	6523	6575	6627	6678	6729	6780	6831	6881	0.6931	5 10 15 20 26 31 36 41 46 51
2.0	0.6931	6981	7031	7080	7129	7178	7227	7275	7324	7372	0.7419	5 10 15 20 24 29 34 39 44 49
2.1	0.7419	7467	7514	7561	7608	7655	7701	7747	7793	7839	0.7885	5 9 14 19 23 28 33 37 42 47
2.2	0.7885	7930	7975	8020	8065	8109	8154	8198	8242	8286	0.8329	4 9 13 18 22 27 31 36 40 45
2.3	0.8329	8372	8416	8459	8502	8544	8587	8629	8671	8713	0.8755	4 9 13 17 21 26 30 34 38 43
2.4	0.8755	8796	8838	8879	8920	8961	9002	9042	9083	9123	0.9163	4 8 12 16 20 24 29 33 37 41
2.5	0.9163	9203	9243	9282	9322	9361	9400	9439	9478	9517	0.9555	4 8 12 16 20 24 27 31 35 39
2.6	0.9555	9594	9632	9670	9708	9746	9783	9821	9858	9895	0.9933	4 8 11 15 19 23 26 30 34 38
2.7	0.9933	9969	1.0006	0043	0080	0116	0152	0188	0225	0260	1.0296	4 7 11 14 18 22 25 29 32 36
2.8	1.0296	0332	0367	0403	0438	0473	0508	0543	0578	0613	1.0647	4 7 11 14 18 21 25 28 32 35
2.9	1.0647	0682	0716	0750	0784	0818	0852	0886	0919	0953	1.0986	3 7 10 14 17 20 24 27 31 34
3.0	1.0986	1019	1053	1086	1119	1151	1184	1217	1249	1282	1.1314	3 7 10 13 16 20 23 26 30 33
3.1	1.1314	1346	1378	1410	1442	1474	1506	1537	1569	1600	1.1632	3 6 10 13 16 19 22 26 29 32
3.2	1.1632	1663	1694	1725	1756	1787	1817	1848	1878	1909	1.1939	3 6 9 12 15 19 22 25 28 31
3.3	1.1939	1969	2000	2030	2060	2090	2119	2149	2179	2208	1.2238	3 6 9 12 15 18 21 24 27 30
3.4	1.2238	2267	2296	2326	2355	2384	2413	2442	2470	2499	1.2528	3 6 9 12 15 17 20 23 26 29
3.5	1.2528	2556	2585	2613	2641	2669	2698	2726	2754	2782	1.2809	3 6 8 11 14 17 20 23 25 28
3.6	1.2809	2837	2865	2892	2920	2947	2975	3002	3029	3056	1.3083	3 5 8 11 14 16 19 22 24 27
3.7	1.3083	3110	3137	3164	3191	3218	3244	3271	3297	3324	1.3350	3 5 8 11 13 16 19 22 24 27
3.8	1.3350	3376	3403	3429	3455	3481	3507	3533	3558	3584	1.3610	3 5 8 10 13 16 18 21 23 26
3.9	1.3610	3635	3661	3686	3712	3737	3762	3788	3813	3838	1.3863	3 5 8 10 13 15 18 20 23 25
4.0	1.3863	3888	3913	3938	3962	3987	4012	4036	4061	4085	1.4110	2 5 7 10 12 15 17 20 22 25
4.1	1.4110	4134	4159	4183	4207	4231	4255	4279	4303	4327	1.4351	2 5 7 10 12 14 17 19 22 24
4.2	1.4351	4375	4398	4422	4446	4469	4493	4516	4540	4563	1.4586	
4.3	1.4586	4609	4633	4656	4679	4702	4725	4748	4770	4793	1.4816	2 5 7 9 12 14 16 18 21 23
4.4	1.4816	4839	4861	4884	4907	4929	4951	4974	4996	5019	1.5041	2 5 7 9 11 14 16 18 21 23
4.5	1.5041	5063	5085	5107	5129	5151	5173	5195	5217	5239	1.5261	2 4 7 9 11 13 15 18 20 22
4.6	1.5261	5282	5304	5326	5347	5369	5390	5412	5433	5454	1.5476	
4.7	1.5476	5497	5518	5539	5560	5581	5602	5623	5644	5665	1.5686	2 4 6 8 11 13 15 17 19 21
4.8	1.5686	5707	5728	5748	5769	5790	5810	5831	5851	5872	1.5892	2 4 6 8 10 12 14 16 19 21
4.9	1.5892	5913	5933	5953	5974	5994	6014	6034	6054	6074	1.6094	
5.0	1.6094	6114	6134	6154	6174	6194	6214	6233	6253	6273	1.6292	2 4 6 8 10 12 14 16 18 20
5.1	1.6292	6312	6332	6351	6371	6390	6409	6429	6448	6467	1.6487	
5.2	1.6487	6506	6525	6544	6563	6582	6601	6620	6639	6658	1.6677	2 4 6 8 10 11 13 15 17 19
5.3	1.6677	6696	6715	6734	6752	6771	6790	6808	6827	6845	1.6864	2 4 6 8 9 11 13 15 17 19
5.4	1.6864	6882	6901	6919	6938	6956	6974	6993	7011	7029	1.7047	2 4 5 7 9 11 13 14 16 18

n	1	2	3	4	5	6	7	8	9	10
$\ln 10^n$	2.3026	4.6052	6.9078	9.2103	11.5129	13.8155	16.1181	18.4207	20.7233	23.0259
$\ln 10^{-n}$	$\bar{3}.6974$	$\bar{5}.3948$	$\bar{7}.0922$	$\overline{10}.7897$	$\overline{12}.4871$	$\overline{14}.1845$	$\overline{17}.8819$	$\overline{19}.5793$	$\overline{21}.2767$	$\overline{24}.9741$

	0	1	2	3	4	5	6	7	8	9	10	Mean Proportional Parts 1	2	3	4	5	6	7	8	9	10
5.5	1.7047	7066	7084	7102	7120	7138	7156	7174	7192	7210	1.7228	2	4	5	7	9	11	13	14	16	18
5.6	1.7228	7246	7263	7281	7299	7317	7334	7352	7370	7387	1.7405										
5.7	1.7405	7422	7440	7457	7475	7492	7509	7527	7544	7561	1.7579										
5.8	1.7579	7596	7613	7630	7647	7664	7681	7699	7716	7733	1.7750	2	3	5	7	9	10	12	14	15	17
5.9	1.7750	7766	7783	7800	7817	7834	7851	7867	7884	7901	1.7918	2	3	5	7	8	10	12	14	15	17
6.0	1.7918	7934	7951	7967	7984	8001	8017	8034	8050	8066	1.8083										
6.1	1.8083	8099	8116	8132	8148	8165	8181	8197	8213	8229	1.8245										
6.2	1.8245	8262	8278	8294	8310	8326	8342	8358	8374	8390	1.8405	2	3	5	6	8	10	11	13	14	16
6.3	1.8405	8421	8437	8453	8469	8485	8500	8516	8532	8547	1.8563										
6.4	1.8563	8579	8594	8610	8625	8641	8656	8672	8687	8703	1.8718										
6.5	1.8718	8733	8749	8764	8779	8795	8810	8825	8840	8856	1.8871										
6.6	1.8871	8886	8901	8916	8931	8946	8961	8976	8991	9006	1.9021	2	3	5	6	8	9	11	12	14	15
6.7	1.9021	9036	9051	9066	9081	9095	9110	9125	9140	9155	1.9169	1	3	4	6	7	9	10	12	13	15
6.8	1.9169	9184	9199	9213	9228	9242	9257	9272	9286	9301	1.9315										
6.9	1.9315	9330	9344	9359	9373	9387	9402	9416	9430	9445	1.9459										
7.0	1.9459	9473	9488	9502	9516	9530	9544	9559	9573	9587	1.9601										
7.1	1.9601	9615	9629	9643	9657	9671	9685	9699	9713	9727	1.9741	1	3	4	6	7	8	10	11	13	14
7.2	1.9741	9755	9769	9782	9796	9810	9824	9838	9851	9865	1.9879										
7.3	1.9879	9892	9906	9920	9933	9947	9961	9974	9988	2.0001	2.0015										
7.4	2.0015	0028	0042	0055	0069	0082	0096	0109	0122	0136	2.0149										
7.5	2.0149	0162	0176	0189	0202	0215	0229	0242	0255	0268	2.0281										
7.6	2.0281	0295	0308	0321	0334	0347	0360	0373	0386	0399	2.0412	1	3	4	5	7	8	9	10	12	13
7.7	2.0412	0425	0438	0451	0464	0477	0490	0503	0516	0528	2.0541	1	3	4	5	6	8	9	10	12	13
7.8	2.0541	0554	0567	0580	0592	0605	0618	0631	0643	0656	2.0669										
7.9	2.0669	0681	0694	0707	0719	0732	0744	0757	0769	0782	2.0794										
8.0	2.0794	0807	0819	0832	0844	0857	0869	0882	0894	0906	2.0919										
8.1	2.0919	0931	0943	0956	0968	0980	0992	1005	1017	1029	2.1041										
8.2	2.1041	1054	1066	1078	1090	1102	1114	1126	1138	1150	2.1163										
8.3	2.1163	1175	1187	1199	1211	1223	1235	1247	1258	1270	2.1282	1	2	4	5	6	7	8	10	11	12
8.4	2.1282	1294	1306	1318	1330	1342	1353	1365	1377	1389	2.1401										
8.5	2.1401	1412	1424	1436	1448	1459	1471	1483	1494	1506	2.1518										
8.6	2.1518	1529	1541	1552	1564	1576	1587	1599	1610	1622	2.1633										
8.7	2.1633	1645	1656	1668	1679	1691	1702	1713	1725	1736	2.1748										
8.8	2.1748	1759	1770	1782	1793	1804	1815	1827	1838	1849	2.1861										
8.9	2.1861	1872	1883	1894	1905	1917	1928	1939	1950	1961	2.1972										
9.0	2.1972	1983	1994	2006	2017	2028	2039	2050	2061	2072	2.2083	1	2	3	4	6	7	8	9	10	11
9.1	2.2083	2094	2105	2116	2127	2138	2148	2159	2170	2181	2.2192	1	2	3	4	5	7	8	9	10	11
9.2	2.2192	2203	2214	2225	2235	2246	2257	2268	2279	2289	2.2300										
9.3	2.2300	2311	2322	2332	2343	2354	2364	2375	2386	2396	2.2407										
9.4	2.2407	2418	2428	2439	2450	2460	2471	2481	2492	2502	2.2513										
9.5	2.2513	2523	2534	2544	2555	2565	2576	2586	2597	2607	2.2618										
9.6	2.2618	2628	2638	2649	2659	2670	2680	2690	2701	2711	2.2721										
9.7	2.2721	2732	2742	2752	2762	2773	2783	2793	2803	2814	2.2824										
9.8	2.2824	2834	2844	2854	2865	2875	2885	2895	2905	2915	2.2925										
9.9	2.2925	2935	2946	2956	2966	2976	2986	2996	3006	3016	2.3026	1	2	3	4	5	6	7	8	9	10

Example: To find ln 6.543, from the row 6.5 and the main column 4 the table gives 1.8779. As no values are entered in the mean proportional parts column opposite 6.5 the nearest values (for 6.6) are taken and from column 3 the value is 5. So ln 6.543 = 1.8779 + 0.0005 = 1.8784.

APPENDIX D

VALUES OF EXPONENTIAL FUNCTIONS

x	e^x	e^{-x}	x	e^x	e^{-x}	x	e^x	e^{-x}
0.00	1.0000	1.0000	0.50	1.6487	0.6065	1.00	2.718	0.3679
0.01	1.0101	0.9900	0.51	1.6653	0.6005	1.01	2.746	0.3642
0.02	1.0202	0.9802	0.52	1.6820	0.5945	1.02	2.773	0.3606
0.03	1.0305	0.9704	0.53	1.6989	0.5886	1.03	2.801	0.3570
0.04	1.0408	0.9608	0.54	1.7160	0.5827	1.04	2.829	0.3535
0.05	1.0513	0.9512	0.55	1.7333	0.5769	1.05	2.858	0.3499
0.06	1.0618	0.9418	0.56	1.7507	0.5712	1.06	2.886	0.3465
0.07	1.0725	0.9324	0.57	1.7683	0.5655	1.07	2.915	0.3430
0.08	1.0833	0.9231	0.58	1.7860	0.5599	1.08	2.945	0.3396
0.09	1.0942	0.9139	0.59	1.8040	0.5543	1.09	2.974	0.3362
0.10	1.1052	0.9048	0.60	1.8221	0.5488	1.10	3.004	0.3329
0.11	1.1163	0.8958	0.61	1.8404	0.5434	1.11	3.034	0.3296
0.12	1.1275	0.8869	0.62	1.8589	0.5379	1.12	3.065	0.3263
0.13	1.1388	0.8781	0.63	1.8776	0.5326	1.13	3.096	0.3230
0.14	1.1503	0.8694	0.64	1.8965	0.5273	1.14	3.127	0.3198
0.15	1.1618	0.8607	0.65	1.9155	0.5220	1.15	3.158	0.3166
0.16	1.1735	0.8521	0.66	1.9348	0.5169	1.16	3.190	0.3135
0.17	1.1853	0.8437	0.67	1.9542	0.5117	1.17	3.222	0.3104
0.18	1.1972	0.8353	0.68	1.9739	0.5066	1.18	3.254	0.3073
0.19	1.2092	0.8270	0.69	1.9937	0.5016	1.19	3.287	0.3042
0.20	1.2214	0.8187	0.70	2.0138	0.4966	1.20	3.320	0.3012
0.21	1.2337	0.8106	0.71	2.0340	0.4916	1.21	3.353	0.2982
0.22	1.2461	0.8025	0.72	2.0544	0.4868	1.22	3.387	0.2952
0.23	1.2586	0.7945	0.73	2.0751	0.4819	1.23	3.421	0.2923
0.24	1.2712	0.7866	0.74	2.0959	0.4771	1.24	3.456	0.2894
0.25	1.2840	0.7788	0.75	2.1170	0.4724	1.25	3.490	0.2865
0.26	1.2969	0.7711	0.76	2.1383	0.4677	1.26	3.525	0.2837
0.27	1.3100	0.7634	0.77	2.1598	0.4630	1.27	3.561	0.2808
0.28	1.3231	0.7558	0.78	2.1815	0.4584	1.28	3.597	0.2780
0.29	1.3364	0.7483	0.79	2.2034	0.4538	1.29	3.633	0.2753
0.30	1.3499	0.7408	0.80	2.2255	0.4493	1.30	3.669	0.2725
0.31	1.3634	0.7334	0.81	2.2479	0.4449	1.31	3.706	0.2698
0.32	1.3771	0.7261	0.82	2.2705	0.4404	1.32	3.743	0.2671
0.33	1.3910	0.7189	0.83	2.2933	0.4360	1.33	3.781	0.2645
0.34	1.4049	0.7118	0.84	2.3164	0.4317	1.34	3.819	0.2618
0.35	1.4191	0.7047	0.85	2.3396	0.4274	1.35	3.857	0.2592
0.36	1.4333	0.6977	0.86	2.3632	0.4232	1.36	3.896	0.2567
0.37	1.4477	0.6907	0.87	2.3869	0.4190	1.37	3.935	0.2541
0.38	1.4623	0.6839	0.88	2.4109	0.4148	1.38	3.975	0.2516
0.39	1.4770	0.6771	0.89	2.4351	0.4107	1.39	4.015	0.2491
0.40	1.4918	0.6703	0.90	2.4596	0.4066	1.40	4.055	0.2466
0.41	1.5068	0.6637	0.91	2.4843	0.4025	1.41	4.096	0.2441
0.42	1.5220	0.6570	0.92	2.5093	0.3985	1.42	4.137	0.2417
0.43	1.5373	0.6505	0.93	2.5345	0.3946	1.43	4.179	0.2393
0.44	1.5527	0.6440	0.94	2.5600	0.3906	1.44	4.221	0.2369
0.45	1.5683	0.6376	0.95	2.5857	0.3867	1.45	4.263	0.2346
0.46	1.5841	0.6313	0.96	2.6117	0.3829	1.46	4.306	0.2322
0.47	1.6000	0.6250	0.97	2.6379	0.3791	1.47	4.349	0.2299
0.48	1.6161	0.6188	0.98	2.6645	0.3753	1.48	4.393	0.2276
0.49	1.6323	0.6126	0.99	2.6912	0.3716	1.49	4.437	0.2254

x	e^x	e^{-x}	x	e^x	e^{-x}	x	e^x	e^{-x}
1.50	4.482	0.2231	2.00	7.389	0.1353	2.5	12.18	0.0821
1.51	4.527	0.2209	2.01	7.463	0.1340	2.6	13.46	0.0743
1.52	4.572	0.2187	2.02	7.538	0.1327	2.7	14.88	0.0672
1.53	4.618	0.2165	2.03	7.614	0.1313	2.8	16.44	0.0608
1.54	4.665	0.2144	2.04	7.691	0.1300	2.9	18.17	0.0550
1.55	4.711	0.2122	2.05	7.768	0.1287	3.0	20.09	0.0498
1.56	4.759	0.2101	2.06	7.846	0.1275	3.1	22.20	0.0450
1.57	4.807	0.2080	2.07	7.925	0.1262	3.2	24.53	0.0408
1.58	4.855	0.2060	2.08	8.004	0.1249	3.3	27.11	0.0369
1.59	4.904	0.2039	2.09	8.085	0.1237	3.4	29.96	0.0334
1.60	4.953	0.2019	2.10	8.166	0.1225	3.5	33.12	0.0302
1.61	5.003	0.1999	2.11	8.248	0.1212	3.6	36.60	0.0273
1.62	5.053	0.1979	2.12	8.331	0.1200	3.7	40.45	0.0247
1.63	5.104	0.1959	2.13	8.415	0.1188	3.8	44.70	0.0224
1.64	5.155	0.1940	2.14	8.499	0.1177	3.9	49.40	0.0202
1.65	5.207	0.1920	2.15	8.585	0.1165	4.0	54.60	0.0183
1.66	5.259	0.1901	2.16	8.671	0.1153	4.1	60.34	0.0166
1.67	5.312	0.1882	2.17	8.758	0.1142	4.2	66.69	0.0150
1.68	5.366	0.1864	2.18	8.846	0.1130	4.3	73.70	0.0136
1.69	5.419	0.1845	2.19	8.935	0.1119	4.4	81.45	0.0123
1.70	5.474	0.1827	2.20	9.025	0.1108	4.5	90.02	0.0111
1.71	5.529	0.1809	2.21	9.116	0.1097	4.6	99.48	0.0101
1.72	5.585	0.1791	2.22	9.207	0.1086	4.7	109.95	0.00910
1.73	5.641	0.1773	2.23	9.300	0.1075	4.8	121.51	0.00823
1.74	5.697	0.1755	2.24	9.393	0.1065	4.9	134.29	0.00745
1.75	5.755	0.1738	2.25	9.488	0.1054	5.0	148.41	0.00674
1.76	5.812	0.1720	2.26	9.583	0.1044	5.1	164.02	0.00610
1.77	5.871	0.1703	2.27	9.679	0.1033	5.2	181.27	0.00552
1.78	5.930	0.1686	2.28	9.777	0.1023	5.3	200.34	0.00499
1.79	5.989	0.1670	2.29	9.875	0.1013	5.4	221.41	0.00452
1.80	6.050	0.1653	2.30	9.974	0.1003	5.5	244.69	0.00409
1.81	6.110	0.1637	2.31	10.074	0.09926	5.6	270.43	0.00370
1.82	6.172	0.1620	2.32	10.176	0.09827	5.7	298.87	0.00335
1.83	6.234	0.1604	2.33	10.278	0.09730	5.8	330.30	0.00303
1.84	6.297	0.1588	2.34	10.381	0.09633	5.9	365.04	0.00274
1.85	6.360	0.1572	2.35	10.486	0.09537	6.0	403.43	0.00248
1.86	6.424	0.1557	2.36	10.591	0.09442	6.1	445.86	0.00224
1.87	6.488	0.1541	2.37	10.697	0.09348	6.2	492.75	0.00203
1.88	6.554	0.1526	2.38	10.805	0.09255	6.3	544.57	0.00184
1.89	6.619	0.1511	2.39	10.913	0.09163	6.4	601.85	0.00166
1.90	6.686	0.1496	2.40	11.023	0.09072	6.5	665.14	0.00150
1.91	6.753	0.1481	2.41	11.134	0.08982	6.6	735.10	0.00136
1.92	6.821	0.1466	2.42	11.246	0.08892	6.7	812.41	0.00123
1.93	6.890	0.1451	2.43	11.359	0.08804	6.8	897.85	0.00111
1.94	6.959	0.1437	2.44	11.473	0.08716	6.9	992.27	0.00101
1.95	7.029	0.1423	2.45	11.588	0.08629	7.0	1096.6	0.000912
1.96	7.099	0.1409	2.46	11.705	0.08543	7.1	1212.0	0.000825
1.97	7.171	0.1395	2.47	11.822	0.08458	7.2	1339.4	0.000747
1.98	7.243	0.1381	2.48	11.941	0.08374	7.3	1480.3	0.000676
1.99	7.316	0.1367	2.49	12.061	0.08291	7.4	1636.0	0.000611

APPENDIX E

AREAS UNDER STANDARD NORMAL CURVE, $\Phi(z)$

z	0.00	0.01	0.02	0.03	0.04	0.05	0.06	0.07	0.08	0.09
0.00	5000	5040	5080	5120	5160	5199	5239	5279	5319	5359
0.10	5398	5438	5478	5517	5557	5596	5636	5675	5714	5753
0.20	5793	5832	5871	5910	5948	5987	6026	6064	6103	6141
0.30	6179	6217	6255	6293	6331	6368	6406	6443	6480	6517
0.40	6554	6591	6628	6664	6700	6736	6772	6808	6844	6879
0.50	6915	6950	6985	7019	7054	7088	7123	7157	7190	7224
0.60	7257	7291	7324	7357	7389	7422	7454	7486	7517	7549
0.70	7580	7611	7642	7673	7704	7734	7764	7794	7823	7852
0.80	7881	7910	7939	7967	7995	8023	8051	8078	8106	8133
0.90	8159	8186	8212	8238	8264	8289	8315	8340	8365	8389
1.00	8413	8438	8461	8485	8508	8531	8554	8577	8599	8621
1.10	8643	8665	8686	8708	8729	8749	8770	8790	8810	8830
1.20	8849	8869	8888	8907	8925	8944	8962	8980	8997	9015
1.30	9032	9049	9066	9082	9099	9115	9131	9147	9162	9177
1.40	9192	9207	9222	9236	9251	9265	9279	9292	9306	9319
1.50	9332	9345	9357	9370	9382	9394	9406	9418	9429	9441
1.60	9452	9463	9474	9484	9495	9505	9515	9525	9535	9545
1.70	9554	9564	9573	9582	9591	9599	9608	9616	9625	9633
1.80	9641	9649	9656	9664	9671	9678	9686	9693	9699	9706
1.90	9713	9719	9726	9732	9738	9744	9750	9756	9761	9767
2.00	9772	9778	9783	9788	9793	9798	9803	9808	9812	9817
2.10	9821	9826	9830	9834	9838	9842	9846	9850	9854	9857
2.20	9861	9864	9868	9871	9875	9878	9881	9884	9887	9890
2.30	9893	9896	9898	9901	9904	9906	9909	9911	9913	9916
2.40	9918	9920	9922	9925	9927	9929	9931	9932	9934	9936
2.50	9938	9940	9941	9943	9945	9946	9948	9949	9951	9952
2.60	9953	9955	9956	9957	9959	9960	9961	9962	9963	9964
2.70	9965	9966	9967	9968	9969	9970	9971	9972	9973	9974
2.80	9974	9975	9976	9977	9977	9978	9979	9979	9980	9981
2.90	9981	9982	9982	9983	9984	9984	9985	9985	9986	9986
3.00	9987	9987	9987	9988	9988	9989	9989	9989	9990	9990
3.10	9990	9991	9991	9991	9992	9992	9992	9992	9993	9993
3.20	9993	9993	9994	9994	9994	9994	9994	9995	9995	9995
3.30	9995	9995	9995	9996	9996	9996	9996	9996	9996	9997
3.40	9997	9997	9997	9997	9997	9997	9997	9997	9997	9998
3.50	9998	9998	9998	9998	9998	9998	9998	9998	9998	9998
3.60	9998	9998	9999	9999	9999	9999	9999	9999	9999	9999
3.70	9999	9999	9999	9999	9999	9999	9999	9999	9999	9999
3.80	9999	9999	9999	9999	9999	9999	9999	9999	9999	9999
3.90	10000	10000	10000	10000	10000	10000	10000	10000	10000	10000

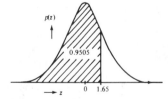

Example: Area under standard normal curve to the left of 1.65 is $\Phi(1.65) = 0.9505$.

APPENDIX F

UPPER TAIL CRITICAL VALUES FOR STUDENT'S t-DISTRIBUTION

One Tail Areas	0.100	0.050	0.025	0.010	0.005	0.0025	0.001	0.0005
$\nu=1$	3.078	6.314	12.71	31.82	63.66	127.3	318.3	636.6
2	1.886	2.920	4.303	6.965	9.925	14.09	22.33	31.60
3	1.638	2.353	3.182	4.451	5.841	7.453	10.21	12.92
4	1.533	2.132	2.776	3.747	4.604	5.598	7.173	8.610
5	1.476	2.015	2.571	3.365	4.032	4.773	5.893	6.869
6	1.440	1.943	2.447	3.143	3.707	4.317	5.208	5.959
7	1.415	1.895	2.365	2.998	3.499	4.029	4.785	5.408
8	1.397	1.860	2.306	2.896	3.355	3.883	4.501	5.041
9	1.383	1.833	2.262	2.821	3.250	3.690	4.297	4.781
10	1.372	1.812	2.228	2.764	3.169	3.581	4.144	4.587
11	1.363	1.796	2.201	2.718	3.106	3.497	4.025	4.437
12	1.356	1.782	2.179	2.681	3.055	3.428	3.930	4.318
13	1.350	1.771	2.160	2.650	3.012	3.372	3.852	4.221
14	1.345	1.761	2.145	2.624	2.977	3.326	3.787	4.140
15	1.341	1.753	2.131	2.602	2.947	3.286	3.733	4.074
16	1.337	1.746	2.120	2.583	2.921	3.252	3.686	4.015
17	1.333	1.740	2.110	2.567	2.898	3.222	3.646	3.965
18	1.330	1.734	2.101	2.552	2.878	3.197	3.610	3.922
19	1.328	1.729	2.093	2.539	2.861	3.174	3.579	3.883
20	1.325	1.725	2.086	2.528	2.845	3.153	3.552	3.850
21	1.323	1.721	2.080	2.518	2.831	3.135	3.527	3.819
22	1.321	1.717	2.074	2.508	2.819	3.119	3.505	3.792
23	1.319	1.714	2.069	2.500	2.807	3.104	3.485	3.768
24	1.318	1.711	2.064	2.492	2.797	3.091	3.467	3.745
25	1.316	1.708	2.060	2.485	2.787	3.078	3.450	3.725
26	1.315	1.706	2.056	2.479	2.779	3.067	3.435	3.707
27	1.314	1.703	2.052	2.473	2.771	3.057	3.421	3.690
28	1.313	1.701	2.048	2.467	2.763	3.047	3.408	3.674
29	1.311	1.699	2.045	2.462	2.756	3.038	3.396	3.659
30	1.310	1.697	2.042	2.457	2.750	3.030	3.385	3.646
50	1.299	1.676	2.009	2.403	2.678	2.937	3.261	3.496
100	1.290	1.660	1.984	2.364	2.626	2.871	3.174	3.390
∞	1.282	1.645	1.960	2.326	2.576	2.807	3.090	3.291

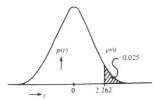

Example: For an area of 0.025 in the upper tail, and $\nu = 9$, the critical value for t is 2.262.

APPENDIX G

LOWER TAIL CRITICAL VALUES FOR MANN-WHITNEY TEST

For each pair of n_x and n_y values, the three values shown give the lower tail values at significance levels:

*** 1%

** $2\frac{1}{2}\%$

* 5%

Upper tail critical value is given by $n_x n_y$ − (lower tail value).

The Mann-Whitney test statistic must be LESS than the lower tail value or GREATER than the upper tail value for significance at the quoted level.

n_y \ n_x	3	4	5	6	7	8	9	10	11	12	13	14	15	16	17	18	19	20
3	0	0	0	0	0	0	1	1	1	1	3	3	4	4	5	5	5	6
	0	0	1	2	2	3	3	4	4	5	5	6	6	7	7	8	8	9
	1	1	2	3	3	4	5	5	6	6	7	8	8	9	10	10	11	12
4			0	1	2	2	3	4	5	6	6	7	8	8	9	10	10	11
		1	2	3	4	5	5	6	7	8	9	10	11	12	12	13	14	15
		2	3	4	5	6	7	8	8	10	11	12	13	15	16	17	18	19
5			2	3	4	5	6	7	8	9	10	11	12	13	14	15	16	17
			3	4	6	7	8	9	10	12	13	14	15	16	18	19	20	21
			5	6	7	9	10	12	13	14	16	17	19	20	21	23	24	26
6				4	5	7	8	9	10	12	13	14	16	17	19	20	21	23
				6	7	9	11	12	14	15	17	18	20	22	23	25	26	28
				8	9	11	13	15	17	18	20	22	24	26	27	29	31	33
7					7	8	10	12	13	15	17	18	20	22	24	25	27	29
					9	11	13	15	17	19	21	23	25	27	29	31	33	35
					12	14	16	18	20	22	25	27	29	31	34	36	38	40
8						10	12	14	16	18	21	23	25	27	29	31	33	35
						14	16	18	20	23	25	27	30	32	35	37	39	42
						16	19	21	24	27	29	32	34	37	40	42	45	48
9							15	17	19	22	24	27	29	32	34	37	39	41
							18	21	24	27	29	32	35	38	40	43	46	49
							22	25	28	31	34	37	40	43	46	49	52	55
10								20	23	25	28	31	34	37	39	42	45	48
								24	27	30	34	37	40	43	46	49	53	56
								28	32	35	38	42	45	49	52	56	59	63
11									26	29	32	35	38	42	45	48	51	54
									31	34	38	41	45	48	52	56	59	63
									35	39	43	47	51	55	58	62	66	70
12										32	36	39	43	47	50	54	57	61
										38	42	46	50	54	58	62	66	70
										43	48	52	56	61	65	69	73	78
13											40	44	48	52	56	60	64	68
											46	51	55	60	64	68	73	77
											52	57	62	66	71	76	81	85
14												48	52	57	61	66	70	74
												56	60	65	70	75	79	84
												62	67	72	78	83	88	93
15													57	62	67	71	76	81
													65	71	76	81	86	91
													73	78	84	89	95	101
16														67	72	77	83	88
														76	82	87	93	99
														84	90	96	102	108
17															78	83	89	94
															88	94	100	106
															97	103	110	116
18																89	95	101
																100	107	113
																110	117	124
19																	102	108
																	114	120
																	124	131
20																		115
																		128
																		139

APPENDIX H

LOWER TAIL CRITICAL VALUES FOR WILCOXON SIGNED RANKS TEST

One Tail Areas	0.005	0.01	0.025	0.05	0.10	0.50
$n = 4$	0	0	0	0	1	5
5	0	0	0	1	3	7.5
6	0	0	1	3	4	10.5
7	0	1	3	4	6	14
8	1	2	4	6	9	18
9	2	4	6	9	11	22.5
10	4	6	9	11	15	27.5
11	6	8	11	14	18	33
12	8	10	14	18	22	39
13	10	13	18	22	27	45.5
14	13	16	22	26	32	52.5
15	16	20	26	31	37	60
16	20	24	30	36	43	68
17	24	28	35	42	49	76.5
18	28	33	41	48	56	85.5
19	33	38	47	54	63	95
20	38	44	53	61	70	105
21	43	50	59	68	78	115.5
22	49	56	66	76	87	126.5
23	55	63	74	84	95	138
24	62	70	82	92	105	150
25	69	77	90	101	114	162.5
26	76	85	99	111	125	175.5
27	84	93	108	120	135	189
28	92	102	117	131	146	203
29	101	111	127	141	158	217.5
30	110	121	138	152	170	232.5

Upper tail critical value is given by:

2 × (value under column 0.50) − (lower tail value).

The Wilcoxon test statistic must be LESS than the lower tail value or GREATER than the upper tail value for significance.

APPENDIX I

CRITICAL VALUES FOR χ^2-DISTRIBUTION

One Tail Areas	0.20	0.10	0.05	0.01	0.001
$\nu=1$	1.64	2.71	3.84	6.63	10.83
2	3.22	4.61	5.99	9.21	13.82
3	4.64	6.25	7.81	11.34	16.27
4	5.99	7.78	9.49	13.28	18.47
5	7.29	9.24	11.07	15.09	20.52
6	8.56	10.64	12.59	16.81	22.46
7	9.80	12.02	14.07	18.48	24.32
8	11.03	13.36	15.51	20.09	26.12
9	12.24	14.68	16.92	21.67	27.88
10	13.44	15.99	18.31	23.21	29.59
11	14.63	17.28	19.68	24.72	31.26
12	15.81	18.55	21.03	26.22	32.91
13	16.98	19.81	22.36	27.69	34.53
14	18.15	21.06	23.68	29.14	36.12
15	19.31	22.31	25.00	30.58	37.70
16	20.47	23.54	26.30	32.00	39.25
17	21.61	24.77	27.59	33.41	40.79
18	22.76	25.99	28.87	34.81	42.31
19	23.90	27.20	30.14	36.19	43.82
20	25.04	28.41	31.41	37.57	45.31
21	26.17	29.62	32.67	38.93	46.80
22	27.30	30.81	33.92	40.29	48.27
23	28.43	32.01	35.17	41.64	49.73
24	29.55	33.20	36.42	42.98	51.18
25	30.68	34.38	37.65	44.31	52.62
26	31.79	35.56	38.89	45.64	54.05
27	32.91	36.74	40.11	46.96	55.48
28	34.03	37.92	41.34	48.28	56.89
29	35.14	39.09	42.56	49.59	58.30
30	36.25	40.26	43.77	50.89	59.70
40	47.27	51.81	55.76	63.69	73.40
50	58.16	63.17	67.50	76.15	86.66
60	68.97	74.40	79.08	88.38	99.61
70	79.71	85.53	90.53	100.43	112.32
80	90.41	96.58	101.88	112.33	124.84
90	101.05	107.57	113.15	124.12	137.21
100	111.67	118.50	124.34	135.81	149.45
110	122.25	129.39	135.48	147.41	161.58
120	132.81	140.23	146.57	158.95	173.62
130	143.34	151.05	157.61	170.42	185.57

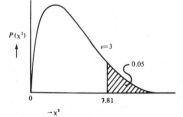

Example: For an area of 0.05 in the tail, and $\nu = 3$, the critical value for χ^2 is 7.81.

Answers to problems

ANSWERS: **Chapter 1**

1.   (i) $x > -1$,   (iv) $x < -1$ or $x > 2$,
       (ii) $x > 2$,   (v) $-1 < x < 2$.
       (iii) $x \geqslant -1/2$,

2.   (i) 0.875,   (iii) 1.222,   (v) 0.033,
       (ii) 2.125,   (iv) 11.667,   (vi) 0.088.

3.   (i) 0.875,   (iii) 1.22,   (v) 0.0333,
       (ii) 2.12,   (iv) 11.7,   (vi) 0.0875.

4.   (i) 1, 1, 3/2, 3;   (iv) 6, 2, 5/2, 2;
       (ii) 3, 1, 1/2, 1;   (v) 0, 0, 1/2, 4.
       (iii) 4, 2, 3/2, 0;

5.   (i) $x < -3$ or $x > 3$,   (iv) $x = -3/2$ or $x = 1/2$,
       (ii) $-3 \leqslant x \leqslant 1$,   (v) $x \leqslant -1/3$ or $x \geqslant 1$.
       (iii) $-1 < x < 3$,

6.   (i) 12,   (iii) 24,   (v) 44,
       (ii) 24,   (iv) 124,   (vi) $n(n + 1)$.

7. $R = k(4 + 3t + t^2)^{\frac{1}{2}}$.

8. $V = k/P$, where k is a constant. Alternatively $PV = k$.

9. $P = kT/V$.

10. $V = kr^3$, $S = Kr^2$.

12. (i) 2,   (iii) $-5/4$,   (v) $-6/5$,
     (ii) $-3/2$,   (iv) $5/4$,   (vi) $-3/8$.

13. (i) $2^{-1} \times 3^{-1}$,   (iii) $2^{-4} \times 3^{5}$,
     (ii) 2×3^{-1},   (iv) $2^{12} \times 3^{2}$.

14. $2^{-1/2} \times 5^{1/6}$.

15. (i) $x = (c - ab)/b$,   (vi) $x = (ab + c)/(a + 1 - b)$,
     (ii) $x = c - ab$,   (vii) $x = (c - ab)/(2a)$,
     (iii) $x = a(c - b)$,   (viii) $x = (a + bc)/(a - c)$,
     (iv) $x = a/(c - b)$,   (ix) $x = -2b/c$,
     (v) $x = (c - a)/b$,   (x) $x = (c - d)/(a - 2b + c)$.

16. (i) $a = b = 1/4$,   (iv) $a = \sqrt{3}, b = 1/3$,
     (ii) $a = 3, b = 1/9$,   (v) $a = 1/3, b = 1/9$.
     (iii) $a = 9, b = 1/3$,

17. (i) 1.6794,   (iii) 3.6794,
     (ii) $\bar{1}.6794 = -0.3206$,   (iv) $\bar{2}.6794 = -1.3206$.

18. (i) 68.5,   (iv) 0.00685,
     (ii) 6850,   (v) 0.685.
     (iii) 0.685,

19. (i) $x + y$,   (iii) $3x + 2y$,
     (ii) $2x - y$,   (iv) $-2x - 2y$.

20. $\alpha = 1, \beta = 2$.

21. 8.

22. (i) 7,   (ii) 24.

23. (i) $1.28a$,   (ii) 14.2 years.

24. $I = I_0 \, 10^{L/10}$, $I = 10^{-6}$.

30. (i) $\pi/3$,   (iv) $\pi/3$,   (vii) $-\pi/4$,
     (ii) $\pi/4$,   (v) $2\pi/3$,   (viii) $\pi/3$.
     (iii) $-\pi/4$,   (vi) $-\pi/6$,

31. (i) $-\pi/4 + n\pi$, where n is an integer;
     (ii) $\pi/3 + 2m\pi$, and $-\pi/3 + 2n\pi$, where m and n are integers;
     (iii) $-\pi/6 + 2m\pi$, and $7\pi/6 + 2n\pi$, where m and n are integers.

35. (i) Increases by 8%,   (ii) $N_n = (1.0715)^n N_0$, $n = 26$.

36. $1/40, 2.5 \times 10^{-2}$.

37. (i) (a) 9.6×10^4,   (b) 2 more hours.

38. (i) Since $\log_5 5 = 1, \log_5 7 > 1$.

39. (i) $2^6 x$,   (ii) 8.75×10^{-2},   (iii) Decreases by 1%.

40. (i) $\log_2 16 = 4$, $\log_2 32 = 5$, hence $4 < \log_2 20 < 5$.
     (ii) 4.32.

ANSWERS: Chapter 2

1. (i) $2, 5, 8, 11$,
    (ii) $1, 8, 27, 64$,
    (iii) x^2, x^4, x^6, x^8,

    (iv) $1, \dfrac{1}{4}, \dfrac{1}{9}, \dfrac{1}{16}$,

    (v) $0, \dfrac{1}{3}, \dfrac{1}{2}, \dfrac{3}{5}$.

2. $5, 10, 20, 40$.

3. $2N, 4N, 8N, 16N$.

4. $k, 2^{3/4}k, 3^{3/4}k, 2^{3/2}k$. $r > 3^{4/3} \cong 4.33$.

5. $2N, 3N, 5N, 8N$.

6. (i) 115.5,   (ii) 48,   (iii) 0.

7. $42, n(n + 1), 9, 32$.

8. 165.

9. 7.

10. (i) 80,   (ii) 40/27,   (iii) 20/27.

11. (i) 3/2,   (ii) 3/4.

12. $4(2^{10} - 1)$.

13. (i) $r = 2$, $S_5 = 124$,   (ii) $r = -2$, $S_5 = 44$.

14. 2/3.

15. 320, 630.

16. 46%.

17. 23%,   14 years.

18. 0.3301, 6 days.

19. (i) 1.016 gm,   (iii) 1.076 gm,
     (ii) 1.025 gm,   (iv) 1.1025 gm.

20. 24, 36, 54 months.

21. 40, 32, 19, 1.

22. (i) 28,   (iii) 126,   (v) 1,
     (ii) 126,   (iv) 1225,   (vi) n.

23.   (i) $1 + 6x + 15x^2 + 20x^3 + 15x^4 + 6x^5 + x^6$,
     (ii) $1 + 12x + 60x^2 + 160x^3 + 240x^4 + 192x^5 + 64x^6$,
     (iii) $1 - 6x + 15x^2 - 20x^3 + 15x^4 - 6x^5 + x^6$,
     (iv) $64 - 192x + 240x^2 - 160x^3 + 60x^4 - 12x^5 + x^6$; 55.0.

24.   (i) $x^4 + 4x^3y + 6x^2y^2 + 4xy^3 + y^4$,
     (ii) $x^4 + 4x^2 + 6 + 4x^{-2} + x^{-4}$,
     (iii) $x^4 - 4x + 6x^{-2} - 4x^{-5} + x^{-8}$,
     (iv) $16x^8 - 96x^5 + 216x^2 - 216x^{-1} + 81x^{-4}$,
     (v) $1 + 4x + 6x^2 + 4x^3 + x^4 + 4y(1 + 3x + 3x^2 + x^3)$
         $+ 6y^2(1 + 2x + x^2) + 4y^3(1 + x) + y^4$,
     (vi) $x^4 + 4x^3 + 2x^2 - 8x - 5 + 8x^{-1} + 2x^{-2} - 4x^{-3} + x^{-4}$.

25. (i) 35,   (iv) 9120,   (vi) 1120,
     (ii) −21,   (v) 210,   (vii) −135/2.
     (iii) −1760,

26. 159%.

28.   (i) $1 + \dfrac{1}{3}x - \dfrac{1}{9}x^2 + \dfrac{5}{81}x^3$, $|x| < 1$,

(ii) $1 - \dfrac{1}{3}x - \dfrac{1}{9}x^2 - \dfrac{5}{81}x^3$, $|x| < 1$,

(iii) $1 + x - \dfrac{1}{2}x^2 + \dfrac{1}{2}x^3$, $|x| < \frac{1}{2}$,

(iv) $1 - x + \dfrac{3}{2}x^2 - \dfrac{5}{2}x^3$, $|x| < \frac{1}{2}$,

(v) $2\left(1 - \dfrac{x}{2^3} - \dfrac{x^2}{2^7} - \dfrac{x^3}{2^{10}}\right)$, $|x| < 4$,

(vi) $\dfrac{1}{2}\left(1 + \dfrac{x}{2^3} + \dfrac{3x^2}{2^7} + \dfrac{5x^3}{2^{10}}\right)$, $|x| < 4$,

(vii) $1 - x + x^2 - x^3$, $|x| < 1$,

(viii) $1 + 2x + 3x^2 + 4x^3$, $|x| < 1$;
    1.9494.

29.   (i) 10 kg,   (iv) 12.66 kg,
      (ii) 11.25 kg,   (v) 11.93 kg.
      (iii) 10.92 kg,

30.   (i) 99.0 kg,   (ii) 94.9 kg,   (iii) 92.4 kg.

31. $(1.05)^r n_0 - 500\{(1.05)^r - 1\}$.

33.   (i) $1 + 7x + 21x^2$,   (ii) $(1.01)^t N_0$,   (iii) Decreases by 25%.

ANSWERS: Chapter 3

1.   (i) $y = 2x$,   (iv) $y = x$,   (vii) $y + 3x = 5$,
      (ii) $y + 3x = 1$,   (v) $y = -2x$,   (viii) $y = 2$,
      (iii) $y + 3x = -1$,   (vi) $y = 2x - 2$,   (ix) $x = 2$.

2. $y + 2x + 1 = 0$.

4.   (ii) $-160/9$,   (iii) $C = F = -40$.

5.   26 cm.

6.   A straight line which cuts the f-axis when $f = 5$ and the n-axis when $n = 100$.
      25%.

7.   (i) $(x - 1)(x - 2)$; $x = 1$, $x = 2$;
      (ii) $(x + 1)(x + 2)$; $x = -1$, $x = -2$;
      (iii) $(2x - 1)(x - 2)$; $x = \tfrac{1}{2}$, $x = 2$;
      (iv) $(3x - 1)(2x + 5)$; $x = \tfrac{1}{3}$, $x = -\tfrac{5}{2}$;
      (v) $(3x + 4)(2x - 3)$; $x = -\tfrac{4}{3}$, $x = \tfrac{3}{2}$;

8.   (i) $x = -1 \pm \sqrt{2}$,   (iii) $x = \tfrac{1}{4}(-3 \pm \sqrt{17})$,
      (ii) $x = -1 \pm \sqrt{3}$,   (iv) $x = \tfrac{1}{3}(1 \pm \sqrt{7})$.

9.   (i) $(x + 1 + \sqrt{2})(x + 1 - \sqrt{2})$,
      (ii) $(x + 1 + \sqrt{3})(x + 1 - \sqrt{3})$,

      (iii) $2\left(x + \dfrac{3}{4} + \dfrac{\sqrt{17}}{4}\right)\left(x + \dfrac{3}{4} - \dfrac{\sqrt{17}}{4}\right)$.

10.   (i) $t = 20/7$,

      (ii) $t = \dfrac{5}{7}(2 \pm \sqrt{2})$,

      (iii) $t = 10/7$, the height is a maximum.

11.   (i) $x = \pm 3i$,   (iv) $x = \tfrac{1}{2}(1 \pm i)$,
      (ii) $x = 1 \pm i$,   (v) $x = \tfrac{1}{2}(-1 \pm i\sqrt{5})$.
      (iii) $x = -1 \pm 2i$,

12.   (i) $(x - 1 - i)(x - 1 + i)$,
      (ii) $(x + 1 + 2i)(x + 1 - 2i)$,
      (iii) $2(x + \tfrac{1}{2} + \tfrac{1}{2}i\sqrt{5})(x + \tfrac{1}{2} - \tfrac{1}{2}i\sqrt{5})$.

13.   (i) $\sqrt{2}\{\cos(\pi/4) + i\sin(\pi/4)\}$,
      (ii) $\sqrt{2}\{\cos(3\pi/4) + i\sin(3\pi/4)\}$,
      (iii) $2\{\cos(-2\pi/3) + i\sin(-2\pi/3)\}$,
      (iv) $\cos(\pi/2) + i\sin(\pi/2)$.

14.   (i) $3 - i$,   (iv) $-1 + 3i$,

      (ii) $-1/2 - 3i/2$,   (v) $\dfrac{1}{10}(9 - 3i)$.

      (iii) $\dfrac{3}{2}(3 - i)$,

15.   (i) $(-1, -4)$, minimum;   (iii) $(1, 1)$,   minimum;
      (ii) $(-1, 4)$,   maximum;   (iv) $(-1, -3)$, maximum.

16. $(-4, 0)$ and $(1, 5)$.

17   (i) $-3, -2, 1$,   (iv) $-\frac{1}{2}, \frac{1}{2}$ (twice),
      (ii) $-1, -\frac{1}{2}, 3$,   (v) -1 (three times).
      (iii) -1 (twice), 2,

18.   (i) $1, -1 \pm \sqrt{2}$,   (iv) $1, \frac{1}{2}(-1 \pm i\sqrt{3})$,
      (ii) $-1, \frac{1}{2}(3 \pm \sqrt{5})$,   (v) $-1, \frac{1}{2}(1 \pm i\sqrt{3})$.
      (iii) $1, \pm 2i$,

19. $-2, -0.414, 2.414$.

20. $(2, -4)$.

21.   (i) $-1, 1, 2, 3$,   (iv) $-1, 1$ (three times),
      (ii) $-1, 1, \pm i$,   (v) $-2, -1, 1, 2, 3$.
      (iii) -1 (twice), 1 (twice),

22.   (i) $\pm \sqrt{2}, \pm \sqrt{3}$,   (iv) $-2, 2$,
      (ii) $1, 2$,   (v) No real roots.
      (iii) $-2, -1, 1, 2$,

23.   (i) $x^5 - 2x^4 + 4x^3 - 8x^2 + 7x - 2$,
      (ii) $-4x^5 + 8x^4 - 13x^3 + 9x^2 - 5x - 3$,

      (iii) $3x^2 + 8x + 7 - \dfrac{10x + 22}{x^2 - 2x + 3}$,

      (iv) $x - 1 + \dfrac{4x - 2}{2x^2 - x - 3}$.

24.   (i) $0.8519, 0.5232$,   (iii) $0.8831, 0.4686$.
      (ii) $0.8793, 0.4756$,

25. 0.356.

26. 101.

27.   (ii) u 0.60 1.20 1.40 (iv) 1.6.

            v 0.00 0.30 0.40 ,

ANSWERS: Chapter 4

1.   (i) $y \leqslant 0$,   (vi) all x, $-1 \leqslant y \leqslant 1$,
      (ii) $-1 < y \leqslant 2$,   (vii) $0 \leqslant y$,
      (iii) all x, $-1 \leqslant y$,   (viii) $x \leqslant 2, 0 \leqslant y$,
      (iv) all x, $0 \leqslant y$,   (ix) all x, $0 \leqslant y < 1$,
      (v) $1 < y$,   (x) all x except $x = -1$,
                                                                all y except $y = 1$.

2. (i) $0 < v < \frac{1}{4}ka^2$, (complete the square),
    (ii) $0 > v$.

3. (i) 2,   (iv) $t^4 + 2t^2 + 3$,
    (ii) 11,   (v) $\cos^2 x - 2\cos x + 3$.
    (iii) $4t^2 - 4t + 3$,

4. $N(0) = 1000$, $N(2) = 1200$, $N(4) = 1800$.

5. $m(0) = \alpha/\beta$, $m(\beta) = \alpha/(2\beta)$, $m(\infty) = 0$.

6. (i) $y = x^{\frac{1}{2}}$, $0 \leqslant x$, $0 \leqslant y$,
    (ii) $y = -x^{\frac{1}{2}}$, $0 \leqslant x$, $y \leqslant 0$.

7. (i) $y = (4 + x^2)^{\frac{1}{2}}$,   all x, $2 \leqslant y$,
    (ii) $y = -(4 + x^2)^{\frac{1}{2}}$, all x, $y \leqslant -2$.

8. (i) $y = (4 - x^2)^{\frac{1}{2}}$, $-2 \leqslant x \leqslant 2, 0 \leqslant y \leqslant 2$,
    (ii) $y = -(4 - x^2)^{\frac{1}{2}}$, $-2 \leqslant x \leqslant 2, -2 \leqslant y \leqslant 0$.

9. $0 \leqslant r \leqslant R$,
    $0 \leqslant v \leqslant \alpha R^2$,

    $r = \left(R^2 - \dfrac{v}{\alpha} \right)^{\frac{1}{2}}$, (note $r > 0$ only),

    $f(R) = 0$,
    $f(\tfrac{1}{2}R) = 3\alpha R^2/4$,
    $f(0) = \alpha R^2$.

10. (i) $f^{-1}(x) = \frac{1}{2}(1 - x)$,
     (ii) $f^{-1}(x) = 1/x$, $x \neq 0$,
     (iii) $f^{-1}(x) = (x - 1)/(x + 1)$, $x \neq -1$,
     (iv) $f^{-1}(x) = \frac{1}{2}(1 - x)^{1/3}$,
     (v) $y_1 = f_1^{-1}(x) = (1 - x)^{\frac{1}{2}}$, $1 \geqslant x$, $0 \leqslant y_1$,
         $y_2 = f_2^{-1}(x) = -(1 - x)^{\frac{1}{2}}$, $1 \geqslant x$, $y_2 \leqslant 0$,
     (vi) $f^{-1}(x) = x^2 - 1$, $0 \leqslant x$.

11. $f^{-1}(x) = \log_{10}(x/\alpha)$.

12. (i) Continuous for all x,
     (ii) Continuous for all x,
     (iii) Discontinuous at $x = 1$,
     (iv) Discontinuous at $x = -2$, $x = +2$,

(v) Continuous for all x,
(vi) Discontinuous at $x = 0$,
(vii) Discontinuous at $x = n\pi$, n an integer,
(viii) Not defined for $-2 < x < 2$.

13.   (i) Continuous for all x,
    (ii) Discontinuous at $x = 1$,
    (iii) Continuous for all x,
    (iv) Continuous for all x,
    (v) Discontinuous at $x = -1$.

14. $t = 4$, 22 animals added to population,
    $t = 12, 24$ animals leave,
    $t = 14$, all animals leave.

15.   (i) 2/3,   (iv) 3,   (vi) 4,
    (ii) 0,   (v) $-1/4$,   (vii) -9.
    (iii) $-1/4$,

16.   (i) $-1/2$,   (iv) 0,   (vi) $4x - 1$,
    (ii) 1/2,   (v) $2x$,   (vii) 6.
    (iii) 2,

17.   (i) 1,   (iv) ∞,   (vi) $-1/3$,
    (ii) 1,   (v) 3,   (vii) 5/3.
    (iii) 0,

18.   (i) $-1/2$,   (iv) 1/2,   (vii) $1/(2x^{\frac{1}{2}})$,
    (ii) 0,   (v) 0,   (viii) $-1/(2x^{3/2})$.
    (iii) $-1/3$ (vi) -1,

19.   (i) 1,   (iii) 1,   (v) 1/2.
    (ii) 1,   (iv) 0,

20.   (i) 4,   (iv) 1,   (vii) $\cos x$,
    (ii) 3,   (v) 1/2,   (viii) $-\sin x$.
    (iii) 1/2,   (vi) 1,

21. $N = \dfrac{b\,10^t}{a + 10^t} = \dfrac{b}{a10^{-t} + 1}$.

ANSWERS: Chapter 5

1. (i) 2.50,   (iv) 4.33,   (vi) 3.10,
    (ii) 8.33,   (v) 1.25,   (vii) 4.55. (all in cm/week).
    (iii) 6.00,

2. (i) 4.9,   (iii) 24.5,   (v) 44.1,
    (ii) 14.7,   (iv) 34.3,   (vi) 24.5 (all in metres/second).

3. (i) 1000,   (iii) 4000,   (v) 3750. (all in cells/hour).
    (ii) 2000,   (iv) 8000,

4. (i) 54,   (iii) 70,
    (ii) 62,   (iv) 62. (all in mg per second).

5. (i) -10,   (iv) -2,
    (ii) -5,   (v) -5. (all in gm per second).
    (iii) -3,   The mass is decreasing.

6. (i) $2x + 2$,   (vi) $-1/(2x^{3/2})$,
    (ii) $1/[2(x + 1)^{\frac{1}{2}}]$,   (vii) $3x^2$,
    (iii) $-1/(x - 1)^2$,   (viii) $1 - 4x$,
    (iv) $-2(a - x)$,   (ix) $4x(x^2 + 1)$,
    (v) $-2(x + 1)^{-3}$;   (x) $\sec^2 x$.

7. (i) $3x^2 + 3$,
    (ii) $6.4x^{0.6}$,
    (iii) $\frac{1}{2}(x^{-\frac{1}{2}} + x^{-\frac{3}{2}})$,
    (iv) $4x(x^2 + 1)$,
    (v) $3(x^2 + 2x + 1) = 3(x + 1)^2$,
    (vi) $6x^2 + 1$,
    (vii) $-1/x^2 - 4/x^3$,
    (viii) $1 - 1/x^2$,
    (ix) $2x - 1$,
    (x) $3(x^2 + x^{-4})$.

8. (i) $3x^2(x^4 - 3x^2 + 2x - 1) + (x^3 + 1)(4x^3 - 6x + 2)$,
    (ii) $-2/(x - 1)^2$,
    (iii) $x(x - 2)/(x - 1)^2$,
    (iv) $(x + 1)^{-2}$,
    (v) $-(1 + x)^{-2}$,
    (vi) $7/(2 - 3x)^2$,
    (vii) $(2 - x^2)/(x^2 + 2)^2$,
    (viii) $(3 - x)/(x + 1)^3$,
    (ix) $-2(x + 2)^{-3}$,
    (x) $4x/(x^2 + 1)^2$.

9. (i) $\cos^2 x - \sin^2 x = \cos 2x$,
    (ii) $1 + 2\cos x$,
    (iii) $-2\cos x \sin x = -\sin 2x$,
    (iv) $\sin x + x \cos x$,
    (v) $2x \cos x - x^2 \sin x$,
    (vi) $3x^2 - \sin x - x \cos x$,
    (vii) $-\cosec^2 x$,
    (viii) $-\cot x \cosec x$,
    (ix) $(\cos x + x \sin x)/\cos^2 x$,
    (x) $\{(1 + x)\cos x - \sin x\}/(1 + x)^2$.

10. $m'(t) = 50 + 8t$,
    $m'(0) = 50$,
    $m'(1) = 58$,
    $m'(2) = 66$,
    $m'(3) = 74$.

11. $m'(t) = -60(t + 2)^{-2}$,
    $m'(0) = -15$,
    $m'(1) = -20/3$,
    $m'(2) = -15/4$,
    $m'(3) = -12/5$,
    $m'(4) = -5/3$.

12. $dR/dx = C - 2x$.

13. $40, 52, 64$.

14. $R = 1/b$, $dR/dI = 0$.

18. $v = 19.6$, $f = 9.8$.

21. $x + y = 2$.

22. $y = -2x$, $y = x - 2$, $\left(\dfrac{2}{3}, -\dfrac{4}{3}\right)$.

23. $(-1, 9)$, $(2, -18)$.

24. $3y - x = -4$.

25.   (i) $(2x + 5)^{-\frac{1}{2}}$,
       (ii) $18x(3x^2 + 1)^2$,
       (iii) $-3/[4(x - 1)^{7/4}]$,
       (iv) $-3\cos^2 x \sin x$,
       (v) $3\cos(3x + \pi/4)$,
       (vi) $3\cos x(1 + \sin x)^2$,
       (vii) $-1/[(1 - x)^{1/2}(1 + x)^{3/2}]$,
       (viii) $x/(1 - x^2)^{3/2}$,
       (ix) $2\tan x \sec^2 x$,
       (x) $(\sin x - 2)/(1 + \sin x)^2$.

26.   (i) 4/3 degrees per hour,   (ii) $0, 2\pi/3$ degrees per hour.

27.   (i) $0.67, 0.80, 0.87, 0.91$,   (ii) 0.83. (all in metres per year).

29. $\dfrac{9\lambda^2(4 - \lambda^3)}{(\lambda^3 + 2)^3}$.

30.   (i) $-x^2/y^2$,
       (ii) $-y/x$,
       (iii) x/y,
       (iv) $-y^{\frac{1}{2}}/x^{\frac{1}{2}}$,
       (v) $x/y - 1/(x^3 y)$,
       (vi) $2\sin x/\cos y$,
       (vii) $-y^2/x^2$,
       (viii) $\dfrac{\cos(x - y) + \sin(x + y)}{\cos(x - y) - \sin(x + y)}$.

31. $5y = x + 24$.

32.   (i) $2/(1 - 4x^2)^{\frac{1}{2}}$,   (iv) $2/(1 + 4x^2)$,
       (ii) $2x/(1 - x^4)^{\frac{1}{2}}$,   (v) $\sin^{-1}x + x/(1 - x^2)^{\frac{1}{2}}$,
       (iii) -1, (why?),   (vi) $2x\tan^{-1}x + x^2/(1 + x^2)$.

33.   (i) $-\cot t$,   (v) $(2t + 1)/(2t - 1)$,
       (ii) $3t/2$,   (vi) $-\tan^2 t$,
       (iii) $\operatorname{cosec} t$,   (vii) $\cot \frac{1}{2}t$.
       (iv) $-1/t^2$,

35.   (i) $-2\sin 2x, -4\cos 2x, 8\sin 2x$,
       (ii) $2x + 3, 2, 0$,
       (iii) $6x^2 - 6x + 1, 12x - 6, 12$,

(iv) $\frac{1}{2}(1 + x)^{-1/2}$, $-\frac{1}{4}(1 + x)^{-3/2}$, $\frac{3}{8}(1 + x)^{-5/2}$,

(v) $(1 - x)^{-2}$, $2(1 - x)^{-3}$, $6(1 - x)^{-4}$,

(vi) $\sin x \cos x$, $2 \cos x - x \sin x$, $-3 \sin x - x \cos x$,

(vii) $2 \sin x \cos x$, $2(\cos^2 x - \sin^2 x)$, $-8 \sin x \cos x$,

(viii) $x(1 + x^2)^{-1/2}$, $(1 + x^2)^{-3/2}$, $-3x(1 + x^2)^{-5/2}$.

ANSWERS: Chapter 6

1. (i) increasing $x < -3$ or $x > 1$; decreasing $-3 < x < 1$,
   maximum $x = -3$, minimum $x = 1$;
   (ii) increasing all x, no stationary points;
   (iii) increasing all x, point of inflexion at $x = -1$;
   (iv) increasing $x < -1$ or $1 < x$; decreasing $-1 < x < 1$,
   maximum $x = -1$, minimum $x = 1$. Discontinuous at $x = 0$;
   (v) decreasing all x, no stationery points. Discontinuous at $x = 0$.

2. (i) minimum $x = -2$,
   (ii) point of inflexion at $x = -2$,
   (iii) none,
   (iv) minimum $x = 2$,
   (v) minimum $x = \frac{1}{2}$, maximum $x = 1$,
   (vi) minimum $x = -1$, maximum $x = 1$.

3. (a) $x > 3$,　　　　(b) $x < 3$.

5. (i) $0, 3, 16, 27, 0$,　　(iv) 16,
   (ii) 12,　　　　　　(v) $x < 3$, $3 < x$.
   (iii) $12x^2 - 4x^3$,

6. $4kR^3/27$.

7. $x = 2$.

8. 10 metres.

9. $C/2$.

10. 2500 at $t = 3$.

11. $l/r = 2$.

12. $\dfrac{4}{9}\pi r^3 \sqrt{3}$.

13. Sides $\frac{1}{4}l$ and $\frac{1}{2}l.$

16. $1/3 < x < 1,$ $y = 5x - 8.$

20. $-2.8\%,$ pressure decreases.

21. $-5\pi\sqrt{2}/6 \cong -3.70\%.$

22. (i) 0.05,   (ii) $\lambda = 2^{2/3}.$

23. (i) $-16\%,$ (ii) $-15\%.$

24. $-10/3\%.$

25. 6%.

27. (i) $\frac{1}{2}$ cm/hr,   (ii) 1/200 cm/hr.

32. $\dot{r} = C/(8\pi r),$ $\dot{V} = 3rC/8.$

33. (i) 4,   (iv) 1/2,   (vii) 1/6,
     (ii) 5,   (v) $-1,$ (viii) 1.
     (iii) 4,   (vi) 1,

ANSWERS: Chapter 7

1.   (i) $x - x^3/3! + x^5/5!,$
     (ii) $1 - x + x^2,$
     (iii) $x + x^3/3 + 2x^5/15,$
     (iv) $1 + \frac{1}{2}x - \frac{1}{8}x^2,$
     (v) $1 - \frac{1}{2}x - \frac{1}{8}x^2,$
     (vi) $1 - \frac{1}{2}x + \frac{3}{8}x^2,$
     (vii) $x - x^3/3 + x^5/5,$
     (viii) $1 + x^2 + x^4.$

2. $15\{2 - t + \frac{1}{2}t^2\},$ 27.3.

3. $1000(1 + t - t^2/15).$

4. $y_1 \cong a\left\{3 - \dfrac{1}{2!}\left(\dfrac{\pi x}{3}\right)^2 + \dfrac{1}{4!}\left(\dfrac{\pi x}{3}\right)^4\right\}$,

$y_2 \cong a\left\{2 - \cos\dfrac{\pi}{5} + \left(\dfrac{\pi x}{5}\right)\sin\dfrac{\pi}{5} + \dfrac{1}{2}\left(\dfrac{\pi x}{5}\right)^2\cos\dfrac{\pi}{5}\right\}$.

5. $3 + \dfrac{2}{3}t + \dfrac{5}{54}t^2$.

ANSWERS: Chapter 8

1. (i) $\pi/3$, $2\pi/3$, $4\pi/3$, $5\pi/3$,
   (ii) $\pi/12$, $5\pi/12$, $13\pi/12$, $17\pi/12$,
   (iii) 0.9046, 5.3786,
   (iv) 0, $\pi/3$, π, $5\pi/3$, 2π,
   (v) $\pi/6$, $5\pi/6$, $7\pi/6$, $11\pi/6$,
   (vi) 0, $\pi/3$, π, $5\pi/3$, 2π,
   (vii) $\pi/4$, $3\pi/4$, π, $5\pi/4$, $7\pi/4$,
   (viii) 0, $\pi/6$, $5\pi/6$, π, $3\pi/2$, 2π.

2. (i) $\theta = 14$, (ii) $\theta = 2$, (iii) $\theta = 10, 18$.

3. (i) 8/3, (ii) 1/6.

4. (i) 0.50, (iv) 0.86, (vii) 1.84,
   (ii) 1.50, (v) 1.08, (viii) 2.29.
   (iii) 0.68, (vi) 1.17,

5. $0.41R$, $0.87R$.

6. 1.29, 2.07.

ANSWERS: Chapter 9

1. (i) $2ax + by$, $bx + 2cy$,
   (ii) $-\sin(x+y) + \cos(x-y)$, $-\sin(x+y) - \cos(x-y)$,
   (iii) $\cos(x+y) + \sin(x-y)$, $\cos(x+y) - \sin(x-y)$,
   (iv) $\cos y - y\cos x$, $-x\sin y - \sin x$,
   (v) $(y^2 - x^2)^{-1/2}$, $-x/[y(y^2 - x^2)^{1/2}]$,
   (vi) $x/(x^2+y^2)^{1/2} - x/(x^2+y^2)^{3/2}$, $y/(x^2+y^2)^{1/2} - y/(x^2+y^2)^{3/2}$,
   (vii) $-y/(x^2+y^2)$, $x/(x^2+y^2)$,
   (viii) $4xy^3 - 9x^2y$, $6x^2y^2 - 3x^3$.

ANSWERS: Chapter 10

1. (a) $\dfrac{5}{7}x^{7/5} + C,$

   (b) $\dfrac{x^2}{2} + \dfrac{1}{x} + C,$

   (c) $\dfrac{x^3}{3} - 2x - \dfrac{1}{x} + C,$

   (d) $x^2 - 2x^{3/2} + 3x^{-1/3} + C,$

   (e) $2x^{1/2} + \dfrac{2}{5}x^{5/2} + C,$

   (f) $\dfrac{x^3}{3} - \dfrac{4}{7}x^{7/2} + \dfrac{x^4}{4} + C.$

2. (a) $\dfrac{3}{4}\sin 4x - \dfrac{2}{3}\cos 3x + C,$

   (b) $-\dfrac{3}{2}\cos 2x - \dfrac{4}{3}\sin 3x + C,$

   (c) $\tan x + \sec x + C,$

   (d) $-\cot x - \operatorname{cosec} x + C.$

3. (a) $\dfrac{x}{2} + \dfrac{1}{12}\sin 6x + C,$

   (b) $-\dfrac{1}{4}\cos 2x + C,$

   (c) $\dfrac{1}{2}\sin x - \dfrac{1}{6}\sin 3x + C,$

   (d) $\dfrac{1}{2}\cos x - \dfrac{1}{6}\cos 3x + C,$

   (e) $-\cot x - x + C,$

   (f) $\dfrac{3}{8}x - \dfrac{1}{4}\sin 2x + \dfrac{1}{32}\sin 4x + C.$

4. (a) $\sin^{-1}\dfrac{x}{3} + C,$

   (b) $\dfrac{1}{3}\sin^{-1}3x + C,$

   (c) $\dfrac{1}{3}\sin^{-1}\dfrac{3x}{2} + C,$

   (d) $\dfrac{1}{6}\tan^{-1}\dfrac{3x}{2} + C,$

   (e) $x - 2\tan^{-1}\dfrac{x}{2} + C,$

   (f) $\dfrac{x}{9} - \dfrac{1}{27}\tan^{-1}3x + C.$

5. (a) 102, (b) 1200.

6. $\dfrac{8}{3}\tan^{-1}\dfrac{t}{12} - \dfrac{t^2}{104} + 1 \qquad 0 \leqslant t \leqslant 12,$

   $\dfrac{4\pi}{3} - \dfrac{23}{13} + \left(\dfrac{3}{26} - \dfrac{\pi}{18}\right)t \qquad 12 < t \leqslant 24.$

7. $\dfrac{65536\pi}{25}.$

8. (a) $\dfrac{2}{9}(3x + 5)^{3/2} + C,$ (d) $\dfrac{1}{2}\sin(2x + \pi/3) + C,$

(b) $\dfrac{1}{6}(2 - 3x)^{-2} + C,$ (e) $-\dfrac{1}{3}\cos\left(3x - \dfrac{\pi}{4}\right) + C,$

(c) $\dfrac{1}{100}(2x - 3)^{50} + C,$ (f) $\dfrac{1}{2}\sec 2x + C.$

9. (a) $-\dfrac{x + 1}{(x + 2)^2} + C,$ (f) $\dfrac{1}{4}\sin^4 x + C,$

(b) $\dfrac{(8x + 1)(x - 2)^{16}}{136} + C,$ (g) $-\dfrac{1}{4}\cos^4 x + C,$

(c) $\dfrac{(3x - 1)(2x + 1)^{3/2}}{15} + C,$ (h) $\dfrac{1}{3}\cos^3 x - \cos x + C,$

(d) $-\dfrac{1}{3}(1 - x^2)^{3/2} + C,$ (i) $\dfrac{2}{3}(1 + \tan x)^{3/2} + C.$

(e) $-(1 - x^2)^{1/2} + C,$

10. (a) $\dfrac{3}{16}(x^4 + 1)^{4/3} + C,$ (d) $-\sin^{-1}\dfrac{1}{x} + C,$

(b) $\dfrac{(5x^2 - 4)(x^2 + 2)^{5/2}}{35} + C,$ (e) $-2\sqrt{2}\cos\dfrac{x}{2} + C.$

(c) $\dfrac{1}{2}\sin^{-1}x - \dfrac{x}{2}\sqrt{(1 - x^2)} + C,$

11. $T = 10 - 6\cos\left(\dfrac{\pi t}{12} - \dfrac{\pi}{4}\right).$

12. (a) $\dfrac{2x}{7}(x - 1)^{7/2} - \dfrac{4}{63}(x - 1)^{9/2} + C,$

(b) $-x\cos x + \sin x + C,$

(c) $\dfrac{x}{2}\sin 2x + \dfrac{1}{4}\cos 2x + C,$

(d) $x^2 \sin x + 2x \cos x - 2\sin x + C,$

(e) $\dfrac{x^2}{2}\tan^{-1}x - \dfrac{x}{2} + \dfrac{1}{2}\tan^{-1}x + C,$

(f) $\dfrac{x}{2}\sec^2 x - \dfrac{1}{2}\tan x + C.$

13. (a) $\dfrac{x}{2(1 + x^2)} + \dfrac{1}{2}\tan^{-1}x + C,$

(b) $-x^4\cos x + 4x^3\sin x + 12x^2\cos x - 24x\sin x - 24\cos x + C,$

(c) $\dfrac{1}{3}\tan x\sec^2 x + \dfrac{2}{3}\tan x + C,$

14. $L = \dfrac{4}{7} - \dfrac{t}{96}(8 - t)^{4/3} - \dfrac{1}{224}(8 - t)^{7/3},\ L = \dfrac{4}{7}.$

ANSWERS: Chapter 11

1. (a) 10,   (c) 2,   (e) 1,

(b) $\dfrac{26}{3}$,   (d) $\dfrac{13}{3}$,   (f) 66.

2. $\dfrac{2}{3}$.

4. (a) 1/12,   (c) $\pi/2$,   (e) $\pi/3$,

(b) $\tan^{-1}\dfrac{1}{\sqrt{2}}$,   (d) $\pi/(4\sqrt{2})$,   (f) $\sqrt{2} - 1$.

5. (a) $\dfrac{8}{15}$,   (b) $24 - 3\pi^2 + \dfrac{\pi^4}{16}$,   (c) $\dfrac{\pi}{4} - \dfrac{2}{3}$.

6. (a) $\dfrac{1}{4}$,   (b) $\dfrac{26}{3}$,   (c) $\sin 1$.

7. (i) 0.48,   (ii) 0.432,   (iii) 0.432.

ANSWERS: Chapter 12

1. (a) 0.2102,   (b) 2.5128,   (c) 4.8154,   (d) -2.0924.

2. (a) $\dfrac{2}{2x - 3}$,   (d) $2\ln x + (\ln x)^2,$

(e) $\operatorname{cosec} x,$

(b) $\dfrac{2}{2x - 3}$,   (f) $\sec x.$

(c) $\ln x,$

3. (a) $\dfrac{x}{x^2 - 1}$,

    (b) $\dfrac{2}{3(1 - x^2)}$,

    (c) $\dfrac{2x}{x^4 - 1}$,

    (d) $\sec x$,

    (e) $\dfrac{1}{x \ln 10}$,

    (f) $\dfrac{x^2 + 2x + 5}{(x + 1)(x^2 + 2x - 3)}$.

4. (a) $\dfrac{(4x^2 + x - 9)(x - 1)(x - 2)^2}{(x + 1)^2}$,

    (b) $\dfrac{(9x^2 + 2x + 1)(x + 1)(x - 1)^2}{2x^{3/2}}$,

    (c) $\dfrac{(x^2 + 32x - 1)(x - 1)^7}{2\sqrt{x}(x + 1)^9}$,

    (d) $\dfrac{(7x + 5)(x - 1)}{3(x + 1)^{2/3}} \cos x - (x - 1)^2(x + 1)^{1/3} \sin x$,

    (e) $(1 + \ln x)x^x$,

    (f) $\left(\dfrac{\cos x}{x} - \sin x .\ln x\right)x^{\cos x}$.

6. (a) 0.009950,   (b) −0.010050.

7. $x - \dfrac{x^2}{2} + \dfrac{x^3}{6}$.

8. $-\dfrac{x^2}{2} - \dfrac{x^4}{12}$.

9. (a) $\dfrac{1}{3} \ln |3x + 2| + C$,

    (b) $-\dfrac{1}{3} \ln |2 - 3x| + C$,

    (c) $\dfrac{1}{2}x + \dfrac{1}{4} \ln |2x + 1| + C$,

    (d) $\dfrac{1}{2} \ln(x^2 + 4) + \dfrac{1}{2} \tan^{-1} \dfrac{x}{2} + C$,

    (e) $-\dfrac{1}{4} \ln |2 - x^4| + C$,

    (f) $- \ln (2 + \cos x) + C$.

10. (a) $\dfrac{1}{2}\ln\dfrac{5}{3}$, (d) $\dfrac{1}{2}\ln 2$, (g) $\ln 2$,

(b) $\dfrac{1}{3}\ln 7$, (e) $\dfrac{1}{3}\ln\dfrac{1}{2}$, (h) $\dfrac{1}{4}\ln 2$,

(c) $\dfrac{1}{4}\ln\dfrac{5}{3}$, (f) $\ln 2$, (i) $\dfrac{1}{2}\ln 2$.

12. (a) $\dfrac{x^2}{2}\ln x - \dfrac{x^2}{4} + C$,

(b) $\dfrac{x^3}{3}\ln x - \dfrac{x^3}{9} + C$,

(c) $-\dfrac{1}{x}\ln x - \dfrac{1}{x} + C$,

(d) $\dfrac{1}{2}(\ln x)^2 + C$,

(e) $\dfrac{x^2}{2}\ln(1+x) - \dfrac{x^2}{4} + \dfrac{x}{2} - \dfrac{1}{2}\ln(1+x) + C$,

(f) $x\ln(1+x^2) - 2x + 2\tan^{-1}x + C$.

14. (i) $(1 + \ln 5)/5$, (ii) $(1 + \ln 2)/2$.

15. (i) $2(1 + 2\ln 5) \cong 8$ cattle,

(ii) $\dfrac{25}{2}(1 + 2\ln 2) \cong 30$ cattle.

16. (a) $e^{\ln 2.x}$, (c) $e^{\ln 3.x}$,
(b) $e^{-\ln 2.x}$, (d) $e^{\ln 10.x}$.

17. (a) $(2x + x^2)e^x$, (d) $-\sin x\, e^{\cos x}$,
(b) $-2x\exp(-x^2)$, (e) $(2\tan 3x + 3\sec^2 3x)e^{2x}$,
(c) $\left(\dfrac{1}{x} + \ln x\right)e^x$, (f) $(e^x + 2 - e^{-x})/(1 + e^{-x})^2$.

18. $\dfrac{3x^2 - 2(y+1)e^{2x}}{3y^2 + e^{2x}}$.

24. (i) $x + x^2 + \dfrac{x^3}{3} + 0$,

(ii) $x + \dfrac{x^2}{2} - \dfrac{5}{24}x^4$,

(iii) $1 + x + \dfrac{x^2}{2} - \dfrac{x^4}{8}$, $x = \dfrac{\pi}{2}$.

25. (a) $-\dfrac{1}{3}e^{2-3x} + C,$ (d) $-\dfrac{1}{2}\exp(-x^2) + C,$

    (b) $\dfrac{1}{3}\exp(x^3) + C,$ (e) $-(x + 1)e^{-x} + C,$

    (c) $e^{\sin x} + C,$ (f) $\dfrac{2^x}{\ln 2} + C.$

26. (a) $\dfrac{e^2 - 3}{4e^2},$ (c) $1,$

    (b) $\dfrac{e - 1}{3e},$ (d) $\dfrac{\pi}{2}.$

28. $e - 1,\ (e - 1)/e,\ (e^2 - 1)/(2e).$

29. (i) $\dfrac{\pi}{2},$ (ii) $4,$ (iii) $\dfrac{1}{2}.$

30. $(1 + \ln 2 - 2^{-t})/\ln 2.$

32. 9082 years.

33. (a) $\dfrac{2}{3}\ln 2$ m,   (b) $\dfrac{2}{3}\ln 10$ m.

34. 14.4%.

35. 19.37.

ANSWERS: Chapter 13

1. (a) $\tan^{-1}(x + 2) + C,$

   (b) $\tfrac{1}{2}\ln(x^2 + 4x + 8) - \tan^{-1}\left(\dfrac{x + 2}{2}\right) + C,$

   (c) $x + \ln(x^2 - 2x + 5) - \dfrac{3}{2}\tan^{-1}\left(\dfrac{x - 1}{2}\right) + C,$

   (d) $-\dfrac{1}{x + 2} + C,$

   (e) $\dfrac{1}{4}\ln|2x - 1| - \dfrac{3}{8x - 4} + C,$

   (f) $x - 2\ln|x + 1| - \dfrac{1}{x + 1} + C.$

2.   (a) $\tan^{-1}4 - \tan^{-1}2$,   (c) $\frac{2}{3}$.   (e) $\frac{1}{2}$,

    (b) $\dfrac{\pi}{4} + \dfrac{1}{2}\ln 2$,   (d) $\ln 2 - 1$,   (f) $\ln 2$.

3.   (a) $\dfrac{5}{x-4} - \dfrac{4}{x-3}$, $5\ln|x-4| - 4\ln|x-3| + C$,

    (b) $\dfrac{3}{x+2} - \dfrac{2}{x-3}$, $3\ln|x+2| - 2\ln|x-3| + C$,

    (c) $\dfrac{3}{x+1} + \dfrac{2}{2-x}$, $3\ln|x+1| - 2\ln|2-x| + C$,

    (d) $\dfrac{3}{2x-1} - \dfrac{1}{x+2}$, $\dfrac{3}{2}\ln|2x-1| - \ln|x+2| + C$,

    (e) $\dfrac{\frac{1}{2}}{1+x} + \dfrac{\frac{1}{2}}{1-x} - 1$, $\dfrac{1}{2}\ln\left|\dfrac{1+x}{1-x}\right| - x + C$,

    (f) $\dfrac{2}{x-4} - \dfrac{1}{x+3}$, $2\ln|x-4| - \ln|x+3| + C$.

4.   (a) $\dfrac{1}{2}\ln 6$,   (d) $\dfrac{1}{5}\ln\dfrac{9}{4}$,

    (b) $\dfrac{1}{6}\ln 5$,   (e) $\dfrac{7}{6}\ln 2 - \ln 3$,

    (c) $2\ln 3 - \dfrac{5}{3}\ln 2$,   (f) $\ln 2 - \dfrac{1}{4}\ln\dfrac{3}{5}$.

6.   (a) $\dfrac{x}{x^2+1} - \dfrac{1}{x+1}$, $\frac{1}{2}\ln(x^2+1) - \ln|x+1| + C$,

    (b) $\dfrac{\frac{1}{2}}{x+1} + \dfrac{\frac{1}{2}}{x-1} - \dfrac{1}{x}$, $\frac{1}{2}\ln|x^2-1| - \ln|x| + C$,

    (c) $\dfrac{1}{x-1} - \dfrac{x}{x^2+x+1}$, $\ln|x-1| - \frac{1}{2}\ln(x^2+x+1)$

                                $+ \dfrac{1}{\sqrt{3}}\tan^{-1}\left(\dfrac{2x+1}{\sqrt{3}}\right) + C$,

    (d) $\dfrac{25/4}{x} - \dfrac{7/2}{x^2} + \dfrac{15/4}{x+2}$, $\dfrac{7}{2x} + \dfrac{25}{4}\ln|x| + \dfrac{15}{4}\ln|x+2| + C$,

    (e) $\dfrac{4}{(x+1)^3} - \dfrac{1}{(x+1)^2}$, $-\dfrac{2}{(x+1)^2} + \dfrac{1}{x+1} + C$,

    (f) $\dfrac{2}{x} - \dfrac{3}{x+1} + \dfrac{1}{x+2}$, $2\ln|x| - 3\ln|x+1| + \ln|x+2| + C$.

7. (a) $\dfrac{1}{2}\ln 2$, (c) $\ln\dfrac{8}{5} - \dfrac{3}{20}$,

(b) $\dfrac{1}{3} - \dfrac{2}{9}\ln 2$, (d) $\dfrac{1}{2}\ln 5 + \ln 3 - 2\ln 2$.

9. $N = \dfrac{AN_0 e^{At/B}}{A + N_0(e^{At/B} - 1)}$.

10. $(a + b)t = \ln\left|\dfrac{b + q}{a - q}\right|$.

ANSWERS: Chapter 14

1. (a) $\dfrac{1}{6}$, (d) $e^2 + e^{-2} - 2$,

(b) $\dfrac{1}{3}$, (e) $\dfrac{32}{3}$.

(c) $2\sqrt{2}$,

2. (a) 0.5, (c) 4,

(b) 8, (d) $(25\pi^2 - 8)/32$.

3. $\dfrac{5}{2}\sin^{-1}\dfrac{1}{\sqrt{5}} = \dfrac{5}{2}\left(\dfrac{\pi}{2} - \sin^{-1}\dfrac{2}{\sqrt{5}}\right)$.

4. $\dfrac{4}{3}$.

5. (a) $\dfrac{3\pi}{2}$, (b) $\dfrac{9\pi}{8}$.

6. 8π mm².

7. (a) $\dfrac{16\pi}{15}$, (b) $\dfrac{\pi}{2}(e^2 - 1)$, (c) $\dfrac{\pi^2}{2}$.

8. (i) $\dfrac{9}{8}$, (ii) $\dfrac{171\pi}{20}$.

10. $\dfrac{64\pi}{5}$.

12. $\dfrac{3}{2} - \ln 2$, (a) $\dfrac{5\pi}{6}$, (b) $\dfrac{8\pi}{3}$.

14. $\dfrac{17}{12}$.

15. $4\sqrt{3}$.

16. $\dfrac{6}{5}$, $\dfrac{12\pi}{7}$, $\dfrac{5692\pi}{1215}$.

17.   (i) $\frac{1}{2}(e - e^{-1})$,   (iii) $\frac{1}{2}(e - e^{-1})$,

       (ii) $\dfrac{\pi}{8}(e^2 + 4 - e^{-2})$,   (iv) $\dfrac{\pi}{4}(e^2 + 4 - e^{-2})$.

18.   (a) $\dfrac{4}{3}$,   (b) $\dfrac{\pi}{16}$,   (c) $\dfrac{1}{2}$,   (d) $\dfrac{1}{4}\ln\dfrac{1}{45}$.

19. $(C = 80)$ $\dfrac{40}{\pi}$, $\dfrac{10}{3}$ mm.

20.   7.5 cm.

ANSWERS: Chapter 15

1.   (a) $\sin y + \cos x = C$,   (e) $\dfrac{1}{2y^2} = \ln|C(x + 1)|$,

     (b) $\cos x \sin y = C$,

     (c) $\dfrac{1}{y} = C - \tan^{-1}x$,   (f) $y^4 = \dfrac{Cx}{4 - x}$,

     (d) $\dfrac{1}{3}(1 + y)^3 = C - \dfrac{x^4}{4}$,   (g) $x^4y^3 = Ce^y$,

                                                (h) $(1 - y)(x^2 - 1)^{\frac{1}{2}} = C$.

2.   (a) $y = \tan(x^2)$,   (c) $\dfrac{x^3}{3} + \dfrac{1}{2}\ln|3 - 2y| = 0$,

     (b) $y^2 = 4 + 2\sin^{-1}x$,   (d) $(1 + x)(1 - y) = 1$.

3. $3W^{1/3} = kt + C$.

4.   slope $= -\dfrac{E}{R \ln 10}$.

5.   (i) $W \cong 27.95H^{2.27}$ or $H \cong 0.230W^{0.441}$,

    (ii) $H \cong 0.959S^{1.26}$ or $S \cong 1.03H^{0.792}$.

6. (i) $Q = \dfrac{4Q_0^3}{(2Q_0 + R_0t)^2}$, $\qquad t = 2(\sqrt{2} - 1)\dfrac{Q_0}{R_0}$,

(ii) $Q = \dfrac{Q_0^2}{Q_0 + R_0t}$, $\qquad t = \dfrac{Q_0}{R_0}$,

(iii) $Q = \dfrac{Q_0^{3/2}}{\sqrt{(Q_0 + 2R_0t)}}$, $\qquad t = \dfrac{3Q_0}{2R_0}$.

7. $\dfrac{dW}{dt} = -kW$, k a positive constant, $40\sqrt{3}$ kg, $\dfrac{30 \ln 0.5}{\ln 0.75} \cong 72$ days.

8. $\dfrac{10 \ln 4}{\ln 1.25} \cong 62$ mins.

9. $\dfrac{10 \ln 3}{\ln 1.5} \cong 27$ mins.

10. (i) $30.4°C$, $\qquad\qquad$ (ii) $26.9°C$.

11. $\dfrac{\ln 5}{\ln 2} \cong 2.32$ hours.

12. (i) $N_0 \exp(-ke^{cT_0})$,

(ii) $N_0 \exp\left[\dfrac{ke^{cT_0}}{\alpha}(1 - e^\alpha)\right]$,

(iii) $N_0 \exp\left[-\dfrac{k\pi e^{cT_0}}{2}\right]$.

13. $N = \dfrac{120}{1 + 3\left(\dfrac{1}{6}\right)^t}$.

14. N tends to A.

15. $N = N_0 \exp\left[\dfrac{6c}{\pi}\sin\left(\dfrac{\pi t}{6} + \pi\right)\right]$.

16. $\dfrac{N}{(K - N)^{b+1}} = \dfrac{N_0}{(K - N_0)^{b+1}}e^{ct}$.

18. $\dfrac{30 \ln 7}{\ln 1.4} \cong 173.5$ minutes.

19. (a) $p(t) = \dfrac{\nu - [\nu-(\mu + \nu)p(0)] \exp[-(\mu + \nu)t]}{\mu + \nu}$,

   (b) $q(t) = \dfrac{\mu - [\mu-(\mu + \nu)q(0)] \exp[-(\mu + \nu)t]}{\mu + \nu}$,

   (c) $\dfrac{1}{\mu + \nu} \ln \left| \dfrac{5(\mu - \nu)}{3\mu - 7\nu} \right|$.

20. $q = \dfrac{100 \left[\left(\dfrac{10}{3}\right)^t - 1 \right]}{20 + \left(\dfrac{10}{3}\right)^t}$ moles/litre.

22. (a) $\frac{1}{2}e^{-x} (\cos x + \sin x) + Ce^{-2x}$,

   (b) $(x + 1)y = \dfrac{x^4}{4} + \dfrac{x^3}{3} + C$,

   (c) $y = \frac{1}{2} + Ce^{-x^2}$,

   (d) $(1 + x)y = x^4 + Cx$,

   (e) $y = C\sin^2 x - \frac{1}{2}$.

23. $N = Ae^{kt} - \dfrac{25kc \sin\left(\dfrac{\pi t}{5} + a\right) + 5\pi c \cos\left(\dfrac{\pi t}{5} + a\right)}{25k^2 + \pi^2}$.

24. $\dfrac{dQ}{dt} = -kQ$, 62 years, $\dfrac{dQ}{dt} = -kQ + Ke^{-ct}$.

27. $\dfrac{ds}{dt} = -as$, $\dfrac{dd}{dt} = bu$, $\dfrac{di}{dt} = cu$.

   Let s_0 be the initial number of susceptible people in the population.

   $s = s_0 e^{-at}$,

   $u = \dfrac{as_0}{b + c - a} [e^{-at} - e^{-(b+c)t}]$,

   $d = bs_0 \left[\dfrac{ae^{-(b + c)t}}{(b + c)(b + c - a)} - \dfrac{e^{-at}}{b + c - a} + \dfrac{1}{b + c} \right]$,

   $i = cs_0 \left[\dfrac{ae^{-(b + c)t}}{(b + c)(b + c - a)} - \dfrac{e^{-at}}{b + c - a} + \dfrac{1}{b + c} \right]$.

28. (a) $y = Ae^{-3x} + \left(B + \dfrac{x}{5}\right)e^{2x}$,

    (b) $y = e^{-2x}(A \cos x + B \sin x) + x - 4/5$,

    (c) $y = e^{-x}(A \cos x + B \sin x) + 1$,

    (d) $y = (A + 2x)e^{-2x} + Be^{-4x}$,

    (e) $y = Ae^{-x} + Be^{2x} + \tfrac{1}{2} - x$,

    (f) $y = A \cos x + \left(B + \dfrac{x}{2}\right) \sin x$.

29. (a) $x = (A + Bt)e^t$, $y = (A - B + Bt)e^t$,

    (b) $x = A + Be^{-2t} + \dfrac{t}{4}(t - 1)$, $y = A - Be^{-2t} + \dfrac{1}{4}(t^2 + t - 1)$,

    (c) $x = A \sin(\sqrt{2}t) + B \cos(\sqrt{2}t)$, $y = (A + B\sqrt{2}) \sin(\sqrt{2}t)$
                                           $+ (B - A\sqrt{2}) \cos(\sqrt{2}t)$.

30. $x = e^t - e^{-t} + 1$, $y = e^t + 2e^{-t}$.

31. $x = 20(\sin t + \cos t) + 30 = 30 + 20\sqrt{2} \sin\left(t + \dfrac{\pi}{4}\right)$.

    $y = 40 \sin t + 5e^{-t} + 60$.

32. $a \ln y - by + c \ln x - dx = C$.

ANSWERS: Chapter 16

1. (a) $u_t = 3(-\tfrac{1}{2})^t$ $t = 0, 1, 2, \ldots$

    (b) $u_t = 2.3^{(t-1)t/2}$ $t = 0, 1, 2, \ldots$

    (c) $u_t = \dfrac{4}{(t - 1)!} (\tfrac{1}{2})^t$ $t = 1, 2, 3, \ldots$

    (d) $u_t = [(t - 1)!]^2 (-\tfrac{1}{2})^{t-1}$ $t = 1, 2, 3, \ldots$

2. $N_t = 10(t + 1)$.

3. (a) $u_t = A(-2)^t + \dfrac{10}{3}$,   (c) $u_t = A(\tfrac{1}{2})^t + t - 2$,

    (b) $u_t = A\left(\dfrac{2}{3}\right)^t + \dfrac{3}{5}$,   (d) $u_t = A\left(\dfrac{2}{3}\right)^t + \dfrac{1}{4} 2^t$.

4.   (a) $u_t = 5$,

    (b) $u_t = 2\left(-\dfrac{1}{3}\right)^t$,

    (c) $u_t = 10t + 2$,

    (d) $u_t = \dfrac{1}{64}(-3)^t + \dfrac{t^2}{8} - \dfrac{t}{16} - \dfrac{1}{64}$,

    (e) $u_t = 1$,

    (f) $u_t = 2^t - \cos\dfrac{\pi t}{3} + \dfrac{1}{\sqrt{3}}\sin\dfrac{\pi t}{3}$,

    All for $t = 0, 1, 2, \ldots$

5.   (a) $N_{t+1} = (k + 1)N_t$; $N_{t+1} = e^k N_t$,

    (b) $N_{t+1} = (Kc + 1)N_t - c(N_t)^2$; $N_{t+1} = \dfrac{Ke^{Kc}N_t}{K + (e^{Kc} - 1)N_t}$.

6. $H_{t+1} = H_t - S_{t+1}$,

    $S_{t+1} = \lambda H_t$,

    $I_{t+1} = I_t + S_t$, $t = 0, 1, 2, \ldots$

7.   (a) $u_t = A\left(\tfrac{1}{2}\right)^t + B2^t - 1$,

    (b) $u_t = A + B2^t + \tfrac{1}{2}(3^t + t + t^2)$,

    (c) $u_t = (A + Bt)(-1)^t + t - 1$,

    (d) $u_t = 2^{t/2}\left[A\cos\dfrac{\pi t}{4} + B\sin\dfrac{\pi t}{4}\right] + 2^{t-1}$,

    (e) $u_t = A + B5^t + t - 2t^2$,

    (f) $u_t = A2^t + B5^t + \left(\dfrac{3}{4} - \dfrac{t}{2}\right)3^t$.

8.   (a) $u_t = 3^t - 2^t + 1$,

    (b) $u_t = 4.2^t - 4 - t$,

    (c) $u_t = \left(2 - \dfrac{3}{2}t\right)2^t + 3^t$,

    (d) $u_t = 2^{t/2}\left[\cos\dfrac{\pi t}{2} + \sin\dfrac{\pi t}{2}\right]$,

    (e) $u_t = (1 - t)3^t + 1$,

    (f) $u_t = \dfrac{4}{3}\cos\dfrac{2\pi t}{3} + \dfrac{4}{3\sqrt{3}}\sin\dfrac{2\pi t}{3} + \dfrac{t}{3} - \dfrac{1}{3}$.

9. $u_t = Ab^t + \left[B + \dfrac{t}{a(a-b)} \right] a^t$ $(a \neq b)$,

    $u_t = \left[A + Bt + \dfrac{t^2}{2a^2} \right] a^t$ $(a = b)$.

10.   (i) $N_t = 8 \left(\dfrac{5}{4} \right)^t$ size of population becomes infinite.

    (ii) $N_t = 12 \left(\dfrac{3}{4} \right)^t - 4 \left(-\dfrac{1}{4} \right)^t$ size of population decreases to zero.

    (iii) $N_t = 9 - (-1)^t$, size of population oscillates between eight and ten.

11. $p_n = \sqrt{3} \left[\dfrac{\sqrt{3}+1}{2\sqrt{3}} \right]^n - \sqrt{3} \left[\dfrac{\sqrt{3}-1}{2\sqrt{3}} \right]^n$,

    p_n tends to zero as n becomes very large.

12.   (a) $u_t = A \left(\dfrac{5}{4} \right)^t + B \left(\dfrac{3}{4} \right)^t$, $v_t = \dfrac{A}{2} \left(\dfrac{5}{4} \right)^t - \dfrac{B}{2} \left(\dfrac{3}{4} \right)^t$,

    (b) $u_t = A2^t - 1$, $v_t = t - A2^t$.

13. $u_t = 20.3^t - 10.2^t$, $v_t = 50.2^t - 10.3^t$;

    $v_t = 0$ when $t = \dfrac{\log 5}{\log 1.5} \cong 3.97$.

14. $u_t = 20 \sin \dfrac{\pi t}{2} + 10 \cos \dfrac{\pi t}{2} + 20$, $v_t = 10 \sin \dfrac{\pi t}{2} + 20$.

15.   (a) $u_t = 2^t - t(-1)^t$,

    (b) $u_t = 2^{t+1} - 3^t$,

    (c) $u_t = A + B \cos \dfrac{\pi t}{2} + C \sin \dfrac{\pi t}{2}$,

    (d) $u_t = [A + Bt + Ct^2 + Dt^3] (-1)^t$.

ANSWERS: Chapter 17

1.   29.97, 0.03337.

2. $0(x^5)$.

3.   (a) 0.687,   (c) 0.594,

    (b) 2.208,   (d) 2.062.

4. (a) 0.687,   (c) 0.594,
    (b) 2.208,   (d) 2.062.

5. 0.204.

6. (a) 0.1667; $1/6 \cong 0.1667$,
    (b) 0.6346; $2 - \ln 4 \cong 0.6137$,
    (c) 0.7788; $\ln(1 + \ln 3) \cong 0.7413$.

7. (i) 0.8317,   (ii) 2.2613.

9. (i) $1/3 \cong 0.3333$,   (ii) 0.3350,   (iii) 0.3333.

10. (a) 1.5   ;   1.5   ,   (d) 0.2691 ;   0.2858,
    (b) -173.67; -219.00,   (e) 1.9118 ;   2.0000,
    (c) 2.00047 ;   2.00000 ,   (f) 3.544; $\pi \cong 3.142$.

11. 16.653; 16.636.

12. $$\int_0^\pi \sqrt{(1 + \cos^2 x)}\,dx.$$

14. 1.045.

15. 7.21 m^3.

16. 38.4 cm^2.

17. 95.6.

18. 21.5 N m.

19. 0.219.

20. $M = 3$, 0.886.

21. When $x = 1.5$ the numerical solution gives $y = 2.766$; exact solution gives
    $y = 3.244$.

22. When $x = 2$,
    (i) Euler's method gives $y = 1.634$,
    (ii) the predictor-corrector method gives $y = 1.646$.
        The exact value of y is 1.646.

23. When $x = 1$,
    (i) $y = 2.336$, (ii) $y = 2.363$, (iii) $y = 2.366$.

24. When $x = 1.4$,
    (i) $y = 8.758$, (ii) $y = 8.762$.
    The exact value of y is 8.762.

25. When $x = 1.6$,
    (i) $y = 2.138$, (ii) $y = 2.335$.

ANSWERS: Chapter 18

1. (a) $\begin{bmatrix} 0 & -12 & 3 \\ 6 & -1 & 6 \\ 9 & 6 & -4 \end{bmatrix}$ (b) $\begin{bmatrix} 8 & 4 & -1 \\ -2 & -3 & -2 \\ -3 & -2 & -4 \end{bmatrix}$

   (c) $\begin{bmatrix} 12 & 0 & 0 \\ 0 & -5 & 0 \\ 0 & 0 & -8 \end{bmatrix}$ (d) $\begin{bmatrix} 16 & 12 & -3 \\ -6 & 0 & -6 \\ -9 & -6 & 0 \end{bmatrix}$

2. $x = 4$, $y = 6$, $z = -11$, or $x = -4$, $y = -6$, $z = -11$.

3. (b) 4, (c) 4, (f) -2.
   (a), (d), (e) do not exist.

4. (a) $\begin{bmatrix} 4 & 3 \\ 4 & 6 \end{bmatrix}$ (b) $\begin{bmatrix} 4 & 6 \\ 2 & 6 \end{bmatrix}$ (c) $\begin{bmatrix} -5 & -6 \\ 6 & -10 \end{bmatrix}$

   (d) $\begin{bmatrix} -5 & -9 \\ -4 & -10 \end{bmatrix}$ (e) $\begin{bmatrix} -8 & 3 \\ 2 & -6 \end{bmatrix}$ (f) $\begin{bmatrix} -8 & 3 \\ 2 & -6 \end{bmatrix}$

   (g) $\begin{bmatrix} -2 & -15 \\ -8 & -18 \end{bmatrix}$ (h) $\begin{bmatrix} -8 & -18 \\ -10 & -12 \end{bmatrix}$ (i) $\begin{bmatrix} -2 & -12 \\ -10 & -18 \end{bmatrix}$

   (j) $\begin{bmatrix} -2 & -15 \\ -32 & 18 \end{bmatrix}$ (k) $\begin{bmatrix} -8 & -15 \\ -12 & -12 \end{bmatrix}$ (l) $\begin{bmatrix} -2 & -12 \\ -10 & -18 \end{bmatrix}$

5. (a) $\begin{bmatrix} 16 & 17 \end{bmatrix}$, (c) $\begin{bmatrix} 5 & -8 & 13 \end{bmatrix}$, (g) $\begin{bmatrix} 0 & -2 & 7 \\ 5 & -8 & 13 \end{bmatrix}$,

(j) $\begin{bmatrix} 20 & 15 \\ 13 & 11 \\ 10 & 10 \end{bmatrix}$,   (k) $\begin{bmatrix} 0 & 6 \\ 6 & 1 \end{bmatrix}$,

(b), (d), (e), (f), (h), (i), (l) do not exist.

6. (a) $\begin{bmatrix} 6 & 0 & 1 \\ 4 & -1 & 3 \\ 19 & 3 & -5 \end{bmatrix}$ (b) $\begin{bmatrix} 6 & -5 & 0 \\ 15 & -14 & -1 \\ -1 & 14 & 8 \end{bmatrix}$

(c) $\begin{bmatrix} 1 & 16 & 11 \\ -7 & 24 & 11 \\ 37 & -4 & 17 \end{bmatrix}$ (d) $\begin{bmatrix} 1 & 16 & 11 \\ -7 & 24 & 11 \\ 37 & -4 & 17 \end{bmatrix}$

(e) $\begin{bmatrix} -5 & 8 & 9 \\ -20 & -20 & 22 \\ -1 & 48 & 13 \end{bmatrix}$ (f) $\begin{bmatrix} 29 & -5 & 6 \\ 13 & -13 & 22 \\ 35 & 21 & -28 \end{bmatrix}$

7. Summer:
    3.36×10^6 eggs, 7.5×10^4 larvae, 4×10^4 pupae, 8.84×10^3 adults.
    Winter:
    6.082×10^5 eggs, 1.494×10^5 larvae, 8×10^3 pupae, 2.21×10^2 adults.

8. $a = b = c = d = 0$.

9. (i) 3,   (ii) 2.

12. 1528. The number of sick individuals each day remains constant.

13. $a = 3, b = 2, c = 1$.

14. (a) $\begin{bmatrix} -6 & 8 & 10 \\ 5 & 4 & -7 \end{bmatrix}$ (d) $\begin{bmatrix} 7 & -6 \\ 8 & 4 \end{bmatrix}$

(e) $\begin{bmatrix} 4 & 4 \\ 5 & -9 \\ -4 & 4 \end{bmatrix}$ (f) $\begin{bmatrix} 3 & 4 & -4 \\ -5 & 12 & 9 \\ 2 & -8 & -4 \end{bmatrix}$

(j) $[-2 \quad -3]$ (k) $\begin{bmatrix} 0 & 8 \\ 8 & -8 \\ 1 & -13 \end{bmatrix}$

(l) $\begin{bmatrix} -2 & 8 & -4 \\ 7 & 0 & -2 \end{bmatrix}$ (m) $\begin{bmatrix} 9 \\ -8 \end{bmatrix}$

(n) $[9 \quad -8]$ (o) 13

(p) $\begin{bmatrix} -3 & 5 & 6 \\ 2 & 1 & -6 \end{bmatrix}$

(b), (c), (g), (h), (i) do not exist.

16. (b)
$$\mathbf{AA^T} = \begin{bmatrix} 10 & 11 & -4 \\ 11 & 21 & -4 \\ -4 & -4 & 17 \end{bmatrix} \qquad \mathbf{A^TA} = \begin{bmatrix} 6 & 4 & -5 \\ 4 & 13 & 2 \\ -5 & 2 & 29 \end{bmatrix}$$

$$\mathbf{BB^T} = \begin{bmatrix} 5 & -5 & 0 \\ -5 & 25 & -10 \\ 0 & -10 & 5 \end{bmatrix} \qquad \mathbf{B^TB} = \begin{bmatrix} 14 & -12 \\ -12 & 21 \end{bmatrix}$$

(c) $\mathbf{A} = \begin{bmatrix} 1 & -\frac{1}{2} & \frac{5}{2} \\ -\frac{1}{2} & 2 & \frac{7}{2} \\ \frac{5}{2} & \frac{7}{2} & -2 \end{bmatrix} + \begin{bmatrix} 0 & \frac{1}{2} & \frac{1}{2} \\ -\frac{1}{2} & 0 & \frac{1}{2} \\ -\frac{1}{2} & -\frac{1}{2} & 0 \end{bmatrix}$

17. (a) $\begin{bmatrix} 1 & 2 & 3 \\ 1 & 1 & 2 \\ 1 & 2 & 1 \end{bmatrix} \begin{bmatrix} x \\ y \\ z \end{bmatrix} = \begin{bmatrix} 1 \\ -1 \\ 1 \end{bmatrix}$ $x = -3, y = 2, z = 0$,

(b) $\begin{bmatrix} 1 & 3 & -1 \\ 2 & 0 & 1 \\ 1 & 2 & 2 \end{bmatrix} \begin{bmatrix} x \\ y \\ z \end{bmatrix} = \begin{bmatrix} 12 \\ 3 \\ 6 \end{bmatrix}$ $x = 2, y = 3, z = -1$,

(c) $\begin{bmatrix} 2 & 3 & -2 \\ 7 & 3 & -3 \\ 1 & -1 & -3 \end{bmatrix} \begin{bmatrix} x \\ y \\ z \end{bmatrix} = \begin{bmatrix} 3 \\ 7 \\ 3 \end{bmatrix}$ $x = \dfrac{16}{25}, y = \dfrac{1}{25}, z = \dfrac{-4}{5}$,

(d) $\begin{bmatrix} 1 & 1 & 1 & -1 \\ 1 & 2 & -1 & -2 \\ 2 & -1 & 1 & 1 \\ 2 & 3 & -3 & 4 \end{bmatrix} \begin{bmatrix} w \\ x \\ y \\ z \end{bmatrix} = \begin{bmatrix} 4 \\ 7 \\ -1 \\ 4 \end{bmatrix}$ $w = 1, x = 2, y = 0, z = -1.$

18. 10, 15, 25.

19. $x = 2, y = 1, z = 3.$

20. (a) $a = 3, b = -1, c = -5, d = 2,$
    (b) $a = 3, b = 5/2, c = 1, d = 1,$
    (c) inverse does not exist,
    (d) $a = 2/3, b = -1/3, c = -5/3, d = 4/3.$

21. (a) $\dfrac{1}{5}\begin{bmatrix} 1 & -3 \\ 1 & 2 \end{bmatrix}$

    (b) $\begin{bmatrix} 3 & 0 & -1 \\ 5 & 3 & -1 \\ -4 & -2 & 1 \end{bmatrix}$

    (c) $\begin{bmatrix} 1 & 1 & 1 \\ 1 & -1 & 2 \\ 2 & 1 & 3 \end{bmatrix}$

    (d) $\begin{bmatrix} 4 & 0 & -5 \\ -18 & 1 & 24 \\ -3 & 0 & 4 \end{bmatrix}$

    (e) $\begin{bmatrix} -6 & -9 & 11 \\ 4 & 6 & -7 \\ 1 & 2 & -2 \end{bmatrix}$

    (f) Inverse matrix does not exist

    (g) $\begin{bmatrix} 3 & 1 & -1 \\ -3 & -3 & 2 \\ 1 & 2 & -1 \end{bmatrix}$

    (h) $\begin{bmatrix} 2 & -1 & 0 & 0 \\ -1 & 2 & -1 & 0 \\ 0 & -1 & 2 & -1 \\ 0 & 0 & -1 & 1 \end{bmatrix}.$

22. (a) (i) $x = 1, y = -1, z = 2,$
    (ii) $x = -3, y = 2, z = 0,$
    (iii) $x = -1, y = 0, z = 1,$
    (b) (i) $x = 4, y = -2, z = 1,$
    (ii) $x = 3, y = 1, z = 2,$
    (iii) $x = 1, y = -1, z = 1,$
    (c) (i) $w = 2, x = 1, y = -\frac{1}{2}, z = \frac{1}{2},$
    (ii) $w = 6, x = 0, y = -5, z = -1,$
    (iii) $w = 8/3, x = -5/3, y = -11/3, z = -8/3.$

23. (a) 2,   (c) 12,   (e) 0,
    (b) −1,   (d) −2,   (f) 10.

24. (a) −200,   (b) 5,   (c) 0,   (d) 3450000.

25. (a) 2, $[\alpha, -\alpha]^T$; 5, $[\alpha, 2\alpha]^T$,
     (b) 2, $[\alpha, -\alpha]^T$; 3, $[5\alpha, -4\alpha]^T$,
     (c) 1, $[4\alpha, 3\alpha, 2\alpha]^T$; 2, $[3\alpha, 3\alpha, 2\alpha]^T$; 3, $[\alpha, \alpha, \alpha]^T$,
     (d) 1, $[\alpha, 13\alpha, 3\alpha]^T$; 2, $[\alpha, 2\alpha, \alpha]^T$; 3, $[\alpha, \alpha, \alpha]^T$,
     (e) 1, $[-2\alpha, \alpha, \alpha]^T$; 2, $[\alpha, \alpha, -\alpha]^T$; 3, $[\alpha, 0, -\alpha]^T$,
     (f) 0, $[\alpha, \alpha, \alpha]^T$; 1, $[\alpha, -\alpha, 0]^T$; 2, $[\alpha, 0, 0]^T$,
     (g) 1 (twice), $[\alpha, \beta, \alpha - 2\beta]^T$; -1, $[\alpha, \alpha, \alpha]^T$,
     (h) -2, $[11\alpha, \alpha, -14\alpha]^T$; 1, $[-\alpha, \alpha, \alpha]^T$; 3, $[\alpha, \alpha, \alpha]^T$.

26. (i) $[324, 432, 216]^T$ (ii) $[324, 404, 244]^T$.

27. The dominant eigenvalue is 2, eigenvector $[20, 5, 2]^T$.
     The population will double each year.

28. (i) $x = 0.83$, $y = 1.08$, $z = 1.04$,
     (ii) $x = 1.000$, $y = 1.000$, $z = 1.000$.

29. 1.190.

30. $\lambda_1 = 1.0127, \lambda_2 = -0.9877, \lambda_3 = 0.0125 + 0.9999i, \lambda_4 = 0.0125 - 0.9999i$.
     Dominant eigenvalue is λ_1 with dominant eigenvector
     $[1038.7, 102.56, 10.127, 1]^T$.
     Ultimate age distribution:
     eggs : larvae : nymphs : adults $= 1038.7 : 102.56 : 10.127 : 1$.

ANSWERS: Chapter 19

1. $y = -24.75x + 377$.

2. $F = 0.12W - 2.13$.

3. (i) 1.5,   (ii) 1.25.

4. $\dfrac{1}{A_{t+1}} = \dfrac{0.4}{A_t} + 0.01$, $A_\infty = 60 = K$.

5. $A = 1.044$, $B = 10.733$, 107 plants.
    Note that the crop failure, when $d = 11$, should be ignored.

6. $k = 0.139$, $\alpha = 0.578$.

7. $A = 1017$, $b = 0.163$; $N = 13803$.

8. $\log(\%A) = 0.0072T - 0.2062$; 3.3.

9. $a = 15$, $b = 2500$, $c = 0.2$.

10. $k = 0.01$, $\alpha = 3.3$.

11. $r = 0.93$, $D = 4.26T + 48.67$, 1st June.

12. $r = 0.9$, $A = 4.2T - 9.6$, 36.6 cm².

13. $\log C = 0.0066t + 1.525$.

14. $a = 2.65$, $b = 0.071$, $r = 0.992$.

15. Normal plasma $t = 51.6 - 11.1 \log C$,
     Diseased plasma $t = 71.6 - 11.0 \log C$,
     Concentration 32.7%.

ANSWERS: Chapter 20

1. (i) 840,   (iv) 35,   (vii) 1,
    (ii) 210,   (v) 1,   (viii) 1.
    (iii) 35,   (vi) 5040,

2. $7 \times 10 \times 4 \times 6 = 1680$.

3. $4^5 = 1024$.

4. $3^3 = 27$.

5. (i) $9 \times 12 \times 11 \times 10 \times 9 = 106920$,
    (ii) $9 \times 8 \times 12 \times 11 \times 10 = 95040$.

6. (i) 15!,   (ii) 6!,   (iii) 9! × 6! .

7. $10 \times 9 \times 8 = 720$.

8. $^{27}P_{10}$.

9. $^{40}C_{10}$.

10. $^7C_2$.

11. $^6C_2$, ^nC_r.

ANSWERS: Chapter 21

1. Discrete; sample space $\{0, 1, 2, \ldots, 18\}$.

2. (i) $(0, \infty)$, continuous,
   (ii) $\{0, 1, 2, \ldots, 20\}$, discrete,
   (iii) $(-\infty, \infty)$, continuous,
   (iv) $\left\{0, \dfrac{1}{N}, \dfrac{2}{N}, \ldots, 1\right\}$ where N is the total number of ticks, discrete.

3. (i) 0.8, (ii) 0.2^5, (iii) $1 - 0.2^5$.

4. (i) $(1/6)^4$, (ii) $4(1/6)(5/6)^3$.

5. $1 - (0.1)^2$.

6. (i) $1 - (8/9)(11/12)(14/15)$, (ii) $(1/9)(1/12)(1/15)$.

7. (i) $3(1/9)(8/9)^2$, (ii) $(1/9)^3$, (iii) $1 - (8/9)^3$.

8. $(1/4)(1/6)(7/8) + (1/4)(5/6)(1/8) + (3/4)(1/6)(1/8)$.

9. (i) $(0.3)(0.7)$, (ii) $(0.3)^3(0.7)$, (iii) $(0.3)^{n-1}(0.7)$.

10. (i) $\dfrac{m}{m + n}$,

    (ii) $\left(\dfrac{m}{m + n}\right)\left(\dfrac{m - 1}{m + n - 1}\right)$, (iii) $\left(\dfrac{n}{m + n - 1}\right)$,

    (iv) $\left(\dfrac{m}{m + n}\right)\left(\dfrac{m - 1}{m + n - 1}\right)\left(\dfrac{m - 2}{m + n - 2}\right)$

    $+ \left(\dfrac{m}{m + n}\right)\left(\dfrac{n}{m + n - 1}\right)\left(\dfrac{m - 1}{m + n - 2}\right)$

    $+ \left(\dfrac{n}{m + n}\right)\left(\dfrac{m}{m + n - 1}\right)\left(\dfrac{m - 1}{m + n - 2}\right)$

    $+ \left(\dfrac{n}{m + n}\right)\left(\dfrac{n - 1}{m + n - 1}\right)\left(\dfrac{m}{m + n - 2}\right)$.

11. (i) (a) $23/64$, (b) $55/64$, (c) $19/32$,
    (ii) $3/8$,
    (iii) (a) $(0.9)^4(0.1)$, (b) 0.3439, (c) 0.59049.

12.   (i) $(1/5)(3/10)$; $(4/5)(3/10) + (7/10)(1/5)$; $1 - (1/5)(3/10)$; $(4/5)(7/10)$,

     (ii) $4(4/5)(1/5)(7/10)(3/10)$;
       $(1/5)^2(3/10)^2 + 4(4/5)(1/5)(7/10)(3/10) + (4/5)^2(7/10)^2$;
       $1 - (1/5)^2(3/10)^2$,

    (iii) $12/19$.

ANSWERS: Chapter 22

1. $1/55$.

2. $4/7$; $26/49$.

3. 0.406; 0.437; 0.088.

4. 1; $1/2$.

5. $e^{-1/3} - e^{-2/3}$.

6.   (i) $(0.8)^6$,
     (ii) $6(0.2)(0.8)^5$,
     (iii) $1 - (0.8)^6 - 6(0.2)(0.8)^5$; 0.0559.

7.   (i) $6(0.2)^5(0.8) + (0.2)^6$,   (ii) 0.9922.

8.   (i) 0.35,   (ii) 0.35^3, $1 - (0.65)^7 - 7(0.65)^6(0.35)$.

9.   (i) $e^{-3.5}$,
     (ii) $1 - e^{-3.5} - 3.5e^{-3.5} - \dfrac{3.5^2 e^{-3.5}}{2} - \dfrac{3.5^3 e^{-3.5}}{6}$.

10. e^{-1}; mean $\leqslant 0.001026$.

11.   (i) 0.8413,   (ii) 0.2206,   (iii) 0.2187.

12.   (i) 0.3085,   (ii) 0.2266; 0.1587 and 0.3085; 5.84.

13.   (i) 46%,   (ii) 54% (iii) approximately 0.

ANSWERS: Chapter 23

1. $\bar{x} = 25.87$, $s^2 = 0.849$.

2. Species A: $\bar{x} = 14.10$, $s = 2.96$; Species B: $\bar{x} = 22.83$, $s = 3.21$.
    Mean foraging heights are very different.
    Species A: median = 13, range = 9; Species B: median = 22, range = 9.
    Unlike the range the standard deviation indicates a slight increase in the
    variability of the foraging heights of species B.

3. $\bar{x} = 38.57$, $s^2 = 215.95$.
    Age is likely to influence length of stay and so it would be better to collect
    data from a specific age group.

4. $\bar{x} = 53$, $s^2 = 144.61$; mode = 55.
    Distribution is skewed and so it is unlikely to be normal.

5. Median = 2, range = 4, mode = 1.

6. (a) $\bar{x} = 1.007$, $s^2 = 1.042$,
    (b) mean = variance $\cong 4.028$.

7. $\bar{x} = 1535.0$, $s^2 = 2310.34$; 0.3783.

8. $\bar{x} = 1.51$, $s^2 = 0.0758$.

9. Small diameter axons: $\bar{x} = 68.33$, $s = 5.60$.
    Large diameter axons: $\bar{x} = 69.98$, $s = 4.20$.
    Impulse speeds have less spread in axons with large diameter.

10. $\bar{x} = 12.9$, $s = 1.1$.

11. Abnormal feet: $\bar{x} = 33.18$, $s = 5.271$.
    Normal feet: $\bar{x} = 14.82$, $s = 2.037$.

12. 0.1359.

13. (i) 0.4260,   (ii) 0.8351.

14. 0.1814; 0.0021; 0.

15. (i) 0.0197,   (ii) 0.0526,   (iii) 0,   (iv) 0.0006.

16. $N(62, 49/40)$.

17. $\bar{x} = 115.5$, $s^2 = 1003.83$; 39.

18. (i) 22,   (ii) 6.

19. 97.

20. 8 Zebu and 32 N'Dama cattle.

21. 25 and 75 from glass-houses A and B respectively.

22. 30 mildly, 30 moderately and 90 severely ill patients.

ANSWERS: Chapter 24

1. One sample z-test, $z = 3.00$, at 5% (two tails) significantly different migration areas.

2. One sample z-test, $z = -2.19$, at 5% (two tails) significant change in contamination counts.

3. One sample t-test, $t = -2.94$, $v = 14$, at 1% (one tail) significantly lower birth weights.

4. One sample t-test, $t = -1.306$, $v = 11$, at 5% (two tails) no significant difference in distance travelled.

5. Two sample z-test, $z = 6.7$, at 1% (one tail) significant difference and so development arrested.

6. Two sample z-test, $z = -1.743$, at 5% (two tails) no significant difference and so blood glucose level not influenced by cadmium concentration.

7. Two sample t-test, $t = 3.45$, $v = 25$, at 5% (two tails) significant difference between diameters.

8. Two sample t-test, $t = -2.74$, $v = 25$, at 5% (two tails) significantly different yields.

9. Two sample t-test, $t = 1.3$, $v = 28$, at 5% (two tails) no significant difference between activities.

10. Paired t-test, $t = 0.54$, $v = 5$, at 5% (one tail) no significant difference between concentrations.

11. Paired t-test, $t = 2.3$, $v = 7$, at 5% (two tails) no significant difference in leaf production.

12. Paired t-test, $t = 4.56$, $\nu = 8$, at 1% (one tail) significant difference between calcium levels.

13. Transformation $x^{0.75}$.

14. Transformation $\ln x$; two sample t-test, $t = -0.20$, $\nu = 10$, at 5% (two tails) no significant difference between parasitaemia levels.

ANSWERS: Chapter 25

1. One sample Sign test, $T = 1$, at 5% (two tails) significant difference between healing times.

2. One sample Sign test, $T = 2$, at 5% (two tails) no significant difference between temperatures.

3. For seedling plants $U = 21\frac{1}{2}$, at 5% (two tails) no significant difference between diameters.

4. Mann-Whitney test, for eggs at $25°C$ $U = 26$, at 5% (two tails) no significant difference between percentages.

5. Mann-Whitney test, for aerated cultures $U = 5$, at 5% (one tail) significant difference between generation times.

6. Wilcoxon paired test, $S_- = 9$, at 5% (two tails) no significant difference between lymphocyte counts.

7. Wilcoxon paired test, $S_- = 0$, at 1% (one tail) significant difference between intensity of infections.

8. Wilcoxon paired test, $S_- = 2$, at 5% (two tails) significant difference between consumption rates.

9. $\chi^2 = 15.60$, $\nu = 2$, at 1% significant difference between treatments.

10. Kruskal-Wallis test, $\chi^2 = 17.2$, $\nu = 3$, at 1% significant difference between groups.

11. Kruskal-Wallis test, $\chi^2 = 0.43$, $\nu = 2$, at 5% no significant difference between diets.

12. $\chi^2 = 0.86$, $\nu = 2$, at 5% no significant difference between treatments.

13. Friedman test, $\chi^2 = 6.5$, $\nu = 2$, at 5% significant difference between eelworm intensities; Friedman test, $\chi^2 = 9$, $\nu = 3$, at 5% significant difference between sterilants.

14. Contingency table, $\chi^2 = 37.0$, $\nu = 2$, at 1% significant relationship between temperature and viability.

15. Contingency table, $\chi^2 = 0.078$, $\nu = 1$, at 5% no significant relationship between colour and surface.

16. Contingency table, $\chi^2 = 12.8$, $\nu = 6$, at 5% significant relationship between size and treatment.

17. (i) Contingency table, $\chi^2 = 16.6$, $\nu = 2$, at 1% significant relationship between type of fumigant and larval mortality;
    (ii) contingency table, $\chi^2 = 6.9$, $\nu = 2$, at 5% significant relationship between type of fumigant and larval and adult mortality.

References

1. Bailey, N. T. J. *The Mathematical Theory of Epidemics*, Griffin (1957).
2. Conover, W. J. *Practical Nonparametric Statistics*, Wiley (1971).
3. Conte, S. D. and Deboor, C. W. *Elementary Numerical Analysis: an Algorithmic Approach*, McGraw-Hill (2nd ed.) (1972).
4. Hamming, R. W. *Introduction to Applied Numerical Analysis*, McGraw-Hill (1971).
5. Hildebrand, F. B. *Advanced Calculus for Applications*, Prentice-Hall (1948).
6. Jeffers, J. N. R. *An Introduction to Systems Analysis: with ecological applications*, Edward Arnold (1978).
7. Leslie, P. H. 'On the use of matrices in certain population mathematics', *Biometrika*, **33**, 183-212, (1945).
8. Leslie, P. H. 'Some further notes on the use of matrices in population mathematics', *Biometrika*, **35**, 213-245, (1948).
9. Lotka, A. J. *Elements of Mathematical Biology*. (Reprinted from *Elements of Physical Biology*, (1925)). Dover Publications (1956).
10. May, R. *Stability and Complexity in Model Ecosystems*, Princeton (1973).
11. Pielou, E. C. *An Introduction to Mathematical Ecology*, Wiley (1969).
12. Protter, M. H. and Morrey, C. B. *Modern Mathematical Analysis*, Addison-Wesley (1964).
13. Quenouille, M. H. *Introductory Statistics*, Pergamon Press (1966).
14. Searle, S. R. *Matrix Algebra for the Biological Sciences (Including Applications in Statistics)*, Wiley (1966).
15. Siegel, S. *Nonparametric Statistics for the Behavioural Sciences*, McGraw-Hill (1956).
16. Smyrl, J. L. *An Introduction to University Mathematics*, Hodder and Stoughton (1978).
17. Sokal, R. R. and Rohlf, F. J. *Introduction of Biostatistics*, W. H. Freeman and Company (1971).
18. Sparks, F. W. and Rees, P. K. *Plane Trigonometry*, Prentice-Hall (1937).
19. Turnbull, H. W. *Theory of Equations*, Oliver and Boyd (1947).
20. Williams, P. W. *Numerical Computation*, Nelson (1972).

Index

MATHEMATICS AND ITS APPLICATIONS
Series Editor: G. M. BELL
Emeritus Professor of Mathematics, King's College London, University of London

series continued from front of book

MATHEMATICS AND ITS APPLICATIONS
Series Editor: G. M. BELL,
Emeritus Professor of Mathematics, King's College London, University of London

Statistics, Operational Research and Computational Mathematics
Editor: B. W. CONOLLY,
Emeritus Professor of Mathematics (Operational Research), Queen Mary College, University of London